AAPG Treatise of Petroleum Geology Reprint Series

The American Association of Petroleum Geologists
gratefully acknowledges and appreciates the leadership and support
of the AAPG Foundation in the development of the
Treatise of Petroleum Geology.

TRAPS AND SEALS I

STRUCTURAL/FAULT-SEAL

AND HYDRODYNAMIC TRAPS

COMPILED BY
NORMAN H. FOSTER
AND
EDWARD A. BEAUMONT

TREATISE OF PETROLEUM GEOLOGY
REPRINT SERIES, No. 6

PUBLISHED BY
THE AMERICAN ASSOCIATION OF PETROLEUM GEOLOGISTS
TULSA, OKLAHOMA 74101, U.S.A.

Library of Congress Cataloging-in-Publication Data

Traps and seals / compiled by Norman H. Foster and Edward A.
Beaumont.   p.   cm. — (Treatise of petroleum geology reprint
series ; no. 6-)
    Includes bibliographies.
    Contents: 1. Structural/fault-seal and hydrodynamic traps.
    ISBN 0-89181-405-1 (v. 1): $34.00
    1. Rock traps (Hydraulic engineering) 2. Petroleum—Geology. 3.
Gas, Natural—Geology. I. Foster, Norman H. II. Beaumont, E. A.
(Edward A.) III.—Series: Treatise of petroleum geology reprint
series: no. 6, etc.
QE611,T73 1988
553.2'8—dc19   88-10471
                                                    CIP

INTRODUCTION

This reprint belongs to a series of reprint volumes which in turn are part of the *Treatise of Petroleum Geology*. The *Treatise of Petroleum Geology* was born during a discussion we had at the Annual AAPG Meeting in San Antonio in 1984. When our discussion ended, we had decided to write a state-of-the-art textbook in petroleum geology, directed not at the student, but at the practicing petroleum geologist. The project to put together one textbook gradually evolved into a series of three different publications: the Reprint Series, the Atlas of Oil and Gas Fields, and the Handbook of Petroleum Geology; collectively these publications are known as the *Treatise of Petroleum Geology*. Using input from the Advisory Board of the Treatise, we designed this entire effort so that the set of publications will represent the state of the art in petroleum exploration knowledge and application. The Reprint Series collects the most up-to-date, previously published literature; the Atlas is a collection of detailed field studies to illustrate all the various ways oil and gas are trapped; and the Handbook is a professional explorationist's guide to the latest knowledge in the various areas of petroleum geology and related fields.

Papers in the various volumes of the Reprint Series are meant to complement the chapters of the Handbook. Papers were selected mainly on the basis of their usefulness today in petroleum exploration and development. Many "classic papers" that led to our present state of knowledge are not included because of space limitations.

We divided the general topic of traps into two volumes: (I) Structural/Fault-Seal and Hydrodynamic Traps, and (II) Stratigraphic/Capillary Traps. Papers in these two volumes deal mainly with the mechanics of trapping hydrocarbons and with trap types. Methods of exploring for traps are covered in the Reprint volume: Methods of Exploration.

Traps are the end product of many processes: the maturation and migration of petroleum, and the formation and preservation of the porous rocks that form the reservoirs and the non-porous rocks that are the seals. There are time and space requirements. A potential trap must have been present during petroleum migration. Pathways must be available from the source to the trap, and reservoir rocks and seal rocks must be in the proper configuration to one another. Without any one of these elements, a trap cannot exist. Primarily, the geometry of a trap is determined by two factors: structure and stratigraphy. The two volumes of the *Traps and Seals* reprint divide papers on traps and seals along these two lines.

Volume I mainly contains papers dealing with structural/fault seal traps, but also includes papers that discuss the controversial subject of hydrodynamic trapping. Volume II contains papers that describe different stratigraphically controlled trap types, the preservation of porosity, and the importance of capillarity in trapping hydrocarbons.

Edward A. Beaumont
Tulsa, Oklahoma

Norman H. Foster
Denver, Colorado

Treatise of Petroleum Geology
Advisory Board

TABLE OF CONTENTS

TRAPS AND SEALS I

STRUCTURAL/FAULT-SEAL TRAPS

HYDRODYNAMIC TRAPS

TABLE OF CONTENTS

TRAPS AND SEALS II
STRATIGRAPHIC/CAPILLARY TRAPS

STRUCTURAL/FAULT-SEAL TRAPS

American Association of Petroleum Geologists Sidney
Powers Memorial Volume, *Problems of Petroleum Geology,*
edited by W. E. Wrather and F. H. Lahee, copyright © 1934
pp. 1-23.

HISTORICAL DEVELOPMENT OF THE STRUCTURAL
THEORY OF ACCUMULATION OF OIL AND GAS[1]

J. V. HOWELL[2]
Ponca City, Oklahoma

ABSTRACT

Although petroleum has been known since a very early date, its geological occurrence became of interest only after discovery of its economic value, and active exploitation began. First mention of the occurrence of oil on anticlines was by William Logan, in 1844, and within a few months after completion of the Drake well both T. Sterry Hunt and H. D. Rogers recalled Logan's earlier observation and noted also that the Drake well was located on an anticline. Numerous other workers confirmed these observations, but it appears that little practical use was made of the theory until I. C. White, in 1885, began to apply it in search for natural gas in Pennsylvania and surrounding states. Coincidently Edward Orton, in Ohio, brought out the importance of geologic structure and by every possible means endeavored to impress on the public the value of geology in prospecting.

During the years following 1900 the United States Geological Survey published a series of studies of oil fields in the eastern states which further supported the theory and served also to give it wider publicity and greater prestige, until finally little or no criticism was heard. When the huge expansion in use of petroleum began, about 1910 the geologist became, for the first time, an important factor in the search for new fields,

The mineral, petroleum, has been known since very remote time, and under diverse names. It has been variously termed "pitch," "slime," "rock oil," "Seneca oil," and perhaps many other designations. First observed in Asia and Europe in natural occurrences, it found but limited uses as a water-proofing substance, as a medicine, and to a certain extent as a lubricant. In the early settlements of America oil and gas springs were soon observed by the white man, or were pointed out to him by the Indians, who had long known them. But not for many years did their usefulness become apparent.

The first settlements in North America were near the Atlantic coast where salt, a prime necessity, was easily obtainable. But as the population gradually spread inland through the thick forests and isolated valleys of the Appalachians, it became increasingly difficult to transport salt

[1] Manuscript received, September 10, 1932.

[2] Consulting geologist, 300 North Fourth Street.

from the coast, and recourse was had to the many salt springs, or "licks," of the mountain valleys. Too often, however, these springs were found to be contaminated by the presence of petroleum, occasionally in considerable amount, and, as its value was unknown, it was looked upon as a nuisance.

Many of the early settlements owed their locations to the proximity of salt licks, where the necessary supplies could readily be obtained. As the population increased and villages multiplied, efforts were made to increase the natural flow of the saline springs by digging shallow wells adjacent to them, and finally wells were drilled to greater depths, as had been done for many years in Europe and in China. It is not surprising that in the Virginias many of the first salt wells encountered oil or gas, for it was shown by W. B. Rogers as early as 1842 (1),[3] that most of the warm saline springs of the Appalachians occurred along the crests of anticlines.

Until about 1855 petroleum found little use in North America, although sold in limited amount as "Seneca oil," and recommended as a liniment for "man or beast." In this year, however, it was found that, by distillation, there could be made from petroleum a light oil similar to coal oil, and much superior to whale oil as an illuminant. Again there began a search for springs, but this time for those which produced oil, as well as salt. The yield of such springs generally was small, and attempts to increase the flow by trenching the adjacent ground, and digging shallow wells in the saturated stream gravels, met with limited success. Then in 1859, Colonel Edward Drake, applying the methods of the salt producers, drilled near an oil spring on Oil Creek, near Titusville, Pennsylvania. At a depth of 69 feet, he obtained a flow of oil, estimated at 35 barrels per day.

Three more wells were completed in that year, 175 in 1860 and 340 in 1861. Despite the onset of the Civil War, the number of wells drilled annually increased to 937 in 1864, remained about stationary until 1867 and then rose rapidly to 1,653 in 1870. Operations spread rapidly over Pennsylvania, the Virginias, Ohio, New York and Ontario, and the production obtained grew much greater than the market could absorb.

Recent search of the literature of that period shows that almost from the date of completion of the Drake well, geologists were attempting to find some means of predicting the results of drilling, and for determining the best locations for prospecting. The first developments were carried out along streams where oil springs were found, and this custom has been

[3] See Bibliography at end of paper.

carried down even to the present day, when "creekology" is not entirely unknown.

Until petroleum had been shown to be of commerical value there was no incentive to search for or exploit it, and speculation concerning its nature and origin was largely academic. Most of the earlier writers emphasize the chemical aspects of the problem, while those geologists who studied the occurrences in the field generally confined their efforts to identification of the associated beds. The physical causes of accumulation did not assume importance until it had been shown by Drake that large underground stores could be obtained by drilling, and that the detection of such deposits in advance of drilling would be of great commercial advantage.

As early as 1842 Sir William Logan, director of the Geological Survey of Canada, visited Gaspé, at the mouth of the St. Lawrence River, and noted the oil seepages on the hills near the shores. The fact that these seepages and springs were located on anticlines was pointed out by him (2) in 1844, but only incidentally to his discussion of stratigraphy. Logan seems not to have been greatly impressed by the matter at the time, but on hearing of the Drake well his earlier observations must have been recalled, either by himself or his associates.

Probably the earliest definite statement of the fact that the newly discovered oil fields of Pennsylvania were located on anticlines is contained in a paper by Professor Henry D. Rogers, of the University of Glasgow, who had previously conducted a geological survey of Pennsylvania. This statement is contained in a paper (5) read before the Philosophical Society of Glasgow on May 2, 1860, nine months after completion of the Drake well. It is apparent that this paper must have been inspired by news of the well, and that while Rogers had not visited it, he was thoroughly acquainted with the locality by reason of many years spent on his survey of the state. In later papers (18, 19) he amplifies his earlier statement and recalls the fact that the Oil Creek (Pennsylvania) fields are located on an anticline discovered many years before and described in his *Geology of Pennsylvania.* (3) Rogers emphasizes the fact that almost all of the warm springs of the Appalachian region are found on the crests of anticlines and that both oil and gas commonly occur with the warm saline water.

Without doubt the first clear statement of the structural or anticlinal theory of accumulation was made by T. Sterry Hunt, of the Geological Survey of Canada. His ideas were first set out in a public lecture in Montreal, and published in the *Montreal Gazette* on March 1, 1861. A more detailed paper by Hunt appeared in the *Canadian Naturalist* for July, 1861 (11), and was reprinted in the *Annual Report* of the Smithsonian Institution for that year, as well as in *Chemical News*, of London. Obviously the paper was considered to be significant, and the fact that it was published in Canadian, American, and British journals renders it improbable that many geologists of the time failed to see it.

Hunt believed that oil was formed chiefly in limestones or in black shales; that it was liberated by folding and fracturing which occurred chiefly along the axes of anticlines. The oil

being lighter than the water which everywhere penetrates these rocks below the water level, naturally rises and accumulates along the crown of these anticlinals. This process is favored by the fact that the strata on either side of the anticlinal dip in opposite directions. (26)

He recognized also that either overlying impervious beds must be present, or that fissures must be sealed by clayey material.

Early in 1861, Professor E. B. Andrews (19), of Marietta, Ohio, traced a line of uplift from eastern Washington County, Ohio, to a point beyond the oil wells of Little Kanawha River, and showed that along this fold were found all of the oil fields, and the oil and gas springs. In this report he presents what appears to be the first published map and cross section of an oil field (Fig. 1). His idea of occurrence was stated as follows:

In the broken rocks along the central lines of a great uplift, we meet with the largest quantity of oil. It would appear to be a law, that the quantity of oil is in a direct ratio to the amount of fissures.

Obviously Andrews believed that the fissures, supposedly most numerous along the anticlinal axes, were the primary cause of accumulation, rather than the anticline itself, and in his subsequent writings he continued to adhere to this view.

In 1863 Henry D. Rogers published two papers, one in Europe (18), and one in the United States (19) in which he stated without qualification that the oil fields of North America occur on the crests and flanks of anticlines. He says: (19)

[It is] ... a very striking fact, that throughout Western Pennsylvania, Northwestern Virginia, Southwestern Ohio, and Eastern Kentucky, or, in other words, throughout all the western borders of the great coal field, where the general flatness of the coal rocks is only at wide intervals interrupted by narrow, but long and sometimes rather sharp anticlinal waves, the more copious emission of the rock oil and the native gases is found to be chiefly restricted to the tracts occupied by the crests and sides of these local billows in the strata.

That the theory had already found support abroad is indicated by a paper by the then French consul in Canada, M. Gauldree-Boileau, in 1862. (12) He credits to Logan the first statement of the anticlinal theory, and discusses the relation of the known occurrences of Ontario to such

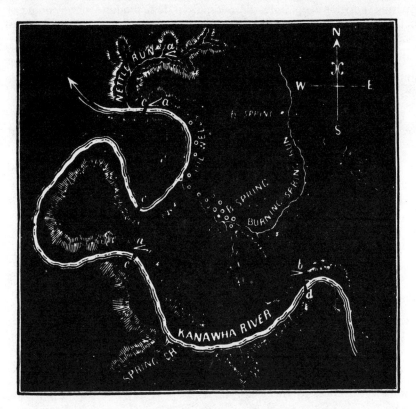

Fig. 1A.—Map of Little Kanawha oil region, Wirt County, Virginia. Scale, about 1.5 inches to a mile. E. B. Andrews, 1861. From a woodcut.

Fig. 1B.—Section on Little Kanawha in vicinity of Burning Spring Run. E. B. Andrews, 1861. *A*, Axis of uplift. *A–B*, Area of greatest number of fissures and great oil wells.

axes. Much further prospecting is predicted along the anticline extending from Burlington Bay, on Lake Ontario, southwestward to Lake St. Clair, and roughly following Thames River, the information being taken from reports by Logan. The following prophetic lines merit quotation.

L'axe anticlinal de sir William Logan est donc indiqué à l'avance pour les explorations à faire en vue d'abtenir de nouveaux approvisionments du petrole. Rien ne donne à penser d'ailleurs que les réservoirs déjà connus soient à la vielle de s'épuisser. Il existe dans le monde des sources de bitumen exploitées depuis des temps fort anciens et qui ne sont pas encore taries. Les phénomènes auxquels on attribue l'origine des pétroles d'Amérique ont dû se produire sur une immense échelle, et leurs résultats ont été sans doute proportionnés à la grandeur des opérations accomplies par le nature.

In 1863 also, William Logan and T. Sterry Hunt presented a re-statement of the anticlinal theory, with additional evidence from the extensive drilling which had taken place in the preceding two years. By study of the well records Hunt had been able to add greatly to knowledge of structure in southern Ontario. This period has been well described in a recent paper by Harkness (85), and is of great interest by reason of the fact that from that time onward, the Canadian Survey stood squarely behind the structural theory.

In 1865 Hunt states that the following conditions are necessary for the accumulation of oil. (26)

1. A source bed [Hunt insists on limestone]
2. Proper attitude of the strata. Anticlines
3. Suitable fissures to act as reservoirs
4. Such impermeability of surrounding and overlying strata as will prevent escape of the accumulated oil

In 1867 Hunt states: (46)

This opinion [that oil occurs only on anticlines] which I have advocated since 1861, is confirmed not only by my work in Upper Canada, but equally by the observations of great numbers of geologists in the United States.

Some opposition to the theory naturally arose, but the opponents, notably J. Peter Lesley and his associates, were able to advance nothing better. The completeness of its acceptance may be judged by the fact that a great majority of the papers on petroleum, written by geologists between 1861 and 1880, refer to the work of either Hunt or Andrews, both advocates of the anticlinal theory.

Hunt and Andrews at first considered that fissures were necessary, and that the oil accumulated in such openings. Its common occurrence in sandstones was explained as being due to the ability of the sandstone to maintain open fissures to a greater extent than either limestones or shales. Alexander Winchell, in 1860, had suggested (7) that sandstones themselves were sufficiently porous to contain oil, even without fracturing, but it was not until several years later that Hunt became converted to this view, while Andrews seems never to have adopted it.

About 1866 E. W. Evans brought forward a modification of the anti-clinal theory known as the "Break theory" (40) which assumed that accumulation takes place along lines of sharp increase of dip (Fig. 2) on

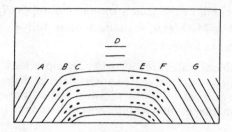

FIG. 2.—Occurrence of oil on "breaks." E. W. Evans, 1866. *B–C* and *E–F* are breaks.

one or both flanks of the anticline, these zones of steep dips being termed "breaks," but the idea did not find many supporters. He discussed fully the influence of water in bringing the oil and gas up into the anticline, and in one paper (40) devotes considerable space to arguing successfully that the oil in West Virginia occurs on an anticline, rather than on a buried hill, as had been suggested by others.

In the preceding year Evans had discussed (20) the mode of occurrence of oil in the reservoir. He explained clearly the prevailing ideas regarding gravitational separation of oil, gas, and water in crevices, not yet having adopted the view that the porosity of the sandstone was sufficient to account for its occurrence therein. His idea is made sufficiently clear by referring to the diagram in Figure 3.

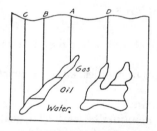

FIG. 3.—Gravitational separation in crevice reservoir. E. W. Evans, 1865.

Evans restated his views in the following year, as follows: (40)

This large accumulation of oil in the crevices of the anticlinal (Burning Springs) would seem to be owing, not solely to a direct connection by a continued line of fractures with the original sources of the oil in strata beneath, but in part also to the collection, from a wider area, of oil that has come up elsewhere, as through the crevices of the adjacent synclines. For, being lighter than water, it would gradually work up between the strata of the slopes.

The foregoing seems to be the first direct statement of a belief in extensive lateral migration.

Alexander Winchell, in 1865, in a commercial report (31) on an area in St. Clair County, Michigan, explained more fully his earlier statements on reservoirs, as follows:

Where the bituminous shales are covered by an impervious layer, as of shale or plastic clay, the oil and gas elaborated are retained in the rocks, filling cavities by driving the water out by elastic pressure, and saturating porous strata embraced in the formation or intervening between it and the impervious cover above.

Winchell specifically states that both gas and oil are formed in the same source beds and migrate upward into the crest of the anticline under the same factors.

Hunt again restated his views in 1868, as follows: (15)

This anticlinal structure appears to be a necessary condition of the occurrence of abundant oil wells; the petroleum being lighter than water, accumulates in *porous strata* or in fissures in the higher part of the anticlinal, and in obedience to a hydrostatic law, rises through openings to heights considerably above the water level of the region. . . . I do not conceive that the gas has any necessary connection with the oil, since large quantities of it are found in rocks which *underlie* the Corniferous [believed by Hunt to be the source of the oil]. If however, as is not improbable, portions of it were generated and now exist in a condensed state in the oil bearing strata, its elasticity would help to raise the oil to the surface.

In 1867 Charles H. Hitchcock summarized (37) the conditions of occurrence of petroleum as follows:

Petroleum may occur:
1. Sometimes in synclinal basins, like the subterranean streams of water in artesian wells.
2. May occur in fissures or cavities, either in synclinal basins or on anticlinal slopes.
3. Along the lines of faults.
4. May exist in *great quantities* beneath anticlinal arches. The roof acts as an impervious cover to confine the fluids.

Besides petroleum, brine and gas are commonly, if not universally, discharged from the orifice of the well, and we may suppose that before the tapping of the cavity they were arranged according to their specific gravities, the gas uppermost, and the brine beneath the others.

Hitchcock still adheres to the view that fissures are essential, and advises that search be made for areas where the strata are much folded and broken, thus producing open cavities. In a paper (29) published two years earlier he had described the occurrence of albertite at Albert Mines, New Brunswick, and called attention to the fact[4] that it occurs in a vein along the axis of an anticline. The albertite is regarded as having been injected into the fissures in a liquid condition, and to have subsequently hardened, so that the albertite veins are analogous to veins of petroleum. He states that

The borings for petroleum in Ohio and West Virginia are most successful along lines of fracture, especially anticlinal axes.

It is interesting to note here that at a meeting of the American Philosophical Society in the spring of 1865, a Mr. Briggs made an oral communication[5] in which he suggested that oil wells flowed because

petroleum is an intimate mechanical mixture of the gases into which petroleum decomposes, with the petroleum fluid, like that which exists between the carbonic acid and the water in a soda fountain.

A French engineer, F. Foucou, visited this country in 1866, and on his return published a short report (43) on the petroleum fields. Particularly noteworthy is his statement that he has fully verified the fact, *well known to American geologists and operators*, that the oil fields are definitely related to anticlines. From the statement that his observations included ". . . relevant à la boussole un grande nombre de puits," it is apparent that Foucou was an early convert to subsurface methods.

From a date at least as early as 1865 and until his death in 1889, J. P. Lesley, director of the Second Geological Survey of Pennsylvania, was a bitter opponent of the anticlinal theory, and attacked it vigorously in his writings and in oral discussions. He adhered to the view that oil accumulated in sandstones and was indigenous thereto; that the most important governing factor was the original character of the rock, and that structure itself was of no importance. In none of the many reports published under his direction is the structural theory given place, but his insistence on the importance of porous sandstones as reservoirs seems gradually to have resulted in abandonment of the older idea of accumulation in crevices.

In 1871 a Canadian geologist, Henry Youle Hind, was employed by local parties to examine an area near Cheverie, Nova Scotia, where oil had been found in joints and crevices in a gypsum quarry near that place.

[4] Hitchcock credits this discovery to Robb and Taylor, *Proc. Amer. Philos. Soc.*, Vol. 5 (1851), p. 242.

[5] *Proc. Amer. Philos. Soc.*, Vol. 10 (1865), p. 136.

His report (49) of 16 pages is not generally accessible, but has been quoted extensively in a previous paper by the writer. (84) It emphasizes the importance of anticlinal structure; the necessity for the presence of an impervious shale, or other sealing beds overlying the reservoir; and states that the oil is of local origin. Hind believed that fissures were necessary to provide a reservoir. Several anticlines near the seepages are described, and it is recommended that tests be drilled on certain of them, but it appears that his advice was not followed. After describing the productive anticlines of Ohio and West Virginia, he states, on p. 9:

It is also to be observed that in some important instances no surface indications were visible, and the finding of the oil was the result of borings for brine-springs, coal, etc., or for oil, guided by the geological structure and age of the rocks. Strata of a certain geological age, and of a particular horizon, and distinctly arranged in an Anticlinal form so that fissure receptacles may permit of the accumulation, being the only guides which led to very successful results.

By 1875, sixteen years after the drilling of the Drake well, the following facts regarding petroleum seem to have become rather generally known and accepted by geologists:

1. It is of organic origin, and to be sought for only in sedimentary rocks.

2. In the regions then being prospected the fields, almost without exception, were located on anticlines.

3. The rocks generally were saturated by water, and through these beds the oil and gas, by reason of their lighter specific gravity, tended to rise to the higher portions of the reservoir.

4. An impervious bed overlying the reservoir rock was essential.

5. Sandstones constituted the most common reservoirs and by this time most geologists believed that fissures were not necessary for accumulation, though many still believed that they were necessary to permit the oil and gas to enter the reservoir from the source beds.

6. That gas and oil usually occur together and that the gas is effective in causing the oil to flow. Solution of gas in petroleum was recognized, but its full significance not appreciated.

Usually a new theory of any sort is subjected to prompt criticism or discussion, but only one voice seems to have been raised against the original statement of the anticlinal theory. J. Peter Lesley expressed his opposition as follows, at a meeting of the American Philosophical Society in 1865. (25)

Stress has been laid by some geologists of note, upon a supposed genetic connection between the accumulation of Petroleum and anticlinal axes. But there are no anticlinal axes in the Pennsylvania oil regions of the French and Oil Creek wells, nor in the Pennsylvania and Ohio Oil regions of the Beaver River, nor in the East Kentucky oil region of the Sandy and Licking waters. The only well defined anticlinal among oil fields is a mere upsqueeze crossing the Ohio River near Marietta, bringing the oil rocks near enough to the surface to be tapped, and thus, only, affecting the finding of oil. . . . The true relationship of petroleum with surface springs seems to be one of simple hydrostatics. Every natural oil spring is an artesian spring, without regard to the existence of anticlinals or profound earth-crust faults.

Again in 1880 he reiterates his continued opposition, in spite of the great amount of additional evidence that has accumulated: (61)

The supposed connection of petroleum with anticlinal and synclinal axes, faults, crevices, cleavage planes, etc., is now a deservedly forgotten superstition. Geologists well acquainted with the oil regions never had the slightest faith in it, and it maintained its standing in the popular fancy only by being fostered by self assuming experts, who were not experienced geologists.

Lesley continued to oppose the theory throughout his life, and it seems to have been his attitude that in large measure caused it to decline in popularity between 1875 and 1885. The Second Geological Survey of Pennsylvania, under the direction of Lesley, existed from 1874 to 1887, and during this time made a very thorough study of the geology and mineral resources of the state. Five volumes were devoted entirely to oil and gas, and in addition the various county reports contain much information on the individual fields. Great numbers of logs were assembled and studied, the stratigraphic positions of the sands determined, and all of the data necessary for a structural study obtained, but seemingly no effort was made to utilize this material in developing any theory to account for the occurrence of the pools. That such a study was not made, and that the Survey was thus deprived of the opportunity to lay the foundation for modern petroleum geology, can be the result only of the tenacity with which Lesley adhered to his early prejudice, first expressed in 1865, repeated in 1880, and again in 1885, when he issued a final broadside against the recently announced view of I. C. White:

It is impossible, . . . that any arrangement of water, oil and gas can occur in the deep oil rocks, such as occurs in a bottle. It therefore seems to me irrational to assign any importance whatever to the extremely gentle anticlinals of the gas-oil region. (67)

The full story of the domination of the Survey by Lesley has never been written, but it is apparent that none of his eminently capable assistants was able to proceed against the prejudice of the director. Of these men, J. F. Carll, the petroleum specialist, and Franklin Platt, seem never

eral hundred feet (five hundred to twenty five hundred feet); (c) Probably very few or none of the grand arches along mountain ranges will be found holding gas in large quantity, since in such cases the disturbance of the stratification has been so profound that all the natural gas generated in the past would long ago have escaped into the air through fissures that traverse all the beds. Another limitation might possibly be added, which would confine the area where great gas-flows may be obtained to those underlaid by a considerable thickness of bituminous shale.

. . . .

The reason why natural gas should collect under the arches of the rocks is sufficiently plain, from a consideration of its volatile nature. Then, too, the extensive fissuring of the rock, which appears necessary to form a capacious reservoir for a large gas-well, would take place most readily along the anticlinals where the tension in bending would be greatest.

It will be noted (paragraph 4) that White states he was employed "with the special object of determining *whether or not it was possible to predict the presence or absence of gas from geological structure.*" In other words, he appears to have been employed to determine whether the anticlinal theory applied also to gas. In the same paragraph he acknowledges his indebtedness to Earseman for the suggestion that in western Pennsylvania the principal gas wells were situated close to the anticlinal axes. Thus there is apparent no disposition on White's part to claim originality for the theory, but he clearly is open to the charge of carelessness in not having examined the earlier literature and giving credit to the many who had suggested such a relationship many years previously. This error he later corrected, (77) giving due credit to Hunt, Andrews, and Hoefer. He then continues:

Thus it appears that the theory had long been recognized and its essential elements published, but the practical oil men had never heard of it in a way to make any impression upon them, and the authors of the theory had made but slight attempts to apply its principles practically in the location of new oil or gas fields. This is the work which the writer has especially accomplished, and in the doing of it so enforced the lessons of geology upon the minds of the men engaged in the practical work of drilling for oil, that the acceptance of the structural theory is now almost universal among them as well as among geologists. In this the writer has been ably assisted by Dr. Edward Orton.

As in years past, Lesley rushed again to the attack (67), this time seconded by his associates, Ashburner (64) and Chance. (70) Lesley advanced no new arguments to support his position, and those offered by Ashburner were readily shown by White (65) to be based on faulty interpretation of structure. By this time the number of oil and gas fields which were demonstrably anticlinal had greatly increased, and White was able to cite an overwhelming number in support of the theory.

At about this time Edward Orton became state geologist of Ohio, and began promptly to study the conditions of occurrence of the oil and gas fields of northwestern Ohio, which first became important in 1884. He soon determined that accumulation here was related to a broad, flattened anticline (Cincinnati arch), and thus further confirmed the anticlinal theory. In this area, however, not all of the oil and gas was found on the crest of the fold, but occurred also in more or less scattered pools along the flanks, so that some modification of the original theory seemed necessary. Orton explained these flank pools as being localized at points of "arrested dip" (73) or terraces, (75) as he later called such features. On a terrace the upward moving oil would, he argued, tend to accumulate on the outer, or lower edge, while downward moving oil would accumulate on the inner, or upper, edge of the terrace. Lateral migration of oil was assumed by Orton and by White, as well as by most of the earlier writers who mention the matter at all.

In his various papers on the Trenton fields of Ohio (73, 75) and Indiana, (74) Orton laid much stress on the importance of terrace accumulation, and in this he has been followed by Phinney (76) and practically all later writers. This type of structure is figured in all textbooks, though the known demonstrable terraces which have produced oil are not numerous outside the area described by Orton. It is quite possible that many of these supposed terraces really are noses, or are locally porous areas in the Trenton. Also it is true that the term terrace has been loosely used in Indiana and even recently it has been applied to structures which most geologists would unhesitatingly classify as noses or plunging anticlines.[9]

During the ten years following the appearance of White's and Orton's studies, little advance was made in the study of oil accumulation, and the literature of the period is meager. But about 1900 the United States Geological Survey, whose attention previously had been given chiefly to the exploration of the western states and to study of ore deposits, began the examination of a number of oil-producing areas in the Appalachian states, particularly Ohio and Pennsylvania. The results of this work have been well summarized by Campbell (79) who points out that the most important discovery made by these workers was that of the extensive occurrence of oil in dry sands, and the recognition that conditions of

[9] An example of this is found on page 195 of "Petroleum and Natural Gas in Indiana," *Indiana Dept. Conservation Pub. 8* (1920). The "Glenzen terrace," shown on this map, is really a nose.

to have committed themselves. J. J. Stevenson, who was primarily a stratigrapher, showed some sympathy toward a structural explanation, but avoided any definite statement. Charles A. Ashburner and H. M. Chance were entirely in agreement with Lesley's view. Only I. C. White, although he deferred to the opinions of his chief while associated with him, seems to have retained an open mind on the question, and within two years after leaving the Survey, had satisfied himself of the truth of the theory and published his conclusions. (63, 65) It would seem probable that the long eclipse of the anticlinal theory was due very largely to the active opposition of the Lesley group, for during the 13 years of its existence this Survey was the largest official geological organization, and the only one operating in the oil-producing areas.

APPLICATION OF ANTICLINAL THEORY PRIOR TO 1885

It has been shown that within two years after the completion of the Drake well in August, 1859, the relation of oil fields to anticlinal axes had been pointed out by Hunt, and Logan in Canada, and by H. D. Rogers, E. W. Evans and E. B. Andrews in the United States. Their conclusions were published in the leading scientific journals of Canada, the United States, England, and Scotland, and it is but fair to assume that these papers were seen by a majority of the geologists of that time, particularly since there were few such journals to be read, and little likelihood of them being overlooked.

Immediately following its first statement, the theory was accepted by Alexander Winchell, who set about to apply it in the examination of areas in Sanilac and St. Clair counties, Michigan; (31) Lambton County, Ontario; (33) Knox and Conshohocton counties, Ohio; (35) and Wetzel County, West Virginia. (36) In Canada, acting for the Provincial Government of Quebec, T. Sterry Hunt reported on the oil possibilities of Gaspé, and accompanied the report (26) with one of the first published maps showing the occurrence of oil on anticlines, and the location of areas to be tested.[6]

Meanwhile, according to F. W. Minshall (quoted by Peckham): (62)

About 1865, Gen. A. J. Warner and Prof. E. B. Andrews, of Marietta, became interested in the White Oak area, and these gentlemen soon began to draw geological inferences which led to the abandonment of the old policy of following the beds of the streams, and to a recognition of the fact that the oil was confined to the crest of an anticlinal; hence the White Oak section, and that alone, has been thoroughly and systematically tested.

[6] This map was reproduced in a paper by R. B. Harkness. (85)

However, so many dry holes were found along the axis of this anticline, that in 1878 Minshall made a careful barometric survey to determine the reason. This resulted in his determination that the production occurred in "local undulations" (domes) along the main fold.

H. Hoefer, a prominent Austrian geologist, in 1876 published *Die Petroleumindustrie Nord-Amerikas* in which he accepted, but apparently without credit, the theory of anticlinal accumulation.

In the Tenth Census Report, 1880, after quoting extensively from published reports of Hunt, Carll, Minshall, and others, Peckham gives, on page 52, the following summary of knowledge of oil accumulation available at that time. (62)

Assuming that Messrs. Hunt, Carll and Minshall have observed correctly, and stated their observations correctly, petroleum occurs in crevices only to a limited and unimportant extent. It occurs saturating porous strata and overlying superficial gravels; it occurs beneath the crowns of anticlinals in Canada and West Virginia, and does not so occur in Pennsylvania, but in the latter region it occurs saturating the porous portions of formations that lie far beneath the influence of superficial erosion, like sand bars in a flowing stream, or detritus on a beach. . . . The Bradford field, in particular, resembles a sheet of coarse-grained sandstone 100 square miles in extent, by from 20 to 80 feet deep, lying with its southwestern edge deepest and submerged, and its northeastern edge highest and filled with gas under extremely high pressure.

It is apparent that Peckham's summary is based entirely on published information, and that his accuracy depends wholly on the accuracy of his sources. Following Carll and Lesley he excludes Pennsylvania from the ranks of the states whose production is anticlinal. But his information on Canada and West Virginia having come from supporters of the theory, he also follows them without questioning the discrepancy in their views. Peckham was a chemist, unfamiliar with geology, and therefore unable critically to examine the geological material which he compiled.

Thus the record shows that within fifteen years after its promulgation by Hunt, the structural theory had been accepted and referred to in published reports by Logan, Hitchcock, H. D. Rogers, E. W. Evans, E. B. Andrews, Alexander Winchell, and Henry Hind, in North America, and by H. Hoefer and probably others in Europe. Lesley alone had seriously attacked it.[7] But it must be remembered that at this period there were few geologists resident in the oil regions and that travel was difficult

---

[7] In a biographical sketch of Lesley (*Proc. Amer. Philos. Soc.*, Vol. 45, 1906) Chance states that during the early sixties Lesley so overworked while doing private geological surveys in the oil fields that he was forced to spend two years abroad to regain his health.

and time consuming. Most were engaged in teaching at colleges and universities in the east, and there was no national and but few state geological surveys. Probably few geologists had opportunity to observe the occurrence of oil in the field, or occasion to record their opinions concerning it, other than in class-room lectures. Unfortunately no records of these are available.[8]

RENAISSANCE OF ANTICLINAL THEORY, 1885–1890

About the year 1880 it became apparent that the natural gas, which hitherto had been a useless and troublesome by-product of the oil fields,

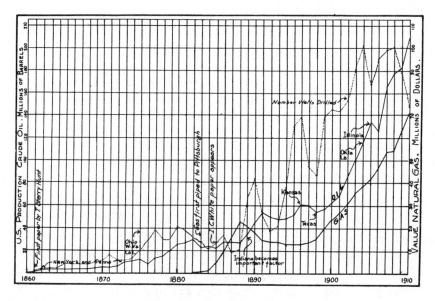

FIG. 4.—Chart showing production of oil and gas, and number of wells drilled each year, 1859–1910.

could be used commercially to replace coal in the steel mills and glass factories of the great industrial centers such as Pittsburgh. During the next decade intensive drilling ensued, in an attempt to keep pace with the growing demands for the new fuel (FIG. 4).

Early in 1883, I. C. White, an assistant on the Second Geological Survey of Pennsylvania since 1875, severed his connection with that organization, and entered commercial work. From the beginning he seems to have been engaged chiefly in search for gas on behalf of a syndicate

[8] Joseph LeConte (*Autobiography*, p. 240) notes that doing the winter of 1866–7 he gave six lectures on Coal and Petroleum at the Peabody Institute, in Baltimore, but it does not appear that these were published.

which was supplying fuel to industrial plants. (77) After two years of this work he published the short paper (63) which has led to his being credited with discovery of the anticlinal theory, an honor which he many times denied. Since reference often is made to this paper, the essential parts are here given.

.This new fuel (gas) . . . has not received that attention from the geologist which its importance demands. So far as the writer is aware, nothing has been published on the subject which would prove of any value to those engaged in prospecting for natural gas, and it is the existence of this blank in geological literature that has suggested the present article.

.   .   .   .

The writer's study of this subject began in June, 1883, when he was employed by Pittsburgh parties to make a general investigation of the natural gas question, with the special object of determining whether or not it was possible to predict the presence or absence of gas from geological structure. In the prosecution of this work, I was aided by a suggestion from Mr. William A. Earseman of Allegheny, Penn., an oil-operator of many years' experience, who had noted that the principal gas-wells then known in western Pennsylvania were situated close to where anticlinal axes were drawn on the geological maps. From this he inferred that there must be some connection between the gas-wells and the anticlines. After visiting all the great gas-wells that had been struck in western Pennsylvania and West Virginia, and carefully examining the geological surroundings of each, I found that every one of them was situated either directly on, or near, the crown of an anticlinal axis, while wells that had been bored in the synclines on either side furnished little or no gas, but in many cases large quantities of salt water. Further observations showed that the gas wells were confined to a narrow belt, only one fourth to one mile wide, along the crests of the anticlinal folds. These facts seemed to connect gas territory unmistakably with the disturbance in the rocks caused by their upheaval into arches, but the crucial test was yet to be made in the actual location of good gas territory on this theory. During the last two years, I have submitted it to all manner of tests, both in locating and condemning gas territory, and the general result has been to confirm the anticlinal theory beyond a reasonable doubt.

But while we can state with confidence that all great gas wells are found on the anticlinal axes, the converse of this is not true; viz., that great gas-wells may be found on all anticlinals. In a theory of this kind the limitations become quite as important as, or even more so than, the theory itself; and hence I have given considerable thought to this side of the question, having formulated them into three or four general rules (which include practically all the limitations known to me, up to the present time, that should be placed on the statement that large gas-wells may be obtained on anticlinal folds), as follows:

(a) The arch in the rocks must be one of considerable magnitude; (b) A coarse or porous sandstone of considerable thickness, or, if a fine grained rock, one that would have extensive fissures, and thus in either case rendered capable of acting as a reservoir for the gas, must underlie the surface at a depth of sev-

accumulation are very different in saturated and unsaturated rocks, and in totally dry rocks the oil may lie in or near the axes of the synclines, or at points where its downward progress has been stopped by interrupted or reversed dips, by increased density of the reservoir rock, or by basinward thinning. To explain such occurrences Clapp and Stone[10] postulated an original condition of saturation during accumulation, followed by draining out of the water as a result of later diastrophic movements. They showed also that the water level in different basins may differ widely, even in the same sand, although that sand be completely saturated. The importance of changes in water level during the course of geologic time was here brought out clearly for the first time.

In his classic report, "The Trenton Limestone as a Source of Petroleum and Inflammable Gas in Ohio and Indiana," (74) Orton brought out the fact that the porosity of the Trenton is dependent on the extent of dolomitization, and that this factor largely governs location of the productive areas scattered over the great arch. In this country, previous to the work of Orton, structure had been shown on maps by means of lines showing axes of folds, and by dip and strike symbols, supplemented by cross sections (Fig. 1). Orton introduced the use of contours.

The contour method of representing the structure of sedimentary rocks appears to have been introduced into the United States by J. P. Lesley, about the year 1858. It was used by him in private reports on coal and iron properties for many years, but the first published structure contour map of an oil field is found in a report on the Punjab oil lands (India) by Benjamin Smith Lyman, a nephew and student of Lesley. This report, (48) published in 1870, contains eight structure contour maps, in more or less detail, an example of which is shown in Figure 5. Lyman used the method again, in 1878, in his report on the oil lands of Japan, (59) but no further instance of its use has been found until the appearance of a map by Orton (73) in 1888. Meantime however, it had been used in showing the structure of the Panther Creek anthracite field of Pennsylvania,[11] the structure having been determined by a combination of surface and subsurface observations. Lyman may therefore be said to have originated the method of using contour lines to delineate oil field structure, but following Lesley, he considered structure to be of little consequence, and even refers to upward folds as "saddles" to avoid use of the word anticline (cross section, Fig. 5). A fuller discussion of the early use

[10] R. W. Stone and F. G. Clapp, *U. S. Geol. Survey Bull. 304* (1907), p. 82.

[11] *Second Geol. Survey Pennsylvania*, Vol. AA, pt. 2, Atlas (1883).

of structure contours may be found in papers by Ashburner and B. S. Lyman[12] from which many of the foregoing data are taken.

In the extensive series of Bulletins and Folios which resulted from the work of the United States Geological Survey from 1900 to 1910, structure contour maps were an essential feature. They did much to popularize

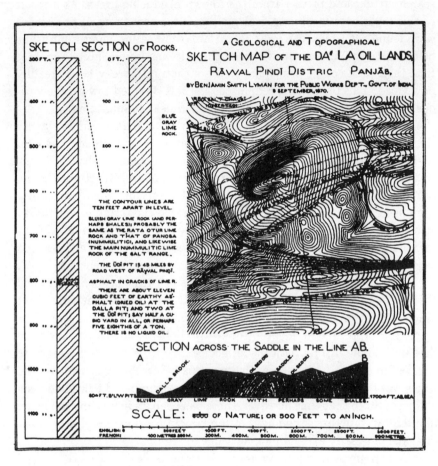

FIG. 5.—First published structure contour map, 1870.

the method, as well as to emphasize the fact that structure is a most important factor in a majority of producing fields, and must be studied in the greatest detail.

[12] C. A. Ashburner, *Second Geol. Survey Pennsylvania 1st Rept. Prog.* (1883), p. 8–AA.
Benj. Smith Lyman, *Amer. Inst. Min. Eng. Trans.*, Vol. 1, pp. 183–92.

PERIOD OF ASCENDENCY OF ANTICLINAL THEORY

During the years 1910 to 1915 the anticlinal theory enjoyed high standing in the scientific world, and acquired a growing prestige among operators. The rapidly growing demand for oil, resulting from rapid increase in number and use of automobiles, caused the prospector for oil to search farther and farther from the known producing areas. As a result of this experience in diverse geologic provinces new types of occurrence were found, of which the most important were saline domes, fault-line fields, and sand lenses. But in all of these it was immediately apparent that the fundamental principles of the structural theory could readily be applied. This being accepted, discussion was directed toward the various modifications of structure which might cause accumulation.

In 1916 F. G. Clapp made the first (80) of several attempts to classify the oil fields of the world into structural groups, which scheme was later revised and modified. (81) Of the hundreds of papers published during this period none seriously questioned the truth of the structural theory, but the great number of unusual types of reservoir which constantly were being demonstrated called for considerable expansion in the theory as originally stated; and the drilling of many dry anticlines recalled forcibly to geologists the truth of White's statement that while "all great gas wells are on anticlines, not all anticlines will produce gas wells."

The great growth of petroleum geology since 1915 has led to gradual changes in the interpretation of structure. It is extremely difficult to set exact dates at which new ideas came in, for many were known a long time before they appeared in the literature. It probably is safe to say, however, that until about 1917 a majority of geologists in the United States were engaged in search for closed anticlines, this being the type of structure which was best known and on which the largest number of fields had been obtained. Since that time however, there has been an increasing tendency to interpret structure in the light of fundamental principles of accumulation, and of greatly expanding knowledge of regional geology.

Munn pointed out (78) in 1909, that the term "anticlinal theory" with its restricted implication, was an unfortunate one, and there has been a growing tendency to replace it with the more inclusive term "structural theory." The important rôle of structure in accumulation of both gas and oil long since has been proved, but the anticline is now considered only as one type of structural reservoir, the simplest, earliest recognized, and undoubtedly the most common.

BIBLIOGRAPHY

In the following pages an effort has been made to list all papers previous to 1890, which have any bearing on the origin, utilization, and growth of the anticlinal theory. Those of greatest importance are referred to in the text, and the others are included to give a better idea of the extent of the literature of that period, and to assist anyone who may care to study further the early use of geology in search for petroleum. Papers are listed in chronological order.

1. W. B. Rogers, "On the Connection of Thermal Springs in Virginia with Anticlinal Axes and Faults," *Trans. Assoc. Amer. Geol. and Nat.*, 1840–42 (1842). Reprint, *Geology of the Virginias* (Appleton, 1884), pp. 577–97.

2. Wm. E. Logan, *Canada Geol. Survey Rept. of Prog.*, 1844 (1846), p. 41.

3. H. D. Rogers, *Geology of Pennsylvania* (1858).

4. J. S. Newberry, "The Rock Oils of Ohio," *Ohio Agricul. Rept., 1859* (1860).

5. H. D. Rogers, "On the Distribution and Probable Origin of the Petroleum or Rock Oil of Western Pennsylvania, New York, and Ohio," *Proc. Philos. Soc. Glasgow*, Vol. 4 (1860), pp. 355–59.

6. J. S. Newberry, "The Oil Wells of Mecca (Ohio)," *Canadian Naturalist*, Vol. 5 (1860), pp. 325, 326.

7. Alexander Winchell, *Geol. Survey Michigan 1st Bienn. Rept. Prog. 1859–60* (1861), p. 73.

8. J. S. Newberry, "The Oil Wells of the Mississippi Valley," *Sci. Amer.*, ser. 2, Vol. 3 (1861), pp. 132, 133.

9. E. B. Andrews, "Rock Oil, Its Geological Relations and Distribution," *Amer. Jour. Sci.* (2), Vol. 32 (1861), pp. 85–93.

10. T. Sterry Hunt, *Montreal Gazette* (March 1, 1861).

11. ———, "Notes on the History of Petroleum or Rock-Oil," *Canadian Naturalist*, Vol. 6 (1861), pp. 241–55. Reprint, *Smith. Inst. Ann. Rept.* (1861), pp. 319–29. Also reprinted in *Chem. News* (London) during same year.

12. M. Gauldrée-Boileau, "Rapport sur l'exploitation de l'huile minérale dans l'Amerique du Nord," *Annales des Mines Mém.*, ser. VI, t. 2, (1862), pp. 95–122.

13. M. Wagner, "Das Petroleum oder Steinöl in Canada," *Zeits. Allg. Erdk.*, Bd. 12 (1862), pp. 279–92.

14. Wm. E. Logan, *Geology of Canada* (1863), pp. 41, 379.

15. T. Sterry Hunt, "Contributions to the Chemical and Geological History of the Bitumens, etc.," *Amer. Jour. Sci.*, Vol. 35 (1863), pp. 157–71.

16. J. P. Lesley, "The Petroleum Vein of Northwestern Virginia," *Proc. Amer. Philos. Soc.*, Vol. 9 (1863), p. 185.

17. T. S. Ridgeway, "Report on the Oil District of Oil Creek, in the State of Pennsylvania," *Jour. Franklin Inst.*, Vol. 45 (1863), pp. 269–73.

18. H. D. Rogers, "Coal and Petroleum," *Good Words* (London, May, 1863), pp. 374–79.

19. ———, "Coal and Petroleum," *Harper's New Monthly Mag.*, Vol. 27 (1863), pp. 259–64.

20. E. W. Evans, "On the Action of Oil Wells," *Amer. Jour. Sci.* (2), Vol. 38 (1864), pp. 159–66.

21. David Murray, "Petroleum, its History and Properties," *Trans. Albany Inst.*, Vol. 4 (1864), pp. 149–66.

22. J. P. Lesley, "On the Mode of Existence of the Petroleum in the Eastern Coal Field of Kentucky," *Proc. Amer. Philos. Soc.*, Vol. 10 (1865), pp. 57, 59.

23. Eli Bowen, *Coal and Coal Oils*, 12mo. (Philadelphia, 1865).

24. J. P. Lesley, *Petroleum, Geological Report on Lands on Paint Lick Fork of Sandy River in Eastern Kentucky*, 8vo. (Philadelphia, 1865).

25. ———, Discussion at meeting of December 1, 1865, *Proc. Amer. Philos. Soc.*, Vol. 10 (1865), p. 190.

26. T. Sterry Hunt, *Petroleum; Its Geological Relations Considered with Especial Reference to its Occurrence in Gaspé* (Quebec, 1865), 19 pp., map.

27. Albert R. Leeds, "The Geography and Geology of Petroleum," *Jour. Franklin Inst.* (3), Vol. 49 (1865), pp. 347–56.

28. Alexander Winchell, "On the Oil Formation of Michigan and Elsewhere," *Amer. Jour. Sci.* (2), Vol. 39 (1865), pp. 350–53.

29. Charles H. Hitchcock, "The Albert Coal, or Albertite, of New Brunswick," *Amer. Jour. Sci.* (2), Vol. 39 (1865), pp. 267–75.

30. Theodore D. Rand, "The Occurrence of Petroleum in Canada," *Jour. Franklin Inst.*, Vol. 80 (1865), pp. 59–65.

31. Alexander Winchell, *Geological Report on Certain Oil Lands Lying in the Counties of Sanilac and St. Clair, Mich.*, 8vo. (Detroit, 1865).

32. ———, "Note on the Geology of Petroleum in Canada West," *Amer. Jour. Sci.* (2), Vol. 41 (1866), pp. 176–78.

33. ———, *Report on the Oil Lands of Lambton County, Canada West*, 8vo. (Detroit, 1866).

34. ———, *Report on the Bruce Oil Lands at Oil Springs, Canada West* (Detroit, 1866).

35. ———, *Geological Report on the Lands of Neff Petroleum Company, in Knox and Conshohocton Counties, Ohio*, 8vo. (Detroit, 1866).

36. ———, *Report on Certain Oil Lands in Wetzel County, West Virginia*, 12mo. (Detroit, 1866).

37. Charles H. Hitchcock, "The Geological Distribution of Petroleum in North America," *Rept. Brit. Assoc.* (1866), pp. 55–57.

38. J. S. Newberry, *Mineral Oil. Prospectus of the Indian Creek and Jack's Knob (Cumberland and Clinton Counties, Ky.) Coal, Salt, Oil Etc. Company; with a Geological Report on the Lands*, 8vo. (Cincinnati, 1866). 20 pp. Reviewed in *Amer. Jour. Sci.* (2), Vol. 41 (1866), p. 284.

39. H. D. Rogers, "On Petroleum," *Proc. Philos. Soc. Glasgow*, 6 (1866), pp. 48–60.

40. E. W. Evans, "On the Oil Producing Uplift of West Virginia," *Amer. Jour. Sci.* (2), Vol. 42 (1866), pp. 334–43.

41. E. B. Andrews, "Petroleum in its Geological Relations," *Amer. Jour. Sci.* (2) Vol. 42 (1866), pp. 33-43.

42. T. Sterry Hunt, "On Petroleum," *Proc. Amer. Assoc.*, Vol. 15 (1866), pp. 29, 30.

43. F. Foucou, "Occurrence and Exploitation of Petroleum in North America," *Mem. Soc. Ing. Civ.* (1867), pp. 82–86.

44. Charles H. Hitchcock, "The Geological Occurrence of Petroleum," *Geol. Mag.*, Vol. 4 (1867), p. 34.

45. Alexander Winchell, "Petroleum in Tennessee," *Min. Manuf. Jour.* (Pittsburgh, January 2, 1867).

46. T. Sterry Hunt, "Sur les pétroles de l'Amerique du Nord," *Bull. Soc. Geol. France*, ser. II, Vol. 24 (1867), pp. 570–73.

47. ———, *Canada Geol. Survey 17th Rept. Prog. 1863–66* (1867), p. 233.

48. B. S. Lyman, *General Report on the Punjab Oil Lands*, folio (Lahore, 1870). (With 8 structure contour maps.)

49. Henry Y. Hind, *Report on the Petroleum Indications at Cheverie, Hants County, Nova Scotia, in Reference to the Probability of a Permanent Supply Being Reached by Boring* (Windsor, Nova Scotia, 1871).

50. T. Sterry Hunt, "On the Oil Wells of Terre Haute, Ind.," *Amer. Naturalist*, Vol. 5 (1871), pp. 576–77.

51. J. S. Newberry, "On the Gas Wells of Ohio and Pennsylvania," *Proc. Lyc. N.H.N.Y.*, 1 (1871), pp. 266–70.

52. ———, "On Ohio and Other Gas Wells," *Amer. Jour. Sci.* (3), Vol. 5 (1873), pp. 225–28.

53. ———, "Geology of Cuyahoga County," *Geol. Survey Ohio*, Vol. 1 (1873), p. 193.

54. H. Hoefer, *Die Petroleumindustrie Nord-Amerikas in geschichtlicher, wirtschaftlicher, geologischer und technischer Hinsicht*, 8vo. (Vienna, 1876).

55. B. S. Lyman, *Geological Survey of the Oil Lands of Japan*, 8vo. (Tokio, 1877).

56. J. S. Newberry, "On the Origin of Petroleum," abstract, *Amer. Naturalist*, Vol. 10 (1876), pp. 316–17.

57. E. S. Nettleton, "On the First Systematic Collection and Discussion of the

Venango County Oil Wells of Western Pennsylvania," *Proc. Amer. Philos. Soc.*, 15 (1877), pp. 429-95.

58. H. Hoefer, *Die geologischer Verhaltnisse der Nord Amerikanischer Petroleumlager* (Ausland, 1878), p. 341.

59. B. S. Lyman, *Geological Survey of Japan (Oil Lands)*, 2nd year's work, 8vo. (Tokio, 1878).

60. Nathaniel S. Shaler, "Petroleum," *Kentucky Geol. Survey Bull. 1* (1879), pp. 5-12.

61. J. Peter Lesley, "Annual Report of the State Geologist," *2nd Geol. Survey of Pennsylvania (3)*, (1880), p. xvi.

62. S. F. Peckham, *Tenth Census Rept.*, Vol. 10 (1880), pp. 38-52.

63. I. C. White, "The Geology of Natural Gas," *Science*, Vol. 5 (1885), pp. 521-22. Reprint, "The Virginias," Vol. 6 (1885), pp. 100-01.

64. Charles A. Ashburner, criticism of White's paper, *Science*, Vol. 6 (1885), pp. 42-43, 184-5.

65. I. C. White, reply to Ashburner's criticism, *Science*, Vol. 6 (1885), pp. 43-44.

66. _____, "The Criticism of the Anticlinal Theory of Natural Gas," *Petrol. Age*, Vol. 5 (1886), pp. 1263-67, pp. 1464-65.

67. J. P. Lesley, "Geology of the Pittsburgh Region," *Trans. Amer. Inst. Min. Eng.*, Vol. 14 (1886), pp. 654-55.

68. Edward Orton, "Preliminary Report upon Petroleum and Inflammable Gas," *Ohio Geol. Survey* (1886), 75 pp.

69. Samuel Aughey, *Ann. Rept. Terr. Geologist Wyoming* (1886), pp. 33-83.

70. H. M. Chance, "The Anticlinal Theory of Natural Gas," *Trans. Amer. Inst. Min. Eng.*, Vol. 15 (1887), pp. 3-13. *Petrol. Age*, Vol. 5 (1886), pp. 1309-12.

71. Charles A. Ashburner, chapter on "Natural Gas" in *A Practical Treatise on Petroleum*, by Benj. S. Crew (1887), pp. 467-78.

72. F. Hue, "Petroleum in America," *Gorn. Jour.*, t. 2 (1888), pp. 100-60.

73. Edward Orton, *Geol. Survey Ohio*, Vol. 6 (1888), p. 94.

74. _____, "The Trenton Limestone as a Source of Petroleum and Inflammable Gas in Ohio and Indiana," *Eighth Ann. Rept.*, *U. S. Geol. Survey*, pt. 2 (1889), pp. 475-662.

75. _____, *Geol. Survey Ohio (3) 1st Ann. Rept.* (1890), p. 51.

76. A. J. Phinney, "The Natural Gas Field of Indiana," *Eleventh Ann. Rept. U. S. Geol. Survey* (1891), pp. 579-742.

77. I. C. White, *West Virginia Geol. Survey*, Vol. 1 (1899), pp. 175-76. (Good summary of his early use of the anticlinal theory.) See also paper by White in *Bull. Geol. Soc. Amer.* (1892).

78. M. J. Munn, "Studies in the Application of the Anticlinal Theory of Oil and Gas Accumulation," *Econ. Geol.*, Vol. 4 (1909), pp. 141-47.

79. M. R. Campbell, "Historical Review of Theories Advanced by American Geologists to Account for the Origin and Accumulation of Oil," *Econ. Geol.*, Vol. 6 (1911), pp. 363-95.

80. I. C. White, "The Anticlinal Theory," *Proc. Amer. Min. Cong. 19th Ann. Sess.* (1917), pp. 550-56.

81. F. G. Clapp, "Revision of the Structural Classification of Petroleum and Natural Gas Fields," *Bull. Geol. Soc. Amer.*, Vol. 28 (1916), pp. 553-602.

82. _____, "The Occurrence of Petroleum," in *Handbook of the Petroleum Industry*, by David T. Day, Vol. 1 (1922), pp. 1-166.

83. E. W. Shaw, discussion, *Econ. Geol.*, Vol. 13 (1918), pp. 207-22.

84. J. V. Howell, "How Old is Petroleum Geology?" *Bull. Amer. Assoc. Petrol. Geol.*, Vol. 14, No. 5 (May, 1930), pp. 607-16.

85. R. B. Harkness, "Early Endeavors on the Anticlinal Theory in Canada," *Bull. Amer. Assoc. Petrol. Geol.*, Vol. 15, No. 6 (June, 1931), pp. 597-610.

BULLETIN OF THE AMERICAN ASSOCIATION OF PETROLEUM GEOLOGISTS
VOL. 50, NO. 2 (FEBRUARY, 1966), PP. 363-374, 11 FIGS., I TABLE

THEORETICAL CONSIDERATIONS OF SEALING AND NON-SEALING FAULTS[1]

DERRELL A. SMITH[2]
Metairie, Louisiana

ABSTRACT

Differentiating between sealing and non-sealing faults and their effects on the subsurface is a major problem in petroleum exploration, development, and production. The fault-seal problem has been investigated from a theoretical viewpoint in order to provide a basis for a better understanding of sealing and non-sealing faults. Some general theories of hydrocarbon entrapment are reviewed and related directly to hypothetical cases of faults as barriers to hydrocarbon migration and faults as paths for hydrocarbon migration. The phenomenon of fault entrapment reduces to a relation between (1) capillary pressure and (2) the displacement pressure of the reservoir rock and the boundary rock material along the fault. Capillary pressure is the differential pressure between the hydrocarbons and the water at any level in the reservoir; displacement pressure is the pressure required to force hydrocarbons into the largest interconnected pores of a preferentially water-wet rock. Thus the sealing or non-sealing aspect of a fault can be characterized by pressure differentials and by rock capillary properties.

Theoretical studies show that the fault seal in preferentially water-wet rock is related to the displacement pressure of the media in contact at the fault. Media of similar displacement pressure will result in a non-sealing fault to hydrocarbon migration. Media of different displacement pressure will result in a sealing fault, provided the capillary pressure is less than the boundary displacement pressure. The trapping capacity of a boundary, in terms of the thickness of hydrocarbon column, is related to the magnitude of the difference in displacement pressures of the reservoir and boundary rock. If the thickness of the hydrocarbon column exceeds the boundary trapping capacity, the excess hydrocarbons will be displaced into the boundary material. Dependent on the conditions, lateral migration across faults or vertical migration along faults will occur when the boundary trapping capacity is exceeded. Application of the theoretical concepts to subsurface studies should prove useful in understanding and in evaluating subsurface fault seals.

INTRODUCTION

Faults are recognized as an important control in the distribution of hydrocarbons in many hydrocarbon provinces. Some faults are known to be sealing to the migration of hydrocarbons under present subsurface conditions; other faults are indicated to be non-sealing to hydrocarbons and commonly are postulated to have provided the path for migration of hydrocarbons in the geologic past. Differentiating between sealing and non-sealing faults and their effects in the subsurface is a major problem for the petroleum geologist and engineer in petroleum exploration, development, and production.

This paper treats the fault-seal problem from a theoretical viewpoint in order to provide the basis for a better understanding of sealing and non-sealing faults. Some general theories of hydrocarbon entrapment are reviewed and related directly to hypothetical cases of faults as barriers to hydrocarbon migration and faults as paths for hydrocarbon migration. Application of the theo-

retical concepts to subsurface studies is considered.

GENERAL THEORY AND EQUATIONS

Hubbert (1953), in deriving the theoretical concepts of a petroleum trap, has shown that the boundary of a reservoir rock is a barrier to hydrocarbon migration because of its capillary properties. Specifically, the boundary represents a difference in the displacement pressure of the reservoir rock and the boundary rock. Displacement pressure is defined as the pressure required to force hydrocarbons into the largest interconnected pores of a hydrophilic (preferentially water-wet) rock. Below this displacement pressure, no hydrocarbons can enter the water-wet rock. The boundary rock necessarily has a higher displacement pressure than the reservoir rock.

Some calculations by Hubbert (1953) give an order of magnitude of the displacement pressures for sediments of various grain sizes in the oil-water system, as shown in Table I. Clay, for example, is a good boundary rock because a large differential pressure is required to force oil into its very minute pores. Clean, uncemented sand is a poor boundary rock (and thus a good reservoir rock) because a very small differential pressure is

[1] Presented before the 15th Annual Meeting of the Gulf Coast Association of Geological Societies, Houston, Texas, October 28, 1965. Manuscript received, February 3, 1965.

[2] Shell Oil Company. (ERP Publication 393).

TABLE I. CAPILLARY DISPLACEMENT PRESSURE FOR
SEDIMENTS OF VARIOUS GRAIN SIZES IN THE
OIL-WATER SYSTEM
(Modified after Hubbert, 1953)

Sediment	Grain Diameters d (Millimeters)	Capillary Displacement Pressure p_d (Atmospheres)
Clay	Less than 1/256*	Greater than 1
Silt	1/256 to 1/16	1 to 1/16
Sand	1/16 to 2	1/16 to 1/500
Granules	2 to 4	1/500 to 1/1000

* The value 1/256 mm. for clay particles is maximum; much finer clays are known. In clay with particle size of 10^{-4}, for example, p_d would be about 40 atmospheres.

required to force oil into the pores. Regardless of the rock type, hydrocarbons must be subjected to a pressure equal to or greater than the displacement pressure of the rock before they will enter the pores of the rock. To state this in different terms, hydrocarbons will be trapped unless they are subjected to a differential pressure equal to or greater than the displacement pressure of the rock. This principle of entrapment of hydrocarbons has been pointed out by Hobson (1954, p. 7–10), Levorsen (1954, p. 436–438), Hill et al. (1961), and Roach (1965, p. 133).

In the common subsurface situation in which water is the wetting phase and hydrocarbons the non-wetting phase, Hill et al. (1961) have shown that hydrocarbons will be trapped if the capillary pressure is less than the displacement pressure of the reservoir boundary material. Capillary pressure is defined as the differential pressure between the hydrocarbons and the water at any level in the reservoir. Pressure in the hydrocarbons will exceed the pressure in the water at any given level because of the difference in density between the hydrocarbons and the water (Hubbert and Rubey, 1959). The differential pressure or capillary pressure increases at increasing elevations above the free water level of the accumulation. Hill et al. (1961) have shown that a boundary rock with a finite displacement pressure has a finite trapping capacity in terms of the thickness of hydrocarbon column. A rock that serves as a boundary to a certain thickness of hydrocarbon column may not necessarily be a boundary to a greater thickness of hydrocarbon column.

Capillary pressure may be expressed by several different equations as shown by Leverett

(1941), Thornton and Marshall (1947), Pirson (1950), Levorsen (1954), and Cole (1961). It is commonly expressed by the equation

$$P_c = (\rho_w - \rho_h)gz \qquad (1)$$

where P_c is capillary pressure in dynes per square centimeter, ρ_w and ρ_h are, respectively, the densities of the water and the hydrocarbons in the reservoir in grams per cubic centimeter, g is the acceleration of gravity in centimeters per second squared, and z is the height in centimeters above the free water level or zero capillary pressure plane of the hydrocarbon accumulation (Leverett, 1941). Expressed in units of measurement commonly used in oil-field practice, the equation for capillary pressure also may be written as

$$P_c = (\rho_w - \rho_h)0.433z \qquad (2)$$

where 0.433 is a derived constant of suitable dimensions to allow an expression of capillary pressure in pounds per square inch, density in grams per cubic centimeter, and height in feet.

An example of capillary pressure in a hypothetical oil reservoir is given in Figure 1. In this example, the maximum capillary pressure is 20 pounds per square inch; i.e., the pressure in the oil phase at the top of the reservoir exceeds the pressure in the water phase by 20 pounds per square inch. Based on the concepts of Hill et al. (1961), the boundary rock overlying this hypothetical oil reservoir must have a displacement pressure of at least 20 pounds per square inch in order to trap the thickness of oil column shown. In more general terms, the hydrocarbons will be trapped if the capillary pressure at all points is less than the displacement pressure of the reservoir boundary material. This concept of entrapment of hydrocarbons is directly applicable to the problem of sealing and non-sealing faults.

For convenient reference, the discussions of sealing and non-sealing faults will deal with hypothetical examples of faulting in an alternating sequence of sandstone and shale such as that found in the Tertiary section of the Gulf Coastal Plain. However, the significant parameter is the capillary property of the rock and not the rock type. The theoretical concepts are equally applicable to other rock sequences such as carbonate rocks which exhibit a wide variation in capillary properties, as shown by Archie (1952), Murray (1960), and others. Stout (1964), for example,

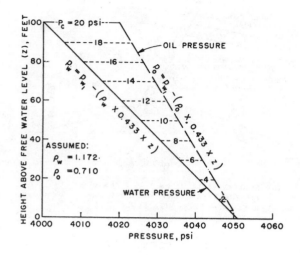

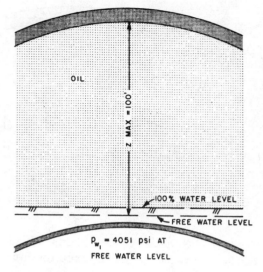

WHERE

P_c = DIFFERENCE BETWEEN THE OIL PRESSURE AND THE WATER PRESSURE AT ANY POINT IN THE RESERVOIR IN psi

P_o = OIL PRESSURE AT ANY POINT IN psi

P_w = WATER PRESSURE AT ANY POINT IN psi

P_{w_i} = PRESSURE OF THE WATER AT THE FREE WATER LEVEL IN psi

ρ_o = DENSITY OF OIL IN THE RESERVOIR IN gm/cc

ρ_w = DENSITY OF WATER IN THE RESERVOIR IN gm/cc

z = HEIGHT ABOVE THE FREE WATER LEVEL IN FEET

Fig. 1 - Capillary pressure in a hypothetical oil reservoir under hydrostatic conditions.

has applied these concepts in his work on pore geometry as related to carbonate stratigraphic traps.

In theory, there are two general ways in which a boundary to the lateral migration of hydrocarbons might result from faulting: (1) by juxtaposed sedimentary lithologic types of different capillary properties and (2) by emplaced fault-zone material formed by mechanical or chemical processes related directly or indirectly to faulting. Sandstone in contact with shale is an obvious example of the first; perhaps less obvious is the second situation, which is illustrated in Figure 2a. Here, two sandstone bodies of different capillary properties are juxtaposed by faulting. These two bodies are termed here Sand 1 and Sand 2. If the displacement pressure of Sand 1 (p_{dR}) is less than the displacement pressure of Sand 2 (p_{dB}), hydrocarbons are trapped, provided that the capillary pressure (P_c) at all points in Sand 1 along

the interface of the two beds is less than the displacement pressure (p_{dB}) of Sand 2. Similarly, hydrocarbons can be trapped against fault-zone material of high displacement pressure, as illustrated in Figure 2b. In both situations the fault must be considered to be a sealing fault because hydrocarbons are trapped at the fault.

From previous discussions, the trapping-capacity limit of the boundary material is determined by the elevation at which the capillary pressure equals the displacement pressure of the boundary, i.e., where

$$P_c = p_{dB}. \qquad (3)$$

The maximum height above the free water level at which hydrocarbons can be trapped by the boundary material is found by substituting equation (2) into equation (3) and solving for z:

$$z\,\text{max} = \frac{p_{dB}}{(\rho_w - \rho_h) \times 0.433}. \qquad (4)$$

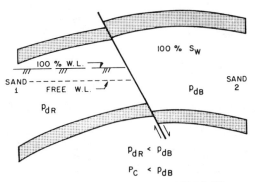

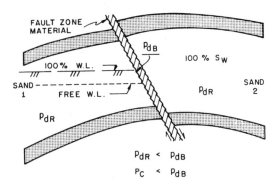

(a) SEALING FAULT FORMED AS A RESULT OF DIFFERENT CAPILLARY PROPERTIES OF JUXTAPOSED SEDIMENTARY LITHOLOGIES (SANDS).

(b) SEALING FAULT FORMED AS A RESULT OF DIFFERENT CAPILLARY PROPERTIES OF RESERVOIR ROCK AND FAULT ZONE MATERIAL.

Fig. 2 - Schematic sections illustrating some hypothetical situations of fault entrapment of hydrocarbons under hydrostatic conditions.

The maximum thickness of hydrocarbon column is somewhat less than z max and can be derived from equation (4). The schematic sections in Figure 3 illustrate some hypothetical examples of fault traps filled to capacity. In these situations, z max in the reservoir can be expressed in terms of the height of the continuous phase of hydrocarbon column (h_o max) and the height from the free water level to the 100 per cent water level (h_w) by

$$z\ \text{max} = h_o\ \text{max} + h_w. \qquad (5)$$

Substituting equation (5) into equation (4) gives

$$h_o\ \text{max} + h_w = \frac{p_{dB}}{(\rho_w - \rho_h) \times 0.433}. \qquad (6)$$

The height from the free water level to the 100-per cent water level (h_w) is determined by the

point at which the capillary pressure (P_c) equals the displacement pressure of the reservoir rock (p_{dR}). For the hydrostatic case, this is expressed by

$$P_c = p_{dR} = (\rho_w - \rho_h) \times 0.433 h_w, \qquad (7)$$

and the equation for h_w is

$$h_w = \frac{p_{dR}}{(\rho_w - \rho_h) \times 0.433}. \qquad (8)$$

Substituting equation (8) into equation (6) and solving for h_o max yields

$$h_o\ \text{max} = \frac{p_{dB} - p_{dR}}{(\rho_w - \rho_h) \times 0.433}. \qquad (9)$$

The maximum thickness of hydrocarbons that can be trapped by the boundary is related to the

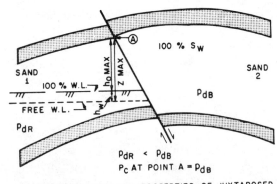

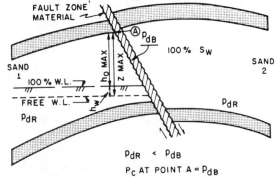

(a) DIFFERENT CAPILLARY PROPERTIES OF JUXTAPOSED SEDIMENTARY LITHOLOGIES (SAND).

(b) DIFFERENT CAPILLARY PROPERTIES OF RESERVOIR ROCK AND FAULT ZONE MATERIAL.

Fig. 3 - Schematic sections illustrating some hypothetical situations of fault traps filled to capacity under hydrostatic conditions.

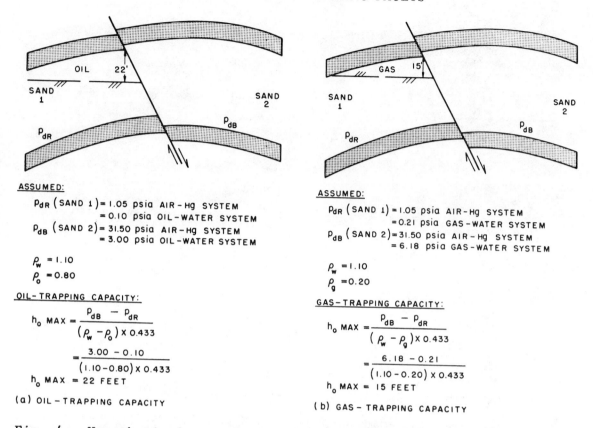

ASSUMED:

p_{dR} (SAND 1) = 1.05 psia AIR-Hg SYSTEM
        = 0.10 psia OIL-WATER SYSTEM
p_{dB} (SAND 2) = 31.50 psia AIR-Hg SYSTEM
        = 3.00 psia OIL-WATER SYSTEM

ρ_w = 1.10
ρ_o = 0.80

OIL-TRAPPING CAPACITY:

$$h_o \text{ MAX} = \frac{p_{dB} - p_{dR}}{(\rho_w - \rho_o) \times 0.433}$$

$$= \frac{3.00 - 0.10}{(1.10 - 0.80) \times 0.433}$$

h_o MAX = 22 FEET

(a) OIL-TRAPPING CAPACITY

ASSUMED:

p_{dR} (SAND 1) = 1.05 psia AIR-Hg SYSTEM
        = 0.21 psia GAS-WATER SYSTEM
p_{dB} (SAND 2) = 31.50 psia AIR-Hg SYSTEM
        = 6.18 psia GAS-WATER SYSTEM

ρ_w = 1.10
ρ_g = 0.20

GAS-TRAPPING CAPACITY:

$$h_o \text{ MAX} = \frac{p_{dB} - p_{dR}}{(\rho_w - \rho_g) \times 0.433}$$

$$= \frac{6.18 - 0.21}{(1.10 - 0.20) \times 0.433}$$

h_o MAX = 15 FEET

(b) GAS-TRAPPING CAPACITY

Fig. 4 - Hypothetical examples illustrating the hydrocarbon-trapping capacity of a boundary sand under hydrostatic conditions.

difference in the displacement pressures of the boundary rock and the reservoir rock. Hydrocarbons can not be trapped unless there is a difference in displacement pressures at the interface of the media. Further, if fault-zone material is present with a displacement pressure less than that of the reservoir sandstones, or if the fault consists of a series of connected, open fractures of low displacement pressure, hydrocarbons will migrate up the fault rather than accumulate in the reservoir beds.

Some hypothetical examples are given in Figures 4 and 5 to illustrate the trapping capacity of a boundary sandstone and boundary fault-zone material, based on the principles discussed. The distribution of the boundary material as well as its displacement pressure will govern the maximum height of hydrocarbon column that can be trapped at a fault. If the boundary material is distributed uniformly along the fault, the displacement pressure of the boundary material will be the controlling factor. In some situations, the distribution of the boundary material may be the

controlling factor. For instance, if the fault-zone material in Figure 5 were distributed across only the upper 20 feet of Sands 1 and 2, it is obvious that only 20 feet of hydrocarbon column could be trapped against the fault in Sand 1.

If the thickness of the hydrocarbon column exceeds the boundary trapping capacity, the excess hydrocarbons will be displaced into the boundary material. Figure 6 illustrates two hypothetical situations in which the trapping capacity of a boundary sandstone is exceeded; capillary pressure equilibrium is assumed in these and succeeding examples. In Figure 6a, hydrocarbons will be trapped until the capillary pressure in Sand 1 at point A equals the displacement pressure of Sand 2. The additional migrating oil is displaced into Sand 2 and will migrate away from the fault. During the period of migration, the fault can be considered to be a non-sealing fault, because oil is migrating across the fault. Ultimately, after migration ceases, the fault will be sealing to some height of hydrocarbon column. The final height of the hydrocarbon column trapped in Sand 1

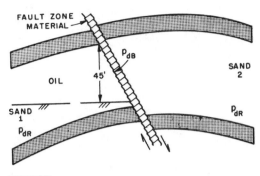

OIL-TRAPPING CAPACITY:

$$h_o \text{ MAX} = \frac{p_{dB} - p_{dR}}{(\rho_w - \rho_o) \times 0.433}$$

$$= \frac{5.95 - 0.10}{(1.10 - 0.80) \times 0.433}$$

$$h_o \text{ MAX} = 45 \text{ FEET}$$

(a) OIL-TRAPPING CAPACITY

GAS-TRAPPING CAPACITY:

$$h_o \text{ MAX} = \frac{p_{dB} - p_{dR}}{(\rho_w - \rho_g) \times 0.433}$$

$$= \frac{12.25 - 0.21}{(1.10 - 0.20) \times 0.433}$$

$$h_o \text{ MAX} = 31 \text{ FEET}$$

(b) GAS-TRAPPING CAPACITY

Fig. 5 - Hypothetical examples illustrating the hydrocarbon-trapping capacity of boundary fault zone material under hydrostatic conditions.

will be equal to the trapping capacity as determined by use of equation (9), regardless of the volume of hydrocarbons that might have migrated into Sand 1. With a suitable trapping environment in the downthrown block, a non-sealing fault to lateral migration could result, as illustrated in Figure 6b. The difference in water levels between the two blocks is related to the

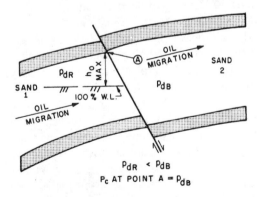

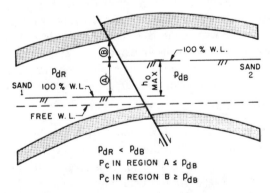

Fig. 6 - Schematic sections illustrating some hypothetical situations in which the trapping capacity of a boundary sand is exceeded under hydrostatic conditions.

difference in displacement pressures of Sands 1 and 2. The fault is non-sealing to lateral migration because there is a continuous phase of hydrocarbons between the fault blocks, and hydrocarbon phase exists. A hypothetical migration, accumulation as long as the continuous hydrocarbons phase exists. A hypothetical migration, accumulation, and production history of such a faulted reservoir is illustrated in Figure 7.

Figure 8 illustrates some hypothetical situations in which the trapping capacity of the fault-zone material is exceeded. In Figure 8a, the additional migrating oil is displaced into the fault-zone material. The oil would then tend to be expelled into Sand 2, because the capillary forces on the oil, as discussed by Hubbert (1953, p. 1977), would be less in Sand 2 than in the fault-zone material. In Figure 8b, the oil displaced into the fault-zone material would migrate vertically through the fault-zone material because the shale is a boundary to lateral migration of the oil displaced into the fault-zone material.

The equations have been developed for determining the boundary trapping capacity under hydrostatic conditions. The principles of entrapment by differences in displacement pressure are valid, however, for hydrocarbons in both hydrostatic and hydrodynamic environments. As shown in Figure 9, hydrocarbons will be trapped in a hydrodynamic situation if the capillary pressure at all points does not exceed the boundary displacement pressure. The determination of capillary pressure for the hydrodynamic case is more difficult than for the hydrostatic case. Pressure in the oil phase can be determined by the equations for a hydrostatic situation; however, determination of pressure in the water phase must be based on a detailed knowledge of the vertical pressure distribution in the dynamic water, which can differ significantly from the vertical distribution in a static situation.

For the traps shown in Figure 9, the trapping capacity of the boundary sandstone, Sand 2, (the maximum thickness of the hydrocarbon column directly below point A) would be the same for the hydrostatic case and for the hydrodynamic case in which water flow across the interface is horizontal. However, horizontal water flow would in very few instances, if ever, be associated with a hydrocarbon reservoir in the subsurface. Water usually will move nearly parallel with the bedding through permeable rocks, as has been pointed out by Hubbert (1953, p. 1970). The dip of the stratum will produce a vertical component of flow in the dynamic water phase. The presence of a hydrocarbon accumulation in a permeable stratum also contributes to some vertical component of flow in the water phase, because the accumulation will act as a restriction to water flow, and water will tend to divert around it. The maximum thickness of the hydrocarbon column below point A (Fig. 9) trapped by the boundary will be greater than in the hydrostatic case if there is a downward component of water flow at the interface, and less than in the hydrostatic case if there is an upward component of water flow at the interface. These examples illustrate that a fault can be non-sealing to water movement without necessarily being non-sealing to hydrocarbon movement.

APPLICATION

Application of the theories of entrapment could resolve many aspects of the fault-seal problem if data were available on the capillary properties of media in contact at the fault. There is, however, a paucity of data on the character of fault zones, fault-zone materials, and associated features in sediments near faults in the subsurface. This presents a considerable problem because a determination of the materials in contact at the fault is not usually possible in the subsurface. The determination that can be made is one of juxtaposed sedimentary lithologic types which would be in contact if no fault-zone material were present.

An indirect approach to the problem is possible. Data are available on the capillary properties of sediments from cores obtained for reservoir and petrophysical determinations. Evaluations can be made from these data to determine if known fault traps are the likely result of a difference in capillary properties of sedimentary lithologic types juxtaposed at the fault. The presence or absence of boundary fault-zone material along the fault might be inferred from these investigations. In addition, the thickness of hydrocarbons trapped at a fault provides a means for estimating the minimum displacement pressure of the fault boundary material, which might be useful for evaluating faults.

The capillary properties of rocks usually are

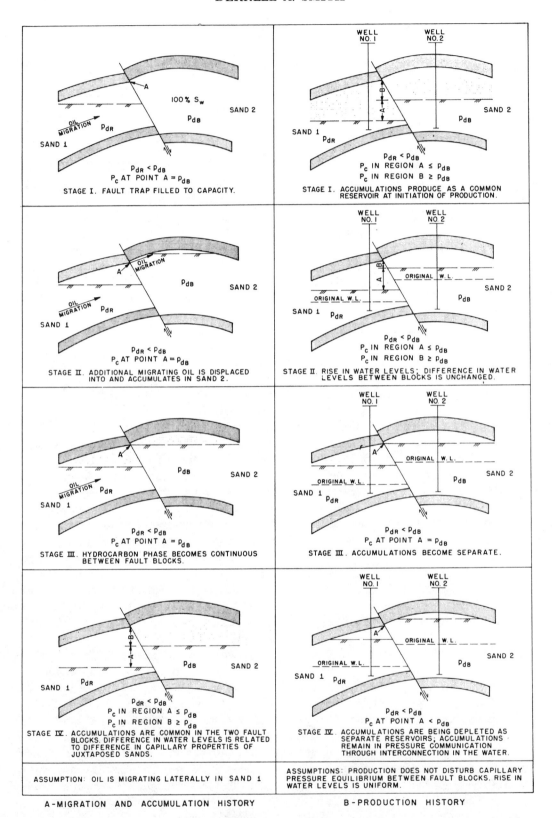

A-MIGRATION AND ACCUMULATION HISTORY   B-PRODUCTION HISTORY

Fig. 7 - Schematic sections illustrating a hypothetical migration, accumulation, and production history of juxtaposed reservoir sands of different capillary properties.

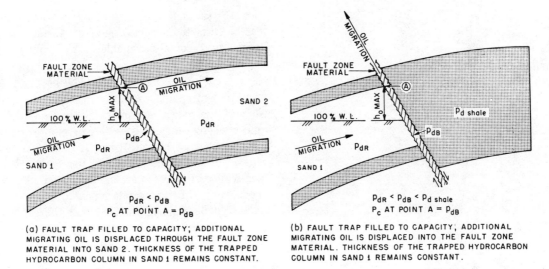

(a) FAULT TRAP FILLED TO CAPACITY; ADDITIONAL MIGRATING OIL IS DISPLACED THROUGH THE FAULT ZONE MATERIAL INTO SAND 2. THICKNESS OF THE TRAPPED HYDROCARBON COLUMN IN SAND 1 REMAINS CONSTANT.

(b) FAULT TRAP FILLED TO CAPACITY; ADDITIONAL MIGRATING OIL IS DISPLACED INTO THE FAULT ZONE MATERIAL. THICKNESS OF THE TRAPPED HYDROCARBON COLUMN IN SAND 1 REMAINS CONSTANT.

Fig. 8 - Schematic sections illustrating some hypothetical situations in which the **trapping capacity** of fault zone material is exceeded under hydrostatic conditions.

measured in the laboratory by injecting mercury into dry rock samples (Purcell, 1949). The displacement pressure can be determined from the air-mercury capillary pressure curve (Fig. 10). In subsurface evaluations, data from the air-mercury system of the laboratory must be converted to the gas-water and oil-water systems encountered in the reservoir rock. Approximate conversion factors are given for the gas-water system as

$$P_c \text{ (gas-water)} = \frac{P_c \text{ (air-mercury)}}{5.1}, \quad (10)$$

and for the oil-water system as

$$P_c \text{ (oil-water)} = \frac{P_c \text{ (air-mercury)}}{10.5}. \quad (11)$$

The conversion factors are based on interfacial tensions of 35 dynes per centimeter for the oil-water system, 72 dynes per centimeter for the gas-water system, and 480 dynes per centimeter for the air-mercury system. The interfacial tension is affected by the composition of the fluid as well as by other factors, including pressure and temperature; therefore, the conversion factors

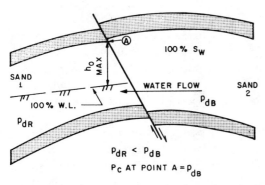

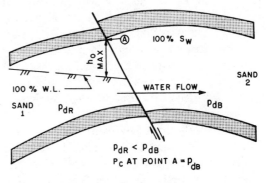

(a) WATER FLOW FROM THE BOUNDARY SAND INTO THE RESERVOIR SAND.

(b) WATER FLOW FROM THE RESERVOIR SAND INTO THE BOUNDARY SAND.

Fig. 9 - Schematic sections illustrating some hypothetical situations of fault entrapment of hydrocarbons under hydrodynamic conditions with sands juxtaposed across the fault.

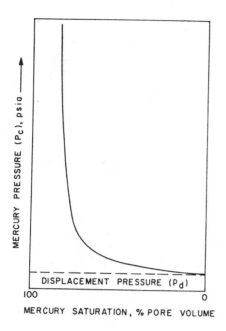

MERCURY PRESSURE (P_c), psia

DISPLACEMENT PRESSURE (P_d)

100   0

MERCURY SATURATION, % PORE VOLUME

Fig. 10 - Idealized air-mercury capillary pressure curve.

given above may be only an approximation in specific cases.

As shown in Table I, sand generally has low displacement pressures and a small hydrocarbon-trapping capacity, whereas clay has high displacement pressures and a large hydrocarbon-trapping capacity. Not all sandstone bodies have low displacement pressures; shaly sandstone, poorly sorted sandstone, and sandstone with calcite or silica cement can have high displacement pressures and a large hydrocarbon-trapping capacity. Figure 11 shows the displacement pressures of some cores from Gulf Coast Tertiary sandstones. Many of the sandstone cores have displacement pressures of sufficient magnitude to trap significant thicknesses of hydrocarbons. A comparison of these observed sandstone displacement pressures with the data used in the hypothetical examples in Figures 4 and 5 will give a general idea of the significance of the data.

Other data necessary to determine the trapping capacity commonly are available from measurements made on a routine basis in field operations. Reservoir-pressure data are available from drill-stem tests and from pressure measurements made during the producing life of a reservoir. PVT analyses provide data on the density of the hydrocarbons under reservoir conditions. In addi-

tion, the subsurface density of oil, gas, and water can be estimated in some cases from surface measurements commonly available on well effluent.

SUMMARY AND CONCLUSIONS

1. In hydrophilic rock, the displacement pressure of the media in contact at the fault determines whether a fault is sealing or non-sealing to the migration of hydrocarbons. If the media have similar displacement pressures, the fault will be non-sealing to hydrocarbon migration. If the media have different displacement pressures, the fault will be sealing, provided that the capillary pressure is less than the boundary displacement pressure.

2. The trapping capacity (the thickness of the trapped hydrocarbon column) of a fault boundary is related to the difference in displacement pressures of the reservoir rock and the boundary rock. If the thickness of the hydrocarbon column exceeds the boundary trapping capacity, the excess hydrocarbons will be displaced into the boundary rock. Dependent on the conditions, lateral migration across faults or vertical migration along faults will occur when the boundary trapping capacity is exceeded.

3. In theory, hydrocarbons can be trapped at a fault where sedimentary lithologic types of different displacement pressure are in contact or where fault-zone material with a displacement pressure greater than that of the reservoir beds is present along the faults. The distribution and displacement pressure of the boundary material along the fault are factors controlling the thickness of hydrocarbon column that can be trapped at a fault.

4. A fault can be non-sealing to vertical migration if fault-zone material has a lower displacement pressure than that of the reservoir beds or if the fault consists of a series of connected, open fractures of low displacement pressure.

5. If the capillary properties of media in contact at the fault (including sedimentary lithologic types and fault-zone material) are known, it can be determined whether a fault is sealing or non-sealing, and the trapping capacity of the fault can be calculated. If the presence or absence of boundary fault-zone material is unknown, data usually are available to determine if known fault traps are the likely result of a difference in capillary properties of sedimentary lithologic types

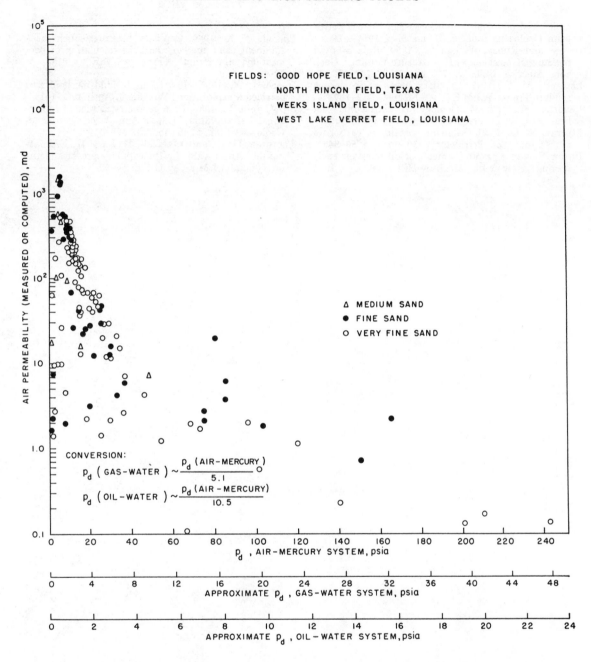

Fig. 11 - Displacement pressure (p_d) from air-mercury capillary pressure curves of some cores from Gulf Coast Tertiary sands.

juxtaposed at the fault. The presence or absence of boundary fault-zone material along the fault can be inferred in some cases from these investigations.

REFERENCES

Archie, G. E., 1952, Classification of carbonate reservoir rocks and petrophysical considerations: Am. Assoc. Petroleum Geologists Bull., v. 36, no. 2, p. 278–298.

Cole, Frank W., 1961, Reservoir engineering manual: Houston, Gulf Publishing Co., p. 5–9.

Hill, Gilman A., Colburn, William A., and Knight, Jack W., 1961, Reducing oil-finding costs by use of hydrodynamic evaluations, *in* Economics of petroleum exploration, development, and property evaluation: International Oil and Gas Educational Center, Southwest Legal Foundation: Englewood Cliffs, N. J., Prentice-Hall, Inc., p. 38–69.

Hobson, G. D., 1954, Some fundamentals of petroleum geology: London, Oxford University Press.

Hubbert, M. King, 1953, Entrapment of petroleum

under hydrodynamic conditions: Am. Assoc. Petroleum Geologists Bull., v. 37, no. 8, p. 1954–2026.

——— and Rubey, William W., 1959, Role of fluid pressure in mechanics of overthrust faulting: Geol. Soc. America Bull., v. 70, no. 2, p. 149–150.

Leverett, M. C., 1941, Capillary behavior in porous solids: Trans. A.I.M.E., v. 142, p. 152–169.

Levorsen, A. I., 1954, Geology of petroleum: San Francisco, W. H. Freeman and Co., p. 433–439.

Murray, R. C., 1960, Origin of porosity in carbonate rocks: Jour. Sed. Petrology, v. 30, no. 1, p. 59–84.

Pirson, Sylvan J., 1950, Elements of oil reservoir engineering: New York, McGraw-Hill Book Co., Inc., p. 245–272.

Purcell, W. R., 1949, Capillary pressures—their measurement using mercury and the calculation of permeability therefrom: Trans. A.I.M.E., v. 186, p. 39–48.

Doach, J. W., 1965, How to apply fluid mechanics to petroleum exploration: World Oil, April, p. 131–134.

Stout, John L., 1964, Pore geometry as related to carbonate stratigraphic traps: Am. Assoc. Petroleum Geologists Bull., v. 48, no. 3, p. 329–337.

Thornton, O. F., and Marshall, D. L., 1947, Estimating interstitial water by the capillary pressure method: Trans. A.I.M.E., v. 170, p. 69–80.

The American Association of Petroleum Geologists Bulletin
V. 64, No. 2 (February 1980), P. 145-172, 29 Figs.

Sealing and Nonsealing Faults in Louisiana Gulf Coast Salt Basin[1]

DERRELL A. SMITH[2]

Abstract Fault-controlled accumulations in the hydropressured Tertiary section were studied in 10 Louisiana Gulf Coast salt basin fields located on low-relief structures. Investigations were limited to traps associated with faults which restrict vertical migration of hydrocarbons; that is, where an accumulation is in contact with the fault. The fault-lithology-accumulation relations observed are (1) fault sealing, with hydrocarbon-bearing sandstone in lateral juxtaposition with shale; (2) fault nonsealing to lateral migration, with parts of the same sandstone body juxtaposed within the hydrocarbon column; (3) fault nonsealing to lateral migration, with sandstone bodies of different ages juxtaposed within the hydrocarbon column; and (4) fault sealing, with sandstone bodies of different ages juxtaposed within the hydrocarbon column. In some places, these four relations are present at different levels along the same fault.

In the examples studied, only faults nonsealing to lateral migration were observed where parts of the same sandstone body are juxtaposed across a fault. With sandstone bodies of different ages juxtaposed, some faults are sealing and others are nonsealing to lateral migration, but sealing faults are the most common. The fault seal apparently results from the presence of boundary fault-zone material emplaced along the fault by mechanical or chemical processes related directly or indirectly to faulting.

INTRODUCTION

Faulting in conjunction with structural closure provides the trapping mechanism for large volumes of hydrocarbons in the Louisiana Gulf Coast salt basin. A close interrelation exists between faulting and hydrocarbon distribution in many of the oil and gas fields in the basin. This study was undertaken (1) to determine the different situations of fault entrapment of hydrocarbons in Tertiary sediments of the Gulf Coast salt basin and (2) to investigate the role of juxtaposed sediments in a sandstone-shale sequence in creating sealing and nonsealing faults.

GENERAL APPROACH, LOCATION, AND METHOD OF STUDY

Relations of faulting to hydrocarbon accumulation were studied in 10 Louisiana Gulf Coast fields located as shown in Figure 1. The fields are associated with faulted, deep-seated structures having relatively low structural dips. Study was restricted to hydrocarbon accumulations in the hydropressured Tertiary sandstone-shale section. These conditions of structure, stratigraphy, and accumulation represent the least complicated geologic conditions suitable ·for investigation of sealing and nonsealing faults.

The faults investigated are normal faults. Growth took place along some of them during the time of deposition of the stratigraphic intervals of interest. Drag flexure of beds is not indicated by well control for the faults investigated; instead, as in the more typical Gulf Coast relation, beds in the upthrown block tend to rise toward the fault and beds in the downthrown block tend to dip downward toward the fault. Sediments are interpreted as juxtaposed on the basis of their relative structural positions across a fault. However, the normal sedimentary lithologies cannot be proved to be in contact at the faults. The juxtaposed lithologies would be in contact if no fault-zone material were present along the fault to separate them.

Criteria for analyzing the fault seal are based mainly on theoretical studies (Smith, 1966) of sealing and nonsealing faults showing that the displacement pressure of media in contact at the fault determines whether a fault is sealing or nonsealing to the movement of hydrocarbons. Displacement pressure is defined as the pressure required to force hydrocarbons into the largest interconnected pores of preferentially water-wet rock. Hydrocarbons can be trapped at a fault where sedimentary lithologic types of different displacement pressures are in contact or where fault-zone material with a displacement pressure greater than that of the reservoir beds is present along the fault (Fig. 2).

Data are usually available to evaluate without speculation only those faults which now restrict vertical migration, that is, where a hydrocarbon

[1]Manuscript received, January 10, 1979; accepted, July 26, 1979.

[2]Shell Oil Co., New Orleans, Louisiana 70160.

This paper is based on results of work conducted at Bellaire Research Center, Shell Development Co. (a division of Shell Oil Co.), Houston, Texas.

Article Identification Number
0149-1423/80/B002-0001$03.00/0

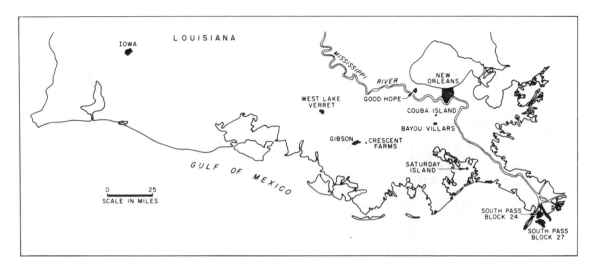

FIG. 1—Index map showing location of fields studied.

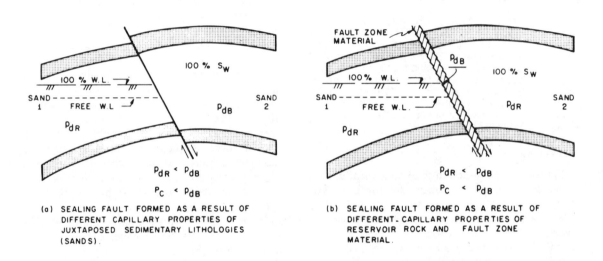

(a) SEALING FAULT FORMED AS A RESULT OF DIFFERENT CAPILLARY PROPERTIES OF JUXTAPOSED SEDIMENTARY LITHOLOGIES (SANDS).

(b) SEALING FAULT FORMED AS A RESULT OF DIFFERENT CAPILLARY PROPERTIES OF RESERVOIR ROCK AND FAULT ZONE MATERIAL.

FIG. 2—Schematic sections illustrating hypothetical situations of fault entrapment of hydrocarbons under hydrostatic conditions (after Smith, 1966). P_{dR}, displacement pressure of reservoir rock; P_{dB}, displacement pressure of boundary rock; P_c, capillary or differential pressure between hydrocarbon phase and water phase in rock pores. If media in contact at fault have similar displacement pressures, fault will be nonsealing to hydrocarbon migration. If media in contact at fault have different displacement pressures, fault will be sealing, provided that capillary pressure, P_c, is less than boundary displacement pressure.

accumulation is in contact with the fault over any vertical distance. Figure 3 illustrates the analyses of the fault seal for hypothetical situations as a basis for determinations made in the study. In all situations, a fault is considered sealing to vertical migration if the hydrocarbon accumulation is trapped against the fault. A fault is sealing to lateral migration if the entire hydrocarbon-bearing sand interval is in lateral juxtaposition with shale or with water-bearing sand (Fig. 3a, b). Common hydrocarbon contacts in the interval of

sandstone juxtaposition indicate that a fault is nonsealing to lateral migration (Fig. 3c). An oil accumulation with free gas caps would not necessarily have a common gas-oil contact if the gas caps are entirely controlled by structural closure or contact of sandstone against shale. A common gas-oil contact with different oil-water contacts is generally indicative of a fault nonsealing to lateral migration (Fig. 3e); this situation might be present in sandstones with lateral and vertical variations in lithology and capillary properties.

HYPOTHETICAL SITUATION	ANALYSIS OF FAULT SEAL	
	VERTICAL MIGRATION	LATERAL MIGRATION
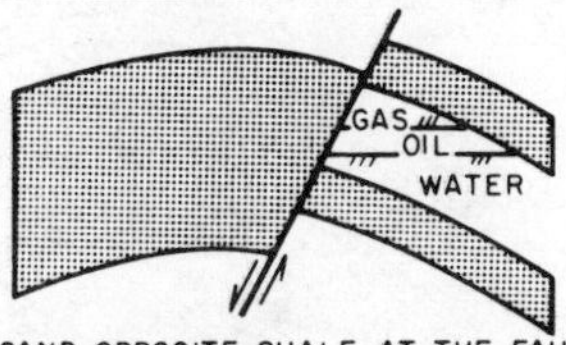 (a) SAND OPPOSITE SHALE AT THE FAULT. HYDROCARBONS JUXTAPOSED WITH SHALE.	SEALING	SEALING RESERVOIR BOUNDARY MATERIAL MAY BE THE SHALE FORMATION OR FAULT ZONE MATERIAL.
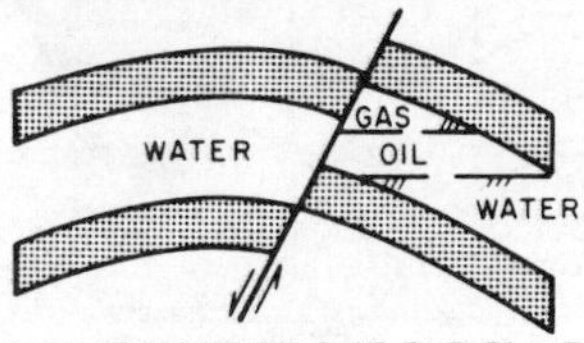 (b) SAND OPPOSITE SAND AT THE FAULT. HYDROCARBONS JUXTAPOSED WITH WATER.	SEALING	SEALING SEAL MAY BE DUE TO A DIFFERENCE IN DISPLACEMENT PRESSURES OF THE SANDS OR TO FAULT ZONE MATERIAL WITH A DISPLACEMENT PRESSURE GREATER THAN THAT OF THE SANDS.
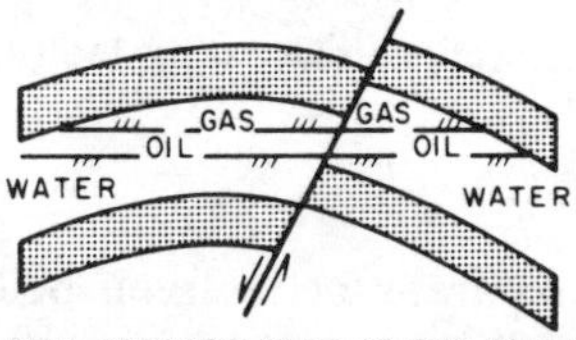 (c) SAND OPPOSITE SAND AT THE FAULT. COMMON HYDROCARBON CONTENT AND CONTACTS.	SEALING	NONSEALING POSSIBILITY IS REMOTE THAT FAULT IS SEALING AND THE RESERVOIRS OF DIFFERENT CAPACITY HAVE BEEN FILLED TO EXACTLY THE SAME LEVEL BY MIGRATING HYDROCARBONS.
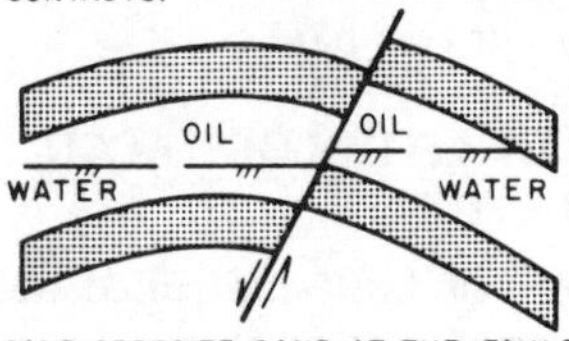 (d) SAND OPPOSITE SAND AT THE FAULT. DIFFERENT WATER LEVELS.	SEALING	UNKNOWN NONSEALING IF WATER LEVEL DIFFERENCE IS DUE TO DIFFERENCES IN CAPILLARY PROPERTIES OF THE JUXTAPOSED SANDS. SEALING IF WATER LEVEL DIFFERENCE IS NOT DUE TO DIFFERENCES IN CAPILLARY PROPERTIES OF THE JUXTAPOSED SANDS.
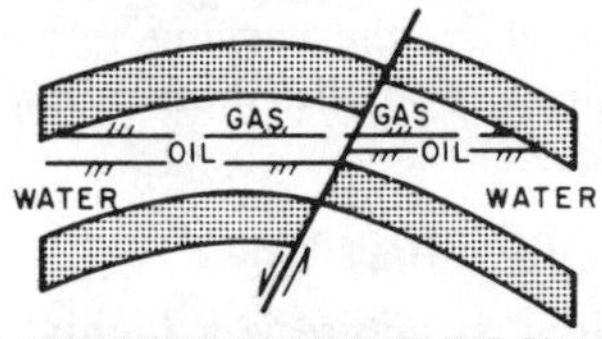 (e) SAND OPPOSITE SAND AT THE FAULT. COMMON GAS-OIL CONTACT, DIFFERENT OIL-WATER CONTACT.	SEALING	NONSEALING POSSIBILITY IS REMOTE THAT FAULT IS SEALING AND MIGRATING GAS HAS FILLED THE RESERVOIRS OF DIFFERENT CAPACITY TO EXACTLY THE SAME LEVEL.
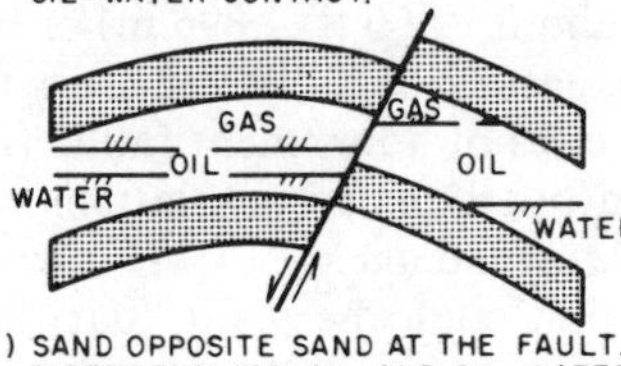 (f) SAND OPPOSITE SAND AT THE FAULT. DIFFERENT GAS-OIL AND OIL-WATER CONTACTS.	SEALING	SEALING A DIFFERENCE IN BOTH GAS-OIL CONTACT AND OIL-WATER CONTACT INFERS THE PRESENCE OF BOUNDARY FAULT ZONE MATERIAL ALONG THE FAULT.
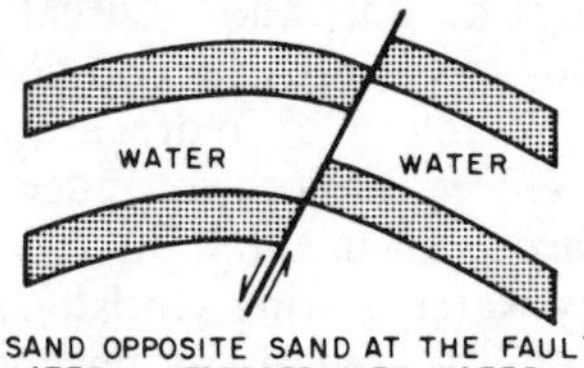 (g) SAND OPPOSITE SAND AT THE FAULT. WATER JUXTAPOSED WITH WATER.	UNKNOWN	UNKNOWN

FIG. 3—Analysis of fault seal for some hypothetical fault-lithology-accumulation relations.

If different hydrocarbon contacts occur at the fault in juxtaposed sandstones, the fault may be either sealing or nonsealing to lateral migration (Fig. 3d, f). A fault is nonsealing to lateral migration if the difference in hydrocarbon levels is attributable to differences in displacement pressures of the juxtaposed sandstones. If reservoir rock with high displacement pressure is in contact with reservoir rock with low displacement pressure, both the oil-water and the gas-oil contact might be different in the juxtaposed sandstones, but both contacts would be at a higher level in the rock with the higher displacement pressure. In this case, the difference in gas-oil contacts would be less than the difference in oil-water contacts in the juxtaposed sandstones. If the difference in water levels is not due to a difference in displacement pressures of the sandstones, the fault is sealing and the presence of boundary fault-zone material can be inferred.

Data are seldom available to evaluate the fault seal where water-bearing sandstones are juxtaposed at a fault (Fig. 3g). The fault could be either sealing or nonsealing to hydrocarbon migration. A fault that is sealing to water movement is also sealing to hydrocarbons in water-wet rock. However, a fault that is nonsealing to water movement is not necessarily nonsealing to hydrocarbon movement (Hill et al, 1961; Smith, 1966).

The evaluations of the fault seal are valid even if some hydrocarbons are now migrating, because hydrocarbons could not be distributed in the observed patterns unless restricted to some extent in vertical (and sometimes lateral) movement by boundary material. However, the analysis of the fault seal is valid only for the thickness of hydrocarbons present in the reservoir prior to initiation of production. For example, a fault that is laterally or vertically sealing to 50 ft (15 m) of hydrocarbon column cannot be assumed to have a capacity to trap even one additional foot of hydrocarbon column. However, a fault that is nonsealing to lateral migration with 50 ft (15 m) of hydrocarbon column juxtaposed could conceivably become sealing to lateral migration under some conditions as the thickness of the hydrocarbon column is reduced during production operations (Smith, 1966). The analyses of the fault seal based on reservoir performance are discussed for some faulted reservoirs included in the study.

The relations of faulting and accumulation derived from the field studies are presented on the basis of the lithologies juxtaposed at the fault, as illustrated schematically in Figure 4. In the Tertiary section of the Gulf Coast salt basin, the beds of nonreservoir rock that serve as boundary mate-

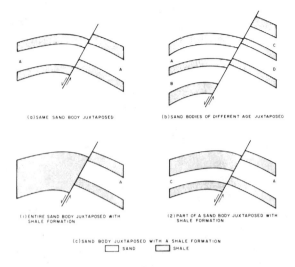

(a) SAME SAND BODY JUXTAPOSED

(b) SAND BODIES OF DIFFERENT AGE JUXTAPOSED

(1) ENTIRE SAND BODY JUXTAPOSED WITH SHALE FORMATION

(2) PART OF A SAND BODY JUXTAPOSED WITH SHALE FORMATION

(c) SAND BODY JUXTAPOSED WITH A SHALE FORMATION

☐ SAND   ☐ SHALE

FIG. 4—Schematic sections illustrating fault-lithology relations discussed.

rial to cap and separate the reservoir beds in the normal stratigraphic sequence are composed chiefly of shale and siltstone. These beds are referred to as shale in this paper.

OBSERVED FAULT LITHOLOGY ACCUMULATION RELATIONS

A large number of fault-controlled accumulations were evaluated in this study. Some of the more informative and interesting field examples are presented to show the varying influence of faulting on hydrocarbon distribution and reservoir performance.

Good Hope Field

The Good Hope structure is a highly faulted, domal closure associated with a salt dome which reaches within about 9,500 ft (2,896 m) of the surface. The fault pattern is a central graben-horst system with a series of adjustment faults (Fig. 5). The main hydrocarbon accumulations are in sandstones of Miocene age.

A determination could be made that the fault was sealing or nonsealing to lateral migration in 41 fault-lithology-accumulation situations in Good Hope field, where each fault is considered to seal vertical migration of hydrocarbons. The investigations involved 9 Miocene sandstones and 12 faults with throws from about 30 to 300 ft (10 to 100 m). Many water-bearing sandstones juxtaposed across the faults are present, but the nature of the fault seal cannot be analyzed. The faults are nonsealing to lateral migration in all 20 cases analyzed in which parts of the same sandstone

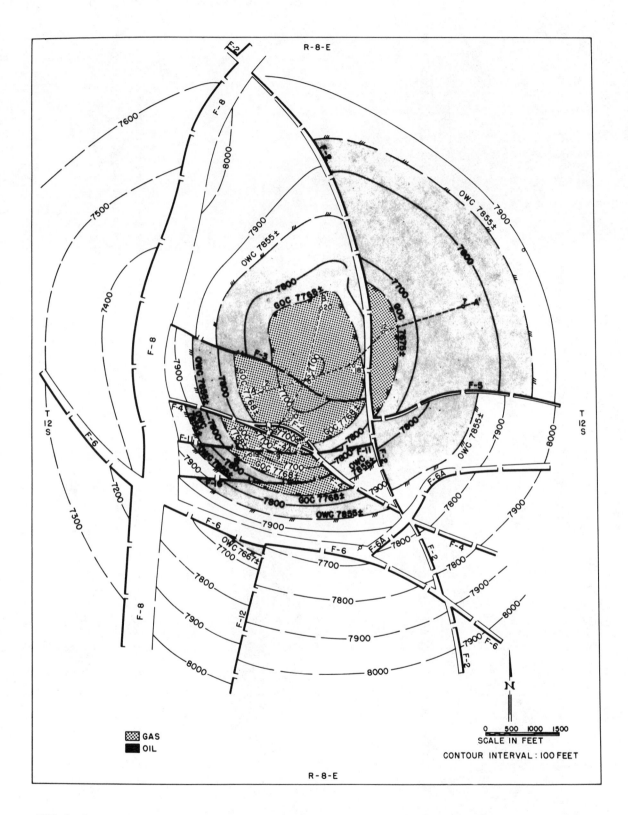

FIG. 5—Structure map of "P" sandstone, Good Hope field. *AA'* and *BB'* are lines of section shown in Figure 6.

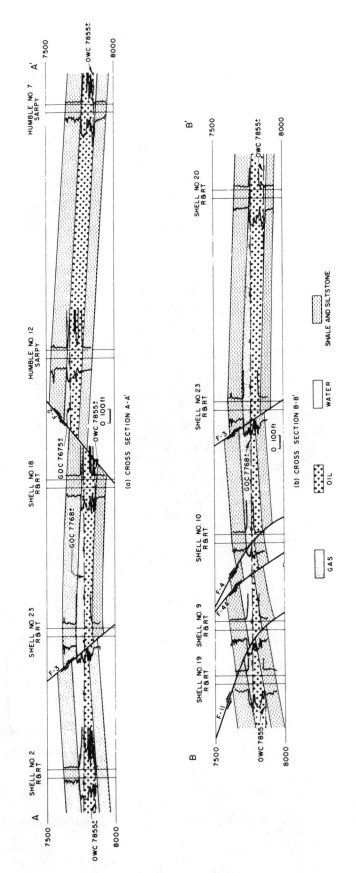

FIG. 6—Cross sections showing accumulations and structural relations in "P" sandstone, Good Hope field. See Figure 5 for locations.

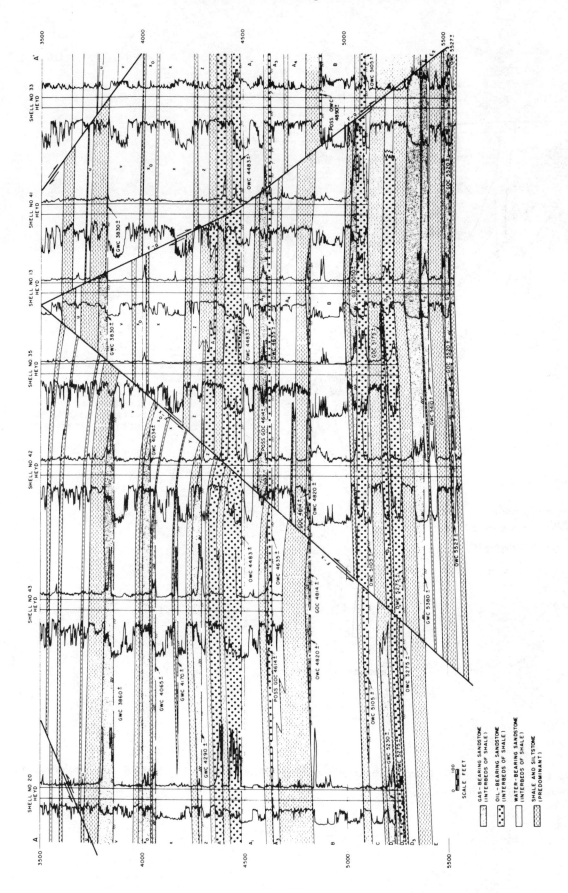

FIG. 7.—Cross section showing accumulations associated with faults F-a and F-x, Iowa field.

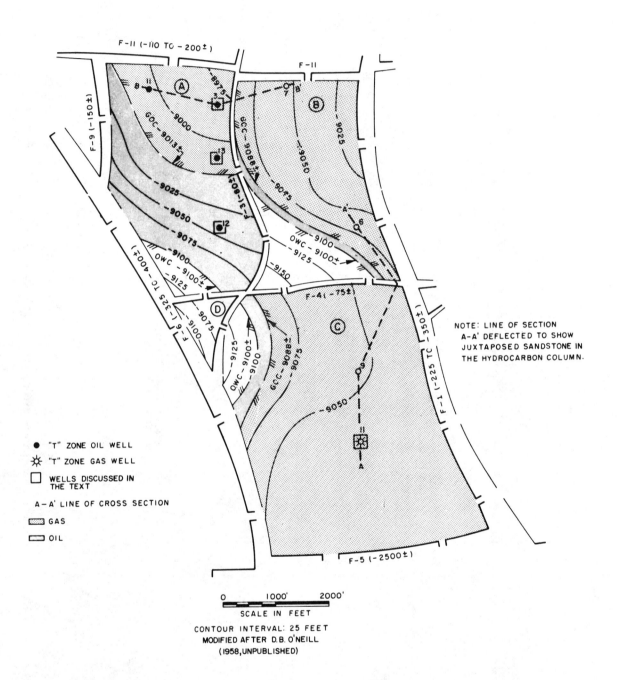

FIG. 8—Structure map of "T" zone in part of South Pass Block 24 field. *AA'* and *BB'* are lines of section shown in Figures 9, 10.

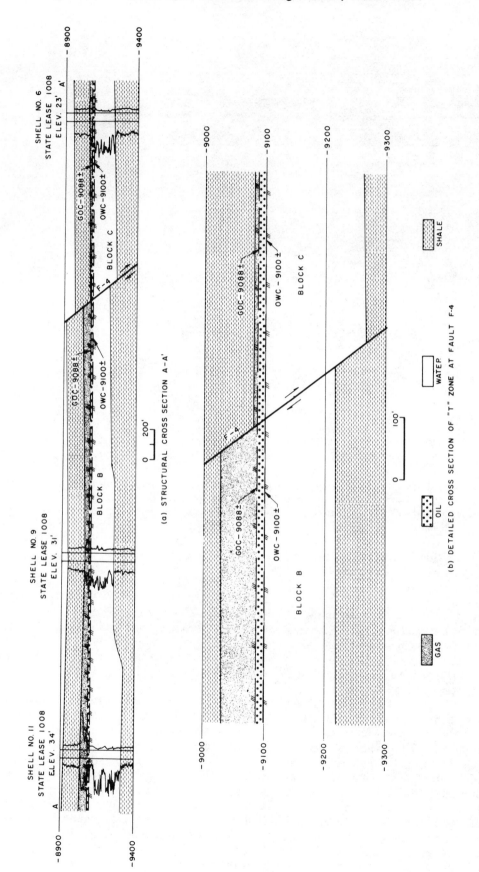

FIG. 9.—Cross sections showing structural relations of some "T" zone accumulations, South Pass Block 24 field. Location shown in Figure 8.

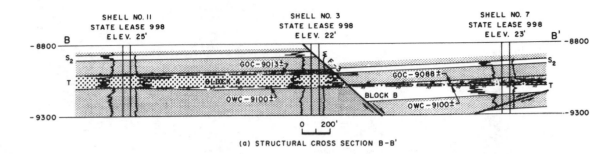

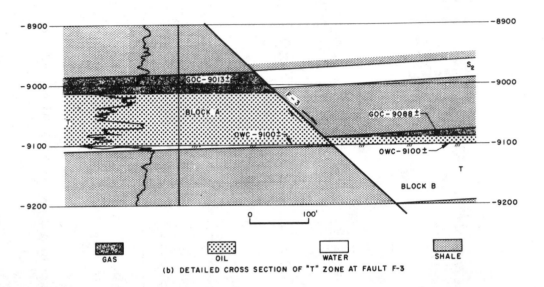

FIG. 10—Cross sections showing structural relation of some "T" zone accumulations, South Pass Block 24 field. Fault F-3 is nonsealing to lateral migration in interval of "T" zone juxtaposition, but is sealing where "T" zone in upthrown block is juxtaposed with "S₂" zone in downthrown block. Location shown in Figure 8.

body are juxtaposed. Some faults are indicated to be sealing to lateral migration at one level and nonsealing at another level with sandstones of different ages juxtaposed.

The "P" zone illustrates some of the observed situations. Parts of the sandstone are juxtaposed within the hydrocarbon column across all faults in the central closure area, except the north segment of fault F-2 (Fig. 6). An oil column with a common water level has been found in the tested blocks in the central closure. The original gas-oil contact is different in the upthrown and downthrown blocks of fault F-2 but is common to the downthrown segments. The downthrown gas-oil contact appears to be controlled by the structural closure above the highest point of sandstone juxtaposition across fault F-2. The accumulations are in capillary pressure equilibrium between fault blocks, and the faults must have been nonsealing to lateral migration during the accumulation history to allow equalization of hydrocarbons between blocks.

Iowa Field

The Iowa field is associated with a salt dome which reaches within about 9,000 ft (2,743 m) of the surface. The locations of the Miocene accumulations associated with faults F-a and F-x are shown in Figure 7. Some shallow gas accumulations in the "v" through "z" zones in the downthrown block of fault F-x are in close proximity to the fault, but are apparently controlled by structural closure. Their distribution does not provide direct data on the fault seal and these zones are not included in the evaluations. In all 11 accumulations in which an evaluation is possible, faults F-a and F-x are indicated to be nonsealing to lateral migration with parts of the same sandstone body juxtaposed. Associated with these faults are five sand-fault situations in which the fault is indicated to be sealing and one situation in which the fault is indicated to be nonsealing to lateral migration with sandstones of different ages juxtaposed.

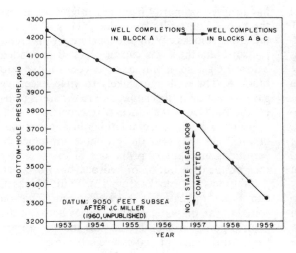

FIG. 11—Bottom-hole pressure history of "T" zone fault-block reservoirs, South Pass Block 24 field.

South Pass Block 24 Field

Accumulations in the "T" zone (upper Miocene) in the South Pass Block 24 field provide some data on the influence of faulting on hydrocarbon movement during the periods of both accumulation and production in a situation where parts of the same sandstone body are juxtaposed. The structural interpretation and the locations of the accumulations in the "T" zone in a part of the field are shown in Figure 8.

The "T" zone is juxtaposed within the hydrocarbon column across fault F-4, and blocks B and C have common oil-water and gas-oil contacts in the interval of sandstone juxtaposition (Fig. 9). Block A has a common water level with the other blocks, but the gas-oil contact is different. Within the interval of sandstone juxtaposition between blocks A and B across fault F-3, the "T" zone has common content (Fig. 10). The gas cap in block B (and thus block C) appears to be exactly con-

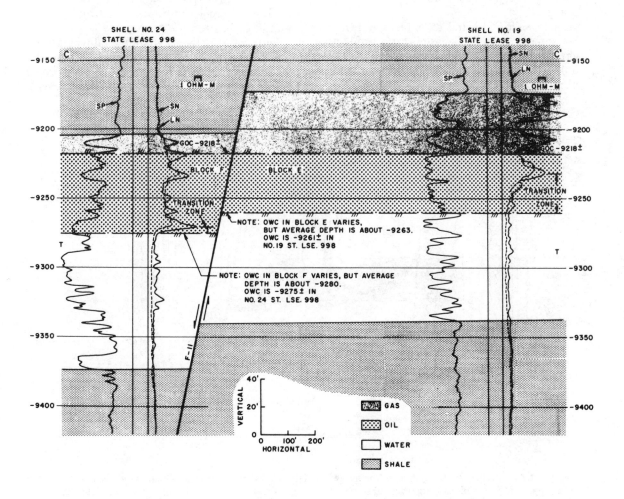

FIG. 12—Cross section showing structural relation of some "T" zone accumulations, South Pass Block 24 field.

Derrell A. Smith

trolled by structural closure above the highest point of juxtaposed sandstone across fault F-3. The interval of sandstone juxtaposition is a spill-point for hydrocarbons and the accumulations are in capillary pressure equilibrium between the segments. Faults F-3 and F-4 must have been nonsealing to lateral migration late in the accumulation history to allow equilization of hydrocarbons between blocks.

Some reservoir performance data indicate that faults F-3 and F-4 are also nonsealing to lateral movement of hydrocarbons during production operations. Well completions were made initially in the thick oil column in block A. Reservoir pressure declined steadily from initial production (Fig. 11). In mid-1957, well 11 State Lease 1008 was the first completion in block C. Bottom-hole pressure was essentially the same as the pressure in block A, which was then about 500 psia (3447 kPa) lower than initial reservoir pressure. Blocks A and C continued to show a similar pressure decline. Later pressure data indicate that block B is also in pressure communication with the other blocks.

Three wells (3, 12, and 13) in block A near fault F-3 showed early high gas-oil ratio production, whereas other wells at a greater distance from fault F-3 produced with normal or solution gas-oil ratios. The excess gas production in block A

has resulted primarily from gas migration across the nonsealing faults from the thick gas caps in the other blocks. With the decline in reservoir pressure, the thick gas cap in block B expanded below the gas spillpoint, and gas could move into block A. The wells most likely to yield excessive gas production in this situation are the wells nearest the fault as in the block A reservoir.

Some "T" zone accumulations in another part of the field illustrate the probable influence of variations in capillary properties of the reservoir rock on the distribution of hydrocarbons between juxtaposed fault blocks (Fig. 12). Fault blocks E and F are indicated to have a common gas-oil contact. The oil-water contact varies between the control points in each block, and there is an apparent difference in water levels between blocks E and F across fault F-11. The upper part of the "T" zone is a delta-fringe deposit in which lateral and vertical changes in lithology are to be expected. The reservoir rock is very fine-grained sandstone, silty sandstone, and sandy siltstone, sometimes shaly. A variation in capillary properties of the reservoir rock can be assumed from the variation in grain size.

Differences in the length of the oil-water transition zone in the two wells shown in Figure 12 tend to confirm the variation in capillary properties of the reservoir rock. Near the water level an

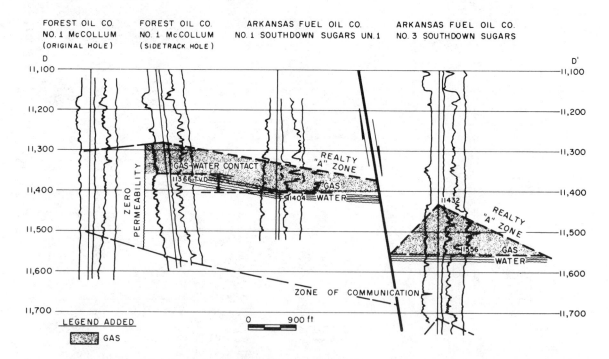

FIG. 13—Cross section showing fault-lithology-accumulation relations, Crescent Farms field. Known structural closure in downthrown block is represented schematically. Downthrown closure is full to fault-sandstone spillpoint and only water-bearing part of Realty "A" zone is juxtaposed at fault. Redrawn from Simmons (1961).

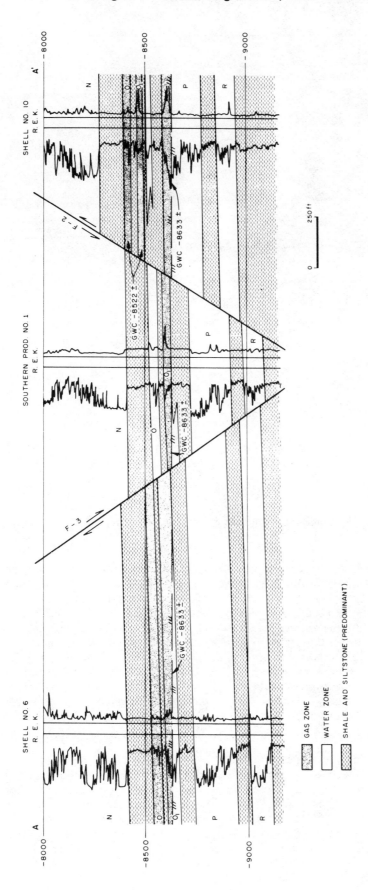

FIG. 14—Cross section showing accumulations and structural relations, Gibson field.

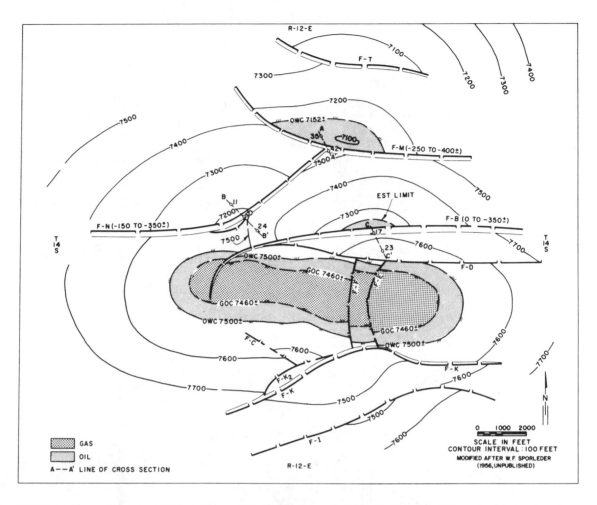

FIG. 15—Structure map of "N" sandstone, West Lake Verret field. *AA'*, *BB'*, and *CC'* are locations of sections shown in Figures 16, 17.

interval of reservoir rock with poorer capillary properties in the upthrown block is juxtaposed with an interval of better capillary properties in the downthrown block. Variations in capillary properties could account for the difference in water levels between fault blocks; thus, fault F-11 is interpreted to be nonsealing to lateral migration. Pressure histories prior to initiating a pressure-maintenance water-injection project indicated that the "T" zone accumulations in blocks E and F produce in pressure communication.

Crescent Farms Field

Crescent Farms field illustrates a special hydrocarbon distribution and reservoir performance with parts of the same sand body juxtaposed. A structural cross section of the Realty zone (upper Miocene) is shown in Figure 13. Only the water-bearing part of the Realty zone is juxtaposed at the fault. The accumulation in the downthrown block appears to be controlled by structural clo-

sure, and the lowest water level position in the downthrown block coincides with the spillpoint of sandstone juxtaposition across the fault. Because the gas accumulation fills the downthrown closure to the spillpoint, the fault must have been nonsealing to lateral migration at the present depth of burial. Reservoir pressure data (Simmons, 1961) indicate that the accumulations are producing in pressure communication, probably through the juxtaposed water-bearing part of the sandstone body.

Gibson Field

The Gibson field provides an example of a fault that is nonsealing to lateral migration of hydrocarbons with sandstones of different ages juxtaposed. The "P" zone (upper Miocene) in the upthrown block of fault F-2 is juxtaposed with the "O_1" zone in the downthrown block of fault F-2, and the zones have a common gas-water contact in the interval of juxtaposed sandstone

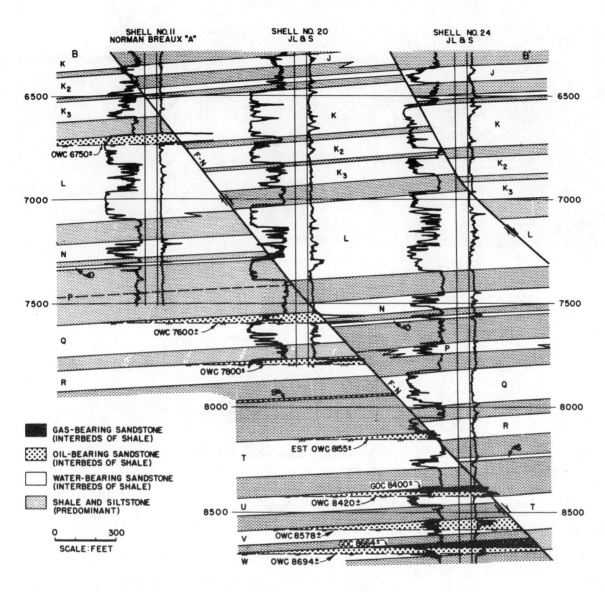

FIG. 16—Cross section showing accumulations associated with fault F-N, West Lake Verret field.

(Fig. 14). These data indicate that equalization of hydrocarbons has occurred across fault F-2 and that the fault was nonsealing during a period of the accumulation history. Reservoir performance data are not definitive, but some data indicate that the fault was nonsealing during production operations.

West Lake Verret Field

The West Lake Verret field is associated with a complexly faulted, anticlinal structure attributed to deep-seated salt movement (Fig. 15); 18 oil and gas accumulations trapped by postdepositional faults F-M, F-N, and F-B in the Miocene section were investigated (Figs. 16, 17). Fourteen accumulations in the upthrown sandstones are juxta-

posed with water-bearing sandstones of a younger age. The possibilities for the existence of the traps with sandstones of different ages juxtaposed are (1) sedimentary variations and facies changes at or near the faults in one or both of the juxtaposed zones, (2) differences in capillary displacement pressures of the juxtaposed sandstones, and (3) boundary fault-zone material along the faults. These possibilities have been investigated, primarily for the situations shown in Figure 17.

With postdepositional faults, stratigraphic traps at or near the faults would not be expected unless faulting occurred along an existing litho-logic boundary. A comparison of the fault positions with the sandstone distribution patterns shows that the faults are not preferentially

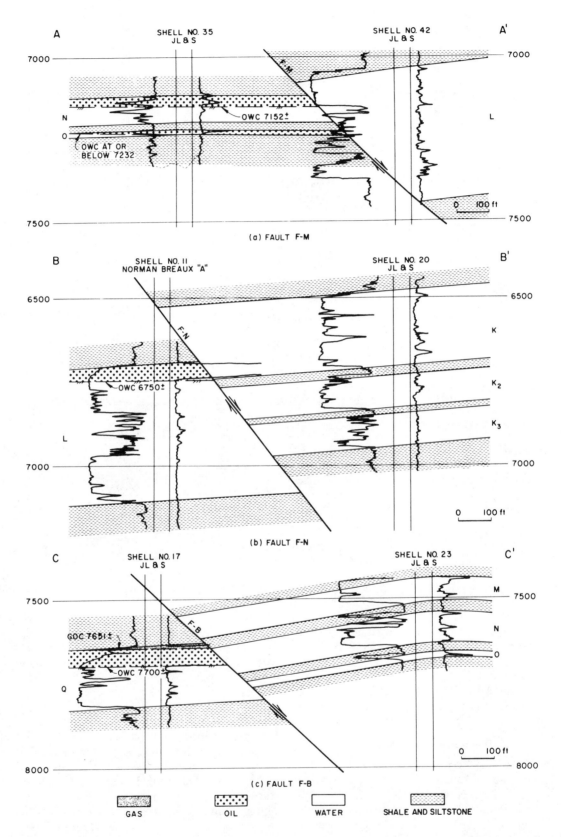

FIG. 17—Detailed cross sections showing structural relations of some hydrocarbon- and water-bearing sandtsones across faults F-M, F-N, and F-B, West Lake Verret field.

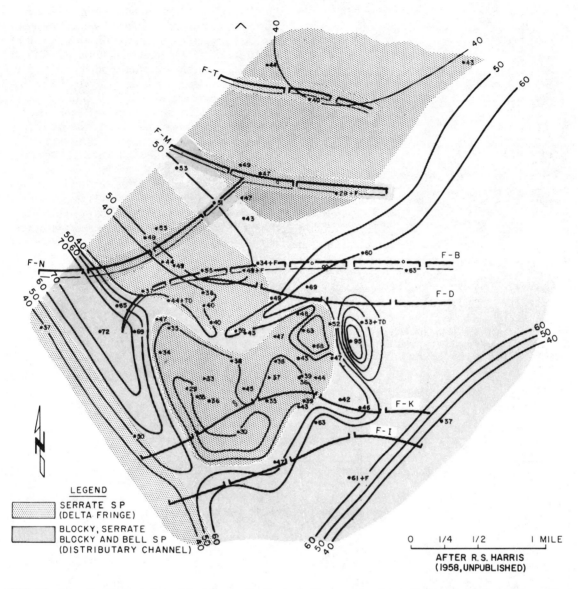

FIG. 18—Map showing relation of faulting to distribution of log types and net thickness in "N" sandstone, West Lake Verret field.

aligned along sedimentary boundaries (Fig. 18). There is no evidence of sedimentary changes in the juxtaposed sandstones that would preferentially trap hydrocarbons along the faults.

Evaluations of the trapping capacity of the juxtaposed sandstones were made as outlined by Smith (1966). Typical air-mercury capillary pressure curves are shown in Figure 19. Air-mercury displacement pressures of the sandstones as determined from the capillary pressure curves range from about 1.5 to 7.0 psia (10.3 to 48.3 kPa). Some thin beds of siltstone within the sand bodies had displacement pressures up to 40.0 psia (275.8 kPa). The capillary properties are rather typical of Gulf Coast reservoir sandstones.

The "N" zone in the upthrown block of fault F-M contains a 50-ft (15 m) oil column juxtaposed with the water-bearing "L" zone (Fig. 17a). Average displacement pressure is similar for the "L" and "N" sandstone cores, and the "L" sand could not provide the trap for the "N" zone accumulation. If "N" sandstone with the lowest air-mercury displacement pressure (1.5 psia; 10.3 kPa) were assumed to be in contact with "L" sandstone with the highest displacement pressure (7.0 psia; 48.3 kPa), the maximum thickness of hydrocarbon column that could be trapped against the "L" zone would be about 3 ft (1 m).

The "K" and "L" sandstones also have similar displacement pressures, and the entrapment of

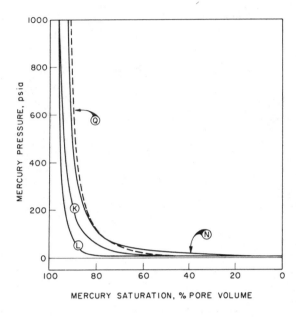

FIG. 19—Typical air-mercury capillary pressure curves for "K," "L," "N," and "Q" zones, West Lake Verret field.

the 50-ft (15 m) "L" zone oil column at fault F-N (Fig. 17b) cannot be explained by a difference in capillary properties. Similarly, the entrapment of the 50-ft (15 m) "Q" zone oil column at fault F-B (Fig. 17c) is not the result of capillary differences in the juxtaposed sandstones. Even if the displacement pressure of the "N" siltstone cores (40.0 psia or 275.8 kPa maximum) were used for computing the trapping capacity, the maximum thickness of oil column that could be trapped in the "Q" zone against the siltstone would be about 23 ft (7 m).

In summary, the evaluations of the fault traps at West Lake Verret field show (1) that the traps are not the result of facies changes or sedimentary variations in either of the juxtaposed sandstones which would preferentially trap hydrocarbons at the faults, and (2) that the traps cannot be explained by a difference in displacement pressures of the normal juxtaposed sandstone lithologies. By elimination, the explanation for the fault traps must be that some type of boundary fault-zone material is present along the faults and is a result, directly or indirectly, of the faulting process.

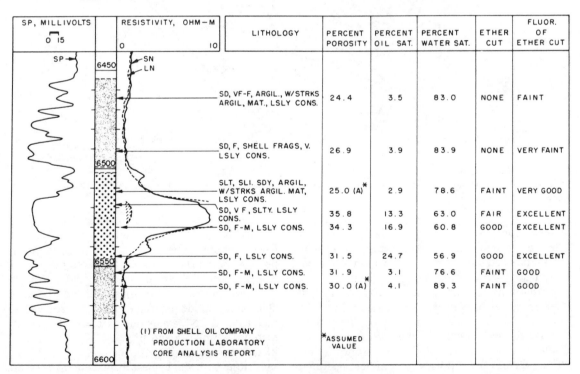

FIG. 20—Electric-log and petrophysical data on well 7, State Lease 1012, South Pass Block 27 field.

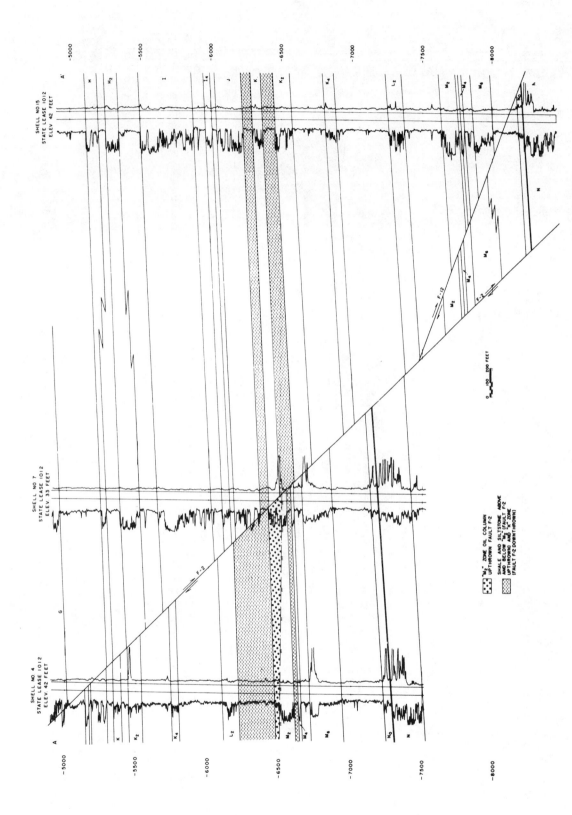

FIG. 21—Cross section showing structural relations across fault F-2, South Pass Block 27 field.

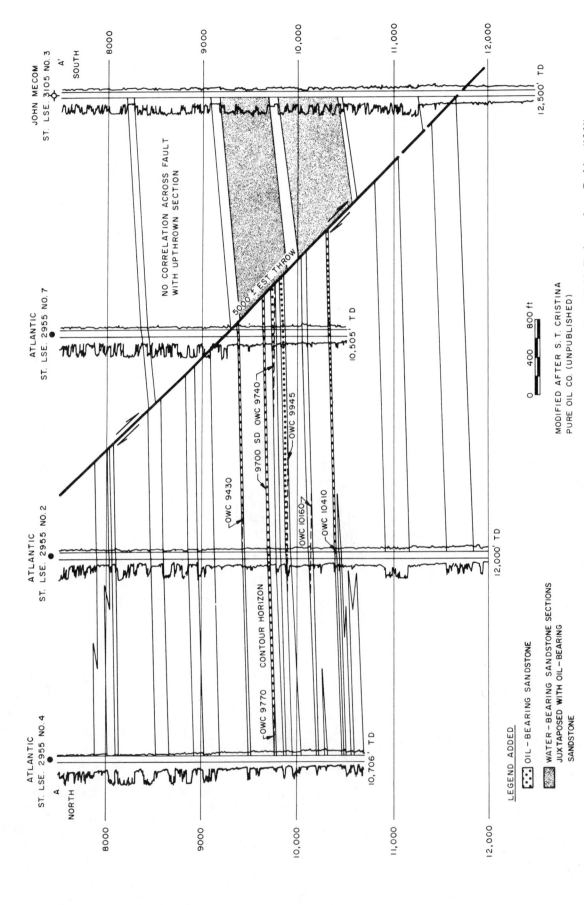

FIG. 22—Cross section showing accumulations and structural relations, Saturday Island field. Redrawn from Perkins (1961).

South Pass Block 27 Field

An example from the South Pass Block 27 field illustrates a sealing fault where a younger sandstone deposited during growth-fault movement is in juxtaposition with an older sandstone deposited prior to growth-fault movement.

The electric log of part of the Pliocene-Miocene section in well 7, State Lease 1012, is shown in Figure 20. The well penetrated a water-bearing sandstone interval separated from an underlying oil-bearing interval by an apparent shale bed about 3 ft (1 m) thick. Major fault F-2 occurs within the apparent shale bed that separates the two intervals, as shown by the structure section in Figure 21. The upper water-bearing sandstone is a part of the "K" zone in the downthrown block of fault F-2 and the lower oil-bearing sandstone is a part of the "M₂" zone in the upthrown block of fault F-2. Well 7 crossed the fault at the point of sandstone juxtaposition and there is little doubt that the fault is sealing.

Fault F-2 originated shortly after deposition of the "L₂" zone (Smith, 1961). The "M₂" zone was deposited prior to the initiation of fault F-2, and the "K" zone was deposited during a period of active movement along the fault.

Bayou Villars, Couba Island, and Saturday Island Fields

Perkins (1961) described these fields and showed that hydrocarbon-bearing sandstones in the upthrown block of large growth faults are juxtaposed with water-bearing zones in the downthrown block. The cross section of the Saturday Island field in Figure 22 typifies the situation. The accumulations in Bayou Villars and Couba Island fields are in the upthrown blocks of growth faults which traverse southwest-plunging anticlinal noses. No dip reversal is indicated in the upthrown blocks in these two fields. In the Saturday Island field, accumulations are in the upthrown block of a large growth fault which strikes east-west across a northeast-plunging nose. Some structural closure is present in the upthrown

block (Perkins, 1961), but the closure is not sufficient to control the accumulations. Most of the hydrocarbon-bearing sandstones are opposite younger water-bearing sandstones at the faults and the faults are sealing to lateral migration of hydrocarbons.

SUMMARY OF OBSERVED FAULT LITHOLOGY ACCUMULATION RELATIONS

The observed fault-accumulation relations are summarized in the following on the basis of lithologies juxtaposed at the fault. The normal sedimentary lithologies cannot be proved to be in contact at the faults because fault-zone material may be present along the faults, separating the juxtaposed lithologies. The presence or absence of boundary fault-zone material can be inferred from available data in some places.

Sandstone in Juxtaposition with Shale

The situation in which hydrocarbon-bearing sandstone is juxtaposed with shale across a fault is common for faults of both very small and very large throw. The schematic sections in Figure 23 illustrate the places where the entire hydrocarbon column is opposite a shale at the fault. The accumulation is trapped at the fault, and the boundary of the accumulation may be the juxtaposed shale or fault-zone material. Shale has a high capillary displacement pressure and a large hydrocarbon-trapping capacity. However, some faults are sealing to lateral migration under conditions other than those involving sandstone opposite shale and no proof is available that the juxtaposed shale is the actual boundary material trapping the accumulations.

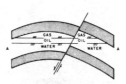

(a) SAND IN JUXTAPOSITION WITHIN THE HYDROCARBON COLUMN. RESERVOIR CONTACTS COMMON. FAULT NONSEALING IN THE INTERVAL OF SAND CONTACT.

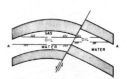

(b) SAND IN JUXTAPOSITION WITHIN THE HYDROCARBON COLUMN. DIFFERENCE IN WATER LEVELS DUE TO DIFFERENCE IN CAPILLARY PROPERTIES OF THE JUXTAPOSED SAND. FAULT NONSEALING IN THE INTERVAL OF SAND CONTACT.

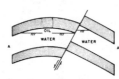

(c) SAND NOT IN JUXTAPOSITION WITHIN THE HYDROCARBON COLUMN. ACCUMULATION IN THE DOWNTHROWN BLOCK CONTROLLED BY STRUCTURAL CLOSURE WITH THE WATER LEVEL AT THE HIGHEST POINT OF SAND CONTACT. ACCUMULATION IN THE UPTHROWN BLOCK CONTROLLED BY CLOSURE OF SAND AGAINST SHALE.

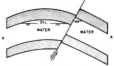

(d) SAND NOT IN JUXTAPOSITION WITHIN THE HYDROCARBON COLUMN. ACCUMULATIONS CONTROLLED BY STRUCTURAL CLOSURE AND CLOSURE OF SAND AGAINST SHALE. THICKNESS OF ACCUMULATIONS LESS THAN MAXIMUM CLOSURE ABOVE THE HIGHEST POINT OF SAND CONTACT.

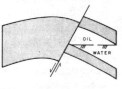

(a) ENTIRE RESERVOIR SAND BODY JUXTAPOSED WITH SHALE; ENTIRE ACCUMULATION JUXTAPOSED WITH SHALE.

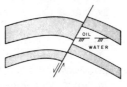

(b) PART OF THE RESERVOIR SAND BODY JUXTAPOSED WITH SHALE; ENTIRE ACCUMULATION JUXTAPOSED WITH SHALE.

FIG. 23—Schematic sections showing observed situations of faulting and accumulation with sandstone in juxtaposition with shale.

FIG. 24—Schematic sections showing observed situations of faulting and accumulation with parts of same sandstone body in juxtaposition.

Some cases were observed in which only a part of hydrocarbon column in a sandstone was laterally juxtaposed with shale. The fault is sealing to lateral migration where sandstone is opposite shale at the fault. However, sandstone opposite sandstone is implied where this shale-accumulation relation is present, and the faults are sometimes sealing and sometimes nonsealing to lateral migration with sandstone juxtaposed.

Parts of Same Sandstone Body in Juxtaposition

Where parts of the same sandstone body are juxtaposed across a fault the fault throw may not be sufficient to displace the entire sand body, and an interval of sandstone juxtaposition remains after faulting (Fig. 24).

In the field examples studied, the two fault segments have a common accumulation where the hydrocarbon column extends below the spillpoint of juxtaposed sandstone (Fig. 24a, b). The faults are nonsealing. In Figure 24b, the oil-water contact of the two blocks is not at the same level, but the accumulation is common to both blocks because the difference in water levels is attributed to a difference in capillary properties of the juxtaposed rock. In Figure 24c the downthrown accumulation fills the structural closure to the spillpoint of juxtaposed sandstone. The hydrocarbon accumulation is only in point contact with the fault and does not provide direct data concerning the fault seal. The fact that the block is full to the spillpoint could indicate that the fault is nonsealing to lateral migration in the interval of sandstone juxtaposition. In Figure 24d, the accumula-

tions do not fill the reservoirs to the spillpoint and are controlled by structural closure and contact of sandstone against shale; data are not available to analyze the lateral fault seal in this situation.

In 35 cases studied in which an accumulation extended below the sandstone spillpoint with parts of the same sandstone body juxtaposed, faults were nonsealing to lateral migration. The upthrown and downthrown blocks have common accumulations which produce as a common reservoir for which production and reservoir performance data were definitive.

Sandstone Bodies of Different Ages in Juxtaposition

Figure 25 illustrates the situation in which a sandstone body in the downthrown block is juxtaposed with an older sandstone body in the upthrown block. The sandstones have been displaced past at least one shale bed (excluding interbeds of shale within a sandstone body) that separates the older bed from the younger bed in the normal stratigraphic sequence. Both sealing and nonsealing faults are present in this situation.

In Figure 25a, the accumulation in the upthrown block extends below the highest point of apparent sandstone contact, but the accumulation is trapped at the fault. Fault traps with sandstones of different ages juxtaposed are associated with both postdepositional faults and growth faults. The trap could be the result of a difference in capillary properties of the juxtaposed beds or the result of the presence of boundary fault-zone material along the fault. Investigations indicate that boundary fault-zone material is the likely reason for the fault seal along the faults analyzed.

In some examples, the accumulation is common to two fault segments with sandstones of different ages juxtaposed, and the fault is nonsealing (Fig. 25b). Sealing faults with sandstones of different ages juxtaposed are much more common than nonsealing faults for the cases that could be analyzed. No example of common accumulation in juxtaposed sandstones of different ages was found in association with faults of greater than 200 ft (61 m) of throw. In addition, faults with associated throw less than 200 ft (61 m) are more commonly sealing than nonsealing to lateral migration where sandstones of different ages are juxtaposed. No statistical significance is attributed to these observations at this time because there are numerous juxtaposed water-bearing sandstones of different ages (Fig. 25c) in which the fault might be nonsealing, but data are not available to evaluate the fault seal.

A major problem is to determine why some faults are sealing and others are nonsealing to lateral migration where sandstones of different

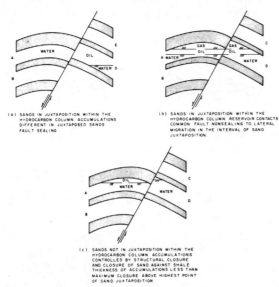

(a) SANDS IN JUXTAPOSITION WITHIN THE HYDROCARBON COLUMN. ACCUMULATIONS DIFFERENT IN JUXTAPOSED SANDS. FAULT SEALING

(b) SANDS IN JUXTAPOSITION WITHIN THE HYDROCARBON COLUMN RESERVOIR CONTACTS COMMON FAULT NONSEALING TO LATERAL MIGRATION IN THE INTERVAL OF SAND JUXTAPOSITION.

(c) SANDS NOT IN JUXTAPOSITION WITHIN THE HYDROCARBON COLUMN. ACCUMULATIONS CONTROLLED BY STRUCTURAL CLOSURE AND CLOSURE OF SAND AGAINST SHALE THICKNESS OF ACCUMULATIONS LESS THAN MAXIMUM CLOSURE ABOVE HIGHEST POINT OF SAND JUXTAPOSITION.

FIG. 25—Schematic sections showing observed situations of faulting and accumulation with sandstone bodies of different ages in juxtaposition.

ages are juxtaposed. The problem appears to be rather complex because the same fault was found to be sealing at one level and nonsealing at another level in some places. Boundary fault-zone material is responsible for the fault traps in many places, but the material could not be identified in the examples studied. The distribution as well as the displacement pressure of the boundary material will govern the maximum thickness of hydrocarbon column that can be trapped. Because boundary fault-zone material is present along some faults, juxtaposition of the reservoir rock with shale is not a prerequisite for a fault trap in the Gulf Coast salt basin.

GULF COAST FAULT ZONES AND BOUNDARY FAULT-ZONE MATERIALS

Much is known about faulting in the Gulf Coast, but few data are available on the character of fault zones, fault-zone materials, and the associated features in sediments near the faults in the relatively soft Tertiary formations of the Gulf Coast subsurface. Some data on Gulf Coast fault zones and fault-zone materials are summarized here because of the importance of this information to the study of sealing and nonsealing faults. Discussions of faulting in published literature are usually hidden in reports of specific fields or geographic areas in the Gulf Coast, and some significant data could have been missed in the literature search.

Faults are generally recognized in the Gulf Coast subsurface by the absence of part of the normal stratigraphic sections in wells. The knowledge that sections are missing provides the basis for an interpretation of faulting when distribution of such gaps and structural discontinuities fit the concepts of faulting developed from outcrop observations and from experimental and theoretical studies. Other data might suggest the presence of faulting, but interpretation of missing stratigraphic intervals usually provides the main evidence for subsurface faulting.

The position of the missing sections (fault cutouts when faulting is interpreted) in wells can be determined if diagnostic electric log markers and other correlation data are available. For example, the position of the major fault shown in Figure 21 can be determined with considerable confidence within an interval of about 3 ft (1 m). This suggests that faulting in the Gulf Coast does not necessarily result in thick zones of highly disturbed or altered sediments along the faults. Typically, the exact position of a fault cutout in a well is difficult to determine because of an absence of closely spaced electric log correlation markers in the normal stratigraphic sequence. Some thick zones of disturbed sediments in which correla-

tion-marker characteristics have been destroyed by faulting may be a possibility.

Some published data give an indication of the thickness of Gulf Coast fault zones. Subsurface faults have apparently been recognized in cores from Raccoon Bend field, Austin County, Texas; Teas and Miller (1933) stated that part of the evidence of faulting in the field consists of "actual fault planes cut in coring, so that sand may lie against shale." Lahee (1925) described a surface exposure of the Wortham fault in Tertiary sediments in Texas in which sandy shales and sandstones of probably Wilcox age are juxtaposed with shales of Midway age. The fault displacement is probably several hundred feet, and a brecciated zone 12 to 15 in. (30 to 38 cm) thick is present along the fault. Stenzel (1946) described the features accompanying the Antrim faults in Eocene sediments of northwestern Houston County, Texas. The faults, where exposed, occur as fault surfaces or as thin fault zones less than 1 ft (0.3 m) thick. The outcrop exposure of two faults in Eocene sediments of the Tyler basin in east Texas is shown by photographs in Figures 26 and 27. The fault in Figure 26 is represented by a fault surface along which sandstone is in contact with shale. In Figure 27, sandstones are juxtaposed, but the two zones are separated by a clay-filled fault zone about 3 ft (1 m) wide.

Other fault exposures in the Gulf Coast are reported in the literature, but descriptions are inadequate to evaluate the thickness of the fault zones. It is clear that some Gulf Coast faults are associated with relatively thin fault zones and in some places they may be represented by discrete surfaces. Although positive evidence is lacking, thick zones of highly disturbed or altered sediments may be present along some faults.

Data are not available to identify the boundary fault-zone materials that form hydrocarbon traps in the Gulf Coast subsurface. Some speculations are possible from the general knowledge of fault-zone materials present in outcrops in the Gulf Coast and other provinces. The possibilities include materials resulting from both mechanical and chemical processes related directly or indirectly to faulting.

Fault gouge, the finely crushed and granulated wall rock along a fault, is frequently mentioned as a possible boundary material. Handin et al (1963, p. 732) reported gouge along faults produced experimentally in rock-deformation tests, and gouge is sometimes present in fault outcrops. If subsurface faulting occurs in sandstone under relatively high effective confining pressure (the difference between total overburden and formation fluid pressures), granulation due to friction along the fault might produce a gouge zone capable of re-

FIG. 26—Photograph of surface exposure of fault of Mt. Enterprise fault system, Tyler basin, east Texas. Middle Weches shale is downthrown about 300 ft (90 m) into juxtaposition with Carrizo sandstone. Carrizo sandstone shows minor drag flexure within about 3 ft (1 m) of fault. Foot wall is marked by thin ledge of indurated sandstone. No gouge is evident, and fault is represented by surface along which sandstone is in contact with shale. Photograph courtesy of H. B. Stenzel.

stricting hydrocarbon movement. In the field examples studied, the faults are nonsealing to lateral migration with the same sandstone body juxtaposed across small faults. In these places, gouge material, if present, did not create a hydrocarbon seal.

Cementation of cavities and fractures along fault zones and cementation of porous fault-zone material such as fault breccia by secondary mineral deposits from subsurface waters might produce a zone that is impermeable to hydrocarbons. Cement-filled openings or fractures along faults are common in outcrop exposures of faults (Trenchard and Whisenant, 1936, p. 636; Reaser, 1961, p. 1759). Vein calcite is present locally in drill cuttings from intervals of faulting in the Gulf Coast, suggesting that cementation of fractures along some fault zones may be present in the subsurface.

Ledges of tightly cemented sandstone from a few inches to a few feet thick in an otherwise loosely consolidated sandstone sometimes parallel faults at the surface. The indurated sandstone ledges along the Mt. Enterprise fault exposures

shown in Figures 26 and 27 are from 1 to 2 in. (2.5 to 5 cm) thick. However, at other places along the same faults, the indurated sandstone forms conspicuous walls up to several feet thick, traceable along the fault uninterruptedly for 1,000 ft (305 m) or more (H. B. Stenzel, personal commun.). The ledges and walls are composed of rusty brown, ferruginous sandstone and are thought to be chiefly, if not entirely, surface features. Ferruginous cementation occurs near the faults as the result of the presence of circulating groundwaters not far below the surface. Similar patterns of cementation can take place at depth, involving other types of cement. Cemented sandstone along a fault might show essentially undisturbed sedimentary bedding and grade laterally into noncemented, loosely consolidated sandstone away from the fault. The cementation is not fault-zone material in the strictest sense, but is nonetheless closely associated with faulting, and faulting has produced the conditions under which such patterns of cementation can occur. Teas and Miller (1933, p. 1473-1475) and Michaux and Buck (1936, p. 806) suggested that cementation of

FIG. 27—Photograph of surface exposure of fault of Mt. Enterprise fault system, Tyler basin, east Texas. Basal Sparta sandstone is downthrown about 600 ft (180 m) into juxtaposition with Carrizo sandstone. No drag flexure is present. Foot wall and hanging wall are marked by thin ledge of indurated sandstone. Fault zone is about 3 ft (1 m) wide and is filled with soft clay. Juxtaposed sandstones are separated by fault-zone clay. Photograph courtesy of H. B. Stenzel.

sandstones has occurred near subsurface faults in the Raccoon Bend and Conroe fields of Texas.

Perkins (1961) postulated that sealing faults in the Gulf Coast may result from soft shale smearing and impregnating the faulted sandstone face during fault movement. He states, "In essence, a natural 'mudcake' is formed over the sand interface and an impervious barrier is created against which migrating hydrocarbons may accumulate." In addition to the postulated shale smear on the faulted sandstone face, another mechanism is possible for emplacing shale along the faults to provide a seal to hydrocarbon migration. Shale can be deformed significantly more than sandstone before faulting occurs (Handin et al, 1963). In the hydropressured stratigraphic sequence of the Gulf Coast, faults might occur in the sandstone beds whereas the shale beds are merely deformed along the potential fault zone (Fig. 28). Continued deformation would ultimately fault the shales, but a zone of deformed shale might become greatly attenuated and trapped along the fault, separating sandstones juxtaposed by the

fault. Rettger (1935, p. 290) referred to such rock material as fault-plane filling, which serves to distinguish the rock from fault gouge.

The Mt. Enterprise fault shown in Figure 27 was suggested by H. B. Stenzel (personal commun.) as an example of shale as a fault-plane filling. The juxtaposed Carrizo and Sparta sandstones are separated by a clay-filled fault zone about 3 ft (1 m) wide. Shales of the Weches formation and of the Marquez member of the Reklaw formation are present between the Sparta and Carrizo sandstones in the stratigraphic sequence. The fault-zone clay has a mineral composition similar to that of some Weches shale samples collected in the area. The clay is not fault-gouge material, but is apparently a part of a shale formation that has become stretched and trapped in the fault zone.

In the hydrocarbon accumulations studied, some examples of sealing faults were observed in which a well crossed the fault at the point of juxtaposition of sandstones of different ages (Fig. 29). In each example, the data indicate that a

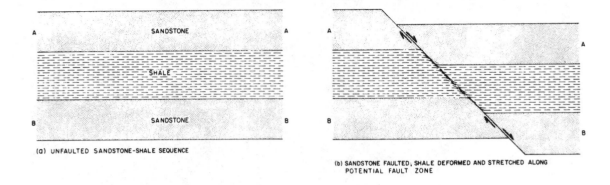

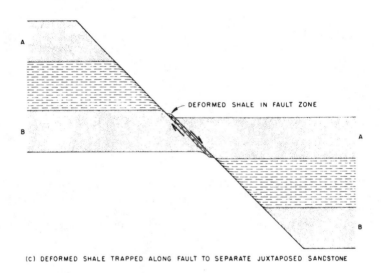

FIG. 28—Schematic sections illustrating emplacement of fault-zone shale postulated from experimental data on relative ductilities of sandstone and shale.

fault zone is present within an apparent shale break that separates the water-bearing and hydrocarbon-bearing sandstone. It is not unreasonable to suggest that this thin shale is aligned along the fault and is a fault-plane filling which serves as a barrier to hydrocarbon migration.

If the fault-zone shale provides the seal, the thickness and physical properties (soft or indurated) of the shale at the time of faulting may be factors which govern whether or not shale will form boundary fault-zone material for hydrocarbon entrapment. Growth faults, relatively near the surface in soft sediments, may thus have a different capacity to trap hydrocarbons from postdepositional faults at depth in more indurated sediments. The fault throw may also be important because the occurrence of a sealing fault seems more likely if the juxtaposed sandstones have moved past a greater thickness of shale.

CONCLUSIONS

Investigations of fault-lithology-accumulation relations in the hydropressured Tertiary section of 10 low-relief structures in the Louisiana Gulf Coast—investigations were limited to faults which restrict vertical migration of hydrocarbons, that is, where an accumulation is in contact with the fault over some vertical distance—have led to the following conclusions:

1. Many fault traps exist in which hydrocarbon-bearing sandstone is in lateral juxtaposition with a shale formation. Proof is not available that the reservoir rock is actually in contact with the juxtaposed shale formation. The reservoir boundary may be the juxtaposed shale formation or fault-zone material.

2. The faults are nonsealing to lateral migration where parts of the same sandstone are juxtaposed

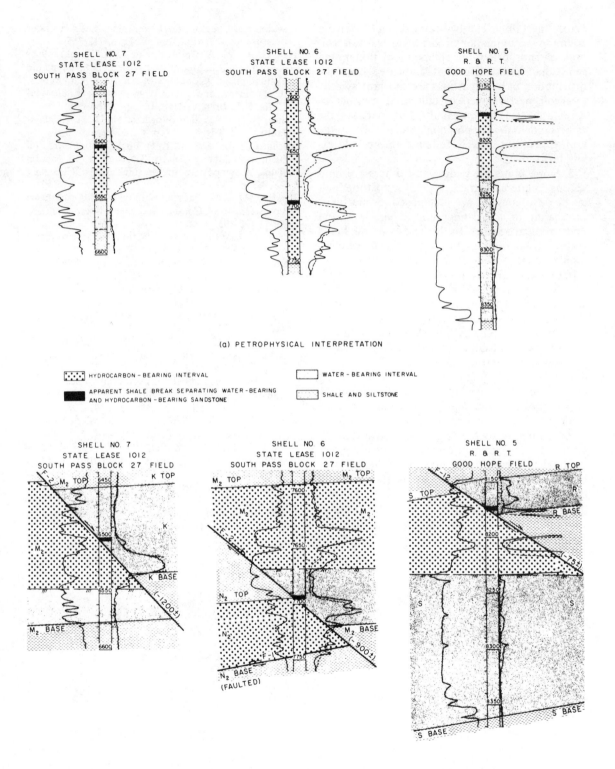

FIG. 29—Observed situations in which well intersects fault at point of juxtaposition of water-bearing sandstone in downthrown block and hydrocarbon-bearing sandstone in upthrown block.

across the faults. The two fault segments have a common accumulation if the hydrocarbon column extends below the spillpoint of juxtaposed sandstone. Nonsealing faults indicated from the distribution of hydrocarbons are substantiated by reservoir performance in a limited number of cases analyzed. Sealing faults with parts of the same sandstone juxtaposed may be present in the Gulf Coast, but none were found in the field examples studied.

3. Some faults are sealing and others are nonsealing to lateral migration where sandstone bodies of different ages are juxtaposed. Some faults are sealing to lateral migration at one level and nonsealing at another. In the fields studied, faults sealing to lateral migration are much more common than nonsealing faults where sandstones of different ages are juxtaposed.

4. Juxtaposition of reservoir rock with shale across a fault is not necesary to form a fault trap. Boundary fault-zone material is indicated along some faults, trapping hydrocarbon accumulations in upthrown sandstones that are juxtaposed with younger water-bearing sandstones.

5. The boundary fault-zone materials creating the sealing faults have not been identified, but the possibilities include materials resulting from both mechanical and chemical processes related directly or indirectly to faulting. The distribution as well as the displacement pressure of the boundary material will govern the maximum thickness of hydrocarbon column that can be trapped at a fault.

REFERENCES CITED

Handin, J., et al, 1963, Experimental deformation of sedimentary rocks under confining pressure: pore pressure tests: AAPG Bull., v. 47, p. 717-755.

Hill, G. A., W. A. Colburn, and J. W. Knight, 1961, Reducing oil-finding costs by use of hydrodynamic evaluations, in Economics of petroleum exploration, development, and property evaluation: Englewood Cliffs, N.J., Prentice-Hall, Inc., p. 38-69.

Lahee, F. H., 1925, The Wortham and Lake Richland faults: AAPG Bull., v. 9, p. 172-175.

Michaux, F. W., Jr., and E. O. Buck, 1936, Conroe oil field, Montgomery County, Texas, in Gulf Coast oil fields, a symposium on the Gulf Coast Cenozoic: AAPG, p. 789-832.

Perkins, H., 1961, Fault closure-type fields, southeast Louisiana: Gulf Coast Assoc. Geol. Socs. Trans., v. 11, p. 177-196.

Reaser, D. F., 1961, Balcones fault system: its northeast extent: AAPG Bull., v. 45, p. 1759-1762.

Rettger, R. E., 1935, Experiments on soft-rock deformation: AAPG Bull., v. 19, p. 271-292.

Simmons, F. E., Jr., 1961, Variations in gas-water contacts of Crescent Farms field, Terrebonne Parish, Louisiana: Gulf Coast Assoc. Geol. Socs. Trans., v. 11, p. 203-212.

Smith, D. A., 1961, Geology of South Pass Block 27 field, offshore, Plaquemines Parish, Louisiana: AAPG Bull., v. 45, p. 51-71.

——— 1966, Theoretical considerations of sealing and nonsealing faults: AAPG Bull., v. 50, p. 363-374; errata in AAPG Bull., v. 51 (1967), p. 1427.

Stenzel, H. B., 1946, Faulting in northwestern Houston County, Texas: Univ. Texas Pub. 4301, p. 19-27.

Teas, L. P., and C. R. Miller, 1933, Raccoon Bend oil field, Austin County, Texas: AAPG Bull., v. 17, p. 1459-1491.

Trenchard, J., and J. B. Whisenant, 1936, Government Wells oil field, Duval County, Texas, in Gulf Coast oil fields, a Symposium on the Gulf Coast Cenozoic: AAPG, p. 631-647.

Reprinted by permission of the World Petroleum Congresses
and Applied Science Publishers, London, from *Proceedings
of the Ninth World Petroleum Congress,* 1975, vol. 2, pp. 209-
221 (ASPL).

PETROLEUM GEOLOGY OF THE NIGER DELTA

Abstract

The Niger Delta is a large, arcuate delta of the destructive, wave-dominated type. A sequence of under-compacted marine clays, overlain by paralic deposits, in turn covered by continental sands, is present throughout, built-up by the imbricated superposition of numerous offlap cycles. Basement faulting affected delta development and thus sediment thickness distribution.

In the paralic interval, growth fault associated rollover structures trapped hydrocarbons. Faults in general play an important role in the hydrocarbon distribution. Growth faults may even function as hydrocarbon migration paths from the overpressured marine clays. Depositional environments of reservoir sands strongly influence well productivity as well as recovery efficiency.

Résumé

Le Delta du Niger est un delta très étendu, de forme arquée et de type destructif (par l'influence des vagues). Une séquence d'argiles marines sous-compactées, recouvertes de dépôts paraliques, eux-mêmes surmontés par des sables continentaux, est présente partout; sa génèse est expliquée par la superposition imbriquée de nombreux cycles régressifs. La fracturation du socle a affecté le développement du delta et par conséquent la distribution des épaisseurs des sédiments.

Les hydrocarbures ont été piégés dans l'intervalle paralique de structures de type "rollover" associées à des failles de croissance. Les failles jouent en général un rôle important dans la distribution des hydrocarbures. Les failles de croissance peuvent même déterminer la migration des hydrocarbures à partir des argiles marines sous-compactées. L'environnement de dépôt des sables réservoirs influence fortement la productivité des puits ainsi que le taux de récupération des hydrocarbures.

1. INTRODUCTION

The petroleum industry in Nigeria has a history which goes back to 1908–14 when a German company, the Nigerian Bitumen Company, drilled 14 wells in the coastal region, some 90 km east of Lagos. Several bituminous accumulations were found at shallow levels.

In 1937 Shell and BP under the name of Shell d'Arcy started their joint venture. At first the activities were concentrated north of the Tertiary Niger delta but no promising discoveries were made. Later, with the improvement in seismic techniques, interesting structures were detected further south in the actual delta. The first commercial discovery was the Oloibiri field,

by K. J. WEBER and E. DAUKORU,
*Shell–BP Nigeria Ltd., P.M.B. 2418,
Lagos, Nigeria*

found in 1955, and by 1958 production had started from the Oloibiri and Afam fields. A rapid increase in the number of discoveries followed.

By 1962 Shell–BP had relinquished more than 50% of its prospecting average. At present twelve companies are active in the Niger delta in cooperation with the Nigerian National Oil Corporation which recently has also started exploring on its own.

After a temporary setback during the 1967/70 civil war, production continued to rise rapidly and Nigeria is now the world's seventh largest oil producer with a production in June 1974 of 2·3 million bd. Gas reserves are also large and several LNG projects have been proposed.

The geological knowledge of the delta has reached an advanced stage but a number of problems remain. The following article summarises present ideas on delta development, structural geology, sedimentology, oil migration and trapping.

It is stressed that this article is but a brief summary of some of the studies made by many Shell–BP geologists over a 25-year period.

2. EARLY PHASES OF DELTA DEVELOPMENT

Pre-Santonian Basin evolution

It seems fairly well established that the oldest pre-Tertiary sedimentary basin, the Benue-Abakaliki Trough, originated as an arm of the triple-junction rift-ridge system that initiated the separation of South America from Africa in the Aptian/Albian.[1] The three arms of the system opened up at different times and different rates. In the South Atlantic, the opening started in the mid-Aptian by crustal stretching and downwarping accompanied by the development of coastal evaporite basins. By Lower Albian it had reached the Gulf of Guinea and extended northeast to form the Benue-Abakaliki Trough.[2] In the North Atlantic, the opening was much earlier, but slower, reaching Senegal by Upper Jurassic. An analogous

triple-junction development has been inferred for the Red Sea–East African Rift System.[1]

By Lower Albian two stable areas could be distinguished on either side of the Benue-Abakaliki Trough, called respectively, the Anambra and Ikpe Platforms (Fig. 1a). On the eastern flank, there were the NW–SE trending Ikang Trough and Ituk High, as well as the Eket Platform, all of which persisted without significant change right into the Tertiary. By contrast, the Benue-Abakaliki Trough was filled in by over 3300 m of sediments of Albian to Coniacian age and then started to close, accompanied by possible crustal subduction.[1,2] The closure was presumably caused by faster seafloor spreading adjacent to northwest Africa than in the south Atlantic,[1] but this has not yet been corroborated by spreading rates from geophysical and stratigraphical data.[2-4]

Santonian–Palaeocene Basin evolution

Consequent to the Campano–Santonian folding, the Benue-Abakaliki Trough was uplifted to form the Abakaliki High, whilst the Anambra Platform was

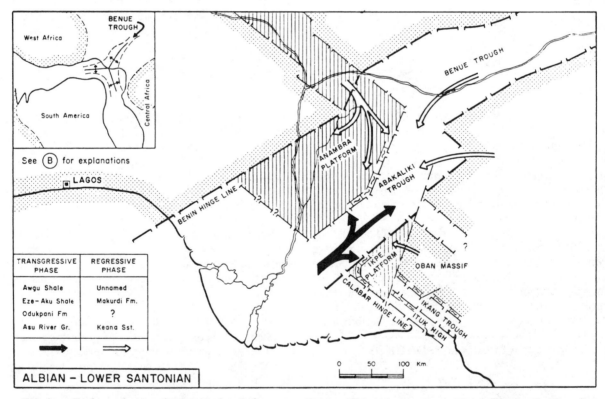

Fig. 1a—Early evolution of Niger Delta Sedimentary Basin, Albian–Lower Santonian (after R. C. Murat).

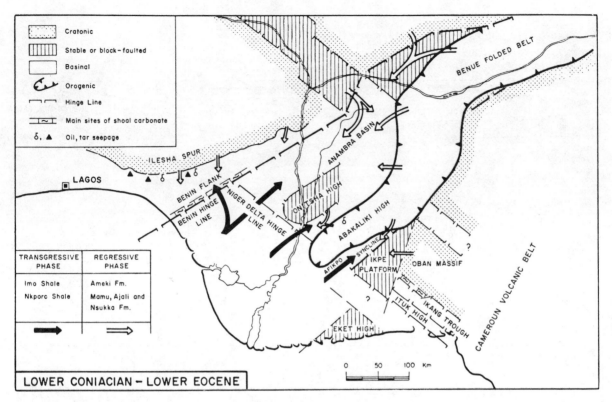

Fig. 1b—Early evolution of Niger Delta Sedimentary Basin, Lower Coniacian–Lower Eocene (after R. C. Murat).

downwarped to form the Anambra Basin (Fig. 1b). The Benin Flank basement, adjoining the Anambra Basin, was then invaded by the sea for the first time.

There existed thus three sedimentary basins from the Campanian to the Palaeocene—the Anambra Basin and the Afikpo Syncline (separated by the Abakaliki High) as well as the thus far, undeformed Ikang Trough. The Onitsha High, a positive anomaly within the Anambra Basin, could have formed at this time by block faulting and tilting.

General sedimentology and hydrocarbon occurrence

The sedimentary fill of the various basins up to the Palaeocene shows a remarkably similar pattern of deltaic cycles. The regressive phases are represented by coarse fluvial and fluvio-marine clastics with the occasional development of coal horizons (in the Maestrichtian), whilst the transgressive phases are mainly represented by shales and calcareous shales. In shelf areas, less affected by clastic influx, fossiliferous pelletal, oncoidal and bioclastic shoal carbonates

developed as lateral equivalents of the basinal marine shales.

Exploration effort in the early days was concentrated mainly in the area fringing the Tertiary delta on the inland side, but only minor hydrocarbon shows were encountered. Some carbonate intervals are oil impregnated but the reservoir characteristics are generally very poor. In addition, several oil and tar seeps have been observed in the Coniacian and Maestrichtian around the nose and western edge of the Abakaliki High as well as in the post Coniacian, along the southern edge of the Ilesha Spur (Fig. 1b)—a stable basement area that forms the north-western limit of the basin.

3. EOCENE TO RECENT DELTA

Eocene–mid-Miocene sub-deltas

The development of Cretaceous proto-deltas was terminated in the Palaeocene by a major transgression, represented by the Imo Shale. This was followed by a

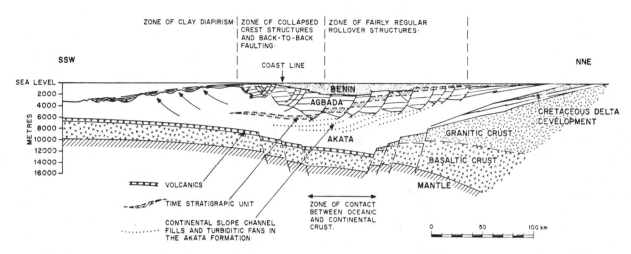

Fig. 2—*Schematic dip section of the Niger Delta (after P. Kamerling).*

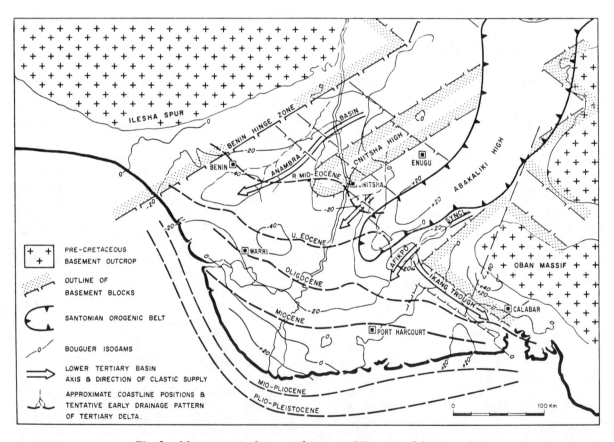

Fig. 3—*Megatectonic frame and stages of Tertiary delta growth.*

regressive phase, represented by the Eocene–Recent delta (Fig. 1b).

The thick wedge of Niger Delta sediments can be considered to consist of three units.[5,6] The basal unit, primarily composed of marine shales, is called the Akata Formation. This unit also comprises some sand beds which are thought to be continental slope channel-fills and turbidites. The Akata Formation ranges in thickness from 600 to probably over 6000 m (Fig. 2).

The overlying paralic sequence, forming the Agbada Formation, consists of interbedded sands and shales with a thickness of 300 up to about 4500 m. The Abgada Formation is built up of numerous offlap cycles of which the sandy parts constitute the main hydrocarbon reservoirs and the shales the cap-rocks.

The topmost unit, the Benin Formation, is composed of fluviatile gravels and sands. This unit is thickest in the central area of the delta (2100 m) where there is maximum subsidence of the basement (Fig. 2).

The formations are strongly diachronous and cut across the time stratigraphic units which are characteristically S-shaped in cross-section (Fig. 2 and Ref. 7).

The important tectonic elements in the post-Palaeocene delta were the Anambra Basin (with a minor sub-basin created by the axially located Onitsha High), the Abakaliki High, the Afikpo Syncline, the Ikang Trough[8] and the contact between continental and oceanic crust (Fig. 1b). These features determined the directions of clastic supply, as well as the areas of maximum deposition, and consequently the outline of the delta front up to the mid-Miocene (Fig. 3).

From the Eocene to the mid-Miocene, the Niger Delta complex progressed along three main sedimentary axes—the Anambra and its subsidiary basin which were fed by the Niger and Benue Rivers, and the Afrikpo Syncline which was fed by the Cross River. Deposition by the Cross River was also along the Ikang Trough for a brief period from the Eocene to the L. Oligocene. Sedimentation at this time was probably rapid, since all clastic debris was transported to these three relatively small areas. The shape of the early deltas could therefore have been lobate, in contrast to the arcuate shape of the present-day delta, in which local outbuilding by individual sub-deltas is prevented by strong longshore currents (Fig. 3).

Growth of post-mid-Miocene delta

After the mid-Miocene, the Niger-Benue and Cross River delta systems merged, and the influence of

Upper Cretaceous tectonic elements was no longer pronounced.

From the mid-Miocene onwards the rate of delta advance was determined by the rate of erosion of newly uplifted blocks in the hinterland, particularly the newly emergent Cameroon Mountains[1,5] as well as by eustatic changes in sea level. Passive subsidence is also suggested by gravimetric and refraction seismic data. These indicate that the maximum overall thickness of sediments in the central delta area was probably accommodated by maximum down-faulting at the junction where the delta built out from continental on to oceanic crust. Negative isostatic anomalies of small magnitude are observed over the central part of the delta which is in line with the concept of a basement yielding under sediment loading. Basement refractions are only recorded along the inland edges of the delta and at the southern apex where the basement comes up (see Fig. 2).

The progressive out-building of the Niger Delta is illustrated by the seaward shift in the inferred positions of the coastline from the Eocene to the present (Fig. 3). The landward position of the present coastline relative to the Pliocene coastline is anomalous and is due to the post-glacial rise in sea level. As a result of this, the continental Benin Formation extends offshore, possibly as far as the shelf break. Mud-filled ancient valleys off the Bonny and other estuaries,[9] and drowned coral thickets have probably also resulted from this transgression.

4. GRAVITY TECTONICS

Growth faults

Rapid sand deposition along the delta edge on top of under-compacted clay has resulted in the development of a large number of synsedimentary gravitational faults (Fig. 4). These so-called "growth faults" are also well known from the US Gulf Coast.[10] Their origin and shape can be explained on the basis of the theory of soil plasticity.

The spacing between successive growth faults decreases with an increase of depositional slope or an increase in the rate of deposition over the rate of subsidence. Growth faults tend to envelop local depocentres at their time of formation. Their trend is thus an indication of the prevailing sedimentological pattern.

The name "growth fault" derives from the fact that after their formation the faults remain active and

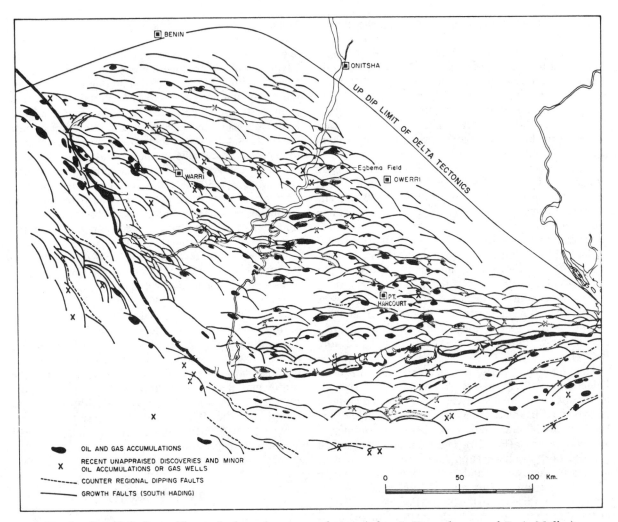

Fig. 4—*Growth faults and known hydrocarbon accumulations (after J. Haremboure and F. A. Molloy).*

thereby allow a faster sedimentation in the down-thrown relative to the upthrown block. The ratio of the thickness of a given stratigraphical unit in the down-thrown block over that of the corresponding unit in the upthrown block is termed the "growth index" which in Nigeria can be as high as 2·5.

The fault throw at the level of the Akata Formation is often as large as several thousand feet. The enhanced sedimentation along the growth fault causes a rotational movement which tilts the beds towards the fault. In this way anticlinal, so-called "rollover" structures are formed along the faults. Some 25 oil-fields in the Niger Delta are basically unfaulted rollover anticlines (Fig. 5a). More common are fields in which one or more additional south hading faults intersect the structure formed by the main fault

(Fig. 5b). Some 70 fields are of this type. About 10 additional fields are simple anticlines with one or more antithetic faults (Fig. 5c).

About 20 fields in the coastal region of the Niger Delta are much more intensely faulted than the above fields. Their fault pattern is of the collapsed crest type, with a series of closely spaced growth faults and a series of antithetic faults (Fig. 5d). The combined effect of the growth faults is a strong rollover of the northern flank. As a result, the upper surface of the Akata Formation also becomes markedly curved and gravitational instability causes the shale bulge to move upward. This in turn leads to the formation of the antithetic faults.

The close spacing of the growth faults in the coastal region is related to the different subsidence pattern

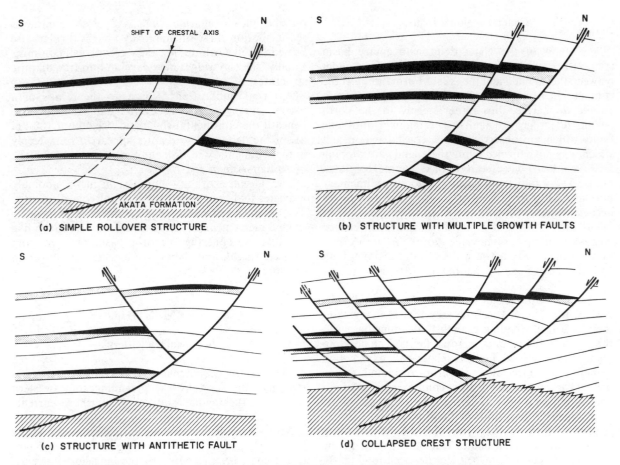

NOTE: ONLY A FEW RESERVOIR SANDS ARE SHOWN IN THE SCHEMATICAL SECTIONS AND THE SAND THICKNESS HAS BEEN ENLARGED

Fig. 5—Principal types of oilfield structures in the Niger Delta with schematical indications of common trapping configurations.

over the oceanic crust in comparison with that over the zone where oceanic and continental crust meet. This latter zone is an area of weakness where the rate of subsidence is higher than in the coastal region (Fig. 2). Consequently in the coastal region, the rate of deposition is higher than the rate of subsidence which leads to a narrowing of the growth fault spacing. A rapid progradation of the delta as a result of a high rate of deposition relative to the subsidence will probably cause a rapid succession of newly formed depocentres, most of which will in turn activate growth faults. In between the zone of collapsed crest structures and that of the simpler structures further inland there is a transitional zone with 6 moderately faulted fields.

The Figs. 5(a–d) also show some examples of hydrocarbon distribution in Nigerian fields. The trapping conditions are discussed in Section 6.

The growth faults die out in the massive continental sands. Near the top of the paralic sequence the dip of the fault planes can be as steep as 55 deg. Downwards, the slope progressively decreases, partly because of the curvature of the fault planes and partly because of the effect of compaction.

An important characteristic of Nigerian rollover structures is the shift of the crestal position with depth (Fig. 5). As a result, a well drilled on the south flank of a shallow horizon often penetrates the north flank of a deep culmination. The north flanks are usually considerably steeper than the south flanks.

Counter regional faults

The shape of the Niger delta sedimentary basin (Fig. 2) is such that in the coastal region the contact between the Agbada and the Akata Formation, *i.e.* the contact which marks the seaward facies change from paralic to marine, dips rather steeply to the north behind major growth faults. The regional antithetic faults which are observed in the coastal region and which may have formed along these landward dipping facies contacts have been termed counter regional faults. These faults can develop into synsedimentary normal faults, acting similar to growth faults. Space needed to accommodate the extra thickness of sediments on the downthrown side of the faults can be created by an upward movement of the Akata Formation marine clay in between the counter regional fault and the nearest south hading growth fault to the north.

The distribution of the major counter regional faults is shown in Fig. 4. It can be seen that in many cases there is a south hading growth fault immediately to the south of the counter regional faults. The development of these growth faults is related to the formation of new depocentres along the front of the advancing delta. The combination of a counter regional fault and an adjacent growth fault to the south constitute what is called "back-to-back faulting".

Back-to-back faulting can also be formed in a different way.[7] Shale bodies can be extruded in the seaward direction from below a depocentre (Fig. 2). In the south-eastern offshore area this type of shale diapirism is known to be common but there are indications that this phenomenon may also occur elsewhere offshore.

The horst-like feature in between the back-to-back faults probably consists mainly of shale. The origin of the back-to-back faulting can thus be inferred from the degree of disturbance of the shale bedding between the faults which is severe in the case of shale diapirism.

Shale ridges and salt diapirs

From the above it can be concluded that the shale upheaval ridges occurring in Nigeria are of three different kinds. First, there are the zones behind major growth faults. Secondly, shale bulges in front of growth faults are often observed and these bulges can sometimes act as positive elements, causing collapsed crest structures and unconformities. Thirdly, along the continental slope shale bodies were extruded in a seaward direction as a result of differential loading

on the plastic marine shale.[7,11] With continued sedimentation these offshore clay upheaval ridges are buried but like salt domes, their growth can continue. Finally, the clay ridges may develop into true diapiric structures.

A recent publication[12] discusses the results of a marine survey which discovered diapiric structures beneath the continental slope and rise, about 100 km SE of the delta. These diapirs appear to root deeply in the sedimentary section and the source layer may be of Aptian–Albian age.

The absence of any notable magnetic anomaly indicates that the structures are not of volcanic origin. The seismic reflection profiles show several pillar shaped piercement structures that look very much like salt diapirs. Thus the evaporite basins, known from offshore Angola and Gabon may extend to the Nigerian offshore area.

5. SEDIMENTOLOGY

Sedimentary patterns

The present geomorphology of the Niger Delta is strongly influenced by the prevailing south-western wind and the regular pattern of longshore currents generated.[13]

Rivers deposit part of their sedimentary load in the coastal plain, whilst the remaining sediments brought to the coast are picked up by the longshore currents. Most of the sand thus accretes along the front of barrier bars (Fig. 6), but a minor portion moves down the slope along submarine channels.[13] This pattern of sedimentation is probably typical for the Miocene to Recent delta.[14,15]

The pre-Miocene delta was probably less smoothly arcuate in shape (Fig. 3) because of the smaller influence of the longshore currents. It is likely, that in the earlier delta stages, the delta front could be lobate in shape in a number of places where major rivers brought in large quantities of sediments. The base of the Agbada Formation undulates markedly and in strike sections fairly large thickness differences are observed especially in the lower part of the formation.

Recently, diapiric shale bodies have been observed in seismic sections from a region north of Port Harcourt. These shale ridges are oriented roughly north–south and are probably associated with differential loading caused by the development of localised depocentres in the pre-Miocene delta.

In lobate shaped deltas, much sand is transported directly to the seashore and this factor, combined with

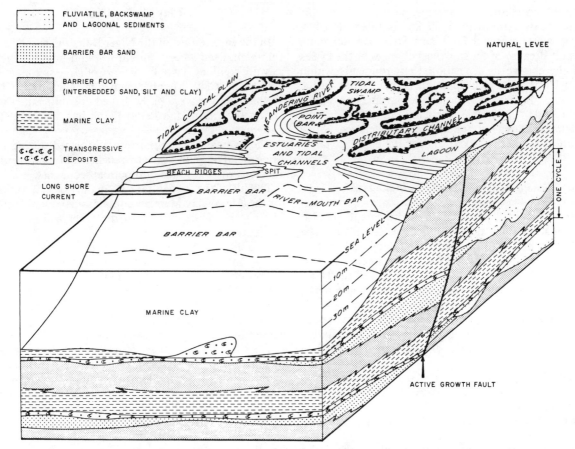

Fig. 6—Block diagram showing coastal geomorphology, cyclic sedimentation and the influence of an active growth fault.

the absence of strong longshore currents, results in sand deposition on the continental slope. It is thought that the sands encountered in the Akata Formation were mainly deposited as fluxo-turbidites in slope-gullies and as submarine fan turbidites in front of the gullies.

Little is known as yet of the distribution of these marine sand deposits because they generally occur at considerable depth underneath thick shale intervals.

Cyclic sedimentation

The paralic sequence is built up by numerous offlap cycles[14] of a thickness of 15–100 m. Most cycles are less than 60 m thick. A complete cycle generally consists of a thin fossiliferous transgressive marine sand overlain by the offlap sequence which commences with a marine shale which passes upward into lamin-

ated fluviomarine sediments. Barrier bar and/or fluviatile sediments follow up to the level where another transgression truncates the offlap sequence (Fig. 6).

Besides barrier bar, point bar and distributary channels sands, the sandy parts of the cycles can also consist of tidal channel fills, river mouth bars, natural levees, and shallow marine sand bars. The sands interfinger with clayey lagoon, marsh, oxbow lake fill and tidal flat deposits. Thin coal intercalations are also found in the upper part of many cycles.

Growth faults have a strong influence on the lateral thickness development of the individual cycles. Especially, barrier bar sands tend to thicken markedly towards growth faults in the downthrown block (Fig. 6).

The individual cycles are very extensive, especially in a direction parallel to the growth faults and within structural units bounded by major growth faults.

Detailed correlation of a large part of the paralic sequence over more than 50 km is often possible. Across a major growth fault, correlation is much more difficult because of the thickness changes of the cycles and the change in depositional environments within the cyclical units in a direction perpendicular to the coastline. In fact correlation across major growth faults is often impossible except for a time stratigraphic correlation based on pollen zones. A detailed pollen zonation has been worked out for the entire delta.

Slope channel-fills and turbidite fans

Underneath the paralic sandy sediments, there also exist marine sand deposits. Sediments of the type known to be found in continental slope gullies were first detected in cores from the Egbema field, situated in the northern part of the delta just east of the Niger (Fig. 4). In this field, these sediments occur at a depth of 1800 m only. Further south however similar sands and probably also some associated turbidite fans have been encountered interbedded with thick marine shales at depths ranging from 3000 to over 4500 m (Fig. 2).

These sands can be over 100 m thick and are characteristically poorly to very poorly sorted, fine clayey sands with some gravel, clay pebbles and plant detritus. The cores show the slumped nature of part of the sands.

In certain of the fields in the north of the Delta these sands sometimes contain important oil reserves.

Reservoir sands

The reservoir quality of the sands is strongly dependent on the depositional environment and the depth.[14] Many reservoirs consist of a single barrier bar or point bar development.

Reservoir sands thicker than 15 m are usually of a complex nature, consisting of a superimposition of sands deposited in the same or different sedimentary environments. Commonly two or three barrier bars develop one on top of the other without the intercalation of a significant marine shale.

Laterally many reservoirs are heterogeneous. A common occurrence is a barrier bar cut by a distributory channel fill at the same stratigraphic level. This lateral heterogeneity is especially striking in the larger fields where the change of the depositional environment with respect to the position of the coastline becomes apparent within each cycle.

Clay filled gullies

In the eastern part of the Niger Delta the paralic and continental sequences contain five and possibly more clay filled gullies. Two of these are of large size, namely the Afam Clay Member and the Kwa Ibo Clay Member. In the west, gully shaped unconformities are also known along the Benin Flank but these are still ill-defined.

The five eastern gullies were all formed during the Miocene. The gullies are entrenched into continental and towards the south into paralic sediments.

The fill of the gullies consists mainly of clay which, for the Afam Clay Member contains a lagoonal fauna. The nature of the clay fill of the other gullies is still being studied.

It is likely that the gullies were initially formed during major transgressions and that existing river valleys guided the axis of the gullies.

Burke[13] explains the gullies as submarine canyons of which the heads cut back into the delta. He relates the location of the gullies to the longshore drift pattern caused by the prevailing south-western wind.

The clay filled gullies are of particular interest where they cut into the paralic sequence as occasionally hydrocarbons are found in truncation traps against the flanks of the gullies. Up to now, no large oil accumulations have been found in truncation traps of this type of land, but offshore underneath the Kwa Iboe Clay Member several important oil discoveries were made, *e.g.* the Ubit field.

Overpressures

Overpressures are encountered in the Tertiary Niger Delta as a result of rapid loading of the undercompacted shales of the Akata Formation by the sandy Agbada and Benin Formations. The Akata Formation is in contact with the sandy paralic Agbada sediments in three different ways. In the first place there is the vertical transition from continuous marine shale into paralic sediments, secondly there are lateral facies transitions and interfingering of sand and clay and thirdly Akata shale is in many places in juxtaposition with Agbada paralic sediments across faults. In each of these cases fluids expelled from the overpressured Akata shales may "inflate" the pressures in the adjacent sands. Consequently overpressures are often encountered before the Akata shale is reached.

Drilling through a major growth fault at considerable depth is particularly hazardous because one may drill from a hydro-pressured interval directly into a strongly overpressured section. In the downthrown blocks a more gradual transition from hydro- to overpressure is often experienced. The overpressures within a given fault block are usually stratigraphically controlled. The levels at which overpressures can be expected are now known over a large part of the delta.

6. OIL MIGRATION AND TRAPPING

Source rock

The main source rock is thought to be the shale of the Akata Formation. A delta-wide study of shale samples indicated that, except for the most eastern part of the delta, the upper part of the Akata Formation can be considered as a mature source rock. The Akata shale has been deposited under anoxic conditions on the continental slope in front of the delta where the nutrient supply for planktonic organisms must have been plentiful.

Although many shale samples from the lower part of the paralic sequence also have source rock characteristics, a recent study has shown that most of these shales are immature.

Migration

Observation of fault zones in unconsolidated sand-shale sequences in outcrops in Germany has shown that, if the throw of a fault reaches several hundred feet, the fault zones develop a laminated character. This consists of streaks of sand, silt and clay, smeared into the fault zones. The thickness of the fault zones in the outcrops ranges from 20 to 60 cm which is similar to the thickness that can be deduced for Nigerian faults in 1:20 scale dipmeter log recordings.

The clay streaks in the fault zone can hamper or prevent flow across a fault whilst the sand streaks can give the fault zones a certain permeability along the fault. In Nigeria the best evidence for the vertical conductivity of the major boundary faults is the fact that in most cases the fault intersection with the upper bedding plane of the reservoir functions as the spill point of the accumulation (Figs. 5 and 7). It is thought likely that these spill points are also the entry points of the hydrocarbons from the fault zone into the reservoir.

At the level of the Akata Formation, the major growth faults offset a thickness of up to several thousand feet of overpressured shale against paralic sediments in the downthrown block. A plausible migration path may thus be from the overpressured shale into and through the fault zone (Fig. 7). The sands juxtaposed against overpressured formations are the only downthrown block reservoirs which occasionally have hydrocarbon accumulations trapped against growth faults. From the conductive fault zones the hydrocarbons appear to flow into the downthrown blocks only. This may be related to the effect of the specific gravity of the hydrocarbons which tends to bring them into that part of the fault zone which is adjacent to the downthrown blocks. A study of the relationship between the level where hydrocarbons are found and the throw of the growth faults indicates that the vertical conductivity of the fault zones for oil probably ceases when the throw drops below 150 m.

Apart from the along-fault migration, other migration routes from the Akata Formation shales must not be ruled out. In this respect the most likely migration path is along regional flanks, *i.e.* from a seaward facies change updip into the south flank of a rollover structure.

Trapping

Although almost all of the roughly 150 oil fields are essentially anticlinal structures, only about one fifth are unfaulted, gently dipping oval anticlines bordered on one side by a growth fault. At deeper levels the rollover movement, combined with regional tilting, can result in structures with their highest point against the back of the next growth fault towards the south (Fig. 5a). The growth faults are nearly always sealing on the upthrown side of the fault zone. Some of the largest hydrocarbon columns are trapped in this manner.

The great majority of Nigerian fields however have at least one fault besides the major growth fault which influences the accumulations (Fig. 5). When the same sands are juxtaposed across a fault, the faults are non-sealing. When the fault throw is larger, the sealing capacity of a fault appears to depend on the amount of shale smeared into the fault zone. This shale smearing is a function of the number and thickness of the clay layers over the throw of the fault.

It has been established statistically that a fault zone is usually sealing at a given level when it has been passed on the downthrown side by an interval consisting of more than 25% of shale. The larger the shale percentage, the larger the trapping capacity of the fault appears to be.

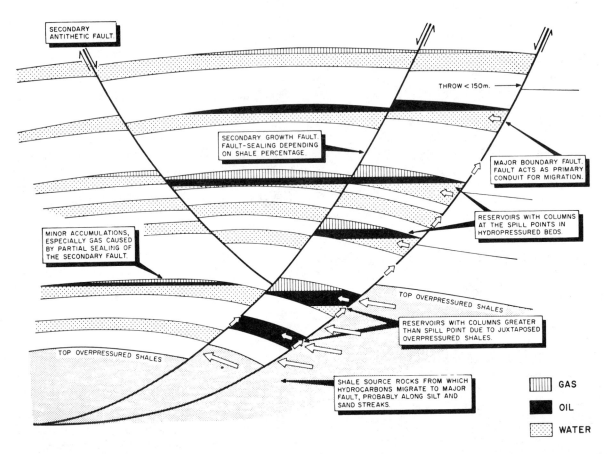

SECONDARY
ANTITHETIC FAULT.

THROW < 150m.

SECONDARY GROWTH FAULT.
FAULT-SEALING DEPENDING
ON SHALE PERCENTAGE.

MAJOR BOUNDARY FAULT.
FAULT ACTS AS PRIMARY
CONDUIT FOR MIGRATION.

RESERVOIRS WITH COLUMNS
AT THE SPILL POINTS IN
HYDROPRESSURED BEDS.

MINOR ACCUMULATIONS,
ESPECIALLY GAS CAUSED
BY PARTIAL SEALING OF
THE SECONDARY FAULT.

TOP OVERPRESSURED SHALES

RESERVOIRS WITH COLUMNS GREATER
THAN SPILL POINT DUE TO JUXTAPOSED
OVERPRESSURED SHALES.

TOP OVERPRESSURED SHALES

SHALE SOURCE ROCKS FROM WHICH
HYDROCARBONS MIGRATE TO MAJOR
FAULT, PROBABLY ALONG SILT AND
SAND STREAKS.

|||| GAS

■ OIL

⦂⦂⦂ WATER

Fig. 7—Schematic section of Nigerian field to show the principal features of the proposed accumulation model
(after R. G. Precious).

Besides trapping by shale smearing, trapping can also occur when a reservoir is juxtaposed against shale.

As outlined above, faults with a throw of 150 m or more must be considered as potential vertical leaks. This fact together with the small dip closure of the structural traps is probably responsible for the small average hydrocarbon column height in the Niger delta. Only about 5% of the oil columns exceed 50 m and hydrocarbon columns in excess of 150 m are very rare.

Minor stratigraphic trapping occurs in many fields as a result of lateral facies changes. Some stratigraphic trapping is also associated with the clay filled gullies. A third form of stratigraphic trapping, resulting from a combination of pinching out of sands and upwarping against the western flank of the Abakaliki High is encountered in several fields situated in the northern delta, just east of the Niger.

7. CONCLUSIONS

The origin of the delta appears to be well understood following the rifting that separated South America from Africa. Basement configuration and structural movements have been deduced from gravimetric measurements and the development of the overlying Tertiary. The various stages of the delta growth have been reconstructed with the aid of the well data.

The most important geological feature is the growth fault which forms the oilfield structures and which may also constitute the migration route of the hydrocarbons from the shales of the Akata Formation into the paralic sand layers of the Agbada Formation. The paralic cyclic sedimentation pattern is well established, but not much is known as yet about the distribution of continental slope channel-fills and turbidite fans.

Both structural and fault trapping are well documented, but the various possibilities of stratigraphic trapping in the delta are still being studied.

Acknowledgement

The authors wish to thank the Shell–BP Petroleum Development Company of Nigeria Limited for permission to publish this paper. Furthermore they stress that much of the ideas presented in the article have been contributed by other geologists for whose co-operation they are very grateful.

References

1. K. C. BURKE, T. F. J. DESSAUVAGIE and A. J. WHITEMAN, African Geology, 1970, Univ. Ibadan, Nigeria, 187–205.
2. E. STONELEY, Geol. Mag., 1966, **103**(5), 385–397.
3. J. B. MINSTER, T. H. JORDAN, P. MOLNAR and E. HAINES, Geophys. J. Royal Astron. Soc., 1974, **36**(3), 541–576.
4. T. D. ALLAN, Earth Sci. Rev., 1969, **5**(4), 217–254.
5. K. C. SHORT and A. J. STÄUBLE, Bull. Am. Ass. Petrol. Geol., 1967, **51**(5), 761–779.
6. E. J. FRÄNKL and E. A. CORDRY, Proc. 7th Wld. Petrol. Congr., 1967, Vol. 2, 196–209.
7. P. J. MERKI, African Geology, 1970, University Ibadan, Nigeria, 635–646.
8. R. C. MURAT, African Geology, 1970, University Ibadan, Nigeria, 251–266.
9. J. R. L. ALLEN, Marine Geol., 1964, **1**, 289–332.
10. D. MALKIN CURTIS, Deltaic Sedimentation, Modern and Ancient, 1970, Soc. Econ. Paleontologists Mineralogists, Spec. Publ. 15, Tulsa, 293–308.
11. J. HOSPERS, The Geology of the East Atlantic Continental Margin, Pt. 4, Africa; Great Britain Inst. Geol. Sci., Rept. 70/16, 1971, 121–142.
12. J. R. MASCLE, B. D. BORNHOLD and V. RENARD, Bull. Am. Ass. Petrol. Geol., 1973, **57**(g), 1672–1678.
13. K. C. BURKE, Bull. Am. Ass. Petrol. Geol., 1972, **56**(10), 1975–1983.
14. K. J. WEBER, Geol. Mijnbouw, 1971, **50**(3), 559–576.
15. J. R. L. ALLEN, Bull. Am. Ass. Petrol. Geol., 1965, **49**(5), 547–600.

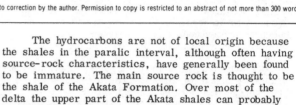

OTC 3356

THE ROLE OF FAULTS IN HYDROCARBON MIGRATION AND
TRAPPING IN NIGERIAN GROWTH FAULT STRUCTURES

by K. J. Weber and G. Mandl, Koninklijke/Shell Exploratie
en Produktie Laboratorium, W. F. Pilaar, Shell Internationale
Petroleum Mij. B.V., F. Lehner, Princeton University, and
R. G. Precious, Sarawak Shell Berhad.

ABSTRACT

A fault zone model is presented which is capable of
explaining the migration, distribution and trapping of
hydrocarbons in Nigerian rollover structures. The main
factor accounting for the occurrence of hydrocarbons in
a rollover structure appears to be the presence of a
large wedge of overpressured marine source-rock shale
on the upthrown side of the structure bounding growth
fault. The driving force for the migration is considered
to be the fluid-pressure differential resulting from the
juxtaposition of overpressured shale against initially
hydropressured paralic sands across the fault.

This mechanism can explain the presence of hydrocarbons in the lowermost series of paralic sands
opposite overpressured shale but not that of shallower
accumulations. Model experiments and outcrop studies
in Germany have indicated that shear zones of normal
faults in sand/shale sequences usually consist of
smeared-in laminae of shale and wedges of sand. In
Nigeria the spillpoints of accumulations in rollover
structures often coincide with the highest point of contact
of bounding growth fault and reservoir. These
observations suggest that migration of hydrocarbons
along major faults may occur along sandy stringers and
wedges in shear zones. In this way, superimposed
reservoirs can be successively filled with hydrocarbons
when connections between hydrocarbon-bearing sand in
the shear zone and the reservoir sand are available.

INTRODUCTION

Throughout the Niger Delta we find a sequence of
undercompacted marine clays (Akata Formation), overlain by paralic deposits (Agbada Form.), in turn
covered by continental sands (Benin Form.). In the
paralic interval growth fault associated rollover
structures trapped hydrocarbons.

The petroleum geology of the Niger Delta[1,2] is
characterised by the occurrence of multiple reservoirs
in the imbricated and superimposed offlap cycles of the
paralic interval. Typically, in a rollover structure
virtually all available sands contain at least some
hydrocarbons over an interval ranging from 1 500 to
6 000 ft, provided the caprock shales overlying the
sands are sufficiently thick.

The hydrocarbons are not of local origin because
the shales in the paralic interval, although often having
source-rock characteristics, have generally been found
to be immature. The main source rock is thought to be
the shale of the Akata Formation. Over most of the
delta the upper part of the Akata shales can probably
be considered mature source rock.

A remarkable feature, common to all fields is,
that nowhere do we observe trapping of hydrocarbons
against the downthrown side of a growth fault, unless
they are juxtaposed against an overpressured zone
across the fault. In many reservoirs the fault intersection
with the top of the reservoir appears to be the spillpoint
of the accumulation. This seems to indicate that the
growth faults form vertical conduits connecting successive
reservoirs.

Several studies have been carried out in Nigeria
to learn more about the composition and structure of
fault zones. A series of 1:20 scale dipmeter logs have
been examined, which show that the widths of shear
zones of normal faults in Nigerian oilfields generally
range from less than a foot to about two feet. The
generally shaly nature of the shear zones is evident
from the high gamma radiation, even where a fault zone
is bounded on both sides by sand. An attempt to take
sidewall samples from a well-defined fault zone was
unsuccessful. The samples from the fault zone proper
were all lost, indicating the probably more indurated or
fragmented nature of the shear-zone material as
compared with the soft friable adjacent sandstones.

At the Koninklijke/Shell Exploratie en Produktie
Laboratorium in Rijswijk (the Netherlands), fault-zone
studies were carried out with the aid of a special ring-
shear apparatus[3]. Experimental shear zones were
produced in an alternating sand/shale sequence.

A team from this laboratory also studied synsedimentary faults exposed in the open-cast lignite
mines near Frechen, west of Cologne (West Germany)[4].
The sand/shale ratio in the Frechen mine (75/25) is
similar to much of the Niger Delta paralic interval. The
observed fault features closely resemble those produced
in the ring-shear apparatus.

Together, the experiments and outcrop studies have provided good insight into the probable composition and structure of the shear zones of the Nigerian growth faults. On this basis, an attempt has been made to explain the migration process which has led to the hydrocarbon distribution observed in the Niger Delta fields.

EVIDENCE OF ALONG-FAULT MIGRATION

a. Evidence from literature

Numerous articles deal with hydrocarbon migration along faults[5-9]. Most articles describe and discuss flow along open fractures in consolidated rocks, e.g. the oilfields in Iraq and Iran[6]. For normal faults, little detail has been published, however, concerning the properties of fault zones in unconsolidated or semi-consolidated sand/shale sequences or of the exact process of the fluid flow along these zones. Two processes that may be applied are discussed in literature, viz hydraulic forces opening a pre-existing fault or even forming fractures[10,11] and the penetration of hydrocarbons into or across a fault zone when the hydrocarbon fluid pressure exceeds the capillary entry pressure of some marginally permeable material in the fault zone[12].

The latter process requires some, albeit marginally permeable, streak or streaks along or across a fault zone and can therefore be applied only if there is evidence of the existence of such streaks. The former process would appear to be generally viable since it only demands that the fluid pressure equals or slightly exceeds the total compressive normal stress that presses the fault walls together ('closing pressure'). At first sight, this subject appears suitable for calculations and quantitative predictions[11].

However, if one considers the complications caused by high fluid pressures, the prospect of making realistic quantitative predictions is less evident. If, for example, the average effective stress (total stress minus pore pressure) is low, as is the case with an overpressured region, the rise in pore pressure that accompanies the accumulation of oil will be of the order of magnitude of the original effective stress. Consequently, it will drastically change the state of effective stress in and around the oil accumulation. Since it is the effective stress that controls the onset of faulting and fracturing of porous rocks, this change of effective stresses may give rise to new shear faulting; it is by no means clear which fault will open first and at what displacement pressure.

Furthermore, when considering a fault in a shale as a mechanical 'plane of weakness' along which the formation can be 'wedged open', one obviously has to account for the access of high fluid pressures to the interior of the fault zone. It is thought that, although the process of hydraulic fracturing is probably one of the main mechanisms allowing along-fault migration, a suitable configuration of the fault zone is essential to this process.

Good evidence of fluid flow along faults in unconsolidated sand/shale sequences comes from an open-cast uranium mine in Texas[13], where mineralisation along a growth fault shows that the mineral-bearing fluids preferentially followed the fault zone. In south Texas there is also some more indirect evidence of flow of saline waters into shallow aquifers via growth faults[13]. The salinities and temperatures encountered in the shallow aquifers indicate that active flow from deeper aquifers takes place. Another interesting instance was reported from the Gulf Coast[14], where, during the drilling of an offshore well, fluids in the uncased part of the borehole entered an intersection with a fault and penetrated the fault zone. Subsequently, a blow-out occurred where the fluids escaped along the subcrop of the fault zone on the sea bed.

These examples all relate to along-fault flow at shallow depth and may not be highly representative of the situations at greater depths.

b. Evidence from the Niger Delta

The main reason for assuming that fluid flow along faults has taken place in Nigeria is the fact that in the downthrown block of the growth faults the intersection of the top of the reservoirs and the faults appear to be the spillpoints of many accumulations (Fig. 1). Frequently the accumulations do not reach the fault but nowhere do we observe an accumulation in the downthrown block trapped against a growth fault, unless it is in juxtaposition with overpressured beds across the fault (Fig. 2).

The basic migration model discussed before[2] is shown in Fig. 3. First, hydrocarbons are expelled into permeable stringers in the upper part of the overpressured Akata shales. From logs there is evidence of the presence of such stringers, especially near the transition zone of the overpressured shales and the paralic sequence. Next, the hydrocarbons flow along these stringers updip towards the growth fault.

At the fault face, highly overpressured hydrocarbons may initially exist in the upthrown block of the fault, partly in juxtaposition with normally pressured water in sandstones in the downthrown block. This juxtaposed interval can attain a thickness of several thousand feet. The idea is that, partly as a result of infiltration into permeable streaks in the fault zone and partly owing to hydraulic fracturing, hydrocarbons and water migrate into reservoirs in part of the downthrown block. Furthermore, mechanical arguments suggest that a growth fault has to change its shape inside the overpressured region as the sediment package is growing. This may cause so-called 'horsetailing' of the fault, which may facilitate penetration of the fault-zone region by the hydrocarbons.

In spite of the above remarks, migration from the overpressured upthrown blocks into the downthrown blocks must be regarded as little known. Nevertheless, there is strong evidence that flow across the growth faults does indeed occur below the top of the overpressured zone on the upthrown side. In many cases the sandstones in the downthrown block opposite the overpressured zone in the upthrown block are also overpressured. However, the pressures are lower than at the same level across the fault. In general, it can be said that the pressure distribution in the lower part of the growth-fault structures is difficult to explain otherwise than by assuming flow across the fault, which inflates the sandstones opposite the overpressured zone in the upthrown block. It is even possible that this flow still continues in some places.

The pressures in the uppermost several hundred feet in juxtaposition with overpressured beds in the upthrown block are usually scarcely higher than hydrostatic. It is thought that the hydrocarbons bleed off upwards along the fault zone to the overlying series of superimposed reservoirs. After filling a reservoir to the spillpoint, the hydrocarbons cascade to the next reservoir. At first, the migrating hydrocarbons mainly consisted of oil but gradually the quantity of gas increased until many of the lowermost reservoirs were flushed by gas and the oil cascaded to higher reservoirs.

The assumption that there exists a continuous permeable zone along the fault zone from which the reservoirs are successively filled is too simple. In that case, all reservoirs would be filled to the spillpoint and there would be a rather systematic change in the type of hydrocarbons filling the superimposed traps. This, however, is not the case. Although virtually all available sands with good caprock shales usually contain some hydrocarbons over intervals of 2 000 - 4 500 ft thickness, the degree of fill of the traps and the oil and gas distribution are much too erratic for such a simple model (Fig. 2).

Where the paralic intervals with hydrocarbons are thin, these intervals are usually overlain by shales or very shaly intervals of some 500 ft or more. Thick hydrocarbon-bearing intervals often reach up to the highest substantial shale layers of the paralic sequence. The base of the hydrocarbon zone is commonly in close proximity to or even below the level of the top of the overpressured Akata shales in the upthrown block of the growth fault.

With the large number of fields available for study in Nigeria, it would appear attractive to study the oil and gas distribution in more detail to reveal clear patterns. Such studies have indeed been carried out, but with little success. This can be understood if one considers that, structurally, few fields are similar and that several processes can lead to secondary migration of hydrocarbons, such as the above-mentioned flushing of oil out of a reservoir by gas which was generated later. Other processes may be: re-activation of faults, differential compaction, regional tilting and lateral flow across semi-sealing faults.

Re-activation of crestal faults is especially important when the caprock shales are thin. In fact, there are probably more small crestal faults of the type shown in Fig. 1 than can be detected in seismic sections or from well evidence.

In many fields, accumulations with different fluid contacts juxtaposed across a fault are observed. It is difficult to judge whether there is still flow across a fault or whether equilibrium with the fault displacement pressure has been reached.

Trapping against faults has been studied in some detail in Nigeria. It was concluded that the trapping capacity of faults is related to smeared-in shale in the shear zones. The fact that hardly any trapping has been observed in the downthrown block against growth faults, but that large columns of hydrocarbons can be trapped in the upthrown block against such faults (Fig. 2b), indicates that the texture of the fault zones must be asymmetric. The asymmetry of the fault zones is a main topic of the following chapters. The only places where trapping against a growth fault occurs in a downthrown block is opposite overpressured beds (Fig. 2b). It is tempting to conclude that active flow into such traps may still take place.

EXPERIMENTAL STUDY OF FAULT ZONES WITH A RING-SHEAR APPARATUS

A special ring-shear apparatus has been designed to study the development of shear zones in granular materials under continued shearing[3]. The ring-shear apparatus consists of a ring-shaped container, the lower half of which can be rotated, whereas the upper half is fixed. The top plate can be loaded up to a pressure of 10 bar (Fig. 4).

Besides experiments with packs of various granular materials, a series of tests was carried out with sand/clay systems to model faulting in a Niger Delta sand/shale sequence. For this purpose, the apparatus was filled with zones of sand alternating laterally with vertical bands of clay. These bands were sheared off under various simulated overburden pressures along a median slip plane.

The clay bands sheared off in this fashion formed one continuous, multi-layered, clay gouge along the median slip plane of the shear zone (Fig. 5). The clay gouge is somewhat contaminated by sand grains derived from the boundary zones of the shear zones but proved to be an effective seal to vertical water flow when tested.

The rate of displacement gradually increased from 5 cm/h, during the initial phase to 50 cm/h during the final stages of the experiments. The clay material contained various natural clays (e.g. from the lignite mines at Frechen) as well as 'non-swelling' pottery clays. The clay was remoulded to about the same plasticity as observed in nature at freshly cut exposures of the lignite mines.

It is interesting to observe that, where shear zones intersect clay beds, sand wedges are formed with their apexes pointing in the direction of shear motion (Fig. 5). Along these wedges clay from unsheared portions of the clay beds is forced into the shear zones. Such wedges may play an important role in hydraulic opening of a clay-sealed fault. Under certain stress conditions, a normal fault - containing one or more planes of weakness - may be 'wedged' open by the fluid intruding via the sand wedge from an adjoining formation. Suggestive, in this respect, was the observed hydraulic opening of experimentally produced shear zones after unloading and during impregnation by liquid plastic under its own hydrostatic load.

Microscopic examination of the granular material along the slip planes showed the presence of a zone of dilatation or pore increase on the downthrown - i.e. active moving - flank of the fault. On the upthrown - i.e. fixed - flank of the fault, compaction or pore decrease occurred. This phenomenon was also observed along small shear faults in nature (Fig. 6) and may facilitate fluid movement along the downthrown side of faults in sandy intervals. The mineralisation along a growth fault in the uranium mine in Texas[13] may well be related to such a zone of dilatation.

OUTCROP OBSERVATIONS OF FAULT ZONES IN FRECHEN

a. General Description

As the ring-shear-apparatus experiments were carried out under rather restricted conditions (e.g. short clay bands, soft clay, strain control by rigid plates), it was thought necessary to test the results against geological reality. Very suitable man-made outcrops of normal faults are present in open-cast lignite mines of 'Rheinische Braunkohlenwerke A.G.' near Frechen, west of Cologne (West Germany)[4].

The faults chosen for this study are situated on the eastern flank of the main trough of the Tertiary Lower Rhine Graben system (Fig. 7) and affect the deltaic sequence that filled this part of the graben. This sequence consists of alternating well-bedded, loose to slightly consolidated sands, silts, shales and gravels, with an intercalation of a thick brown-coal layer in the lower part of the exposed section. Sand predominates over the other lithological constituents. The average sand/shale ratio is about 75/25 in the Frechen mine.

The faults of the Lower Rhine rift valley were active during sedimentation[15]. Synsedimentary fault movements, especially during the Upper Miocene and lowermost Pliocene, have led to considerable growth (some 30%) across faults. Throws up to about 100 m and dips averaging around 70° have been observed. Features commonly associated with this type of normal faults are usually well exposed along the freshly cut slopes of the pits. They include antithetic faults, normal drag, and dip into the fault plane.

b. Tectonic Shear Zones Associated with Normal Faulting

The principal displacement along the fault is usually not confined to a single surface or to conjugate sets of single surfaces. Slip motion appears to have taken place along many surfaces distributed throughout a 'shear zone' of varying width. Such shear zones have definite boundaries within which the style of shearing contrasts with the slip patterns in the adjacent rock.

The widths of these shear zones vary with the lithological compositions of the fault walls over the throw interval. If sand is displaced against sand, the shear zones are usually only a few centimetres wide. However, if sand is displaced against other materials, the zones are often wider and include material from different locations along the fault walls. Such shear zones have a lithologically layered appearance. Clay gouge is present where shale beds are sheared off, either by a minor shear or by a major fault.

The style of shearing within the shear zones appears to vary with the lithology of the gouge material. In sandy material, a style of shearing is often observed that fits into the pattern of the 'classical' shear zone as described by Tchalenko[16,17]. Typical of this style are the fault-parallel 'principal displacement shears' (D-shears) or 'principal slip surfaces' and the 'Riedel shears' (R-shears) which lie en echelon, inclined at 10 - 30° to the D-shears, with the acute angle pointing in the direction of the relative movement of the fault block in which they occur (Fig. 8).

Whereas 'slip planes' are clearly visible in the sandy part of the shear zone, the extent of discontinuous deformation in the clayey part becomes apparent only if the material is somewhat dried out. Where fault-parallel shears exist along the margins of the clay gouge, a very regular pattern of sigmoidal shears appears after some drying (Fig. 9).

c. Deformation Outside the Shear Zones

On either side of the shear zone or major slip plane of the faults, much wider zones with traces of continuous and/or discontinuous (slip) deformation were found. Fault drag folds in shale layers are the most prominent examples of continuous deformation. Nearly everywhere along a fault, coal and sands in particular show traces of discontinuous deformation along slip surfaces. In the downthrown block, drag folds are often intersected by these slips. Figure 10 shows the development of a clay gouge with a pattern of slips in the adjacent sand.

A large number of observations suggest that these slips form two sets of conjugate shears whose attitude with respect to the major slip planes of the fault is the same as that found for Riedel shears (R-shears) and their conjugates (R'-shears) in the type of shear zone discussed above. The material is deformed along these sets of shears by a 'double-gliding' motion, probably alternating along the two sets of slip surfaces. Where such a set of R and R'-shears cut through a sand/shale bedding plane, characteristic graben structures are formed (Fig. 10).

d. Clay Gouges and Sand Wedges

Further away from its source bed, the clay shows no intersection by R or R'-shears and forms a band-like gouge between more or less parallel walls, i.e. the gouge gradually becomes thinner further away from the clay source bed. The clay gouges are rarely thicker than 50 cm, which agrees with the observations of the 1:20 scale dipmeter logs from Nigeria. A gouge usually contains clay material from different beds. These clays show parallel alignment with sharp boundaries. They hardly contain any inclusions of other material from the fault walls.

Sandy material between two different clays was found only where a drag-folded clay bed approached a shear zone which already contained a clay gouge from another bed (Fig. 11). In such cases it was invariably observed, however, that the sand wedged out and the clay gouges on both sides amalgamated to form one continuous gouge.

Clay-gouge material is supplied both from the upthrown and the downthrown part of the clay bed. Continuity of the clay gouge was observed over about 400 m of exposure. The importance of a clay gouge as a barrier against fluid flow across the fault is obvious.

For the migration along the fault, the most important feature is probably the sand wedge between the clay gouge and the overlying drag fold of clay, which extends at an acute angle into the fault zone. Although the sand wedges were observed to terminate in the two-dimensional exposures studied, the distance from the sand in the wedge to sands of the overlying sand bed is often very small. In Fig. 11 it can be seen how the Riedel shears can almost interconnect two successive sand beds near the tip of the sand wedge of the lowermost sand. Along the Nigerian growth faults, with their much larger throws, the effect of the R-shears is probably even more pronounced. The graben-like features as shown in Fig. 10 may virtually interconnect many successive sands. Thus, hydraulic opening[11] of these Riedel shears over short distances would be enough to allow hydrocarbons to flow from one sand to the overlying sand layer.

It logically follows that very thick shales or very shaly intervals will effectively stop upward migration along a fault. In these cases it is unlikely that a sand wedge will come near an overlying sand. This agrees with the distribution of hydrocarbons in the Nigeria fields in cases where such thick shales occur within the paralic sequence.

Wherever observed, the fault zones are asymmetric in their configuration. The shale wedges present on the upthrown side of the fault point downwards, and it can be understood why trapping of hydrocarbons is possible on this side. Furthermore, hydraulic fracturing from the tip of downwards pointing sand wedges appears to be very unlikely.

CONCLUSIONS

The ring-shear-apparatus experiments and the Frechen exposures both show very similar fault zone configurations. Although the fault zones do not appear to contain continuous permeable streaks between successive sands, it is nevertheless, possible that the observed sand wedges, which extend at an acute angle into the fault zones, come close to providing such interconnections. Especially if the fault throw is large and the shales relatively thin, the associated Riedel shears will reduce further the distance between successive sand beds. It is probable that in most cases the actual interconnection is only established with the aid of a certain degree of wedging open of a Riedel-shear by hydraulic forces.

Migration of hydrocarbons along faults will be barred by thick shales. This is in agreement with the field observations in Nigeria. Trapping of hydrocarbons in the upthrown block against growth faults can be explained by the presence of downwards pointing shale wedges continuing into the smeared-in clay of the clay gouges. The capacity of these traps is enhanced further by the fact that the downwards pointing sand wedges in the upthrown blocks are unlikely to become connected with an underlying sand by means of hydraulic fracturing.

It is felt that at least some of the above conclusions may be equally valid in the Gulf Coast growth fault region.

REFERENCES

1. Weber, K.J., "Sedimentological aspects of oil fields in the Niger Delta", Geol. & Mijnbouw, 50 (1971), no. 3, pp. 559-576.

2. Weber, K.J. & Daukoru, E., "Petroleum geology of the Niger Delta", Trans. 9th World Petrol. Congr., Tokyo, 2 (1975), pp. 209-221.

3. Mandl, G., de Jong, L.N.J. & Maltha, A., "Shear zones in granular material. An experimental study of their structure and mechanical genesis", Rock Mechanics, 9 (1977), pp. 95-144.

4. Quitzow, H.W., "Tektonik und Grundwasserstockwerke im Erftbecken", Geol. Jahrbuch, 69 (1954), pp. 455-464.

5. Dufour, J., "On regional migration and alteration of petroleum in South Sumatra", Geol. & Mijnbouw, 19, (1957), pp. 172-181.

6. Dunnington, H.V., "Stratigraphical distribution of oil fields in the Iraq-Iran-Arabia basin", Inst. Petrol. Jour. 53, (1967), no. 520, pp. 129-161.

7. Fowler, W.A.Jr., "Pressures, hydrocarbon accumulation, and salinities - Chocolate Bayou Field, Brazoria County, Texas", Jour. Petrol. Techn, April 1970, pp. 411-423.

8. Gavrilov, V.P., "Relationships between abyssal fractures in young platforms and gas-petroliferous structures in platformal blankets", as in USSR, Internat. Geol. Rev., 14, (1972), no. 9, pp. 917-925.

9. Gavrish, V.K., "Role of the abyssal fractures in fluid migration", Geologiceky Zhurnal, 34 (1974), no. 4, pp. 17-27.

10. Hubbert, M.K. & Rubey, W.W., "Role of fault pressure in mechanics of overthrust faulting. Part 1: Mechanics of fluid-filled porous solids and its application to overthrust faulting", Geol. Soc. Amer. Bull., 70 (1959), pp. 115-166.

11. Secor, D.T.Jr., "Role of fluid pressure in jointing", Am. Jour. of Science, 263 (October 1965), pp. 633-646.

12. Smith, D.A., "Theoretical considerations of sealing and non-sealing faults", Bull. Am. Ass. Petr. Geol., 50 (1966), pp. 363-374.

13. Gabelman, J.W., "Migration of uranium and thorium-exploration significance", Studies in geology no. 3, Am. Ass. Petr. Geol., Tulsa, Oklahoma, 1977.

14. Baird, R.W., "The role of geophysics to assist with marine drilling operations", SPE Paper 6865 (1977).

15. Frange, W., "Tektronik und Sedimentation in den Deckschichten des Niederrheinischen Hauptbraun-kohlenflözes in der Ville, mit Bemerkungen zur Feintektronik der Niederrheinischen Bucht", Fortschr. Geol. Rheinland und Westfalen, 2 (1958), pp. 651-682.

16. Tchalenko, J.S., "The evolution of kink-bands and the development of compression textures in sheared clays", Tectonophysics, 6 (1968), no. 2, pp. 159-174.

17. Tchalenko, J.S., "Similarities between shear zones of different magnitudes", Geol. Soc. Am. Bull., 81 (1970), pp. 1625-1640.

ACKNOWLEDGEMENTS

The authors wish to thank Shell Research B.V. in Rijswijk (the Netherlands) and Shell BP-Nigeria for permission to publish this paper.

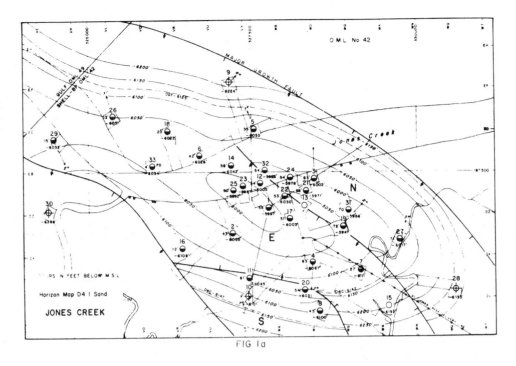

FIG 1a

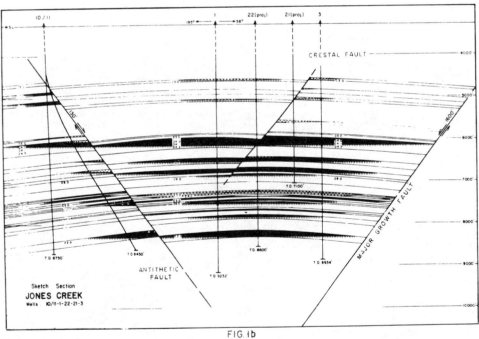

FIG. 1b

Fig. 1 - Map and section of the Jones Creek field showing a series of reservoirs with hydrocarbon-water contacts near the intersection of the top of the reservoir and the major growth fault.

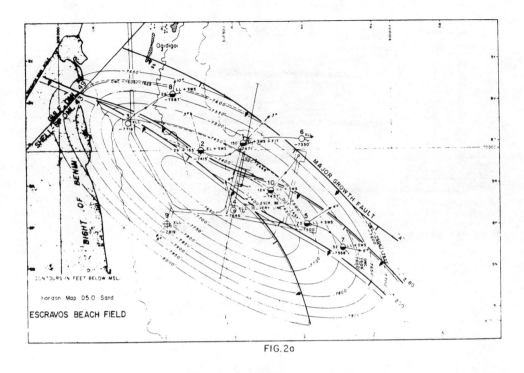

FIG. 2a

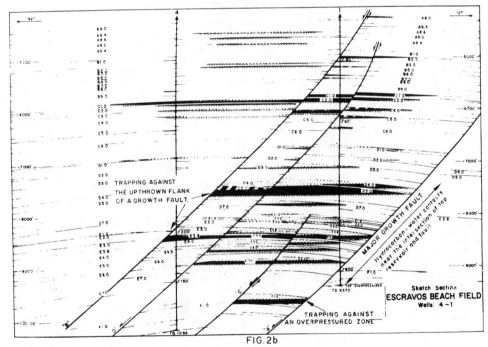

FIG. 2b

Fig. 2 - Map and section of the Escravos Beach field showing the hydrocarbon fault relationships in a field with multiple growth faults.

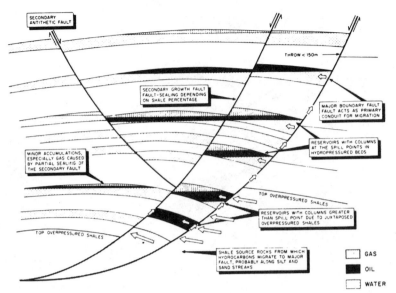

Fig. 3 --Schematic section of Nigerian field to show the principal features of the proposed accumulation model.

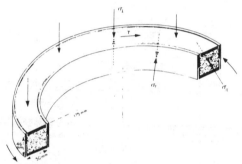

Fig. 4 - Transparent annulus of ring shear apparatus.

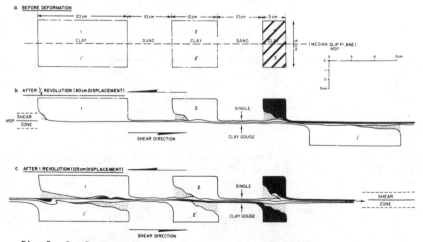

Fig. 5 - Development of a multi-layered, single clay gouge in an alternating sand/shale sequence sheared in a ring-shear apparatus.

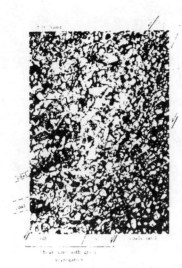

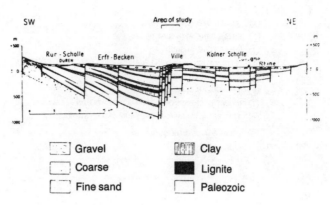

Gravel ☐ Clay
Coarse ☐ Lignite
Fine sand ☐ Paleozoic

Fig. 7 - Cross-section through the lower Rhine Graben
(after H. W. Quitzow).

Fig. 6 - Natural shear in fine-to-
medium grained sand showing segregation, dilation (d) and compaction (c)
inside the shear zone. (Frechen mine,
West Germany); Magn. 15x.

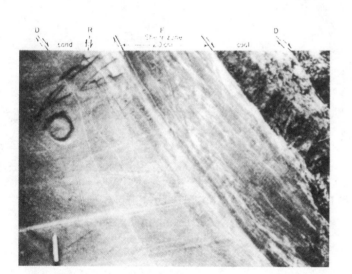

Fig. 8 - Shear zone of major fault (F) (Wiedenfelder
fault, Garsdorf mine, throw approx. 100m) containing
composite gouge of sand, clay and coal material. R-
and D- shears visible inside--forming shear lenses--
and outside shear zone.

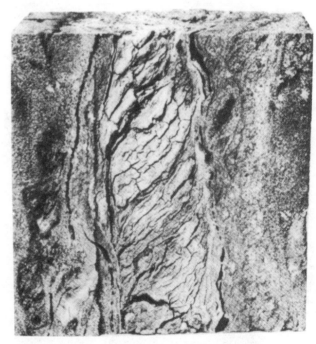

Fig. 9 - Sample from shear zone of "Max Rudolf"
fault (Frechen mine) showing sigmoidal shear pattern
in clay gouge bounded by principal displacement
shears (natural size).

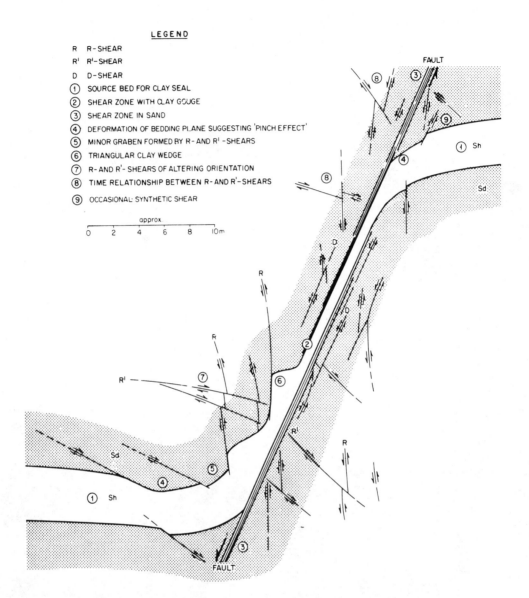

Fig. 10 - Main observations along a normal fault in the Frechen brown-coal mines.

90

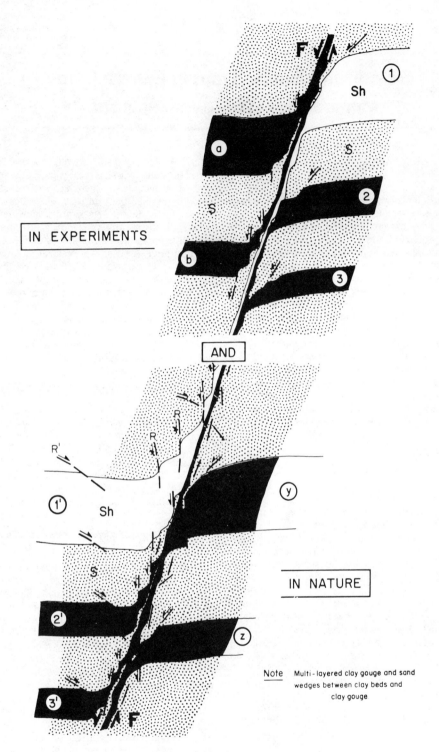

IN EXPERIMENTS

AND

IN NATURE

Note Multi-layered clay gouge and sand
wedges between clay beds and
clay gouge

Fig. 11 - Comparison of fault zones in the Frechen browncoal
pits and as produced in the ring-shear apparatus.

The American Association of Petroleum Geologists Bulletin
V. 69, No. 12 (December 1985), P. 2098-2109, 16 Figs.

Cenozoic Diapiric Traps in Eastern China[1]

WANG XIE-PEI, FEI QI, and ZHANG JIA-HUA[2]

ABSTRACT

Diapiric traps, including diapirs of salt and mud or igneous intrusives, have recently been found in many places in the Cenozoic petroliferous basins in eastern China, and most of them produce oil and gas. During the Eocene–early Oligocene, salt-lake basins evolved extensively. Plastic source materials for diapirism were deposited in the basins in great thickness. We have found that the diapiric traps of salt and mud in eastern China are unpierced or slightly pierced structures. The diapiric materials are a mixture of salt, gypsum, and mudstone, but mudstone is the main component of the plastic bodies. Based on an analysis of the structural features of the diapirs and the regional tectonic setting, we believe that the diapiric traps are caused by a combination of horizontal stress due to regional tectonic movement and vertical stress due to gravitational instability. Some diabase diapirs are arranged in a series of small anticlinal traps along the regional faults in the Subei basin of Jiangsu province. Oil and gas have been found in certain of these diapirs.

INTRODUCTION

The diapiric structure is an important type of hydrocarbon trap. Whether such traps actually exist in eastern China, however, continues to be a subject of much dispute. In the past, most geologists believed that no diapirs developed in eastern China because no strong piercement occurred and no thick plastic strata had been found. Recently, with the development of deep drilling and seismic surveying, diapiric structures have been identified in many places in the Cenozoic basins of eastern China by studying the traps in such basins as Songliao, Bohai Bay, Subei-South Yellow Sea, Jianghan, and Dongpu. Although the intensity of piercement in those areas is rather weak, the flowing and doming of the plastic material at deep levels play an important role in the formation of oil and gas traps in overlying strata.

According to O'Brien (1968), diapirism is broadly defined as a process in which earth materials from deeper

levels have deformed and pierced, or appear to pierce, shallower formations. In the present paper, the term diapirism is used to include nonpiercing, conformable structures cored by mobile, buoyant rock, as well as piercing structures. Based on the temperature at which piercement occurs, diapirism can be an igneous intrusion—piercement at high temperature—or, in the restrictive sense, it can be salt and mud intrusion—piercement at low temperature. The diapirs related closely to oil and gas traps and distributed extensively in petroliferous basins are salt domes and mudlumps, which are the focal point of this paper. In addition, because extrusion occurred frequently in eastern China's petroliferous basins and because genetic relationships between the hydrocarbon traps and the magmatic activities have been found in the Subei–South Yellow Sea basin, we also discuss magmatic diapirism.

DIAPIRIC TRAPS OF SALT AND MUD

Diapiric traps of salt and mud are distributed extensively in eastern China. Typical traps are the Dongying, Shengli-cun-Tuozhuang, and Xinzhen structures in the Dongying depression, Bohai Bay basin; the Wenliu structure in the Dongpu depression, Henan province; and the Wangchang structure in the Jianghan basin, Hubei province. These structures have several features in common.

Regular Patterns in Petroliferous Basins

Diapiric structures are generally arranged in a regular pattern in petroliferous basins. For example, the Dongying depression, where the well-known Shengli oil field is located, is the most productive area in Bohai Bay basin (Figure 1). Dongying is a Tertiary half-grabenlike depression extending east-northeast and divided into three regional structural zones: the north steep slope, the south gentle slope, and the central rise (Figure 2) (Fei Qi and Wang Xie-pei, 1982). The data collected from deep wells and seismic profiles confirm that great thicknesses of strongly deformed Eocene plastic beds consisting of salt, gypsum, and soft mudstone exist at deep levels across the depression (Figures 3, 4). These deformed plastic beds in the shape of a ridge or pillow directly affect the structure of overlying strata. Note that the ridges and pillows are arranged en echelon in several structural zones parallel to the east-northeast regional fault trend. Generally, structures on the central rise are more deformed, with higher amplitude and more complex fault systems on the crest than structures in other zones. Most structures on the north steep slope appear as en echelon anticlines with pillowlike plastic cores at depth and without complex faulting on the crest. These north slope structures form the best traps for oil and gas in the Dong-

[1]Manuscript received, May 21, 1984; accepted, July 12, 1985.

[2]Wuhan College of Geology, Petroleum Geology Teaching and Research Section, Wuhan, People's Republic of China.

Special thanks to the Geological Research Institute of the Shengli Oil Field and the Geological Brigade of East China Bureau of Petroleum Geology for providing much valuable data. We also thank M. P. A. Jackson, S. B. Frazier, and W. W. Tyrrell, Jr., and the geologists of Shengli oil field for their helpful comments and suggestions.

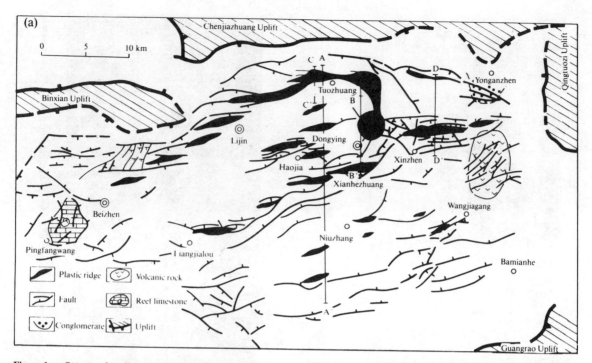

Figure 1a— Structural outline map, Dongying depression, Shengli oil field area of Bohai Bay basin, Shandong province, north China (between long. 117°50′ and 118°45′, and lat. 37°46′ and 37°20′). Locations for seismic profiles in Figures 2, 7, 8c, and 10 are shown.

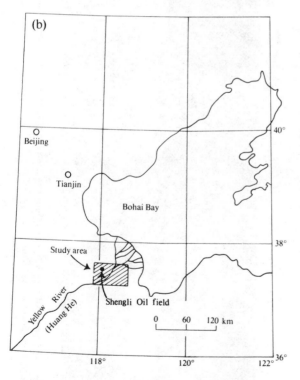

Figure 1b—Index map showing location of Shengli oil field area.

ying depression. The structures on the south gentle slope are small in area and amplitude, but have a clear orientation and arrangement pattern.

Qianjiang depression in Jianghan basin is another example of an Eocene-Oligocene salt basin. The total thickness of the Oligocene formation is more than 3,500 m, and the cumulative thickness of salt is about 1,800 m. A series of salt domes in the depression is arranged concentrically (Figures 5, 6).

Doming and piercing of diapiric materials occur primarily because the density of plastic materials is lower than that of the overlying sediments; i.e., a density inversion causes gravitational instability. Actually, the force caused by this process is related to a tectonic vertical stress, whereas the regular arrangement patterns of these diapiric structures are caused obviously by the horizontal stress of regional tectonic movement. Therefore, we believe that Cenozoic diapiric structures in eastern China are formed by a combination of vertical and horizontal stresses.

Nonpiercing or Slightly Piercing Diapirs

Most of the diapirs are nonpiercing or slightly piercing. Unlike typical strong diapirs, the diapiric structures in eastern China are weakly deformed, and the flanks of these structures generally dip less than 10°.

The Dongying structure on the central rise of the Dongying depression is a well-known example (Figure 7). It is a domal anticline composed of an Oligocene sand-shale

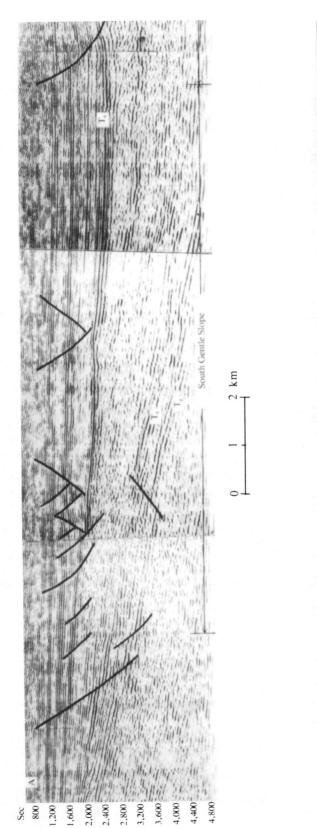

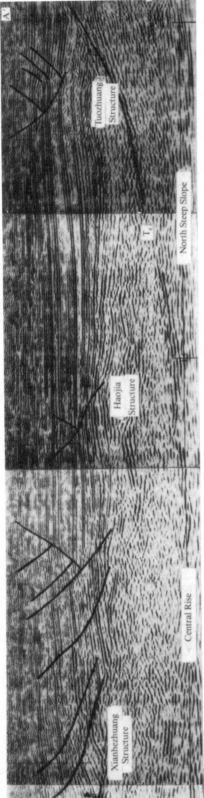

Figure 2—Seismic profile (AA′) across Dongying depression. T_6 = bottom of Eocene Shahejie formation; T_9 = bottom of Eocene Shahejie-2 (Es_2) formation; and T_g = basement of basin. Location shown in Figure 1.

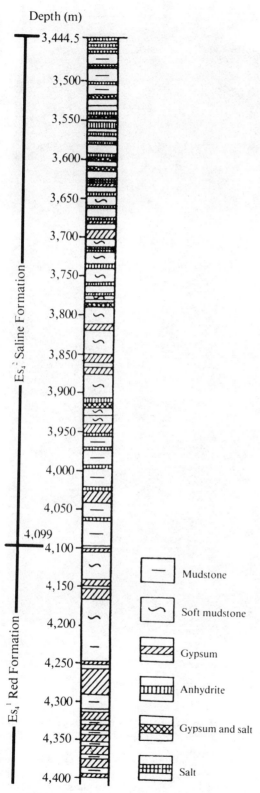

Depth (m)

Es$_4^2$ Saline Formation

Es$_4^1$ Red Formation

— Mudstone

∿ Soft mudstone

▨ Gypsum

▥ Anhydrite

▧ Gypsum and salt

▤ Salt

Figure 3—Column of plastic strata, Shahejie formation (Es$_4$), Dongying depression.

sequence with a 200-m closure. The dip of the flanks is 5°–8° at shallow levels. Many faults on the crest form a complex radial graben system. The Eocene Shahejie formation (Es$_4$) at a deep level of the structure is composed of salt, gypsum, and soft mudstone, and it is more than 1,000 m thick (Figure 3). The seismic profile shows that the Es$_4$ reflective horizons are disordered and discontinuous. A clear unconformity boundary between the Es$_4$ disordered sequence and the overlying Es$_3$ sand-shale stratified sequence suggests that a piercement occurs at the bottom of the Es$_3$ sequence (Figure 7). Stratigraphic correlation shows that a lacuna of about 250 m at the bottom of the Es$_3$ has resulted from piercement (Wang Xie-pei et al, 1981). This "drawbridge" effect—as it is called by Gulf Coast geologists (Hughes, 1968)—is the strongest diapirism we have seen in eastern China.

The Tuozhuang structure located in the north steep slope of the Dongying depression is an en echelon anticline. The Eocene plastic core at a deep level is in the shape of a pillow, without piercement of overlying strata (Figure 8).

The Wangchang oil field in the Qianjiang depression is another example (Figure 9). It is a long, narrow anticline with a typical salt dome. The Oligocene sediments are 1,000 m thick in the flanks, but they thicken abruptly in the core to 2,200 m. The amplitude of the dome may reach 800 m. The flanks dip about 20° at shallow levels and 65°–70° at deep levels. The entire section of Oligocene sediments has high plasticity; therefore, no evidence of piercement is found in the upper part of such plastic strata although the structure is highly domed.

Interbedded Plastic Strata

Cores of the diapiric structures are composed of interbedded plastic strata, including salt, gypsum, and mudstone, but mudstone is predominant. Salt-lake basins evolved extensively during the Eocene–early Oligocene in eastern China. The distribution of salt is primarily limited to a small area—the center of the basin, which forms only one-fifth to one-fourth of the whole basin. Most of the basin is occupied by mudstone. The cumulative mudstone thickness of the entire section is much greater than that of salt and gypsum, with mudstone generally forming 60-80% of the total thickness. These characteristics are important when analyzing the origin of the diapirs because the nature of the plastic materials and the geologic conditions causing plastic flow are different. Generally, the intensity and duration of the plastic movement of salt are stronger and longer than that of mudstone. The paucity of salt is one of the major causes for the lack of strong diapirs in eastern China.

Graben-Fault System Above Diapirs

As is typical in petroliferous basins of the world, a graben-fault system is commonly found above diapirs. The Xinzhen structure in Shengli oil field, Dongying depression, is a typical example (Figures 10, 11). Xinzhen is a long, narrow asymmetric anticline trending nearly west-east. The south flank dips about 10°, and the north flank may dip

Figure 4—Cores of Shahejie unit (Es$_4^1$) in Dongying depression. Scale is in centimeters.

25°–30° at depth. A series of complex graben faults extends parallel to the axial strike and is clustered on the crest of the structure. Having been intersected by two groups of faults trending northeast and northwest, the entire anticline is cut into more than 100 blocks. Therefore, the Xinzhen oil field reservoir has been severely destroyed, and a considerable amount of oil and gas may have disappeared along the faults. Oil field development has confirmed that reserves in the Xinzhen structure are poor even though it is located in the most favorable position—on the central rise of the petroliferous basin.

Experience in prospecting for oil and gas suggests that a graben-fault system on the crest of an anticline indicates diapirism at deep levels, and the complexity of the fault system reflects the intensity and amplitude of the doming of plastic strata.

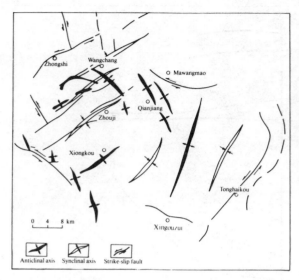

Figure 5—Structural outlines of Qianjiang depression, Jianghan basin, Hubei province (between long. 112° and 113°40′, and lat. 30°50′ and 29°40′). Location shown in Figure 6.

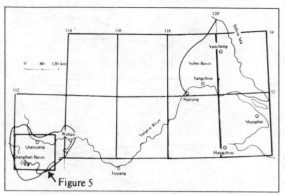

Figure 6—Location of Subei and Jianghan basins.

MAGMATIC DIAPIRIC TRAPS

Magmatic diapiric traps have not been studied thoroughly enough. However, having analyzed the types and mechanism of structural traps in the Subei basin (Figure 6), we found that structural traps are closely related genetically to magmatic activities.

The Subei basin is a complex Tertiary basin strongly divided by inner uplifts (Figure 12). The basin area is 23,000 km². More than 150 anticlinal traps have been found. Although these structures are located in different units of the basin, their structural characteristics are similar.

1. All of these anticlinal traps have a small area. Two-thirds of the structures in the Subei basin are less than 5 km². The smallest one is only 0.5 km².

2. Structures are nearly circular. Axial trends of these dome-shaped anticlines are irregular and show no relation to the regional trend (Figure 12).

3. These domes have a regular spatial arrangement along the regional fundamental faults in north-northeast and northeast trends. A prominent gravitational high is found along these fundamental faults on the map of regional gravitational anomalies, as might be expected for basic intrusions.

4. Graben-fault systems developed on the crests of the anticlines (Figure 13). Diabasic intrusions have been found at deep levels of the structures. Diabase has either intruded along the bedding to form a sill (Figure 14) or has pierced vertically to form a plug (Figure 15).

5. Cenozoic magmatic activities are strong in the Subei basin and can be divided into at least 12 stages (Wang Xie-pei et al, 1982). We have observed many volcanic cones in the western part of the basin. Also, a series of small anti-

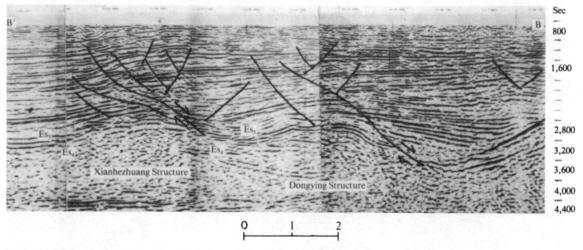

Figure 7—Seismic profile (BB′) across Dongying and Xianhezhuang structures in Dongying depression. Es = Eocene Shahejie formation. Location shown in Figure 1.

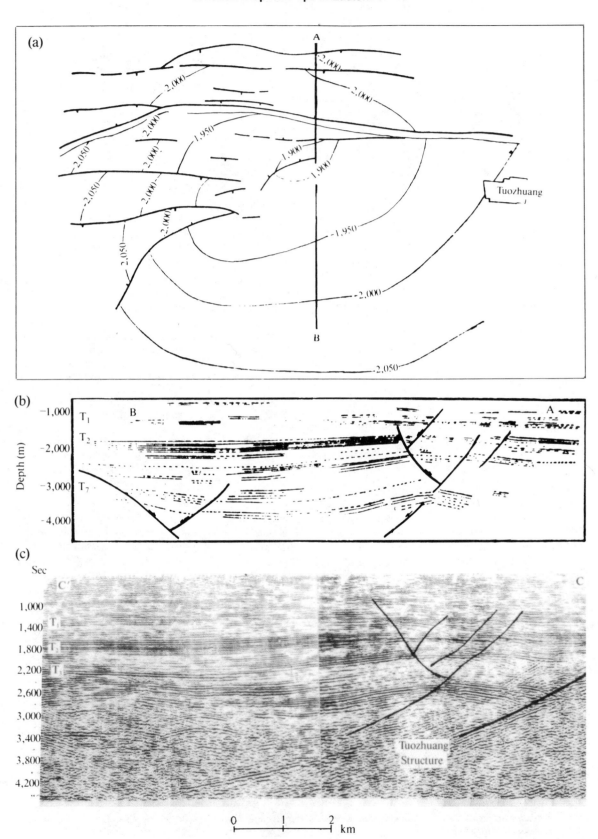

Figure 8—(a) Structural contour map (T_2) (C.I. = 50 m), (b) seismic depth profile, and (c) seismic profile (CC′) of Tuozhuang structure, Shengli oil field. Location of c shown in Figure 1.

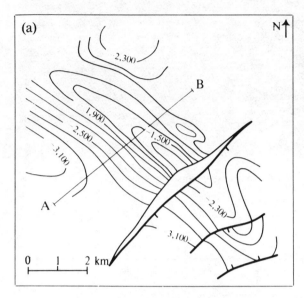

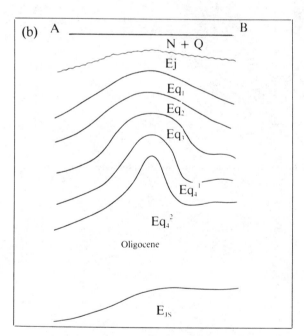

Figure 9—(a) Structural contour map on top of unit Es_3 (top left corner is long. 112°40′, lat. 30°50′), and (b) cross section AB of Wangchang oil field, Jianghan basin, Hubei province. N + Q = Neogene and Quaternary; Ej = Eocene Jinghezhen formation; Eq = Eocene Qianjiang formation; and E_{JS} = Eocene Shashi formation. C.I. = 200 m.

clines is distributed along the major petroliferous fault zone, Wubao, and a series of volcanic cones is regularly arranged at the surface on the strike of the fault zone. These volcanic cones are similar to small anticlines on the fault zone in size, form, and direction of arrangement (Figure 16). This phenomenon indirectly reflects the genetic relation between the anticlinal structures and magmatic activities.

Therefore, we believe that a certain proportion of the small anticlinal traps in the Subei–South Yellow Sea basin are formed by diabasic diapirs.

CONCLUSIONS

Many salt-lake basins developed extensively in eastern China during the Eocene–early Oligocene, and a sedimentary sequence composed of interbedded salt, gypsum, and mudstone was commonly deposited. Plastic deformation and diapirism occurred during the Cenozoic regional tectonic movement.

These diapiric traps are unpierced or slightly pierced structures. The diapiric material is a mixture of salt, gypsum, and mudstone, but mudstone is the main constituent.

Mechanically, diapiric structures in eastern China are caused by a combination of horizontal and vertical stresses.

In the Dongying depression, intense diapirism created three diapiric structural zones, each of which has a different ability to trap oil and gas. With a higher closure and integrated form, the diapiric structures on the steep slope are the most productive. Diapiric traps on the central rise vary in productivity, depending on the degree of trap destruction by faults. Structures on the gentle slope are too small to form favorable traps.

Diabasic diapiric traps are commonly arranged in a series of small anticlines along the regional fundamental faults in the Subei basin. Oil and gas have been found in some of these traps.

REFERENCES CITED

Fei Qi, and Wang Xie-pei, 1982, A preliminary study of diapiric structures in oil and gas-bearing basins in eastern China: Oil and Gas Geology (China), v. 3, no. 2, p. 113-123.

Hughes, D. J., 1968, Salt tectonics as related to several Smackover fields along the northeast rim of the Gulf of Mexico basin: Gulf Coast Association of Geological Societies Transactions, v. 18, p. 320-330.

O'Brien, G. D., 1968, Survey of diapirs and diapirism, in Diapirism and diapirs: AAPG Memoir 8, p. 1-9.

Wang Xie-pei, Fei Qi, and Zhang Jia-hua, 1981, Mechanism of the formation of diapiric structural traps in Dongying depression, Shandong province: Acta Petrolei Sinica (China), v. 2, no. 3, p. 13-22.

—————— Wang Shu-zhen, and Yang Xiang-min, 1981, The tectonic framework and mechanism of formation of local structural traps in the oil-gas-bearing basins in northern Jiangsu: Petroleum Exploration and Development (China), no. 3, p. 1-15.

(Continued)

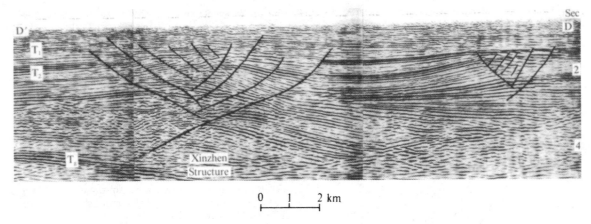

Figure 10—Seismic profile (DD′) of Xinzhen oil field in Dongying depression. T₁ = bottom of Neogene; T₂ = bottom of Eocene Dongying formation; and Tg = basement of basin. Location shown in Figure 1.

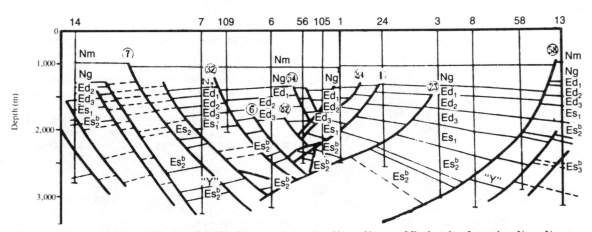

Figure 11—Structure section of Xinzhen oil field in Dongying depression. Nm = Neogene Minghuazhen formation; Ng = Neogene Guantao formation; Ed = Eocene Dongying formation; Es = Eocene Shahejie formation; and circled number = fault number.

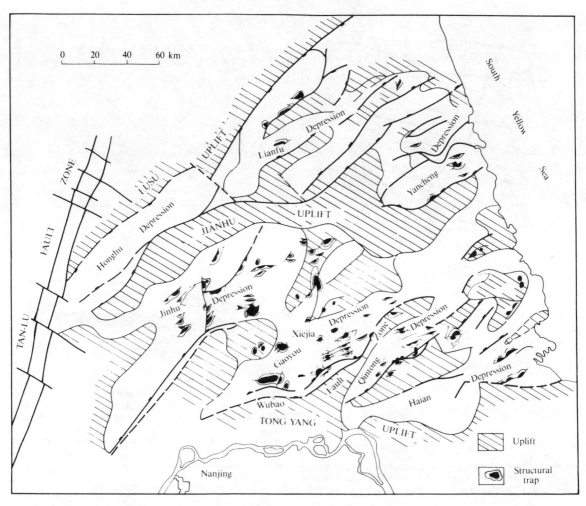

Figure 12—Structural uplifts and depressions with distribution of local structures, Subei basin, Jiangsu province (between long. 117°40′ and 121°, and lat. 34°10′ and 32°20′). (Modified from Wang Xie-pei et al, 1982.) Heavy solid lines define faults; broken lines are inferred faults. Location shown in Figure 6.

(Continued)

101

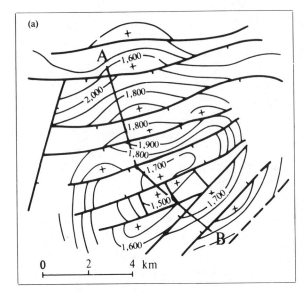

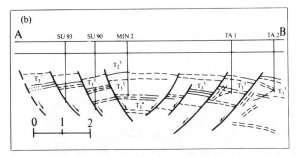

Figure 13—(a) Structural contour map on bottom of T_3^3 unit (top left corner is long. 119°, lat. 33°; C.I. = 100 m), and (b) cross section AB of Minqiao structure, Subei basin, Jiangsu province.

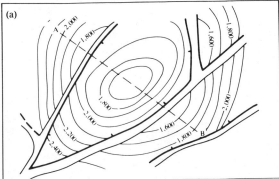

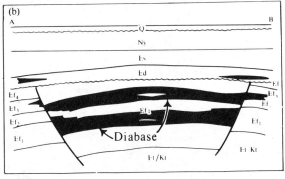

Figure 14—(a) Structural contour map on top of Ef_2 (top left corner is long. 119°10′, lat. 32°50′; C.I. = 100 m), and (b) cross section AB of Matouzhuang structure, Subei basin, Jiangsu province. Q = Quaternary; Ny = Neogene Yangcheng formation; Es = Eocene Shanduo formation; Ed = Eocene Dainan formation; Ef = Eocene Funing formation; and Et/Kt = Eocene or Cretaceous Taizhou formation.

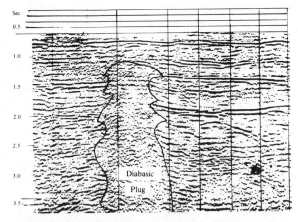

Figure 15—Xiejia diabasic plug, Subei basin, Jiangsu province. Location shown in Figure 12.

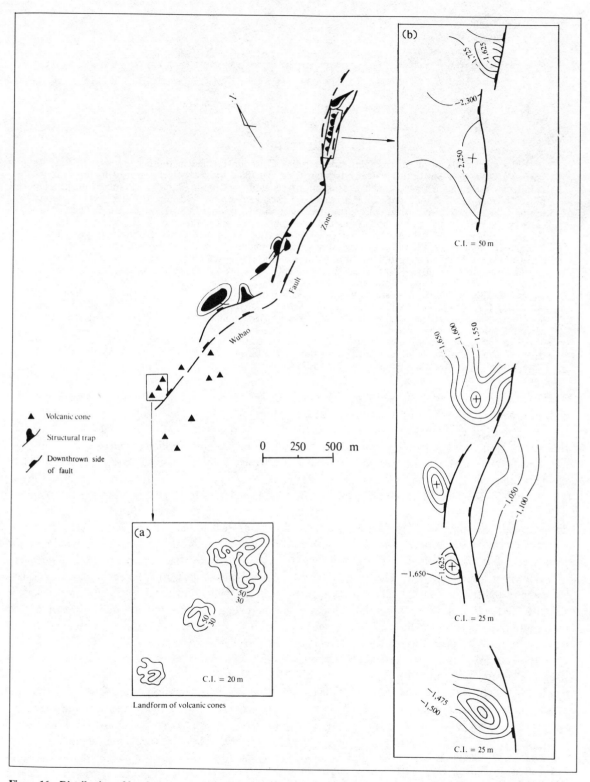

Figure 16—Distribution of local structures along Wubao fault zone, Subei basin, showing genetic relation between small anticlines and magmatic activities. (a) Top right corner is long. 119°, lat. 32°45′; (b) top right corner is long. 120°, lat. 32°55′.

American Association of Petroleum Geologists Memoir 2, *Backbone of the Americas: Tectonic History From Pole to Pole,* edited by O. E. Childs and B. W. Beebe, copyright ©1963 pp. 152-159.

STRUCTURAL DEVELOPMENT OF SALT ANTICLINES OF COLORADO AND UTAH[1]

FRED W. CATER[2] AND D. P. ELSTON[2]
Denver, Colorado

ABSTRACT

The salt anticlines of eastern Utah and western Colorado formed in the deepest part of Paradox basin—a basin developed during Pennsylvanian time and filled by great thicknesses of upper Paleozoic sediments, including a thick sequence of evaporites belonging to the Paradox Member of the Hermosa Formation. The salt anticlines originated either as tectonic folds or as folds over basement faults soon after the evaporites were deposited, probably in middle Pennsylvanian time. These structures were parallel to, and probably formed concomitantly with, the rise of the ancestral Uncompahgre highland, the front of which paralleled closely that of the southwest front of the present day Uncompahgre Plateau. Rapidly accumulating arkosic sediments of the Permian Cutler Formation, derived from this highland, probably buried parts of the salt anticlines; elsewhere along the anticlines the salt rose isostatically as rapidly as the sediments were deposited. In places the Cutler was later intruded by the cores of the buried salt anticlines. Parts of the cores were exposed at the surface at least until the Morrison Formation was deposited in Late Jurassic, so that the formations pinch out along the flanks of the salt cores. Variations in thickness—chiefly thinning—of the Morrison and later Mesozoic formations over the crests of the salt cores indicate that salt flowage was still active after the salt cores were buried.

The salt anticlines attained their present forms—except for modifications imposed by later collapse of the crestal parts of the anticlines—during the early Tertiary when the rocks of the region were folded, and the salt anticlines were accentuated.

The salt anticlines of southwestern Colorado and southeastern Utah are a northwestward-trending group of structures occupying a zone of the same trend about 110 miles long and 25-30 miles wide. The major anticlines are decidedly elongate, the salt cores being from 2 to 5 miles wide, whereas the lengths range from 30 to 70 miles (Fig. 1). The salt anticlines occupy the deep, trough-like, northeasterly part of Paradox basin, a basin which formed during Pennsylvanian time and in which accumulated thick deposits of evaporites, principally salt. The ancestral Uncompahgre uplift, an element of the Ancestral Rockies, was formed concomitantly with the basin, and borders it on the northeast. The southwest front of this uplift approximately coincides with the southwest front of the present Uncompahgre Plateau uplift, but the plateau is areally much less extensive than the original uplift.

Rocks exposed or cut by drill holes in the region range in age from Precambrian to late Ter-

tiary (Fig. 2). Precambrian rocks, dominantly granites and gneisses, crop out along the Uncompahgre Plateau. Paleozoic rocks of Cambrian, Devonian, Mississippian, and early Pennsylvanian age are known only from drill holes. Rocks of the middle Pennsylvanian Hermosa Formation, including the salt-bearing Paradox Member, and rocks of the Rico Formation, which ranges in age from middle Pennsylvanian to Permian? depending on location, are exposed only in the valleys carved from the crests of the salt anticlines. Rocks of later Paleozoic and Mesozoic age are widely exposed, but Tertiary rocks are less extensive and include only intrusive porphyritic rocks and scattered conglomerate of doubtful late Pliocene age.

The Hermosa Formation, especially the Paradox Member, is of great interest, not only because of its economic potential for oil, gas, and potash, but also because of the controlling influence it has exerted, owing to its plasticity, on the stratigraphy, thickness, and structure of succeeding formations. It is probably safe to assume that at no place in the salt anticline region is the Paradox undisturbed by flowage, and, there-

[1] Read before the Association at Denver, Colorado, April 26, 1961. Manuscript received, May 5, 1961. Publication authorized by the Director, U. S. Geological Survey.
[2] U. S. Geological Survey.

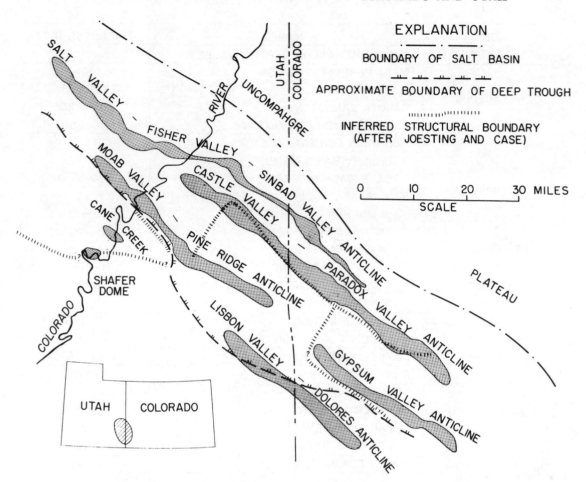

FIG. 1.—Map of salt cores of anticlines, southwestern Colorado and southeastern Utah.

fore, the original thickness is difficult to determine. Various estimates have been made of the original thickness of the Paradox in the deep part of the basin; the highest estimate of local thicknesses along the site of Paradox Valley is 7,000 feet. If one assumes that the cores from each of these anticlines contain salt from the adjacent synclines and that virtually all the salt originally in the synclines has been squeezed out, then a wedge averaging about 3,000 feet thick beneath and adjacent to Gypsum Valley, and 4,000 feet thick beneath and adjacent to Paradox Valley, would have been required to supply the salt presently in the two cores. The continuity of beds in the leached caps of the two salt cores suggests that no great amount of the Paradox Member was lost by extrusion, and there is no

evidence in the surrounding sediments to suggest that large-scale extrusion occurred. Rather, it seems that the salt moved slowly upward at a rate about sufficient to keep pace with deposition, and that the salt presently in the cores, plus the calculable amount represented by the residuum in the cap, represents most of the salt originally deposited in this part of the basin.

The major salt anticlines are not only parallel to each other but also are parallel to the edge of the ancestral Uncompahgre uplift—a fact that has led many geologists to the conclusion that the anticlines were controlled by deep-seated structures. This conclusion now is more defensible. Drilling and geophysical information has dispelled any doubt concerning the existence of deep-seated structures in pre-Paradox rocks, and,

AGE	NAME	THICKNESS (FT.)
LATE TERTIARY(?)	CONGLOMERATE IN GYPSUM VALLEY	
CRETACEOUS	MESA VERDE FORMATION	100+
	MANCOS SHALE	2000(?)
	DAKOTA SANDSTONE	150-220
	BURRO CANYON FORMATION	50-250
JURASSIC	MORRISON FORMATION	
	BRUSHY BASIN MEMBER	300-600
	SALT WASH MEMBER	250-600
	SAN RAFAEL GROUP	
	SUMMERVILLE FORMATION	0-110
	ENTRADA SANDSTONE	0-220
TRIASSIC(?) AND JURASSIC	GLEN CANYON GROUP	
	NAVAJO SANDSTONE	0-400+
	KAYENTA FORMATION	0-250
	WINGATE SANDSTONE	0-450
TRIASSIC	CHINLE FORMATION	0-450
	MOENKOPI FORMATION	0-1300
PERMIAN	CUTLER FORMATION	0-10,000(?)
PENNSYLVANIAN	RICO FORMATION	0-940
	HERMOSA FORMATION	
	UPPER MEMBER	0-2900(?)
	PARADOX MEMBER	0-14,000(?)
	LOWER MEMBER	0-195
——— ? ———	MOLAS FORMATION	0-135
MISSISSIPPIAN		
DEVONIAN	PRE-PENNSYLVANIAN PALEOZOIC ROCKS	2000±
CAMBRIAN		
PRECAMBRIAN		

Fig. 2.—Generalized section of rock formations in salt anticline region.

in a broad way, has even indicated the nature of these structures. Structural displacements of pre-Paradox rocks where no corresponding displacements of surface rocks exist, have been proved by drilling at a number of places. On the southwest flank of the Lisbon Valley anticline, drill holes show that the pre-salt rocks locally have about 2,500 feet of structural relief. Under the southwest flank of the Gypsum Valley anticline the base of salt is about 1,800 feet below sea level; 7 miles to the southwest the base is about 600 feet lower, whereas beneath the northeast flank of the anticline the base is about 2,600 feet lower. On the basis of magnetic anomalies

this ridge in pre-Paradox rocks had been predicted by Joesting and Byerly (1958) prior to drilling. Between a hole drilled in the center of the Paradox Valley anticline and a hole recently drilled in the adjoining syncline 5 miles to the southwest, more than 5,000 feet of structural relief exists in the pre-salt rocks. The gravity data obtained by Joesting and Byerly indicate the dislocation causing this structural relief lies beneath, and parallel to, the southwest flank of the anticline. The steepness of the gravity gradient would seem to indicate a fault, as it is extremely doubtful that any fold could produce so steep a gradient. All in all, the evidence suggests that the pre-salt rocks are cut into a series of fault blocks; some of the larger of these faults probably controlled the loci of the salt anticlines, such as those that probably underlie the southwest flanks of both Paradox Valley and Gypsum Valley anticlines. In both these anticlines the rocks northeast of the faults have been downdropped.

Drilling information and geophysical data also suggest that the individual blocks are tilted so that the strata dip southwest as shown in Figure 3. This, of course, would mean that pre-salt strata are deepest immediately below the salt cores. Such tilted strata with the updip, faulted ends sealed by salt could form efficient oil and gas traps. Most of the folds are asymmetrical, the north flanks of the anticlines being the steeper, and the troughs of the synclines lying near their southwest margins.

Joesting and Byerly (1958) found that in addition to the gravity anomalies resulting from the visible northwestward-trending salt structures, strong northeastward-trending anomalies

also exist, but the nature of these is unknown. By and large, whatever the cause of these anomalies, they have been largely ineffective in exerting any control on visible surface structures. One possible exception may be the offsets which occur at the southeast ends of the Paradox and Gypsum Valley anticlines. Possibly these anomalies may reflect faults beneath the Paradox, but if so, drilling to date has not indicated the direction of the throw on them. Another possibility, of course, is that these anomalies represent northeastward-trending masses of high density, abnormally magnetic rocks in the crystalline Precambrian basement.

The location of the salt anticlines may have been determined before deposition of the salt even began, because it seems probable that the basement structures that localized the anticlines developed prior to Paradox time. R. J. Hite (1960) has shown, as the result of detailed study of the internal stratigraphy of the salt-bearing unit of the Paradox Member, that the sites of the Lisbon Valley, Cane Creek, and northwest

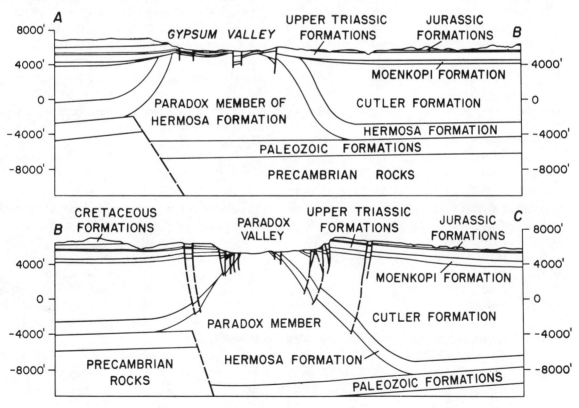

Fig. 3.—Section across Gypsum Valley and Paradox Valley anticlines.

part of the Moab Valley anticlines, along the southwestern margin of the salt anticline region, were sedimentary troughs during most of the time of salt deposition, and that some of the oldest salt beds deposited in the troughs are missing over the adjacent structurally high areas. A. W. Neff (1960) has reported that in a drill hole which penetrated structurally high pre-salt rocks in the syncline between the Gypsum Valley and Dolores anticlines, the lower member of the Hermosa Formation, the Molas Formation, and the upper part of the lower Mississippian rocks apparently are missing. In other words, structural disturbances immediately preceding and perhaps during early Paradox time were so great that some areas projected above levels of deposition and were subjected to erosion. Recent information from the drill hole in the syncline southwest of the Paradox Valley anticline indicates similar conditions exist in the structurally high pre-salt rocks.

Prior to 1958 it was generally thought that the salt anticlines formed largely, if not entirely, by salt intrusion which was controlled, nonetheless, by regional tectonic forces. In that year and in 1959, however, Elston, Shoemaker, and Landis (Elston and Landis, 1960) proved, as the result of detailed mapping of outcrops in Paradox Valley, Gypsum Valley, and Salt Valley, that beds of the Rico and Cutler formations locally rest unconformably on the Paradox and limestone members of the Hermosa Formation. Other evidence, but not so clear-cut, indicates that the Paradox and upper limestone member of the Hermosa are also unconformable. These unconformable relations indicate that the salt anticlines probably started as faults that passed upward into folds at the top of the Paradox, rather than as folds developed because of salt flowing under the stress of a load of overlying rocks.

Theoretical considerations and relations observed in other salt anticlines of the world argue strongly against the possibility that the unconformable relations observed represent deposition of beds of the Rico and the Cutler over exposed intrusive cores. The main argument against this possibility is that the post-Paradox beds would not have been thick enough to have developed the stress differentials necessary to have started

intrusion by the beginning of Rico time.

Balk (1949, p. 1820) has stated that ". . . a mass of salt should be under an excess compression of at least 60 kg/cm² in one direction, in order that slip along favorably oriented glide planes of its crystals may commence." Parker and McDowell (1955, p. 2391-2392), on the basis of Balk's remarks, have suggested that a pressure differential equal to 1,000 feet of beds is necessary to start flowage of salt. With a thickness of 2,000 to 2,500 feet of sediments of the upper part of the Hermosa overlying the salt at the beginning of Rico time, it is unlikely that the stresses developed could have been critical—that is, it seems that it would have been impossible beneath a load of only 2,000-2,500 feet of sediments to develop pressure differentials approaching the quoted 60 kg/cm² necessary to start salt flowage and, thereby, intrusion. And then it should be recalled that there is evidence that the upper part of the Hermosa rests unconformably on the Paradox Member.

There is much geologic evidence, however—such as the rock salt glaciers of the Near East and some of the collapse features developed in the Colorado-Utah salt anticlines—that indicates the strength of salt to be much lower than the value we have cited; but it should be remembered that these geologic features formed under conditions where salt was free to move and where water was sufficiently abundant to permit some solution. Under conditions where incipient solution can occur, salt has virtually no strength, but such conditions are far different from those that probably would have existed in the salt of the Paradox had it been confined beneath a cover of strong limestone 2,000 feet thick.

If the foregoing analysis is correct, it seems unlikely that intrusion of salt played much part in the early development of the anticlines. This is not to say, however, that intrusion of beds of the Paradox into younger beds, notably of the Cutler, has not occurred; rather it is a common phenomenon.

The lithic characteristics of the Cutler indicate very rapid deposition; under these conditions the Cutler deposits could have temporarily buried the rising salt cores. This is illustrated in the third section of the series of diagrammatic

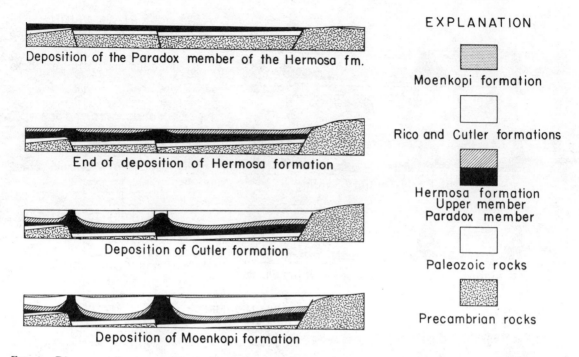

EXPLANATION

Moenkopi formation

Rico and Cutler formations

Hermosa formation
Upper member
Paradox member

Paleozoic rocks

Precambrian rocks

Deposition of the Paradox member of the Hermosa fm.

End of deposition of Hermosa formation

Deposition of Cutler formation

Deposition of Moenkopi formation

Fig. 4.—Diagrammatic cross sections showing development of Paradox Valley and Gypsum Valley anticlines.

cross sections showing the development and collapse of the salt anticline (Figs. 4 and 5). With a decrease in the rate of deposition toward the end of Cutler time, the rising salt at places broke through to the surface again before the basal part of the Moenkopi Formation (Triassic) was deposited; this is indicated by intrusive contacts and by contacts between Paradox and Cutler that are almost vertical through distances of several thousand feet beneath the Moenkopi along the flanks of the salt cores.

It seems probable that much of the available salt in the areas between the salt cores had been squeezed out before the lowermost beds of the Moenkopi were deposited. Because of the irregular nature of the pre-salt surface, however, sizable thicknesses of salt probably were trapped where post-Paradox beds sagged to points where they were in contact with structurally high pre-salt beds, and further escape of salt was thereby prevented. One possible place where considerable salt could have been trapped is in the syncline between the Dolores and Gypsum Valley anticlines where the pre-salt structural high under the southwest flank of Gypsum Valley could have

formed a barrier when post-salt beds came to rest upon it. Later, this salt could have escaped following Tertiary deformation of the region, which may have re-established escape ways. The unusual structural depth of this syncline may be the result of this process. In most places, beds of Mesozoic age show evidence of only relatively slow growth of salt cores, as the rate of upwelling of salt balanced or slightly exceeded the rate of removal of salt by solution and the cap by erosion. Thus, growth about kept pace with the accumulation of sediments in surrounding areas, and the upper surface of the salt maintained nearly constant altitudes as the rocks in the surrounding country gradually settled around them.

With the exhaustion of available salt in the synclines in Late Jurassic time, active rise of salt generally ceased, although minor readjustments of the salt within the cores continued, and the Morrison Formation of Jurassic age very likely completely blanketed the region.

With the exception of the northeast flank of the Sinbad-Fisher Valley anticline, there seems to have been no tendency for linear structures flanking the salt anticlines analogous to the rim

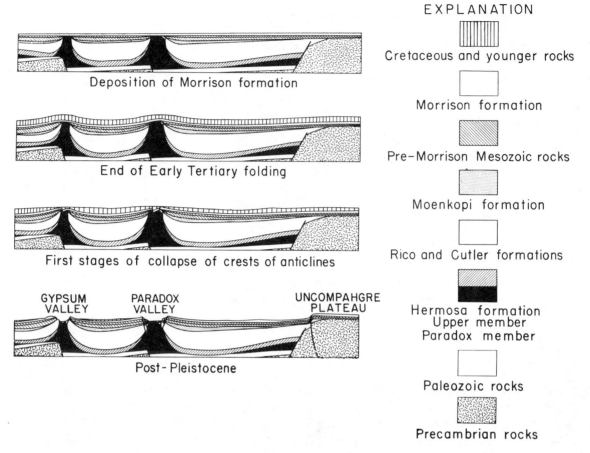

EXPLANATION

Cretaceous and younger rocks

Morrison formation

Pre-Morrison Mesozoic rocks

Moenkopi formation

Rico and Cutler formations

Hermosa formation
Upper member
Paradox member

Paleozoic rocks

Precambrian rocks

Deposition of Morrison formation

End of Early Tertiary folding

First stages of collapse of crests of anticlines

GYPSUM VALLEY PARADOX VALLEY UNCOMPAHGRE PLATEAU

Post-Pleistocene

Fig. 5.—Diagrammatic cross sections showing development and collapse of Paradox Valley and Gypsum Valley anticlines.

synclines of the Gulf Coast salt domes to develop. On the northeast flank of the Sinbad-Fisher Valley anticline, the Moenkopi Formation is nearly three times as thick as normal in the region, but it pinches out abruptly on the flanks of the salt cores, and thins rapidly to the northeast as the Uncompahgre Plateau is approached. Perhaps considerable Paradox material is trapped between this anticline and the Uncompahgre Plateau.

Following deep burial of the salt cores by late Mesozoic and possibly early Tertiary sediments, the region was deformed sometime during the early Tertiary, and the salt anticlines attained approximately their present structural form except for modifications caused by later collapse of their axial parts. Somewhat later, possibly through relaxation of the stresses responsible for

folding, the crests of the anticlines were faulted or dropped as grabens. Still later, following epeirogenic uplift of the Colorado Plateau and breaching of the crests of the anticlines by erosion, the present stage was reached. Uplift of the Uncompahgre Plateau has been a comparatively recent event; in fact, the final 2,000 feet of uplift has occurred since Unaweep Canyon, which cuts completely across the plateau, was abandoned by the Colorado River.

Collapse of the salt anticlines seems to have little real significance economically, but there is one collapse feature—a marginal anticline—that has been drilled for oil several times without success. This structure, called the Dry Creek anticline, lies at the southeast end of the Paradox Valley anticline and along its southwest flank. Most of the crest of the Paradox Valley anti-

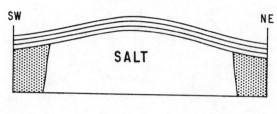

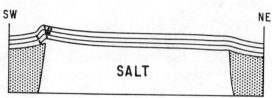

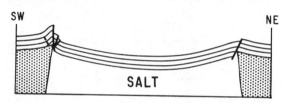

Fig. 6.—Diagrammatic cross sections showing development of the downsag and faulted marginal anticlines at the southeast end of the Paradox Valley anticline.

cline failed by faulting and slumping, but this part failed as the buttressing rocks in the flanks of the anticline were pushed apart during the sagging of the crest, as shown in Figure 6.

Because the confining seal of sediments northwestward along the Paradox Valley anticline had been removed by erosion, the salt in the cores was free to escape; as the crest of this part of the anticline sagged, the salt was squeezed northwestward into Paradox Valley like toothpaste out of a tube, and removed by solution and erosion. Eventually dips in the crestal block of rocks reversed and the block subsided further along the marginal faults. Thus, what had been a single large anticline became two, one over each flank of the salt core. The Dry Creek anticline over the southwest flank of the core is by far the more impressive, and is the one that has been drilled. It is of interest to note that, as the result of this process of collapse, the sharply flexed beds adjacent to the marginal faults here have gained the appearance of drag folds in reverse.

REFERENCES

Balk, Robert, 1949, Structure of Grand Saline salt dome, Van Zandt County, Texas: Am. Assoc. Petroleum Geologists Bull., v. 33, no. 11, p. 1791-1829.

Elston, D. P., and Landis, E. R., 1960, Pre-Cutler unconformities and early growth of the Paradox Valley and Gypsum Valley salt anticlines, Colorado, *in* Short papers in the geological sciences: U. S. Geol. Survey Prof. Paper 400-B, p. 261-265.

Hite, R. J., 1960, Stratigraphy of the saline facies of the Paradox Member of the Hermosa Formation of southeastern Utah and southwestern Colorado, *in* Four Corners Geol. Soc. Guidebook 3d Field Conf.: Geology of the Paradox fold and fault belt, 1960, p. 86-89.

Joesting, H. R., and Byerly, P. E., 1958, Regional geophysical investigations of the Uravan area, Colorado: U. S. Geol. Survey Prof. Paper 316-A, p. 1-17.

Neff, A. W., 1960, Comparisons between the salt anticlines of South Persia and those of the Paradox basin; *in* Four Corners Geol. Soc. Guidebook 3d Field Conf.: Geology of the Paradox fold and fault belt, 1960, p. 56-64.

Parker, T. J., and McDowell, A. N., 1955, Model studies of salt-dome tectonics: Am. Assoc. Petroleum Geologists Bull., v. 39, no. 12, p. 2384-2470.

The American Association of Petroleum Geologists Bulletin
V. 68, No. 3 (March 1984), P. 333-362, 26 Figs.

Graben Hydrocarbon Occurrences and Structural Style[1]

T. P. HARDING[2]

ABSTRACT

Major hydrocarbon occurrences, types of traps, and structural styles have been synthesized from the Sirte basin, the Suez and Viking grabens, and from other normal faulted regions. Hydrocarbons occur in a stacked succession of one or more basins: pregraben, graben fill, and interior sag. Preservation of pregraben accumulations depends on late initiation of crustal arching and limitation of uplift to the graben shoulders. Furthermore, a stable pregraben tectonic environment is required to ensure internal continuity of subsequent fault-block closures.

Hydrocarbon traps in the pregraben and graben-fill sediments are primarily dependent on the multidirectional orientation of normal faults, tilting of fault blocks, and, in many cases, flexing or erosion parallel with block edges. The fault pattern is dominated by longitudinal faults subparallel with the graben axis and, secondarily, by oblique faults. The direction and degree of block rotation are influenced by fault profile (planar or listric), degree of extension, fault pattern (doglegs, junctions, terminations), downwarping of the subsequent sag basin, and isostatic adjustments between large blocks. Fold closures result from the upward termination of faults into forced folds, which may be accentuated by drag during renewed faulting. Passive drape and differential compaction extend the folds to shallower depths. Because they form along block edges, the flexures have trends that mimic the multidirectional fault patterns. Truncation traps are usually developed from the erosional retreat of fault scarps and they can also mimic the fault patterns.

Where the sedimentary succession is complete, an interior sag basin is superimposed across the graben. Its lower part is structured mostly by drape and differential compaction across the underlying graben fault blocks and fault-block topography. The upper part is usually an unsegmented syncline, and hydrocarbon traps at this level include depositional buildups (reefs, turbidite mounds), and salt diapirs or other structures caused by superimposed deformation unrelated to graben tectonics.

[1]Manuscript received, January 28, 1983; accepted, July 13, 1983.
A briefer, less documented version of this article has appeared in Geologie en Mijnbouw, v. 62, p. 3-23 (Harding, 1983a).
[2]Exxon Production Research Co., Houston, Texas 77001.
I am indebted to the exploration staffs of Esso Exploration and Production U.K. and Esso Exploration and Production Norway, Inc., for their assistance in the study of Viking graben hydrocarbon occurrences. Instructive discussion and helpful suggestions for the manuscript were provided by K. T. Biddle, R. P. George, S. H. Lingrey, D. W. Phelps, and R. C. Vierbuchen of Exxon Production Research Co., and G. C. Mudd of Esso Exploration and Production U.K. This work was originally conducted for Exxon Production Research Co., and I am grateful for their permission to publish. Rebecca A. Miller drafted most of the figures.

INTRODUCTION

This work documents the kinds of major hydrocarbon accumulations that have thus far been found in the Sirte basin of Libya, the Gulf of Suez in Egypt, and the Viking graben of the North Sea. Accumulations are described either according to the stratigraphic level of the reservoir or by the kind of trap that dominates a particular trend or region.

Applied in the proper context, the Sirte basin and the Suez and Viking grabens provide models for basin development and models for the major hydrocarbon plays and their prospects (traps) that may be encountered in future exploration ventures. The structural characteristics of hydrocarbon traps and the typical structural style of grabens are summarized from these basins and are corroborated by the writer's subsurface and surface investigations of additional grabens that are not referenced in the present article. Certain kinds of structures and relationships are repeated in most areas, and these similarities are emphasized (see also Harding and Lowell, 1979). A measure of the relative importance of each trap type within its basin is provided by the estimated ultimate recoverable hydrocarbons that they contain. Reserve volumes are indicated on summary figures but are not discussed in the text. The hydrocarbon richness of grabens, of course, is controlled by many factors, and some of these are beyond the scope of the present discussion. The factors that control the sealing properties of normal faults are also not described. The reader is referred instead to Smith (1980) and Weber et al (1978).

The patterns of faulting in the Gulf of Suez have been investigated by Robson (1971) and by El-Tarabili and Adawy (1972). Hay (1978) and Halstead (1975) have studied faulting in the Viking graben. Falvey (1974) and Vierbuchen et al (1982) have presented thermal-mechanical models for the development of rifts (i.e., graben systems) and divergent margins. Basin relationships observed in the three examples of the present paper bear on this subject.

Past accounts of hydrocarbon occurrences in grabens have been treated either in detailed studies of individual fields or in basin overviews that emphasize statistics and stratigraphic levels of production. Hydrocarbons in the North Sea and Sirte basin have been discussed from the latter viewpoint by Ziegler (1979) and Parsons et al (1980), respectively. For the Gulf of Suez, Gilboa and Cohen (1979) have documented the cross-sectional characteristics of hydrocarbon traps.

Discussions of the stratigraphy, reservoir development, and general geology of each basin are limited in the present article. The reader is referred instead to past work such as that of Goudarzi (1970) for Libya, Garfunkel and Bartov (1977) for the Gulf of Suez, and Ziegler (1978, 1980) for

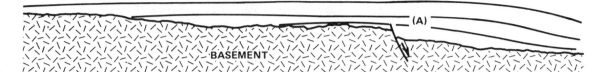

(A) PREGRABEN: INITIAL CRUSTAL STRETCHING(?), CONTINENTAL MARGIN SUBSIDENCE, OTHER

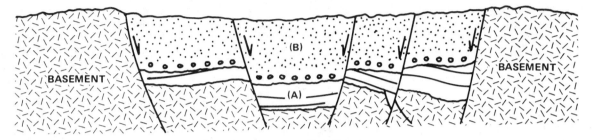

(B) GRABEN-FILL BASIN: CRUSTAL STRETCHING AND DOWNFAULTING

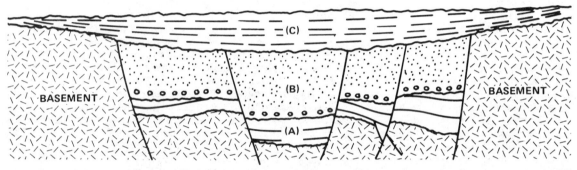

(C) INTERIOR SAG BASIN: CRUSTAL COOLING AND DOWNWARPING

Figure 1—Sequence of sedimentary basins containing major hydrocarbon accumulations in graben settings and their dominant subsidence mechanisms. Not all basin types are present or productive in every graben system, and deposition of pregraben sedimentary cover (A) may be dependent on factors other than graben tectonics.

the North Sea.

In the discussion that follows, I first review in general terms the major hydrocarbon accumulations in the Sirte, Suez, and Viking grabens. Next, the tectonic controls are described for the occurrence of these accumulations and for the development or preservation of the basins in which they occur. Last, the structural styles of grabens and controls for individual traps are summarized from a wider range of examples, and these are presented in the context of profile and map characteristics.

MAJOR HYDROCARBON OCCURRENCES

A tectonic sequence unique to grabens, if allowed to evolve fully, can cause a superimposition of two or possibly three dissimilar sedimentary basins or sites (Figure 1) (see also Falvey, 1974): (A) pregraben sedimentary cover deposited prior to normal block faulting; (B) graben-fill basin deposited during faulting; (C) interior-sag basin deposited after faulting.

In most instances the basins are separated by unconformities that are important to the trapping of hydrocarbons (Falvey, 1974). The pregraben basins may be of several types that are unrelated to the formation of the graben, or they may possibly be subsided by the earliest stage of the graben cycle. The postgraben interior sag is essentially an unfaulted syncline that forms as the final effect of the graben tectonics. Each of the three basins has the capacity to contain major hydrocarbon reserves where there has been deposition of reservoir facies and source rocks, or in some cases, just deposition of reservoir rocks (Figures 2–4).

The basins are sometimes structured by later deformational events, such as compression or strike slip (Illies and Greiner, 1978; Milanovsky, 1981). This structuring can add greatly to the basin's prospectiveness, especially to that of the interior sag, which may otherwise lack abundant closures (Harding, 1983c). Such redeformation, however, is not a direct part of the graben-forming process and is therefore not a subject of the present work.

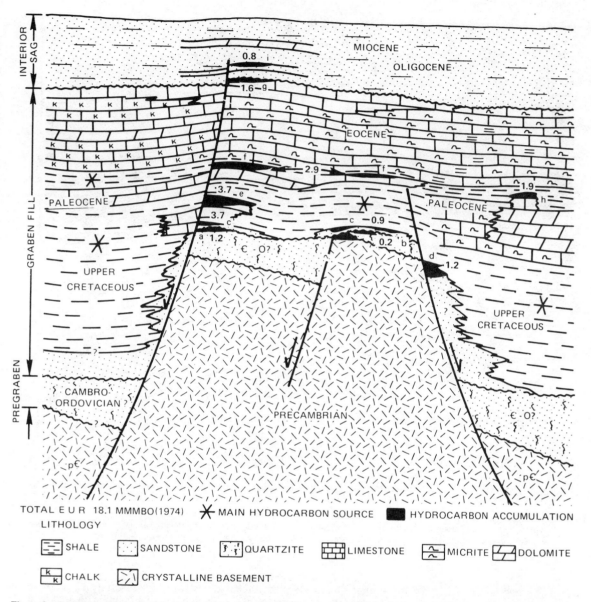

Figure 2—Schematic diagram of cross-sectional character of traps, and stratigraphic level and lithology of hydrocarbon-producing reservoirs in northwest portion of Sirte basin, Libya. Succession of basin types is indicated at left margin. Estimated ultimate recoverable reserves (EUR) are for total liquids and are stated in billions of barrels of oil (MMMBO). Trap types discussed in text indicated by a-i. Numbers on diagram indicate EUR for each trap type and were derived from Parsons et al (1980) and this study.

Sirte Basin Hydrocarbon Occurrences

In the Sirte basin, all three potential sedimentary cycles (pregraben, graben fill, and interior sag) are present and productive. These cycles are found in three broad graben arms—Sirte basin deep and Hagfa trough, Tumayn subbasin, and Sarir-Hameimat trough (Figure 5)—which define a large triple junction within the continental crust of the African plate. Graben formation started in the Late Cretaceous and ended with the downwarping of the interior sag in the middle Late Eocene (Parsons et al, 1980).

The northwestern arm (Sirte basin deep) of the triple junction is the most prolific; its hydrocarbon occurrences are summarized in Figure 2.

Accumulations in Pregraben Reservoirs

During the Paleozoic and Early Cretaceous (Conant and Goudarzi, 1967; Klitzsch, 1970), passive margin-sag and shelf sequences spread southward from the north margin of the ancestral African plate and covered the future site of the Sirte basin. Within these sediments, the fractured

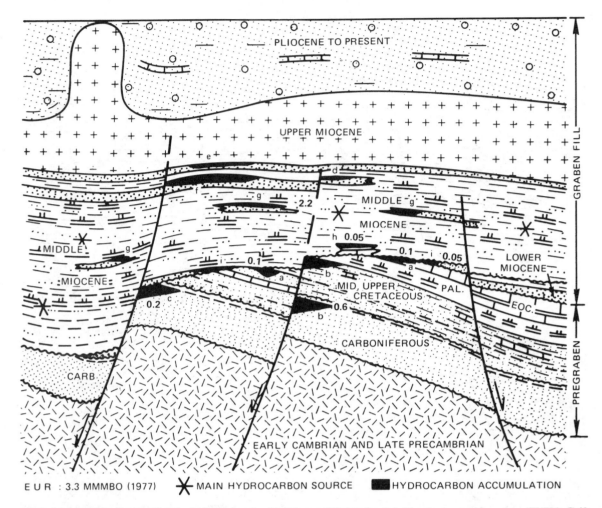

Figure 3—Schematic diagram of major hydrocarbon accumulations and their estimated ultimate recoverable reserves (EUR) in Gulf of Suez, Egypt. Trap types and reservoir lithologic characteristics have been synthesized from published data (see text). Hydrocarbon reserves (from Gilboa and Cohen, 1979) are for liquids only. Trap types discussed in text indicated by a-c. Salt indicated by +, marl by —; other symbols explained in Figure 2.

Cambrian-Ordovician quartzites of the Gargaf Group provide reservoirs for major hydrocarbon accumulations below the Upper Cretaceous and lower Paleogene graben fill (Parsons et al, 1980). Fluvial sandstones of the Lower Cretaceous Nubian Sandstone are the reservoir for a second group of major fields, which are productive mostly in the Sarir-Hameimat trough (not shown in Figure 2) (Sanford, 1970). A fractured and weathered granitic basement is a third subunconformity reservoir, and it produces at one major field on the Amal platform (Figure 5) (Williams, 1972).

In both the Gargaf and Nubian fields, hydrocarbons have accumulated directly under an angular unconformity at the base of the graben fill. Traps have been filled by hydrocarbons derived primarily from Upper Cretaceous marine shales deposited in adjacent graben deeps (Parsons et al, 1980). Gargaf accumulations and most other producing fields in the northwestern arm of the Sirte basin are

located at the updip edge of broad platforms (e.g., fields on the Amal, Zelten, and Beda platforms in Figure 5; platforms in this discussion are large, positive structures, usually formed by an agglomeration of several fault blocks, as in Figure 6). Nubian production is mainly from midtrough structures where the relatively low structural position of the blocks has protected the reservoir from erosion (fields in the Sarir and Hameimat troughs in Figure 5).

Individual closures in both the Gargaf Group (see Cambro-Ordovician quartzite in Figure 2) and Nubian Sandstone are truncated fault blocks, some with vertical closure increased by block-edge drag or drape flexing (a in Figure 2) (Sanford, 1970), and buried topographic highs (b in Figure 2) (Roberts, 1970). The geometries of these and other traps, and controls for their development are discussed in the section on graben structural styles.

One large Nubian field provides a notable exception to the dominance of structure-related traps in the pregraben

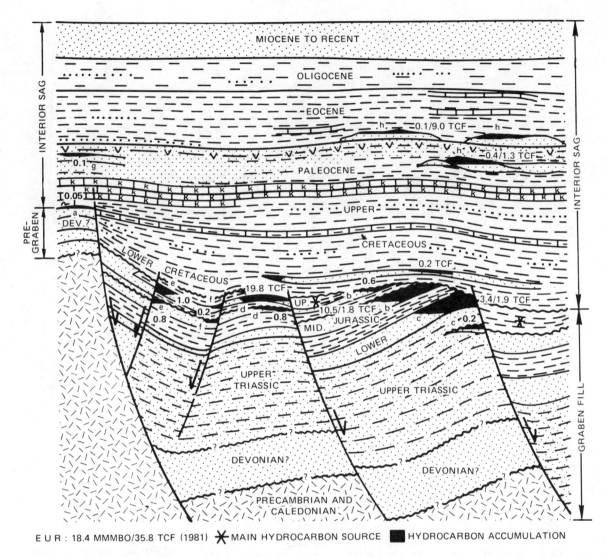

E U R : 18.4 MMMBO/35.8 TCF (1981) ✳MAIN HYDROCARBON SOURCE ◼ HYDROCARBON ACCUMULATION

Figure 4—Schematic diagram of major hydrocarbon accumulations and their estimated ultimate recoverable reserves (EUR) in Viking graben, North Sea, compiled from published data (see text), augmented with data supplied by exploration staffs of Esso Exploration and Production U. K. and of Esso Exploration and Production Norway. Reserves include gas stated in trillions of cubic feet (TCF). Trap types discussed in text indicated by a-h. Several traps related to salt flowage at far south end of graben are excluded. Volcanic tuff indicated by v; other symbols explained in Figure 2.

sedimentary cover. Here hydrocarbons have accumulated in a truncation trap at the base of the Upper Cretaceous shales, but in an off-structure position that is not block faulted (Clifford et al, 1980).

Accumulations Within Graben Fill

The most prolific and varied of the Sirte basin hydrocarbon occurrences, however, lie within the graben-fill cycle. At this level, rich Upper Cretaceous and Paleocene source rocks, Upper Cretaceous to Eocene shallow-marine clastic and carbonate reservoir rocks, and abundant closures are all closely juxtaposed (Figure 2).

Course clastics at the base of the graben fill were derived

from Gargaf quartzites and from crystalline basement rocks that were exposed at the high edges of large platforms during the Cretaceous erosion (c in Figure 2). These basal sands are reservoirs for major accumulations at the west edge of the Zelten platform and at the east side of the Amal platform (Figure 5) (Roberts, 1970; Williams, 1972). Sands with a similar provenance contain major reserves in a low-side fault trap opposite the east boundary of the Amal platform (d in Figure 2) (Williams, 1968).

Prolific carbonate reservoirs occur stratigraphically above the basal sands. Productive Paleocene reefs are localized mostly at high fault-block corners (e in Figure 2; includes several fields at the west side of Dahra, Beda, and Zelten platforms in Figure 5) (Joiner and Myers, 1972).

117

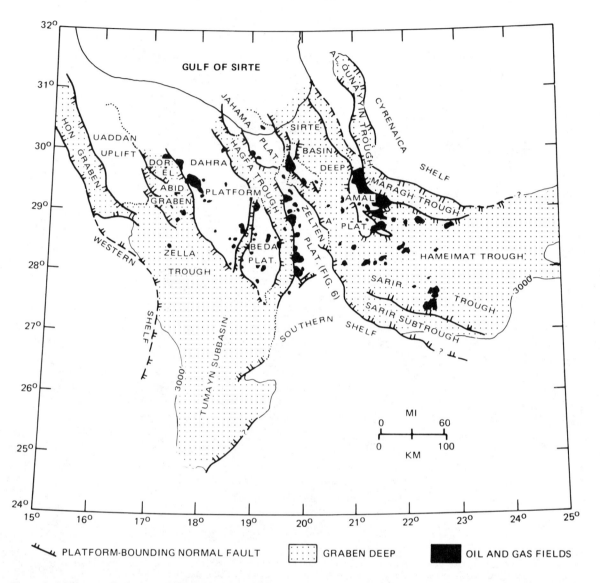

Figure 5—Major tectonic elements of Sirte basin, Libya. Five major northwest-trending grabens have formed an unusually wide basin complex in northwest portion of basin and provide several sites for maturation of hydrocarbons. Accumulations in this area are summarized in Figure 2.

Closures in porous Paleocene carbonate banks (Bebout and Pendexter, 1975) occur where the banks are draped across platform boundaries (f in Figure 2; see also Figure 7) (Fraser, 1967). Porous Eocene carbonates, reworked at the unconformity at the top of the graben sequence (g in Figure 2), are additional carbonate reservoirs, particularly near the south block of the Amal platform. Pinnacle reefs located within the Sirte basin deep provide traps at a level where local structural closures are absent (h in Figure 2) (Brady et al, 1981).

In many cases the hydrocarbons within the graben-fill reservoirs are partially trapped against the upthrown side of the block-bounding faults (Figure 7; this study). Where the normal faults are effective barriers to migration

(Weber et al, 1978; Smith, 1980), this has greatly increased the trap capacities of the associated block-edge flexures or sedimentary buildups. The spillpoint for these accumulations is commonly the highest level of the producing reservoir on the fault's low side.

Accumulations Within Interior-Sag Basin

Since the end of the graben phase—or, approximately, since the end of Eocene—the Sirte basin has subsided as a single tectonic unit. Subsidence has taken the form of a broad, simple downwarp superimposed across the earlier fault-block architecture. Closures are limited at this level and accumulations occur at flexures associated with selec-

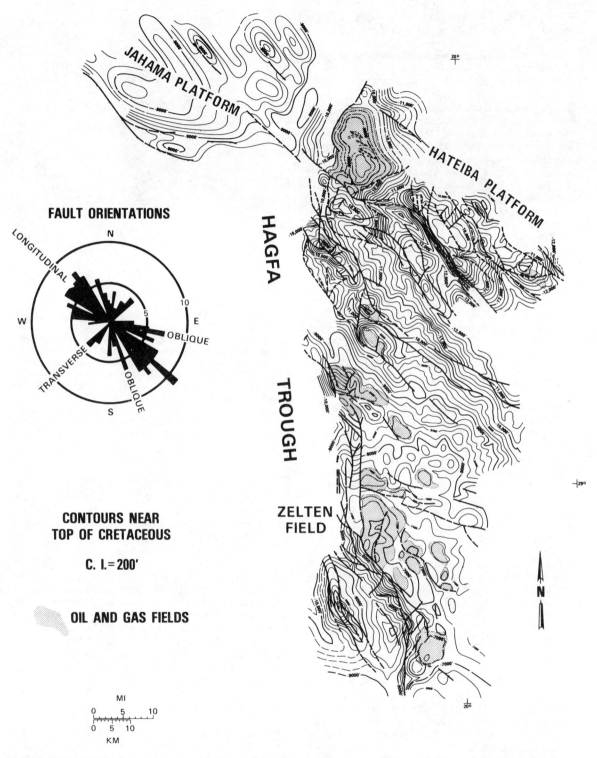

FAULT ORIENTATIONS

**CONTOURS NEAR
TOP OF CRETACEOUS**

C. I. = 200'

OIL AND GAS FIELDS

Figure 6—Structure of Zelten platform, north-central Sirte basin, and plot of numbers of major faults with orientations within each 5° geographic quadrant. Normal faults parallel with regional strike of Sirte basin deep are termed "longitudinal"; those trending at right angles to the deep are "transverse." Faults whose orientations lie in between longitudinal and transverse are termed "oblique." See Figure 5 for orientation of Sirte basin deep and location of Zelten platform.

tive rejuvenation of an older platform-bounding fault (i in Figure 2; south block of Amal platform in Figure 5).

Gulf of Suez Hydrocarbon Occurrences

Accumulations in Pregraben Reservoirs

The Paleozoic and Mesozoic shelves of the north margin of the African plate also extended southward at various times across the location (Figure 8) of the present Gulf of Suez (Soliman and Faris, 1963). The shelfal sediments again provide reservoirs for important hydrocarbon accumulations, here lying beneath a major unconformity at the base of a dominantly marine, Miocene to Holocene graben fill (Figure 3).

Major reserves have been found in nonmarine sandstones of the "Nubian" sandstone, considered by Brown (1980) to be carboniferous and Cenomanian in age in the Gulf of Suez. Nonmarine to marine sandstones of younger Cretaceous age are also productive, and Eocene shelfal carbonates contain minor accumulations (Gilboa and Cohen, 1979). Said (1962) has suggested that in some fields the porosity of these Eocene carbonates has been enhanced by weathering processes at the unconformity at the base of the graben fill. Eocene and Cretaceous shales and marls may be limited hydrocarbon sources for the accumulations in the pre-Miocene section (Said, 1962), but the dominant source is the thick middle Miocene shale and marl of the graben fill itself (Gilboa and Cohen, 1979).

Principal closures in the pregraben reservoirs are provided by (1) the truncation of rotated fault blocks by an unconformity near the base of the graben fill (a in Figure 3) (Morgan and El-Barkouky, 1956), (2) faults (b in Figure 3; see also Figure 9) (Thiebaud and Robson, 1981), (3) fault-associated flexures (present in Figure 9 but not a factor controlling this hydrocarbon accumulation), and (4) combinations of these features (c in Figure 3) (Said, 1962, his Figure 47). All of the structural closures formed during graben deformation.

The pregraben reservoirs are most prolific along a mid-trough zone, which is the central element of three parallel structurally high basement-block trends (Figure 8). A fourth high trend at the southwest margin of the graben corresponds with the Esh Mallaha range where the basement crops out extensively. This trend completes the graben framework but is nonproductive.

Accumulations within Graben Fill

Graben formation began in the Oligocene (Garfunkel and Barkov, 1977). Important middle Miocene reservoirs were derived from Nubian sands and from a crystalline basement, both still exposed in large fault blocks flanking either side of the central producing trend (Figure 10) (El-Ashry, 1972). The graben-fill reservoirs are most prolific along the central "uplift," but large reserves are also contained in the eastern producing trend. Major traps are fault closures (d in Figure 3) (Brown, 1980) or flexures draped along fault-block boundaries (e in Figure 3) (Ayouty, 1961), or a combination of these two (f in Figure

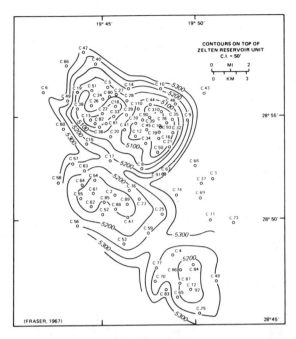

Figure 7—Drape closure of Zelten oil field on western, updip, edge of Zelten platform. Structure's shape reflects an underlying trap-door fault block. Productive Paleocene(?) and lower Eocene, shallow-water limestone facies (Bebout and Pendexter, 1975) are draped over buried junction of northwest-southeast longitudinal fault and north-south oblique fault at field's southwest and west flanks, respectively (see Zelten field on Figure 6) (Fraser, 1967). Reefal buildup in the Paleocene Heira Shale underlies northern portion of pool and accentuates drape flexure there.

3) (Gilboa and Cohen, 1979). Minor stratigraphic accumulations are controlled by the updip pinch-out of sands in both upthrown and downthrown fault blocks (g in Figure 3) (Omar, 1975; Gilboa and Cohen, 1979), by several reefal buildups located at the high edges of fault blocks (h in Figure 3) (Ghorab, 1961), and by the updip wedge-out of lower Miocene sands deposited at the base of the graben fill (i in Figure 3) (Gilboa and Cohen, 1979, their Table 1).

Salt structures are present within the upper portion of the graben fill along the deeper southern segments of the Suez graben. These structures have been growing since the late Miocene, but the deformed, upper Miocene sediments are generally devoid of reservoir facies, and the post-Miocene sediments lack overlying hydrocarbon seals (Moustafa, 1974). Graben faulting is still active, and there is no interior-sag basin.

Viking Graben Hydrocarbon Occurrences

All three basin types—pregraben, graben fill, and interior sag—are present in the Viking graben, and two of the basins have prolific hydrocarbon occurrences (Figure 4). Exploration for reservoirs below the base of the graben fill is still in an early stage, and the limits and significance of the pregraben play are not established.

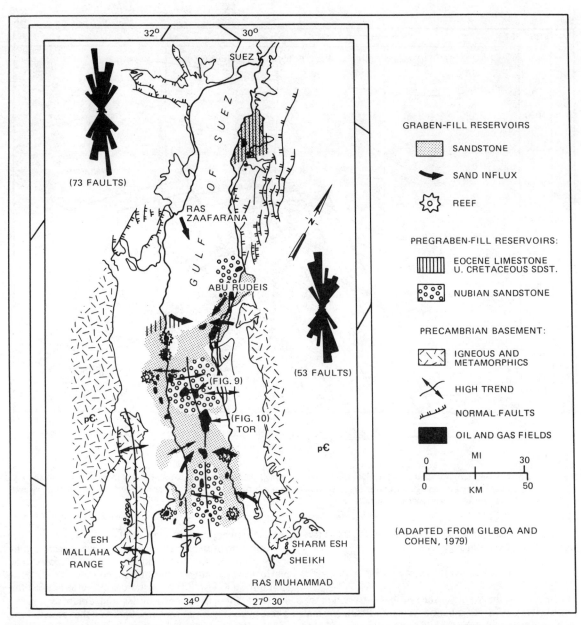

Figure 8—Oil fields and distribution of productive zones in Gulf of Suez. Numbers of surface faults with orientations within each 5° geographic quadrant are plotted in rose diagrams for northeast (after Robson, 1971) and southwest margins of graben.

Accumulations in Pregraben Reservoirs

The basement underlying the northern North Sea sedimentary sequence had its final consolidation at the conclusion of the Caledonian orogeny (Ziegler, 1978, his Table 1). Directly following this event an undefined, intracontinental basin or sag, termed the "Orcadian basin," is thought to have subsided across the region (Ziegler, 1978, 1980). Coarse, nonmarine clastics of the Devonian-age Old Red Sandstone series are exposed on either flank of

the Viking graben and have been encountered under the Mesozoic graben fill at several wells. Subsidence of this basin has been attributed to late orogenic strike slip with extension (Ziegler, 1978; Steel and Gloppen, 1980).

Thus far, the exploration of pregraben reservoirs has resulted in two small oil discoveries in the Quadrant 16 area. Both fields are located at the high side of the western shoulder of the graben. Here, the possibly fractured arkosic sandstone reservoirs of the Old Red Sandstone facies directly underlie anticlinal closures mapped at the uncon-

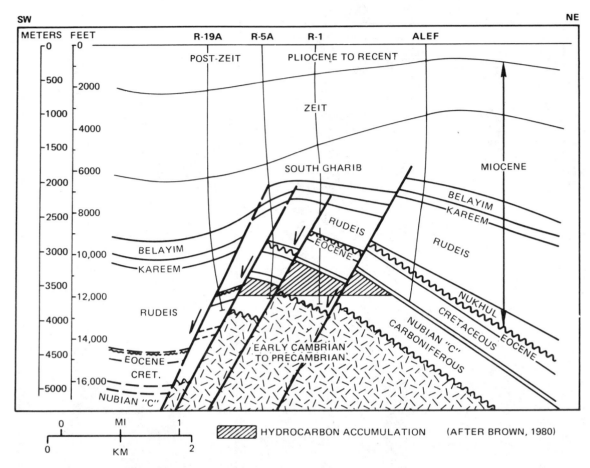

Figure 9—Cross section across Ramadan oil field, Gulf of Suez, showing hydrocarbon trap in pregraben-fill reservoirs (Nubian "C" sands) that is controlled by graben-age normal faults. Associated flexing forms narrow rollover at high side of block boundary and broader syncline in low side. See Figure 8 for location.

formity at the base of the Cretaceous (a in Figure 4) (see Harms et al, 1981, their Figures 2, 3).

Accumulations Within Graben Fill

Formation of the Viking graben probably started in the Triassic (Hay, 1978; Saeland and Simpson, 1982). Normal faulting culminated in the middle Kimmerian (ca. 156 Ma, P. R. Vail, 1982, personal communication), but it continued locally into the Early Cretaceous and rarely into the Tertiary (Figure 11) (Saeland and Simpson, 1982; Kirk, 1980, his Figure 4). Jurassic stages of faulting were intense and established the series of rotated platforms characterizing the northwestern portion of the basin (Figure 12). The upthrown edges of the platforms were eroded at approximately the end of the graben cycle (Vail and Todd, 1981). Closure was produced by a combination of out-of-the-basin structural dip within the platform and basinward-facing erosional scarps. Most major accumulations in the prolific northwestern flank of the Viking graben produce from Middle and Lower Jurassic sands truncated beneath such topographic highs (b in Figure 4, see also Figures 13-

16) (De'Ath and Schuyleman, 1981).

Truncation of the Jurassic reservoir sands is stratigraphically complex. Vail and Todd (1981) have proposed that platform rotation was continuous and that episodic erosion during low stands of eustatic sea level produced the series of angular unconformities and onlap sequences. These unconformities converge updip on the rotated slope of the platforms in rocks ranging in age from Jurassic to Early Cretaceous. Vail and Todd (1981) have recognized a period of maximum truncation in the late Middle Jurassic (ca. 156 Ma).

Hydrocarbon trapping at many of the truncated fault blocks is partially controlled in either the strike (Figure 14) (Bowen, 1975; Davies and Watts, 1977) or updip direction by faulting or flexing (c in Figure 4, see also Figure 17) (Kirk, 1980; Grey and Barnes, 1981). The truncation traps are morphologically similar to those in the Sirte and Gulf of Suez basins, but in the latter areas they occur at the base of the graben fill instead of near or at its top.

Closures controlled solely by structure are not as numerous in the Viking graben but are still important. These traps are forced folds formed along the upthrown edges of

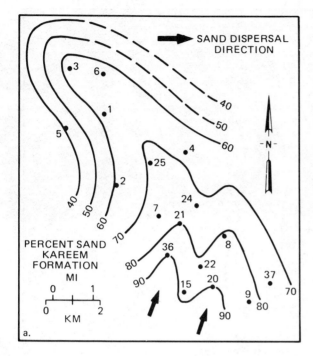

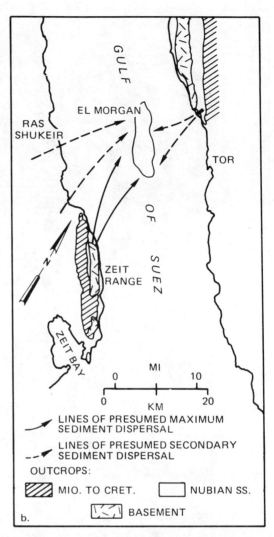

(EL-ASHRY, 1972)

Figure 10—Percent sand map of middle Miocene Kareem reservoir within El Morgan producing area (a) and provenance for Kareem clastics at flanking fault blocks (b), Gulf of Suez. Sand-shale ratios and coarsening directions of sand grains corroborate this interpretation (El-Ashry, 1972). Trap for hydrocarbons is longitudinal horst with block-edge flexure (Brown, 1980, his Figures 6, 7). See Figure 8 for location.

blocks (d in Figure 4) (Harms et al, 1981, their Figure 2), high-side fault closures (e in Figure 4) (Budding and Inglin, 1981), or combinations of the two (similar to Figure 17 but without reservoir truncation) (Harms et al, 1981). Traps closed directly against the downthrown side of faults (f in Figure 4, see also Figures 15, 16) (Albright et al, 1980) and rollovers associated with faults antithetic to platform fault boundaries are also productive, mostly from Upper Jurassic sands along the southwest edge of the Viking graben (Harms et al, 1981). Low-side fault traps, however, do not contain major reserves here or in most other grabens.

Accumulations Within Interior-Sag Basin

Closely spaced, multiple unconformities near the top of

the graben fill prohibit a precise dating of the contact between the fill and the interior-sag basin. On the basis of significantly diminished faulting, the base of the sag basin is considered in this work to be a transitional zone within the Lower Cretaceous.

Broad flexures dominate the structural style within the lowermost portion of the sag fill (Figures 12, 15) (Kirk, 1980, his Figure 4). They are thought to be the result of passive drape and differential compaction across the buried topography, combined with varying amounts of fault offset and forced folding across certain platform boundaries that remained active during the waning stages of faulting (unpublished structural studies). The resulting structural relief is considerably greater than that observed in other sag basins, which mostly merge downward into the graben fill without an intervening structural discor-

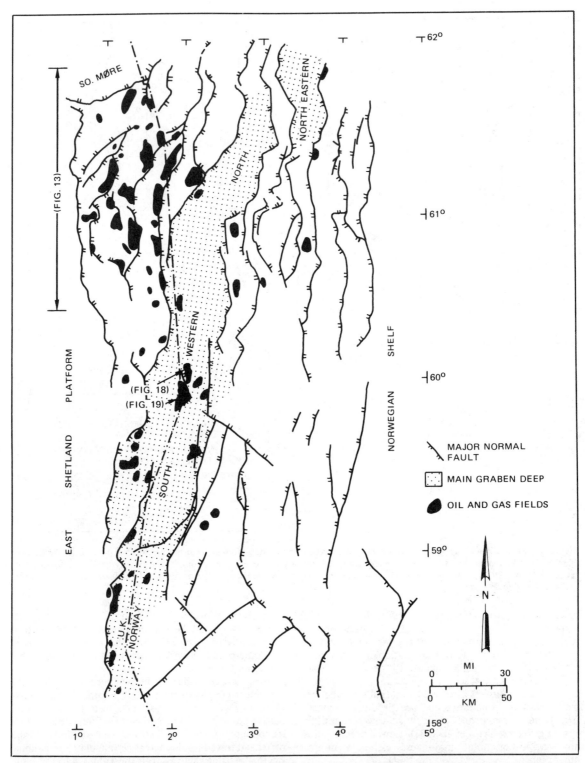

Figure 11—Major faults and oil and gas fields of Viking graben. Merger of north-south fault system with northeast-trending zones at north end of basin defines regional dogleg. Main graben deep shifts across basin axis within this dogleg (see Figure 24 for possible explanation). United Kingdom portion, after maps by Esso Exploration and Production U.K.; Norwegian portion, south of 60° north, after Ronnevik et al (1975); remainder from this study.

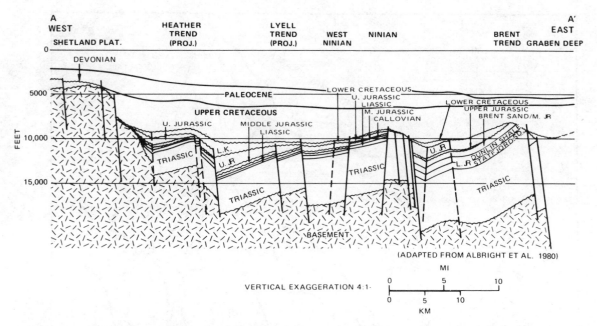

Figure 12—East-west cross section across northwest flank of Viking graben; see Figure 13 for location. Platforms at level of graben fill (i.e., Heather, Lyell, west Ninian, Ninian, and Brent) are rotated westward into large normal faults located at each platform's western boundary. Structural relief within lower sag fill has flexure style expressed by Upper–Lower Cretaceous contact.

dance or angular unconformity. Several folds above the western shoulder of the graben contain hydrocarbons in Paleocene sands (g in Figure 4) (G. C. Mudd, 1982, personal communication). Within the graben proper, a general absence of Cretaceous reservoirs has precluded the occurrence of major hydrocarbon reserves in the lower level of the interior-sag basin.

The upper portion of the sag fill is largely devoid of structural closures within the graben proper (Figure 12) (Kirk, 1980, his Figure 4; Ziegler, 1980, his Figure 32). Paleocene and basal Eocene sands, whose source was platform areas west of the western shoulder of the graben (Morton, 1982), contain major hydrocarbon accumulations. Morton has interpreted the Paleocene sands as gravity flow and younger deltaic deposits. He considered the lower Eocene sands to be of shallow-marine origin. Depositional mounding of the lenticular reservoir sands, in places amplified by gentle drape and differential compaction across buried graben structures, provides the closures (h in Figure 4, see also Figures 18, 19) (Heritier et al, 1981).

To the south, in the Central graben of the North Sea, thick Upper Permian evaporites have provided a mechanism for deforming the upper portions of the interior-sag fill (Ziegler, 1979), and this deformation occurs in other sag basins (Trusheim, 1960). At the Central graben, major hydrocarbon reserves are contained in fractured uppermost Cretaceous and basal Paleocene chalks domed by salt diapirs (Van den Bark and Thomas, 1980). In the vicinity of the diapirs, other closures, which have also resulted from salt flowage, occur at the level of the graben fill. They include turtle structures, residual salt highs, and possibly, nonpiercement salt pillows (unpublished struc-

tural studies; Trusheim, 1960). These latter structures extend into the far south end of the Viking graben and provide traps for several additional fields (not included on Figure 4).

TECTONIC CONTROLS FOR OCCURRENCE OF MAJOR HYDROCARBON ACCUMULATIONS

Accumulations in Pregraben Reservoirs

Grabens superimposed across a prior depositional site have a potential for traps below the graben fill. To be effective for retaining large volumes of hydrocarbons, the older sediments must be devoid of complex pregraben structures that would disrupt the internal continuity of later fault-block closures. In addition, the older sediments must be preserved from erosional removal prior to and during graben formation.

The Sirte and Gulf of Suez grabens were superimposed on continental shelves, and these tectonic settings remained relatively stable until the start of rifting (Conant and Goudarzi, 1967; Garfunkel and Bartov, 1977). Because of this, the graben fault blocks do not contain structures from previous deformation. The history of the pregraben reservoirs in the Viking graben is not well known.

During normal fault deformation, broad arching of the crust is a mechanism potentially destructive to the preservation of the pregraben sediments. According to Vierbuchen et al (1982), graben-inducing mechanisms of crustal and lithospheric stretching—and the consequent lithospheric heating and subsequent cooling—may be balanced

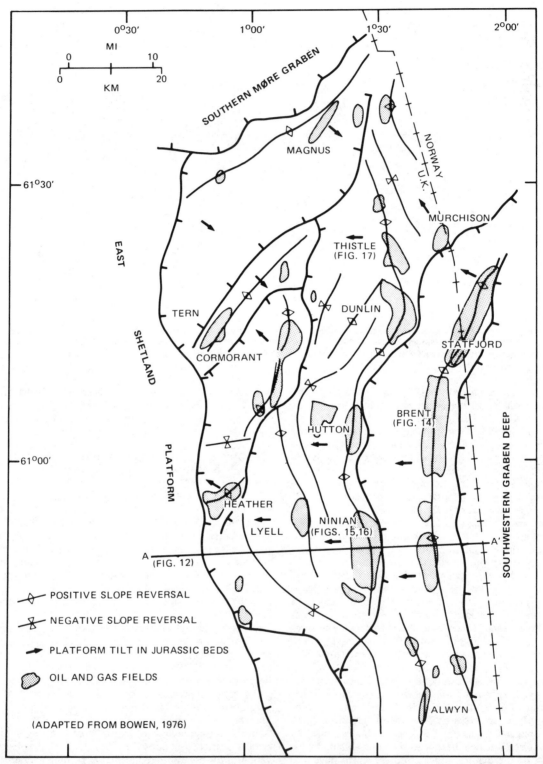

Figure 13—Topographic features, major faults, and main platforms at late Kimmerian unconformity (approximate Jurassic-Cretaceous contact) on northwest flank of Viking graben. Positive slope reversals are combination of platform tilt (mostly west slopes) and erosional "scarps" at platform edges (mostly east slopes). See Figures 11 and 12 for location and geologic cross section, respectively.

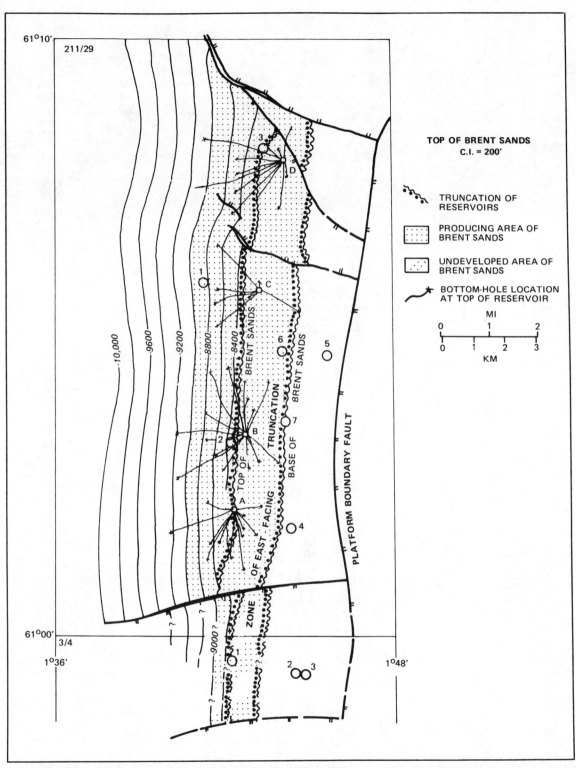

Figure 14—Structure of truncation trap in Middle Jurassic Brent sands at Brent oil field, Viking graben; see Figures 12 and 13 for cross section and field location, respectively. Updip limit of reservoir sands parallels north-south platform boundary fault. This relationship demonstrates control exerted on pattern of reservoir truncations by such faults. Structure in block 211/29 north of southernmost transverse fault from Esso Exploration and Production U.K., 1981.

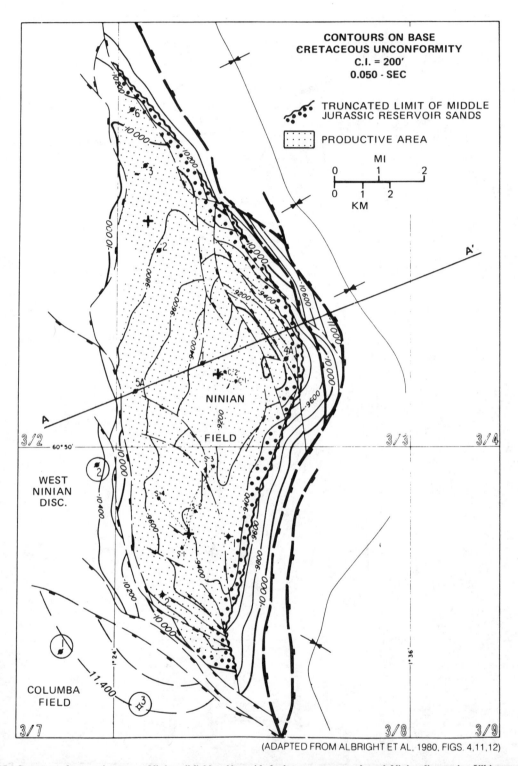

(ADAPTED FROM ALBRIGHT ET AL, 1980, FIGS. 4,11,12)

Figure 15—Structure of truncation trap at Ninian oil field and low-side fault traps at west and south Ninian discoveries, Viking graben (see Figures 12 and 13 for geologic cross section and field location, respectively). Large northwest-southeast and north-south normal faults at field's east side bound a trap-door structure, which is reflected by truncated limits of Middle Jurassic reservoir sand. West Ninian Jurassic discovery appears to be a broad warp of field's west flank that extends across secondary longitudinal normal fault. At Jurassic Columba field, closure occurs where oblique fault cuts across trap door's south plunge.

NINIAN FIELD

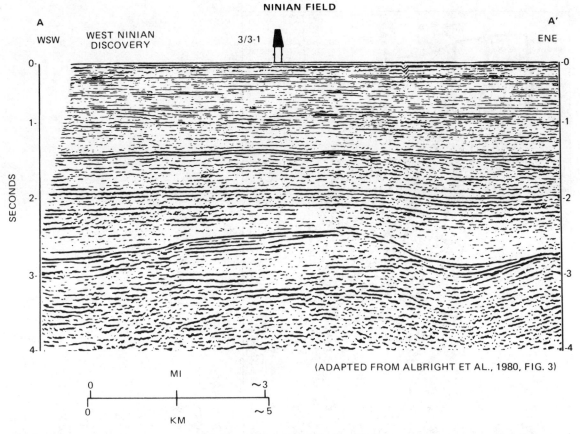

(ADAPTED FROM ALBRIGHT ET AL., 1980, FIG. 3)

Figure 16—East-west seismic profile of Ninian oil field and west Ninian discovery, Viking graben (see Figure 15 for location). Trunca-tion of Jurassic section and dip reversal at base of sag fill is apparent at structure's crest and east side (at 2.4–2.9 sec). Positive flexure caused by drape and differential compaction occurs above unconformity. Below this surface, large normal faults are downthrown to east at structure's east boundary, and smaller normal faults are downthrown to west at west Ninian.

in several ways to give differing sequences of subsidence or uplift. In some areas, the surface of the crust may have been broadly downwarped at the initiation of faulting or perhaps just before (Asmus and Guazelli, 1981; Milanovsky, 1981), preserving the older sediments from erosion. In the Gulf of Suez, arching came after signifi-cant graben subsidence by block faulting (Garfunkel and Bartov, 1977), and this timing was important to the preser-vation of the pregraben play. Cretaceous and older arching preceded or accompanied initiation of faulting in the Sirte basin, however, and the Nubian Sandstone and Gargaf Group are absent under central portions of that basin (Clifford et al, 1980, their Figure 3; Parsons et al, 1980).

Graben-Fill Accumulations

On the northeast side of the Gulf of Suez and in the Viking graben, normal faulting occurred in multiple stages (Garfunkel and Bartov, 1977; Hay, 1978; Grey and Barnes, 1981). An initial, pervasive stretching phase apparently resulted in relatively closely spaced normal faults, mostly with moderate displacements (pre-upper Gharandal faulting in Figure 20). In the Gulf of Suez, the

faulting was accompanied by rotation of fault blocks and deep erosion at some block edges. Extension in a later stage was concentrated on fewer, larger and more widely spaced faults, some of which had had earlier movements as well (faults that offset both the older and younger sedi-mentary sections on Figure 20).

During the early pervasive faulting, the pregraben reser-voirs were preserved from erosion in most blocks across the east flank of the Suez graben (Figure 20). While the entire graben system subsided, the base of the sedimentary cover was apparently downfaulted below the erosional base level at each successive structure (e.g., Eocene and below on Figure 20). The second stage of faulting pro-duced platforms composed of agglomerations of the older and smaller fault blocks, and these outline the major tec-tonic framework (Garfunkel and Bartov, 1977).

Garfunkel and Bartov (1977) noted this structural sequence in the Gulf of Suez and determined that the sec-ond phase of faulting was accompanied by arching of the rift shoulders. This latter faulting did not occur until the latest early Miocene. In the Viking graben, the major faulting that culminated in the present platform architec-ture and the arching, thought by Ziegler (1978, 1980) to

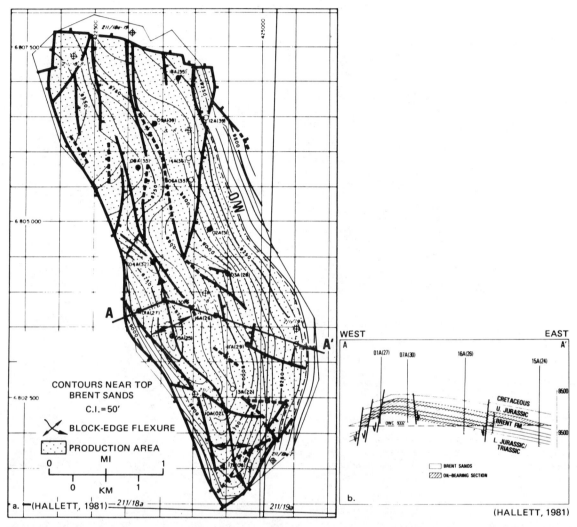

Figure 17—Structure map (a) and cross section (b) of Thistle oil field, Viking graben (see Figure 13 for location). Updip trapping of hydrocarbons in Middle Jurassic sands is by combination of block-edge flexure and fault at south end, and by high side of block boundary fault at north end. Parallelism of flexure and block boundary demonstrates that fold's formation is dependent on block relief.

have occurred south of the graben, were about coeval. The North Sea arching has been dated as Middle Jurassic (Ziegler, 1978, 1980), well after the initiation of graben deformation.

Arching is typically most obvious at the rift "shoulders," the area outside the most external of the boundary faults. Regional cross sections of the Gulf of Suez (Figure 20) (Schurmann, 1966, his Figure 32), the Viking graben (Figure 12) (Kirk, 1980, his Figure 4; Ziegler, 1980, his Figure 32), and the Sirte basin (Parsons et al, 1980, their Figure 5) demonstrate that subsidence by normal faulting has outpaced the broader arching of the crust. Differences in structural relief within these grabens were caused primarily by the rotation and displacement of individual blocks and do not appear to reflect an integrated, graben-wide upwarp.

Interior-Sag Accumulations

Major subsidence of both the Sirte basin and the Viking graben continued well after block faulting ceased (Kirk, 1980, his Figure 4; Parsons et al, 1980, their Figure 5). The final stage of subsidence there has the form of a regional, mostly unfaulted downwarp. The flexural downwarp forms the interior-sag basin, and includes both the graben and the adjacent graben shoulders. The magnitude of this subsidence is indicated by the comparative thickness of the shallower, unfaulted strata on Figures 12 and 21. In the Sirte basin, the interior sag has over 12,000 ft (3,650 m) of relief at the base of the Oligocene.

Subsidence of the interior-sag basin in the Viking graben took two forms: initial differential subsidence of the major structural elements during the Early Cretaceous

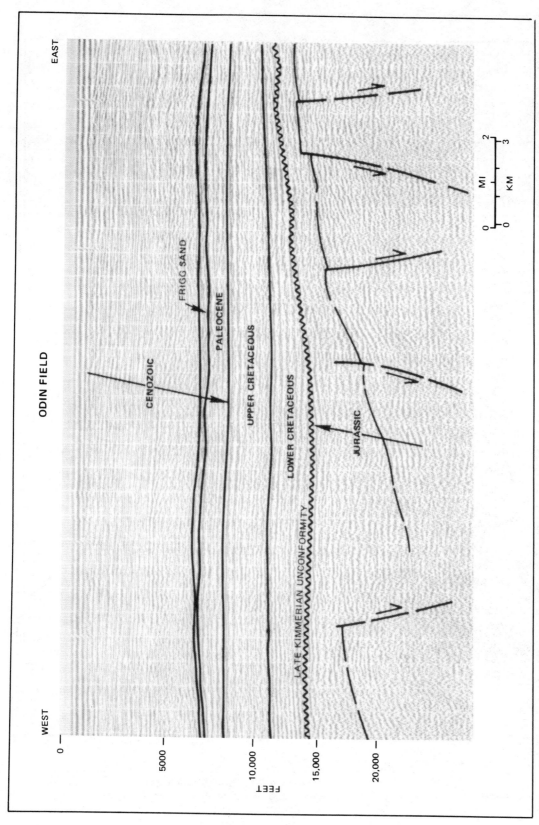

Figure 18—East-west seismic profile across Odin gas field, Viking graben (D. G. Blair, 1982, personal communication). Hydrocarbon production is from lower Eocene Frigg sands deposited within interior-sag fill. Accretionary mounding of these sands provides most of dip reversal for hydrocarbon trap, and there is minimal structural influence from underlying graben-stage fault blocks. See Figure 11 for location of field.

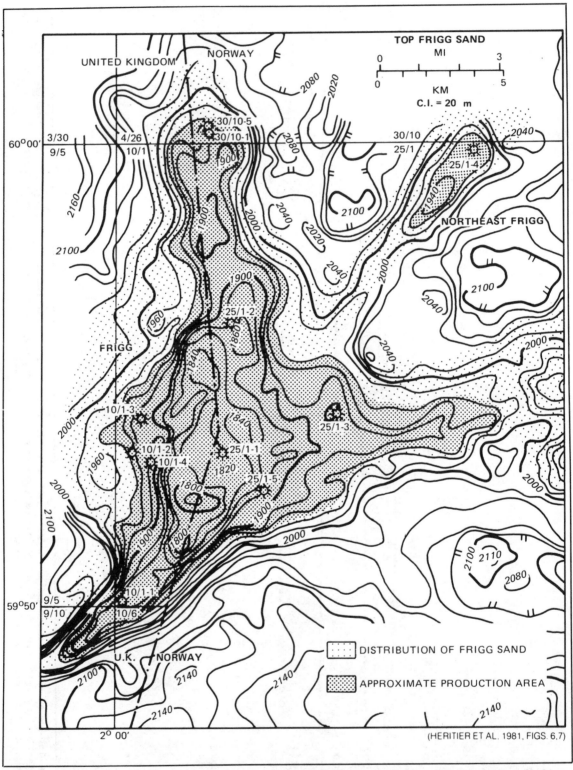

Figure 19—Structure of Frigg and Northeast Frigg gas fields, Viking graben. Fields are just south of Odin (Figure 18), and hydrocarbon traps are due to similar accretionary mounding of lower Eocene Frigg sands. Closure at Frigg has been increased by drape across trap-door fault buried within Jurassic graben fill (Heritier et al, 1981, their Figure 7). See Figure 11 for location of fields.

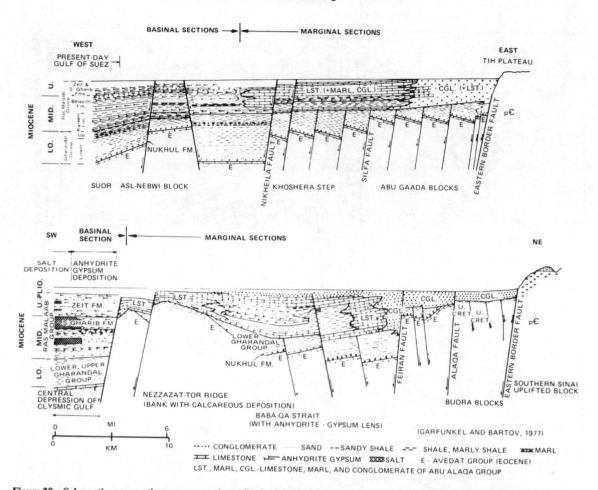

Figure 20—Schematic cross sections across northeast flank of Gulf of Suez graben. Faulting occurred mostly during lower Gharandal deposition (earliest Miocene) and in post-lower Gharandal (late early Miocene to Holocene). Regional arching is pronounced on Landsat imagery of terrain external to graben (this study), but does not appear to affect structures within trough itself. Conglomerate facies, derived from uplifted rift shoulders, mark initiation of arching during second faulting event (Garfunkel and Bartov, 1977).

and into the Late Cretaceous (illustrated by the base of the Upper Cretaceous in Figure 12), followed by integrated subsidence of the basin as a single, unsegmented downwarp in the latest Cretaceous and Tertiary (illustrated by the base of the Paleocene in Figure 12). The early period of differential subsidence caused significant structural relief within the Cretaceous section.

GRABEN STRUCTURAL STYLE AND CONTROLS FOR TRAP DEVELOPMENTS

Closures associated with extensional block faulting are relatively easily interpreted on seismic data. This is partly because most structures repeat the same uncomplex cross-sectional characteristics. The lack of uniqueness in profile expression, however, makes it difficult to correlate block boundaries through successive seismic lines. Consequently, the conversion of profile control to structural maps is more difficult.

Rapid variations in displacement magnitude, abrupt changes in fault orientation, termination of blocks, and the junctions of different fault trends all increase the mapping difficulty. These characteristics usually negate the potential for reasonably constant trend patterns. They cause reconnaissance mapping of fault blocks to be more dependent on control density than are other structural styles.

Furthermore, every graben system has its own set of provincial characteristics and geologic history, but certain structures and types of patterns are repeated. Other investigators have also noted this repeatability in morphology and tectonic development (Illies, 1970). The following description of the structural style of grabens synthesizes relationships observed in the three basins described in the previous sections, and also utilizes investigations of other grabens and broader based topical studies of certain graben structures.

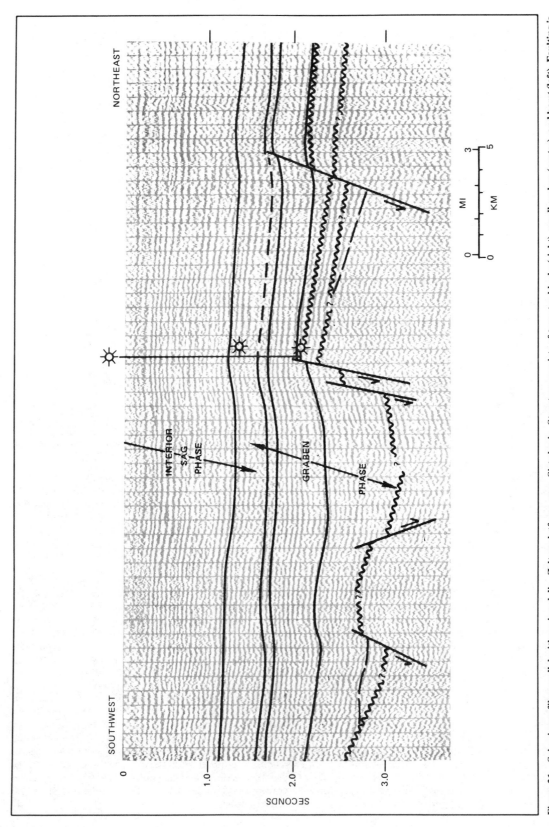

Figure 21—Seismic profile parallel with regional dip, Zelten platform area, Sirte basin. Structure consists of two step blocks (right), small graben (center), and horst (left). Faulting is accompanied by conventional dip-slip drag flexing and forced drape at level of graben-phase deposits. Flexing caused by passive drape and differential compaction becomes progressively more dominant upward within interior-sag deposits. Gas accumulations have been found in both drape-flexure and fault-block traps.

Profile Characteristics

Extensional fault blocks to depths resolvable with exploration data characteristically have a slab-like or internally unflexed profile. In the subsurface, the edges of these blocks commonly have two (Figure 16) or three (Figures 19, 21) vertically stacked structural assemblages that together are distinctive of the fault-block style. Each level has its own particular trap potential. At the deepest mappable level, the structure may be a simple fault closure, as within the pre-Rudeis formations in Figure 9 and within the pregraben reflectors at the middle step block in Figure 21. In these instances, the retention of hydrocarbons is largely dependent on the sealing capacity of the block bounding faults (see Weber et al, 1978; Smith,, 1980). Fold closure becomes progressively more important at shallower structural levels within the sedimentary cover.

Block-Edge Flexures

Most folds in grabens are thought by the writer to be attributable to forced drape and the conventional fault drag associated with dip-slip displacements. Examples of forced folds are considered to be dip reversal at the top of the Belayim Formation in Figure 9 and the upper portion of the graben phase at the center step block of Figure 21. This flexing may be initiated by an upward propagation of the dislocation surface. Seismic and surface data available to this study demonstrate that normal faults commonly die out upward (and along strike) into flexures, which appear to develop prior to brittle rupture. Robson (1971), and El-Tarabili and Adawy (1972) noted this relationship in the Suez graben. Flexure form is concordant with the sense of the underlying offset, and a positive rollover or monoclinal knee develops above the upthrown edge of the block.

As continuing displacements propagate the faults farther upward, the forced folds are accentuated by dip-slip drag, and the structure evolves into the faulted flexure that typifies the intermediate structural level. Examples are the rollovers within the Belayim and Kareem Formations in Figure 9, and near the top of the graben phase at the step block nearest the right margin of Figure 21. If the block bounding faults are an effective barrier to migration at this level, combination flexure and fault traps may develop and part of the hydrocarbon column will be trapped directly against the fault (Figure 17).

Establishment of positive closure from block-edge flexing is dependent on a combination of block tilt and the direction of fault dip. If faults dip away from the direction of tilt, downward warping of beds toward the downdropped block will either flatten or reverse the tilt, depending on the magnitude of the flexing (southwest side of the horst in Figure 21). If the fault dip is similar to the direction of tilt, downward warping of beds toward the low side of the fault will merely increase the degree of tilt. The result is a monoclinal flexure with no structural closure (northeast side of the horst in Figure 21).

In many grabens it has been observed by this study that the low-side upturns (synclines) into the fault are typically broader than the companion high-side positive rollovers.

This is also apparent at the level of the Belayim and Kareem Formations in Figure 9 and within the graben phase at the step block at the right margin of Figure 21. In some places only the syncline is present, as below the base of the interior-sag deposits east of the Ninian field in Figure 16 (below 3.0 sec) and within the lower portion of the graben phase at the middle step block in Figure 21.

Fold traps are common in the Sirte basin (this study) and the Suez graben (El-Tarabili and Adawy, 1972), but they are rare in the prolific northwest flank of the Viking graben (Figure 12). The formation of the positive dip reversals adjacent to the upthrown side of the faults causes them to be vulnerable to large-scale erosion of fault-block edges. This erosion has occurred in the Viking graben region at the level of the hydrocarbon-producing reservoirs (Middle Jurassic and Liassic in Figure 12). High-side flattening of the platform tilt and positive rollover have been preserved at the Thistle oil field (Figure 17) and at the Block 34/10 Delta structure (i.e., Gullfaks oil field; Saeland and Simpson, 1982).

Structures at the shallowest level are dominantly passive drape and differential compaction flexures. They form above and reflect the structural relief of the underlying fault blocks (interior-sag phase in Figure 21) or the residual topography of an unconformity at the top of the fault blocks (above 2.8 sec in Figure 16). Their culminations provide relatively broad, unbroken "anticlinal" closures. Differential compaction can also enhance closure within the lower structural levels.

Block Rotation

Block tilt is consistent in some regions, and this consistency has led to the establishment of prolific hydrocarbon trends and areas with numerous closures (e.g., those in Figures 6, 12, 13). In other places, tilt is random, differing even between adjacent structures. Several block-rotation geometries have been observed in the Basin and Range province of Nevada, where Wernicke and Burchfiel (1982) have related them to different kinds of dislocation surfaces: block-rotating listric (i.e., downward flattening as in Figure 22b) and planar normal faults (i.e., uncurved as in Figure 23), and non-block-rotating, high-angle planar normal faults and low-angle extensional detachments. These faults occur together in some areas studied by Wernicke and Burchfiel, and examples have been observed on seismic profiles (McDonald, 1976). Sets of planar faults are nested in the hanging wall of a listric fault and terminate downward into that fault (see also Angelier and Colletta, 1983). The listric fault continues into a shallow, basal detachment that bounds unrotated sheets.

Application of Wernicke and Burchfiel's (1982) kinds of normal faults to rift settings is being studied. Most investigators of deeply subsided grabens, such as those described in this article, suggest that the basement-involved listric normal faults cut to much deeper levels, possibly merging downward with a subhorizontal zone of increased ductility (Figure 22a) (Proffett, 1977; Bally et al, 1981). In this way, the faults mechanically thin the crust and induce the deep subsidence centered between strands of the fault system. Deep-going planar faults have also been proposed, and they may terminate downward into ductile shear folds (Figure 23) (Morton and Black, 1975). In interpretations by Proffett (1977) and Morton and Black (1975), the fault profiles have been attributed to the changes in displacement mode that occur at depth with increasing temperature and pressure.

Listric faults cause bed rotation because the hanging-wall block maintains an approximately constant angular contact with the curving fault surface (Figure 22b) (Hamblin, 1965). As displacements proceed, this block is progressively rotated. Steepest inclinations occur where displacements have moved the hanging wall down to the flatter segments of the fault. Both larger displacements and more abrupt fault curvature increase the degree of low-side rotation. Magnitudes of tilt differ on either side of the listric faults, and this relationship distinguishes them from the sets of rotating planar faults that bound equally tilted blocks (Wernicke and Burchfiel, 1982; see Bally et al, 1981, for further discussion of listric normal faults).

The extension of terrain containing planar faults causes rotation of both the blocks and the fault planes bounding them (Figure 23) (Morton and Black, 1975; Angelier and Colletta, 1983). The result can be compared geometrically with the incremental slip and rotation generated by tilting a row of dominoes sideways. In Morton and Black's (1975) model, the obvious space problems at the base of the blocks (potential gaps) and at their margin are alleviated by ductile flow.

The similarity between Morton and Black's (1975) scheme for planar faults and the pre-upper Gharandal fault blocks shown by Garfunkel and Bartov (1977) at the northeast side of the Gulf of Suez (especially the upper cross section in Figure 20) tempts one to apply this style to the Suez region. The post-Gharandal faults mostly depressed each platform without rotation. These latter faults could belong to the Wernicke and Burchfiel (1982) category of high-angle, nonrotating planar faults. Angelier and Colletta (1983) have also recognized planar faults in the Gulf of Suez, but the sequence of faulting proposed in their model differs from that indicated in Figure 20.

In other areas, platform tilt appears to be late and imposed primarily by subsidence of the interior-sag basin. In portions of the Sirte basin, for example, platform dips are subparallel with dips in the overlying sag fill (Figure 21) (Parsons et al, 1980, their Figure 5), indicating negligible rotation during faulting.

Block tilt can be affected by many additional factors, including inherited pregraben dips (which affect pregraben reservoirs), ramping at fault terminations, isostatic adjustments, underlying intrusions, impinging uplifts or subsidences from external deformation, and superimposed stress systems not related to graben formation. Regional changes in the graben's trend, such as major junctions and doglegs, can also change the pattern of block rotation. The inconsistency of block rotation is not yet fully explained.

Map Characteristics

Regional Trend Patterns

On a large scale, grabens typically have long, straight, or gently curving segments, termed "straightaways," connected by doglegs (the north and southwestern segments of the Viking graben deep in Figure 11) (Coleman, 1974, his Figure 1) and loops (Illies, 1970, his Figure 11). Adjacent grabens, not always with similar histories, can have widely differing orientations (Milanovsky, 1981, his Figures 18, 20–22). They may join at junctions whose configurations range from low-angle oblique to perpendicular T intersections and include the more idealized triple junctions of plate interiors (Figure 5) (Illies, 1970, his Figure 11; Burke and Dewey, 1973, their Figure 3). In some instances, grabens do not connect directly but, instead, overstep at their ends (the north and northeastern graben deeps in Figure 11) (Illies, 1974, his Figure 7).

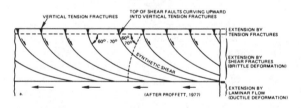

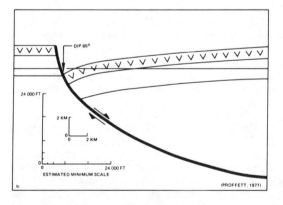

Figure 22—Conceptual diagram of changes in failure mechanisms that are thought to result in deep-going listric fault profile (a) and block rotation induced by this profile (b). Depth at which fault soles out can be highly variable. Differing amounts of rotations result from differences in displacement magnitude, from abruptness of fault curvature, and from rotation superimposed by later generation of listric faults.

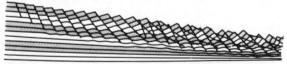

Figure 23—Hypothetical illustration of crustal attenuation and external rotation of planar normal faults and their intervening fault blocks. In this representation, time, progressive attenuation, and renewed faulting increase to right. It is doubtful that hydrocarbon accumulations would survive extreme deformation shown at far right.

The interiors of graben junctions may be complexly structured because of the combined deformation by the different fault sets of each graben arm. They can have complex histories where fault activity and subsidence have shifted from one arm of the junction to another (Illies, 1974). Trap-door closures can be concentrated within these areas, but they may be overly segmented by the intense faulting.

Doglegs form where the boundaries of a graben assume an oblique orientation for a short distance and then return to the regional trend (Figure 24). Major faults within the oblique middle segment may continue across the entire system, taking the trough axis from one side of the basin to the other. This may change the relative positions of shallow- and deep-water facies, and the direction of block rotation can reverse abruptly. Areas with a high concentration of closures, particularly trap-door closures, are formed where faults of the regional and oblique directions intersect at the dogleg bends. The optimally structured

northwest flank of the Viking graben is an example (Figure 13).

If extension proceeds sufficiently, doglegs and the ends of overstepped grabens (Courtillot et al, 1974) can become sites for strike-slip faults. At doglegs, the segments oriented most nearly in line with the regional extensional stress (e.g., the middle portion of the dogleg in Figure 24) will ultimately require a strike-slip fault to accommodate the continued extension. Because of the spatial relationship between graben segments, the direction of strike slip is opposite that of the apparent offset of the connected grabens, similar to transform faults at spreading centers. The spatial relationships also suggest that the strike-slip displacements typically would be accompanied by divergence. If so, compressional folding would be subdued or eliminated by the extensional components. In these instances, the potential wrench zone might inherit the style of the earlier dogleg normal faults, making its identification very difficult. Courtillot et al (1974) have modeled the

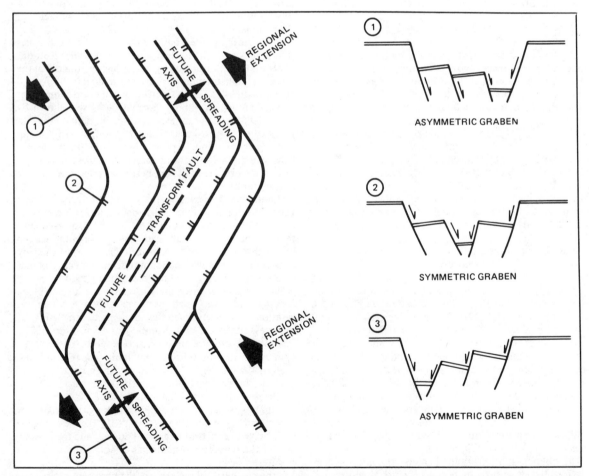

Figure 24—Schematic map and cross sections of dogleg graben; scale ranges from few to many kilometers. Graben segments at map's top and bottom are in regional trend and normally would persist for greater distances than central oblique segment. Faults are arbitrarily treated as listric and deep-going. Termination of right-dipping faults in cross section 1 by oblique fault set of central segment results in dominance of left-dipping faults in cross section 3. Direction of block rotation changes with this reversal of fault dip. Strike slip may occur along oblique segment after large extension (see discussion in text).

fault patterns developed between the overstepped ends of grabens, and Harding (1983b) has described a style of strike-slip deformation associated with extension.

Bathymetry and other data suggest that precursors of transform faults have been initiated within oceanic crust at doglegs in the trend of the northern and central Red Sea (Coleman, 1974, his Figure 1). Structures characteristic of strike slip have not been documented convincingly, however, in the hydrocarbon-producing grabens that I have studied. There are several possible explanations. First, these grabens are underlain by continental crust, and its inhomogeneities would be prone to produce initially multidirectional block faulting in response to crustal extension. The resulting grid of fault blocks may preempt the subsequent lateral propagation of through-going strike-slip faults. These grabens also, presumably, may not have had sufficient extension and strain rates to generate major strike slip. Alternatively, because of the difficulty in differentiating wrench faults and normal faults in extensional settings, the awareness of the role of strike slip in graben deformation may be incomplete.

Multidirectional Normal Faults

Grabens are complicated internally by the bifurcation, junction, interference, and termination of normal faults (Harding and Lowell, 1979). Structural vergence changes abruptly and sporadically because of this interaction between differently oriented faults (Schurmann, 1966, his Figure 32). The occurrence of several orientations of normal faults in grabens has been illustrated by Parsons et al (1980, their Figure 3) and Sanford (1970, his Figures 2, 22) in the Sirte basin, by Robson (1971, his Figures 2–6) and Garfunkel and Bartov (1977, their Figure 2) in the Gulf of Suez, and by Halstead (1975, his Figures 7, 10–12) and Hay (1978, his Figure 5) in the Viking graben. It has been reproduced with tectonic models (e.g., Freund and Merzer, 1976) and has been studied theoretically (e.g., Reches, 1978).

Plotting of fault orientations in the Sirte (Figure 6), Suez (Figure 8), and Viking grabens, and in other grabens studied by the writer demonstrates that there is a degree of organization that is repetitive. This has also been noted by Robson (1971) and El-Tarabili and Adawy (1972) in the Gulf of Suez, by Hay (1978) in the Viking graben, and by Freund and Merzer (1976) in other normal faulted regions. Longitudinal faults subparallel with the graben's regional strike are dominant. Oblique faults trend to either side of the longitudinal faults, and both oblique orientations are equally numerous in most of the grabens studied. The oblique faults decrease in number as their obliquity to the graben trend increases, and transverse elements are the least common. Typically, 90% of the block-bounding faults lie within an 80° quadrant (see rose diagrams in Figures 6 and 8; see also El-Tarabili and Adawy, 1972).

The oblique and longitudinal faults combine to form platform edges with zigzag borders (Figure 6), clusters of two or three-sided blocks termed "trap doors" (Figures 15, 25, respectively), and straightaways where only one trend is developed (King, 1949). Oblique faults commonly strike oblique to the dip of beds; and where the two oblique fault sets junction in an updip direction, the faults or their associated block-edge flexures may provide both updip and strike closure. Many of the productive traps in Figures 6 and 13 are located at such fault intersections. Regional straightaways and individual blocks bounded by a single fault direction do not have as high a propensity for forming traps, because closure parallel with the strike may be lacking.

The oblique faults in Freund and Merzer's (1976) models were initiated by strike slip, but the fault-block architecture rapidly inhibited continuation of these displacements, and normal slip was superimposed on the inclinal fractures. Obliquely oriented faults in the Viking graben and several oblique trends in the Suez graben have been interpreted as wrench or transform faults by Hay (1978) and Abdel-Gawad (1970), respectively. These authors, however, gave no direct stratigraphic or structural evidence for strike slip. According to B. Wernicke (1982, personal communication), slickensides on faults associated with shallow extension detachments in the Basin and Range province of Nevada demonstrate a large component of divergent oblique slip on faults not oriented parallel with the regional trend.

In numerous graben basins, however, seismic profiles (this study) and surface exposures (King, 1965; Robson, 1971; El-Tarabili and Adawy, 1972) demonstrated that the different fault orientations all have a similar "dip-slip," extensional style. That a significant strike-slip component may be absent on oblique normal faults was shown by King's (1965) stratigraphic piercing-point data on the Salt Flat graben of west Texas. King, Robson, and El-Tarabili and Adawy, all minimized the role of strike slip in the grabens they studied. For the purposes of hydrocarbon exploration, the oblique faults are best treated as conventional normal faults.

The timing of displacements within the graben system varies in detail as activity shifts back and forth from one group of faults to another (this study). However, it is thought that, locally, faults with different orientations can be synchronous. This synchroneity is demonstrated by trap-door corners where the total displacement is similar on adjoining sides of the fault block (Figure 25).

The influence of basement anisotropy is critical in block faulting, and in some regions one or more of the fault orientations have been developed along preexisting zones of weakness (Robson, 1971; Ramberg and Smithson, 1975). It is not clear, however, if this is a prerequisite for the development of the multidirectional fault sets (Freund and Merzer, 1976; Illies, 1970). At the north end of the Viking graben, the regional north-south faults turn abruptly northeastward (Figure 13) to parallel the Precambrian and Caledonian basement trend exposed along the strike near Trondheim, Norway (Ramberg et al, 1977). The change in fault strike has formed a regional dogleg, and the interaction of the two fault systems has established a decidedly closure-prone area.

In the northern Viking graben, the basement grain intersected the graben at an optimal oblique angle. It is thought that basement grain parallel with the graben would probably constrain faulting mostly to the longitudinal orientation and thus could diminish closure development.

Figure 25—Oblique air photo of three-sided trap-door fault block at east margin of Salt Flat graben, west Texas. Structure is 3–6 mi (5–10 km) wide. Style and age of boundary faults are similar on all three sides, and trap door appears to be result of simultaneous dip-slip displacement on three faults.

Transverse grain is apparently utilized much less commonly by the graben faulting (e.g., frequency plots in Figures 6, 8; and in other grabens, El-Tarabili and Adawy, 1972). In some areas, the observable basement anisotropy does not correspond to any of the fault orientations, yet the multidirectional fault patterns are still developed (Illies, 1970).

Secondary Normal Faults

Individual blocks may be segmented by secondary normal faults, and these can provide subsidiary closures (Columba field and west Ninian discovery in Figure 15). They may adversely disrupt a potential closure if the faulting is severe. Secondary faults are considered to be faults that are restricted to the interior of a single block and do not bound the block.

At some structures, the secondary faults repeat the trend of the block boundaries (Figure 15), but in other areas they have oblique or transverse orientations (Figure 14) (Robson, 1971; King, 1965). The latter two patterns may result from local stress within the block, and they can be an important source of closure along the structure's strike (Figure 14).

Block-Edge Flexures

Where flexures are developed, their trends repeat the multidirectional orientations of the longitudinal and oblique block bounding faults (Figures 6, 7, 17) (El-Tarabili and Adawy, 1972). This relationship occurs because the flexures are formed by forced drape, drag, or differential compaction at fault-block edges. The fold patterns include parallel, relay (similar to relay fault patterns), dogleg, zigzag, and interference. The patterns are combined in various ways, and they typically present a gridded appearance on flexure maps that is considered to be characteristic of fault-block deformation. In the Viking graben, trends of the block-bounding faults have localized fault-face erosion (Figures 14, 15). Because of this, the distribution of topographic features there also suggests a fault-block style (Figure 13).

Other Graben Styles

Some grabens have structural styles that differ importantly from the examples described in the present study. These include grabens that are distinguished by a unidirectional set of normal faults, or by large vertical tension gashes, which are arranged en echelon along the trough axes (Mohr, 1968; Gibson and Tazieff, 1970). Mohr related such faulting in the main Ethiopian rift to a diffuse, divergent crustal couple. At the Wonju fault belt, the en echelon fractures are younger than the rift boundaries, and are closely spaced and trend obliquely to the axis of subsidence (Figure 26).

Because of their orientation oblique to graben subsidence, such fractures may trend obliquely to the bed strike. Traps associated with this style may be highly seg-

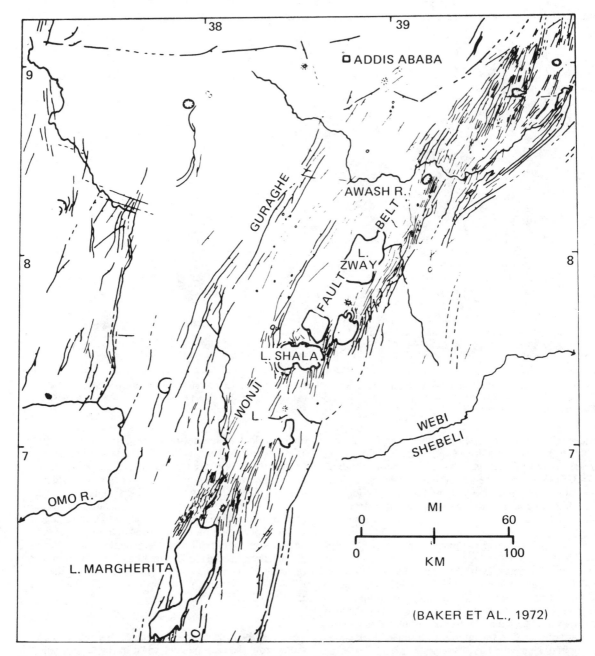

Figure 26—Northeast segment of main Ethiopian rift and Wonji fault belt, Ethiopia. Wonji fault belt is composed of high-angle, north-northeast trending en echelon faults; it extends outward from northeast corner of map to Lake Margherita. Rift's border faults were initiated first (Gibson and Tazieff, 1970). They outline a regional dogleg, only southern portion of which is visible on map.

mented by the close spacing of faults, and the fault strike could cause structures to be open obliquely up the dip of potential reservoirs.

In some regions, the superpositioning of several differently oriented extensional stress systems has apparently resulted in increased structural segmentation. In addition to various longitudinal and oblique faults, faults transverse to the regional trend may be common. In Malta, for example, two-stage faulting has produced a complex trellis-like pattern (Illies, 1981, his Figure 6).

CONCLUSIONS

Two tectonic factors are critical to the occurrence of hydrocarbon accumulations in graben settings, and these have been emphasized in the present work. They are (1) the subsidence and stacking of sedimentary basins, and (2) the internal deformation of these basins by multidirectional normal faults.

Basin subsidence follows several different sequences. Early workers have proposed that grabens result from crestal normal faulting of a rising arch and, therefore, that the initiation of arching must precede basin formation (e.g., Falvey, 1974). This may have been the sequence at the Sirte basin. In the Gulf of Suez, however, the start of graben faulting coincided with broad subsidence. This timing is critical, in that reservoirs predating the graben fill were preserved from erosion and a major hydrocarbon play was thereby established in the older section.

Numerous closely spaced, comparatively small displacement faults occurred first. Their movements later shifted to fewer, more widely spaced, larger displacement zones, and this established the broad platforms. Arching in the Gulf of Suez came during the second stage, after considerable graben deposition. It then affected only the rift shoulders. The continuing graben subsidence established and preserved a second major hydrocarbon play, this one within the graben-fill section. The Viking graben may have had a sequence similar to the Gulf of Suez but events during its earliest stage are not well known.

The zigzag, gridded patterns of multidirectional normal faulting establish many of the hydrocarbon traps. Although the structural style of individual grabens may have different controls, the repetition of patterns suggests that the fault style, and therefore its hydrocarbon traps, do not require unique circumstances for development, as has been proposed in the past. At a number of grabens, including several not discussed in this article, the differently oriented faults are seen to be generally synchronous, to have a similar normal slip, and not to be dependent on corresponding basement grains for each of their multiple orientations. If the conclusions from these observations prove correct, all grabens can be considered to have a high probability of containing abundant structural closures.

REFERENCES CITED

Abdel-Gawad, M., 1970, The Gulf of Suez—a brief review of stratigraphy and structure: Philosophical Transactions of the Royal Society of London, v. A267, p. 41-48.

Albright, W. A., W. L. Turner, and K. R. Williamson, 1980, Ninian field, U. K. sector, North Sea, in Giant oil and gas fields of the decade: 1968-1978: AAPG Memoir 30, p. 173-193.

Angelier, J., and B. Colletta, 1983, Tension fractures and extensional tectonics: Nature, v. 301, p. 49-51.

Asmus, H. E., and W. Guazelli, 1981, Descricao sumaria das estruturas da margem continental Brasileira e das areas oceanicas e continentais, adjaceates—hipoteses sobre o tectonismo cuasador, e implicacoes para os prognosticos do potencial de recursos minerais; in H. E. Asmus, ed., Reconhecimento global de margem continental Brasileria: Petrobras (CENPES), Serie Projecto Remac, n. 9, p. 187-269.

Ayouty, M. K., 1961, Geology of Belayim fields: Third Arab Petroleum Congress, Alexandria, Egypt, p. 1-12.

Baker, B. H., P. A. Mohr, and L. A. J. Williams, 1972, Geology of the Eastern Rift system of Africa: GSA Special Paper 136, 67 p.

Bally, A. W., D. Bernoulli, G. A. Davis, and K. Montadert, 1981, Listric normal faults: Oceanologica Acta, Proceedings of the 26th International Geological Congress, Geology of Continental Margins Symposium, p. 87-101.

Bebout, D. G., and C. Pendexter, 1975, Secondary carbonate porosity as related to early Tertiary depositional facies, Zelten field, Libya: AAPG Bulletin, v. 59, p. 665-693.

Blair, D. G., 1975, Structural styles in North Sea oil and gas fields, in A. W. Woodland, ed., Petroleum and the continental shelf of northwest Europe; v. 1, Geology: London, Applied Science Publishers, p. 327-335.

Bowen, J. M., 1976, The Brent oil field, in A. W. Woodland, ed., Petroleum and the continental shelf of northwest Europe: v. 1, Geology: New York, John Wiley, p. 353-361.

Brady, T. J., N. D. J. Campbell, and C. E. Maher, 1981, Intisar "D" oil field, Libya, in L. V. Illing and G. D. Hobson, eds., Petroleum geology of the continental shelf of northwest Europe: London, Hayden and Son, p. 543-564.

Brown, R. N., 1980, History of exploration and discovery of Morgan, Ramadan and July oil fields, Gulf of Suez, Egypt, in A. D. Maill, ed., Facts and principles of world petroleum occurrence: Canadian Society of Petroleum Geologists Memoir 6, p. 733-764.

Budding, M. C., and H. F. Inglin, 1981, A reservoir geological model of Brent sands in southern Cormorant, in L. V. Illing and G. D. Hobson, eds., Petroleum geology of the continental shelf of northwest Europe: London, Hayden and Son, p. 326-334.

Burke, K., and J. F. Dewey, 1973, Plume-generated triple junctions: key indicators in applying plate tectonics to old rocks: Journal of Geology, v. 81, p. 406-433.

Clifford, H. J., R. Grund, and H. Musrati, 1980, Geology of a stratigraphic giant: Messla oil field, Libya, in Giant oil and gas fields of the decade: 1968-1978: AAPG Memoir 30, p. 507-524.

Coleman, R. G., 1974, Geologic background of the Red Sea, in C. A. Burke and C. L. Drake, eds., The geology of continental margins: New York, Springer-Verlag, p. 743-751.

Conant, L. C., and G. H. Goudarzi, 1967, Stratigraphic and tectonic framework of Libya: AAPG Bulletin, v. 51, p. 719-730.

Courtillot, V., P. Tapponnier, and J. Varet, 1974, Surface features associated with transform faults—a comparison between observed examples and an experimental model: Tectonophysics, v. 24, p. 317-329.

Davies, E. J., and T. R. Watts, 1977, The Murchison oil field: Norwegian Petroleum Society, Proceedings of the Mesozoic Northern North Sea Symposium, Oslo, p. 15-1 to 15-24.

De'Ath, N. G., and S. F. Schuyleman, 1981, The geology of the Magnus oil field, in L. V. Illing and G. D. Hobson, eds., Petroleum geology of the continental shelf of northwest Europe: London, Hayden and Son, p. 380-391.

El-Ashry, M. T., 1972, Source and dispersal of reservoir sands in the El Morgan field, Gulf of Suez, Egypt: Sedimentary Geology, v. 8, p. 317-325.

El-Tarabili, E. S., and N. Adawy, 1972, Geologic history of Nukhul-Baba area, Gulf of Suez, Sinai, Egypt: AAPG Bulletin, v. 56, p. 882-902.

Falvey, D. A., 1974, The development of continental margins in plate tectonic theory: APEA Journal, v. 14, p. 95-106.

Fraser, W. W., 1967, Geology of the Zelten field, Libya, North Africa: 7th World Petroleum Congress Proceedings (Mexico): Amsterdam, New York, Elsevier Publishing Co., v. 2, p. 259-264.

Freund, R., and A. M. Merzer, 1976, The formation of rift valleys and their zigzag fault patterns: Geological Magazine, v. 113, p. 561-568.

Garfunkel, Z., and Y. Bartov, 1977, The tectonics of the Suez rift: Geological Survey of Israel Bulletin 71, 43 p.

Ghorab, M. A., 1961, Abnormal stratigraphic features in Ras Gharib oil field: Third Arab Petroleum Congress, Alexandria, Egypt, v. 1, p. 1-10.

Gibson, I. L., and H. Tazieff, 1970, The structure of Afar and the northern part of the Ethiopian rift: Philosophical Transactions of the Royal Society of London, v. A267, p. 331-338.

Gilboa, Y., and A. Cohen, 1979, Oil trap patterns in the Gulf of Suez: Israel Journal of Earth Sciences, v. 28, p. 13-26.

Goudarzi, G. H., 1970, Geology and mineral resources of Libya—a reconnaissance: U. S. Geological Survey Professional Paper 660, 104 p.

Grey, W. D. T., and G. Barnes, 1981, The Heather oil field, in L. V. Illing and G. D. Hobson, eds., Petroleum geology of the continental shelf of northwest Europe: London, Hayden and Son, p. 335-441.

Hallett, D., 1981, Refinement of the geological model of the Thistle field, in L. V. Illing and G. D. Hobson, eds., Petroleum geology of the continental shelf of northwest Europe: London, Hayden and Son, p. 315-325.

Halstead, P. H., 1975, Northern North Sea faulting: Norwegian Petroleum Society, Proceedings of the Jurassic Northern North Sea Symposium, Stavanger, p. 10-1 to 10-38.

Hamblin, W. K., 1965, Origin of "reverse drag" on the downthrown-side of normal faults: GSA Bulletin, v. 76, p. 1145-1164.

Harding, T. P., 1983a, Graben hydrocarbon plays and structural style:

Geologie en Mijnbouw, v. 62, p. 3-23.

—— 1983b, Divergent wrench fault and negative flower structure, Andaman Sea, in A. W. Bally, ed., Seismic expression of structural styles, v. 3: AAPG Studies in Geology Series 15, p. 4.2-1 to 4.2-8.

—— 1983c, Structural inversion at Rambutan oil field, south Sumatra basin, in A. W. Bally, ed., Seismic expression of structural styles, v. 3: AAPG Studies in Geology Series 15, p. 3.3-13 to 3.3-18.

—— and J. D. Lowell, 1979, Structural styles, their plate-tectonic habitats and hydrocarbon traps in petroleum provinces: AAPG Bulletin, v. 63, p. 1016-1058.

Harms, J. C., T. Tackenberry, E. Pickles, and R. E. Pollock, 1981, The Brae oil field area, in L. V. Illing and G. D. Hobson, eds., Petroleum geology of the continental shelf of northwest Europe: London, Hayden and Son, p. 352-357.

Hay, J. T. C., 1978, Structural development in the northern North Sea: Journal of Petroleum Geology, v. 1, p. 65-77.

Heritier, F. E., P. Lossel, and E. Wathne, 1981, The Frigg gas field, in L. V. Illing and G. D. Hobson, eds., Petroleum geology of the continental shelf of northwest Europe: London, Hayden and Son, p. 380-391.

Illies, J. H., 1970, Graben tectonics as related to crust-mantle interaction, in J. H. Illies and St. Muellar, eds., Graben problems: Stuttgart, Schweizerbart, p. 4-27.

—— 1974, Taphrogenesis and plate tectonics, in J. H. Illies and K. Fuchs, eds., Approaches to taphrogenesis: Stuttgart, Schweizerbart, p. 433-460.

—— 1981, Graben formation—the Maltese Islands—a case history: Tectonophysics, v. 73, p. 151-168.

—— and G. Greiner, 1978, Rhinegraben and the Alpine system: GSA Bulletin, v. 89, p. 770-782.

Joiner, D. S., and C. E. Myers, 1972, Developing a 10 billion barrel reservoir with modern techniques: Society of Petroleum Engineers of AIME, European Spring Meeting, Amsterdam, Preprint, n. SPE-3739, 15 p.

King, P. B., 1949, Regional geologic map of parts of Culberson and Hudspeth counties: U. S. Geological Survey Oil and Gas Investigations, Preliminary Map 90, reprinted 1960.

—— 1965, Geology of the Sierra Diablo region, Texas: U. S. Geological Survey Professional Paper 480, 185 p.

Kirk, R. H., 1980, Statfjord field—a North Sea giant, in Giant Oil and gas fields of the decade: 1968-1978: AAPG Memoir 30, p. 95-116.

Klitzsch, E., 1970, Die structur-geschichte der Zentralsahara: Geologische Rundschau, v. 59, p. 459-527.

McDonald, R. E., 1976, Tertiary tectonics and sedimentary rocks along the transition: Basin and Range province to plateau and thrust belt province, Utah, in J. G. Hilt, ed., Symposium on geology of the Cordilleran hinge line: Rocky Mountain Association of Geologists, p. 281-317.

Milanovsky, E. E., 1981, Aulacogens of ancient platforms—problems of their origin and tectonic development: Tectonophysics, v. 73, p. 213-248.

Mohr, P. A., 1968, Transcurrent faulting in the Ethiopian rift system: Nature, v. 218, p. 938-941.

Morgan, D. E., and A. El-Barkouky, 1956, Geophysical history of Ras Gharib field, in P. L. Lyons, ed., Geophysical case histories: Society of Exploration Geophysicists, v. 2, p. 237-247.

Morton, A. C., 1982, Lower Tertiary sand development in Viking graben, North Sea: AAPG Bulletin, v. 66, p. 1542-1559.

Morton, W. H., and R. Black, 1975, Crustal attenuation in Afar, in A. Pilger and A. Rosler, eds., Afar depression of Ethiopa: Proceedings of the International Symposium on Afar Region and Related Rift Problems, Stuttgart, E. Schweizerbart'sche Verlagsbuchhandlung, p. 55-65.

Moustafa, A. M., 1974, Review of diapiric salt structures, Gulf of Suez: Fourth Exploration Seminar, Egyptian General Petroleum Corp., Cairo, 21 p.

Omar, K. Z., 1975, Genetic sequences and oil occurrence of Miocene strata in Um El Yusr field, Gulf of Suez, A. R. E.: Ninth Arab Petroleum Congress, Dubai, Paper, n. 115 (B-3), p. 1-26.

Parsons, M. G., A. M. Zagaar, and J. J. Curry, 1980, Hydrocarbon occurrences in the Sirte basin, Libya, in A. D. Maill, ed., Facts and principles of world petroleum occurrence: Canadian Society of Petroleum Geologists Memoir 6, p. 723-732.

Proffett, J. M., 1977, Cenozoic geology of the Yerington district, Nevada, and implications for the nature and origin of basin and range faulting: GSA Bulletin, v. 88, p. 247-266.

Rambert, I. B., and S. B. Smithson, 1975, Grided fault patterns in a late

Cenozoic and a Paleozoic continental rift: Geology, v. 3, p. 201-205.

—— R. H. Gabrielsen, B. T. Larsen, and A. Solli, 1977, Analysis of fracture patterns in southern Norway: Geologie en Mijnbouw, v. 56, p. 295-310.

Reches, Z., 1978, Analysis of faulting in three-dimensional strain field: Tectonophysics, v. 47, p. 109-129.

Roberts, J. M., 1970, Amal field, Libya, in Geology of giant petroleum fields: AAPG Memoir 14, p. 438-448.

Robson, D. A., 1971, The structure of the Gulf of Suez (clysmic) rift, with special reference to the eastern side: Journal of Geological Society of London, v. 127, p. 247-276.

Ronnevik, H. C., W. von der Bosch, and E. H. Bandlien, 1975, A proposed nomenclature for the main structural features in the Norwegian North Sea: Norwegian Petroleum Society, Jurassic Northern North Sea Symposium Proceedings, JNNSS/18, p. 1-16.

Saeland, G. T., and G. S. Simpson, 1982, Interpretation of 3-D data in delineating a subunconformity trap in Block 34/10, Norwegian North Sea, in The deliberate search for the subtle trap: AAPG Memoir 32, p. 217-235.

Said, R., 1962, The geology of Egypt: Amsterdam, New York, Elsevier Publishing Co., 377 p.

Sanford, R. M., 1970, Sarir oil field, Libya—desert surprise, in Geology of giant petroleum fields: AAPG Memoir 14, p. 449-476.

Schurmann, H. M. E., 1966, The Precambrian along the Gulf of Suez and the northern part of the Red Sea: Lieden, E. J. Brill, 404 p.

Smith, D. A., 1980, Sealing and nonsealing faults in Louisiana Gulf Coast salt basin: AAPG Bulletin, v. 64, p. 145-172.

Soliman, S. M., and M. I. Faris, 1963, General geologic setting of the Nile delta province and its evaluation for petroleum prospecting: 4th Arab Petroleum Congress, Beirut, Paper 23 (B-3), 11 p.

Steel, R., and T. G. Gloppen, 1980, Late Caledonian (Devonian) basin formation, western Norway: signs of strike-slip tectonics during infilling, in P. F. Ballance and H. G. Reading, eds., Sedimentation in oblique-slip mobile zones: International Association of Sedimentology Special Publication 4, p. 79-103.

Thiebaud, C. E., and D. A. Robson, 1981, The geology of the Asl oil field, western Sinai, Egypt: Journal of Petroleum Geology, v. 4, p. 77-87.

Trusheim, F., 1960, Mechanism of salt migration in northern Germany: AAPG Bulletin, v. 44, p. 1519-1540.

Vail, P. R., and R. G. Todd, 1981, Northern North Sea Jurassic unconformities, chronostratigraphy and sea-level changes from seismic stratigraphy, in L. V. Illing and G. D. Hobson, eds., Petroleum geology of the continental shelf of north-west Europe: London, Hayden and Son, p. 216-235.

Van den Bark, E., and O. D. Thomas, 1980, Ekofisk: first of the giant oil fields in western Europe, in Giant oil and gas fields of the decade: 1968-1978: AAPG Memoir 30, p. 195-224.

Vierbuchen, R. C., R. P. George, and P. R. Vail, 1982, A thermal-mechanical model of rifting with implications for outer highs on passive continental margins, in J. S. Watkins and C. L. Drake, eds., Studies in continental margin geology: AAPG Memoir 34, p. 765-778.

Weber, K. J., G. Mandl, W. F. Pilaar, F. Lehner, and R. G. Precious, 1978, The role of faults in hydrocarbon migration and trapping in Nigerian growth fault structures: Offshore Technology Conference Paper, OTC 3356, p. 2643-2652.

Wernicke, B., and B. C. Burchfiel, 1982, Modes of extensional tectonics: Journal of Structural Geology, v. 4, p. 105-115.

Williams. J. J., 1968, The sedimentary and igneous reservoirs of the Augila oil field, Libya, in Geology and archeology of northern Cyrenaica, Libya: Petroleum Exploration Society of Libya 10th Annual Field Conference Guidebook, p. 197-204.

—— 1972, Augila field, Libya: depositional environment and diagenesis of sedimentary reservoir and description of igneous reservoir, in Stratigraphic oil and gas fields—classification, exploration methods, and case histories: AAPG Memoir 16, SEG Special Publication 10, p. 623-632.

Ziegler, P. A., 1978, Northwest Europe: tectonics and basin development: Geologie en Mijnbouw, v. 57, p. 589-626.

—— 1979, Factors controlling North Sea hydrocarbon accumulations: World Oil, v. 189, November, p. 111-124.

—— 1980, Northwest Europe basin: geology and hydrocarbon provinces, in A. D. Maill, ed., Facts and principles of world petroleum occurrence: Canadian Society of Petroleum Geologists Memoir 6, p. 653-706.

Reprinted by permission of the Geological Society of America, from *Geological Society of America Bulletin*, 1964, vol. 75, no. 4, pp. 307-316.

G. De MILLE

J. R. SHOULDICE   } *Imperial Oil Limited, Calgary, Alberta, Canada*

H. W. NELSON

Collapse Structures Related to Evaporites of the Prairie Formation, Saskatchewan

Abstract: The salt deposits of the Middle Devonian Elk Point Group extend from North Dakota to Canada's Northwest Territories. Subsurface knowledge is best in Saskatchewan. Here the most widespread of these deposits, the Prairie Formation, reaches a maximum thickness of 670 feet and contains the extensive potash ores now being developed. The Prairie salt beds overlie thin carbonates containing reefs. These sediments form part of a depositional sequence starting with thin shales and grading upward to carbonates, followed by evaporites climaxing in potash salts and capped with red beds. This sequence is overlain by similar depositional units in which, however, evaporites are less well developed.

Because of the excellent subsurface control in this vast gently dipping shelf area, it is possible to recognize structural features that might go unnoticed in more complex areas. Structural highs due to salt flowage are unknown. There are, however, many structural lows which are due to removal of salt by subsurface leaching.

The Rosetown low, covering about 144 sections, is one of the larger features of this type.

The Hummingbird trough is a complex structure. Seismic and subsurface information indicate that there have been repeated episodes of salt leaching accompanied by downwarping and compensation in overlying strata.

It is impossible to escape the conclusion that the Prairie evaporites have been severely modified by subsequent solution. The unusual structures resulting from this phenomenon may be present but unrecognized in other salt basins.

CONTENTS

INTRODUCTION

The Middle Devonian Elk Point Group contains the largest salt deposits in Canada. They extend 1200 miles from the northern United States to Canada's Northwest Territories, covering parts of the provinces of Alberta, Saskatchewan, and Manitoba as well as the northeastern corner of British Columbia (Fig. 1). In Saskatchewan, the area to be dealt with, the most extensive of these deposits is called the Prairie Formation, which consists largely of salt and lesser anhydrite beds.

One of the most striking features related to the Prairie Formation is structural "lows" occurring along the margin of the salt and locally within the area of the salt deposit. A recent exploratory well that penetrated one of these lows provides geological data in support of the hypothesis that they originated from postdepositional interstratal solution. We will review these data and attempt to relate the oc-

Geological Society of America Bulletin, v. 75, p. 307–316, 10 figs., 1 pl., April 1964

currence of the collapse features to sources of weakness in the underlying basement.

Saskatchewan has other Devonian evaporite deposits, both older and younger than the Prairie Formation, in which similar collapse phenomena have been observed. These evaporites, however, are less extensive and will not be discussed here.

ness of over 800 feet in Alberta and 670 feet in Saskatchewan. The northern end of the basin was open to the sea but restricted by carbonate banks. The southern part, being farthest from the open sea, contained the most concentrated brines. The well-known potash deposits now being developed are located in Saskatchewan. Although they represent only a small portion

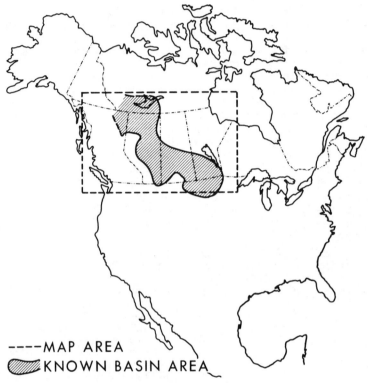

----MAP AREA

KNOWN BASIN AREA

Figure 1. Elk Point salt basin. Shaded area shows the extent of the evaporites. Rectangular area is the area appearing in Figure 3.

STRATIGRAPHY

The sedimentary section in southern Saskatchewan is of moderate thickness (Fig. 2). In general, the lower half of the section consists of competent Paleozoic carbonates, while the upper half consists of Mesozoic clastics. The Prairie Formation occurs at about the middle of the Paleozoic section.

The basin in which the Prairie Formation was deposited was part of a sedimentary shelf bordering the Canadian Shield (Fig. 3). The region to the west was the site of geosynclinal deposition and tectonic activity. The evaporites, however, have undergone only mild epeirogenic movements. They reach a thick-

of the total evaporites, these potash beds constitute a remarkably extensive deposit containing major reserves of high-grade ore.

In Saskatchewan, the evaporites of the Prairie Formation represent the final stage of a major depositional cycle. At the onset of this cycle, the sea spread over the eroded pre-Devonian surface, depositing a thin basal shale. This was followed by fairly clean, shelf-type carbonates, presumably deposited under comparatively stable tectonic conditions. Subsequently, however, a central basin developed, and extensive reef banks formed along the margins of the subsiding trough. In addition, isolated pinnacle reefs grew from the basin deep, obtaining heights of up to 400 feet. Then, as a

144

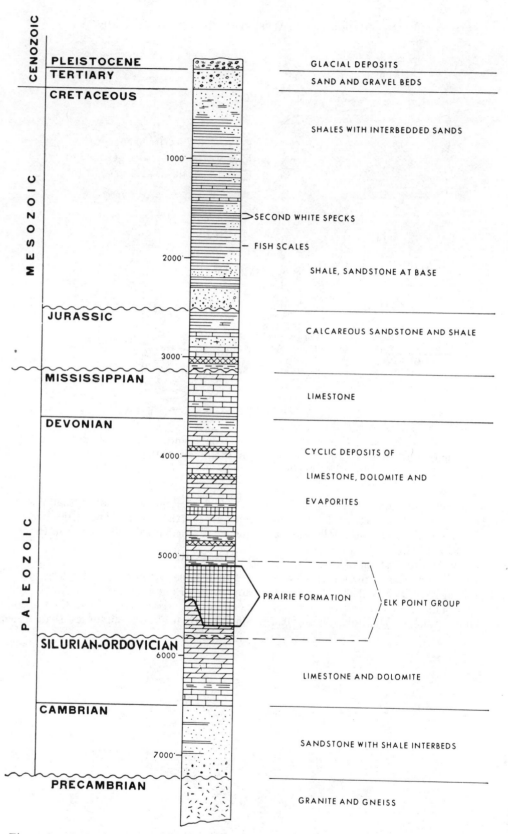

Figure 2.   General stratigraphic column showing position of the Prairie evaporites in the sub-surface of southern Saskatchewan

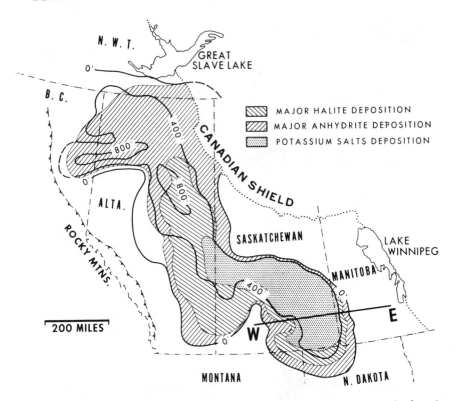

Figure 3. Isopachs of Prairie Formation and equivalents. The potash deposits (stippled) are developed near the top of the Prairie Formation and are confined mainly to Saskatchewan. Section line shown on Figure 4

result of increasing salinity in this inland sea, the reefs died off and were covered by evaporites. Thin anhydrite beds were followed by thick sequences of halite and, as concentration increased, by sylvite and carnallite. As the basin became filled, the retreating sea left a thin deposit of red silts and shales. The evaporites are now buried under a cover of younger rocks 1000 feet thick in Manitoba, about 5000 feet thick at Regina, and more than 8000 feet thick in North Dakota. The depositional sequence is schematically illustrated in Figure 4.

STRUCTURE

The regional structure of this large area is simple. The Prairie evaporites are tilted gently to the south and west, away from the Canadian Shield, with an average dip of less than 30 feet

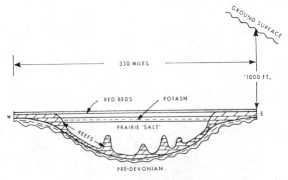

Figure 4. Schematic cross section, illustrating the cyclic change from normal marine to evaporite and red-bed deposition. *See* Figure 3 for location of section.

per mile. Because of the gentle regional structure and the excellent subsurface control, it is possible to distinguish local structural features associated with the evaporites. No structural highs due to salt flowage have been observed in Saskatchewan despite the considerable depth of burial. There are, however, some anomalous lows which appear to be related to the salt deposits. trough, which extends from Regina southward, beyond the International Boundary.

ROSETOWN LOW

The Rosetown low is a structural feature covering about four townships, or 144 square miles. It is well within the limits of the Prairie evaporites, where nearly 400 feet of salt would

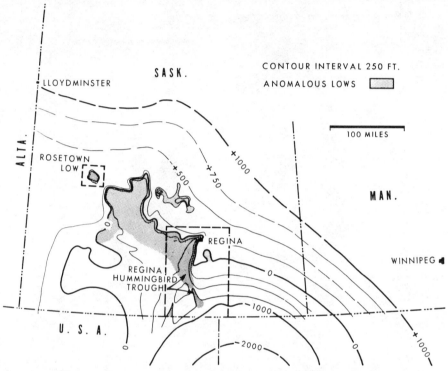

Figure 5.   Structure contours of Upper Cretaceous Second White Specks. The gentle regional southwest tilt is complicated by anomalous structural depressions.

posits. Figure 5 shows the structure drawn on the Upper Cretaceous Second White Specks horizon. Sediments dip gently away from the Canadian Shield south into the Williston Basin and west into the Alberta Basin. This comparatively smooth regional pattern, however, is interrupted by several structural lows indicated by the shaded areas on the map.

Various authors, for example, Bishop (1953) and Edie (1959), have interpreted these lows as being the result of large-scale subsurface leaching of Devonian salt beds. New evidence to support this theory is supplied by recent deep drilling in two separate areas. The first area is the Rosetown low, located in the western part of Saskatchewan. The second area is a more linear feature, the Regina-Hummingbird

normally be expected. The area is drift-covered, and the low was known only from seismic work before it was drilled. Seismic control on several horizons above the Middle Devonian was good, but the attitude of deeper beds could not be reliably determined.

The seismic data indicated a rectangular, flat-bottomed, steep-sided low in an area of very gentle regional dip (Fig. 6). This feature, which has an amplitude of several hundred feet, was believed to be the result of solution of the salt at depth.

Recently a deep test (Imperial Fortune 13–20–30–14–West of 3d Meridian) drilled within this low confirmed that the salt was missing. Moreover, the amount of depression, 380 feet, was nearly equal to the thickness of salt normally expected for this area. Beds underlying the salt were found to be regional structurally and not involved in the Rosetown low.

These facts constitute some of the evidence for subsurface leaching of salt. The only alternate explanation would be to assume that an ancient structural high existed in this area,

Figure 1 of Plate 1 shows rock typical of the lower parts of these breccias, in which much of the material is of a fine nature. Some of it, both fragments and matrix, is believed to represent insoluble matter within the original salt.

Figure 2 of Plate 1 shows higher parts of the breccia. The fragments are much larger, and there is less relative separation. Where core is

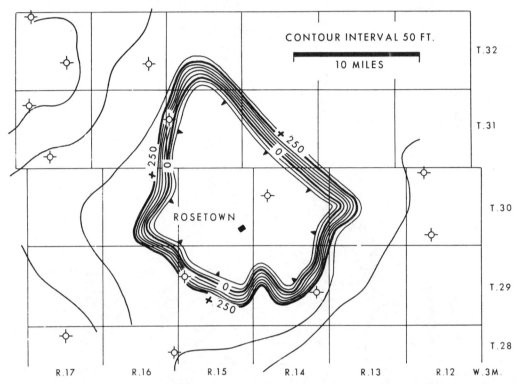

Figure 6. Rosetown low. The Fortune well in T. 30, R. 14–W3, which was drilled in the structural depression, showed salt to be missing and encountered a breccia in the expected position of the salt.

preventing the deposition of salt. Subsequently, in post-Cretaceous times the strata would have returned precisely to their original position with respect to strata underlying the salt. This seems to be an improbable coincidence for this and other similar structures in Saskatchewan.

The deep test revealed a breccia in the beds overlying the position of the missing salt. The breccia is similar to breccias found in other areas of suspected salt-leaching in Saskatchewan. Notable features of the breccias are: (1) their well-defined base coinciding with the top of the strata underlying the Prairie evaporites; and (2) their upward extension through two or more formations of differing lithology. In view of this, there is little doubt that the breccias resulted from solution and collapse.

available it shows these breccias grading upward a hundred or more feet and eventually merging with fractured but otherwise unbrecciated strata.

From a genetic point of view it is interesting to note the location of the Rosetown low with respect to the original salt basin. The southwest flank of this basin consists of an extensive fringe reef which is partially overlapped by the upper beds of the Prairie evaporites. This reef extends in a southeast direction through the Rosetown low. It also coincides with much of the great structural depression between the Rosetown feature and Regina (*See* Fig. 5). This strongly suggests that the reef trend has influenced salt-leaching. Permeability of the reefal beds may have aided the movement of

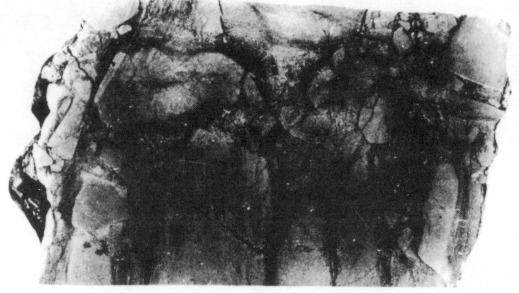

Figure 1. Sample showing the more fragmented nature of the lower part of the brecciated section

Figure 2. Sample showing higher parts of the breccia

SOLUTION BRECCIAS IN AREAS OF SALT REMOVAL

PLATE 1:

149

water which dissolved the overlying salt. Furthermore, irregular compaction of the strata over the reef may have produced fractures which allowed water to circulate downward to the salt.

The possibility of basement movement along the trend of reef growth must also be considered, particularly in view of the remarkably

pian in Late Cretaceous time are also shown. Seismic data and subsurface information from numerous deep tests have been used in these interpretations.

In the southern part of the trough the present structural low on the Mississippian coincides with the one which existed in Late Cretaceous time. This suggests that in this area

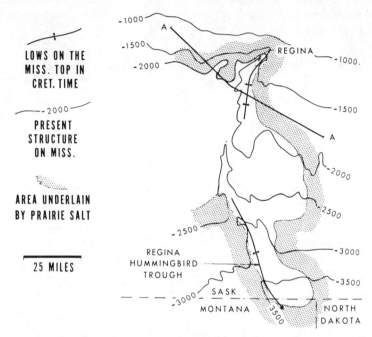

Figure 7. Mississippian structure map of Regina–Hummingbird trough. The present structural trough is the result of salt solution which began in pre-Cretaceous time throughout the trend and recurred in post-Cretaceous time at the north end. *See* Figure 8 for structure along A–A'.

straight northeastern border of the Rosetown low. Such true tectonic movements may have triggered the salt-solution phenomenon and then have become masked by the resulting collapse features.

REGINA-HUMMINGBIRD TROUGH

The Regina-Hummingbird trough extends southward from the city of Regina for more than 100 miles. Like the Rosetown low, it is considered to have resulted from salt solution within the Prairie evaporites and collapse of the overlying strata. The outline of the trough is illustrated in Figure 7 by structure contours on the top of the Mississippian. The shaded areas are underlain by Prairie salt which is approximately 200–300 feet thick in this region. Structural lows which existed on the Mississip-

subsidence occurred between Mississippian and Late Cretaceous time and most likely is related to the post-Mississippian erosional interval. Post-Mississippian subsidence is indicated by the presence, in the depression, of an extra section of Mississippian strata preserved below the unconformity, and by excessive thickness of the overlying red beds at the base of the Mesozoic section. In the vicinity of Regina, however, the present structural low is offset from the low which existed in Late Cretaceous (Second White Specks) time. In this part of the trough, therefore, subsidence must have taken place both before and after Second White Specks time.

It is apparent then, that salt removal occurred throughout the trough area in pre-Cretaceous time and recurred in the north end in Late Cretaceous or Tertiary time. In fact, detailed isopachs of units overlying the Prairie evaporites suggest that salt removal may have begun as early as Late Devonian time and recurred repeatedly, with maximum leaching occurring during the post-Mississippian erosional interval and during the post-Second White Specks time.

The manner of structural development is generalized for illustration in Figure 8, which shows sections drawn along the line A–A' of

permit the fluid movements necessary for salt-leaching. A similar suggestion was made by Grieve (1955) for leaching phenomena observed in Silurian salt beds in southwestern Ontario. The probability of basement activity can be inferred from an examination of the Nemo-Estes trend. (Fig. 9). This structural trend is characterized by gravity and magnetic gradients caused by lithological and structural differences in the basement rocks. The overlying sedimentary section contains folds and

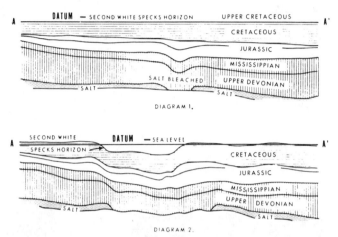

Figure 8. Structural section Regina-Hummingbird trough. The structure along Section A–A' (Fig. 7) is illustrated schematically in diagram 1 for the Late Cretaceous time; diagram 2 shows the present-day, much broader area of salt removal. Length of section 40 miles, thickness 3500 feet.

Figure 7. The first panel reconstructs pre-Cretaceous structure, with the Second White Specks as datum. The area where the salt is absent is marked by a structural depression of the pre-Cretaceous beds. The thickening in the Jurassic corresponds with the amount of salt removal.

The second panel illustrates the present structure and the wider area of salt removal. The broad sag corresponds exactly with the area in which the salt is absent, except in the eastern part where the thicker section of the Jurassic compensates for the missing salt and causes the Second White Specks to lie at the elevation normally anticipated for the salt-bearing areas.

Careful study of the geometry of the rock masses involved in the deformation indicates that in most cases some structural movement must have occurred beyond that compensating for the salt removal. One may speculate whether basement movements fractured the sedimentary section to create permeability and

faults which developed by intermittent movements over a long period of time. The magnetic and gravity trends extend into Canada, where they show general association with the Regina-Hummingbird trough. This relationship suggests, although it does not prove, that basement movement initiated salt-removal which in turn resulted in the greater structural deformation of this area.

CONCLUSIONS

Both of these anomalous structural lows offer conclusive evidence that the Prairie evaporites have been severely affected by underground solution and that leaching took place while the salt body was buried under thousands of feet of sediments.

In both cases the amplitude of the depression below the regional structure is nearly equal to the amount of salt which would normally be expected but, in fact, is missing in these structures.

The structural lows contain solution breccias

CONCLUSIONS

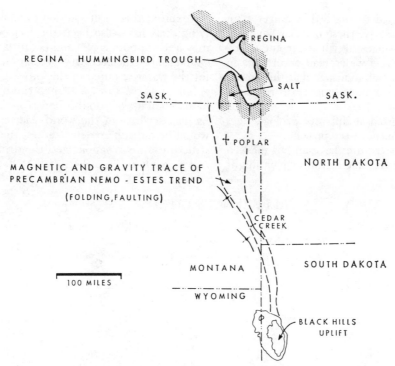

Figure 9. Tectonic sketch, Nemo-Estes trend, showing the coincidence of the Regina-Hummingbird trough and the Nemo-Estes tectonic trend which suggests a relationship between these features.

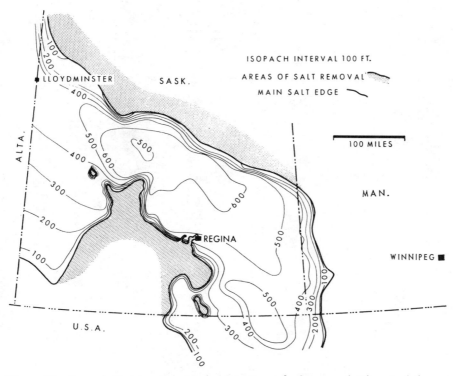

Figure 10. Isopach Prairie evaporite showing areas of salt removal. The stippled areas contain the solution breccias which suggest a greater previous extent of the Prairie salt.

153

which are confined to the beds younger than the salt. Significantly, these breccias have not been observed where the salt is present. They do, however, have a widespread occurrence in adjacent areas, which indicates that the Prairie salt originally had a much broader distribution, as shown on Figure 10.

Solution occurred at different geologic ages from Late Devonian to the present.

The unusual structures have all but masked any more subtle tectonic features which might have initiated the salt-solution process. Underlying reefs have also been the loci of salt removal. Both porous reef and bedrock fractures would supply the necessary circulation channels for the water required in this process.

Solution structures similar to those found in Saskatchewan no doubt occur in evaporite basins elsewhere in the world. However, they would be difficult to detect unless the regional structure were as simple, and the drilling control as good, as in southern Saskatchewan.

REFERENCES CITED

Bishop, R. A., 1953, Saskatchewan exploratory progress and problems: Alberta Soc. Petroleum Geologists News Bull., v. 1, no. 8, p. 3–6

Edie, R. W., 1959, Middle Devonian sedimentation and oil possibilities, central Saskatchewan: Am. Assoc. Petroleum Geologists Bull., v. 43, no. 5, p. 1025–1057

Grieve, R. O., 1955, Leaching (?) of Silurian salt beds in southwestern Ontario as evidenced in wells drilled for oil and gas: Canadian Mining and Metall. Bull., v. 48, p. 12–18

Manuscript Received by the Society, February 25, 1963

THE AMERICAN ASSOCIATION OF PETROLEUM GEOLOGISTS BULLETIN
V. 51, NO. 10 (OCTOBER, 1967), P. 1929-1947, 14 FIGS.

SALT SOLUTION AND SUBSIDENCE STRUCTURES, WYOMING, NORTH DAKOTA, AND MONTANA[1]

JOHN M. PARKER[2]

Denver, Colorado

ABSTRACT

Salt beds with a thickness range of a few feet to 600 ft are common in rocks of Middle Devonian and Permian ages in the northern Rocky Mountain area. Variations in thickness of these salt beds in short horizontal distances require consideration of the hypothesis that the salt-thickness changes are due to post-depositional salt removal by solution and not to original variations in the thickness of salt deposited. Thin, regionally persistent sedimentary units directly above and below the salts prove that, locally, there was little depositional change of salt thickness. Therefore, within the regions of salt deposition, the salt must have been removed after deposition in the present thin or zero salt areas.

Local removal of Middle Devonian salt (Prairie Formation) in the Williston basin occurred during later Devonian time and continued through the time of deposition of the Mississippian Lodgepole and Tilston Formations. Local removal of Permian salt (Minnelusa Formation, Opeche Shale, and Goose Egg Formation) in both the Williston and Powder River basins took place near the close of the Jurassic. Compensating deposition took place in the collapse depressions created by underlying salt removal, thus dating the age of solution.

In the examples of salt removal described in this paper, a few to several hundred feet of sediment was deposited over the salt before salt removal took place. It is postulated that salt removal was accomplished by the formation of local and regional fracture systems which allowed ascending water to escape from regional aquifers below the salt beds, and that this water dissolved the salt and carried it in solution to the ocean floor.

Oil accumulations in some places are related to the formation of local subsidence structures by salt removal. Geologists have postulated the presence of regional lineaments along and above which salt has been removed. Locally, the salt-removal areas may be irregular in size, shape, and distribution, although, as a whole, these areas may occur within a regional band.

INTRODUCTION

Salt beds with a thickness range of a few feet to 600 ft are common in rocks of Middle Devonian, Permian, and Jurassic(?) ages in the northern Rocky Mountain area. These salt beds were deposited in a variety of marine and continental conditions: (1) simple closed basins, (2) clusters of closed basins, and (3) lagoon and basin-arm complexes surrounding environments (1) and (2). Subsurface control obtained in the last 10 years shows that considerable thickness variations take place in these salt beds within short distances. These variations are not compatible with the depositional environments listed above which require (1) low-gradient-ocean- or lake-bottom slopes and (2) continuity of deposition except at the limits of the evaporite environment. Thus, these radical thickness changes in salt sections require consideration of the hypothe-

sis that they were caused by post-depositional salt removal by solution and not by variations in the thickness of salt deposited.

POWDER RIVER BASIN PERMIAN SALT SOLUTION

Figure 1 is a stratigraphic cross section and Figure 2 a structural cross section in the Dillinger Ranch oil field of Campbell County, Wyoming. These sections show a well with approximately 100 ft (30 m) less salt and associated section in the Permian Goose Egg Formation than in an offset well 2,000 ft (600 m) distant. Thin regionally persistent sedimentary units directly above, between, and below the salt beds prove that locally there was little depositional change of salt thickness. Therefore, in the present thin or zero salt areas that are within the regional salt-deposition area, the salt must have been removed after deposition.

Because of the substantial lateral changes of salt thickness due to subsequent salt removal, three structurally different sedimentary units are present: (1) the unit from the base of the salt downward to the base of the sedimentary section; (2) the unit from the top of the salt upward in the section to the base of the first bed

[1] This paper is a revision of the presidential address titled "Salt Beyond the Borehole" given in Denver, Colorado, at the Rocky Mountain Section, 16th Annual Meeting, of The American Association of Petroleum Geologists, October 24, 1966. Manuscript received and accepted, June 16, 1967.

[2] Geologist, Kirby Petroleum Co.

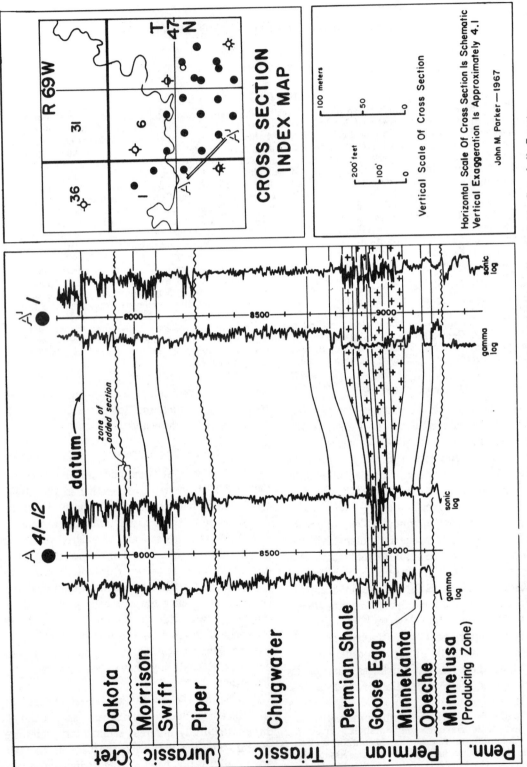

Fig. 1.—Stratigraphic cross section. Datum is top of Dakota, Dillinger Ranch field, Campbell County, Wyoming. Salt is shown in this and all subsequent cross sections by +++ symbol.

deposited after all salt removal ceased; and (3) the unit from the base of the bed that was deposited after all salt removal had occurred to the near-surface rocks. Beds above the salt that have been thickened by additional sedimentation to compensate for the removal of the salt indicate the time of salt removal.

In the Dillinger Ranch field this compensatory thickening is in the Morrison and the lower part of the Dakota Formations. For example, between the eastern and western wells in Figure 1, there is a loss of about 100 ft (30 m) of salt. The Morrison and lower Dakota section between these two wells thickens 38 ft (11.4 m), roughly half the thickness of salt removed. Two well locations north of the west well, the increase in thickness of the Morrison and lower Dakota interval (compared with the east well on the cross section) is 93 ft (27.9 m). Thus, nearly complete compensation has occurred.

In other areas, the compensatory thickening may not appear in the actual well where the salt was removed but the compensatory thickening is present in the adjacent area. In the Powder River basin, salt-removal compensation may appear only in the Swift, the Morrison, the lower Dakota, or in combinations of these three formations.

POWDER RIVER BASIN TECTONIC SEQUENCE AND MINNELUSA OIL MIGRATION AT DILLINGER RANCH FIELD

In the Dillinger Ranch field the writer assumed that the sources of the oil were the Opeche Shale and Minnekahta Limestone. The oil migrated from these source rocks and, during the time of deposition of the lower and middle Chugwater (Late Triassic), was trapped in the tops of erosional highs in the Minnelusa. These highs were preserved at the unconformity surface below the Opeche. During the time of upper Chugwater deposition differential uplift occurred and the oil in the Minnelusa topographic highs remigrated through available sandstone bodies into new highs. The Chugwater isopachous map (Fig. 3) shows that the Dillinger Ranch high had 75 ft (22.5 m) of structural relief. The upper Chugwater (approximately 400–800 ft or 120–240 m) was then planed off and Jurassic deposition occurred. During the late Jurassic, salt of the Goose Egg Formation was dissolved, causing the formation of subsidence structures.

The salt isopachous map (Fig. 3) shows the location of a postulated fracture zone through which the dissolved salt escaped. The brine ascended an approximate vertical distance of 1,085 ft (326 m) as shown on Figure 2. The isopachous map (Fig. 3) also shows a major salt thin west of the fracture zone and a small salt thin east of the fracture zone. The oil pools that had concentrated in the Late Triassic highs now moved, where sandstone continuity permitted, into new highs formed by the regional and local structural adjustments which had caused the fractures and allowed salt solution. The oil accumulations remained in the new highs until the Powder River basin was downwarped during the Laramide orogeny (Late Cretaceous), at which time the Minnelusa oil pools again moved, sandstone continuity permitting, to the updip edges of porous sandstone and dolomite lentils, or to areas of sandstone termination against Opeche channel fill. Figure 4 is a structural map in the Dillinger Ranch field, the eroded top of the Minnelusa Formation being datum.

NORTH DAKOTA DEVONIAN SALT SOLUTION

Bishop (1954) and Baillie (1955) established that solution of Prairie Formation salt had occurred in the past and that it continues in the outcrop areas.

As pointed out in the North Dakota Geological Society's 1961 publication on the Devonian System, the solution of Prairie salt and subsequent collapse occurred from Devonian (early Duperow) to Jurassic time in the United States part of the Williston basin, and from Devonian (Dawson Bay) to Recent time in Canada.

More recently, Anderson and Hunt (1964) discussed the Prairie salt solution in the Hurd area (Figs. 5–7) and the subsequent adjustment by thickening in the Lodgepole, Tilston, and Frobisher-Alida beds of Mississippian age. They also stated that, if migration of oil can occur or continue after collapse of the beds, chances for oil accumulation in these salt collapse structures are excellent.

Fig. 5 is a stratigraphic cross section (datum: top of Silurian Interlake) of the Hurd area, Bottineau County, North Dakota, showing the thinning and complete local absence of the Prairie evaporite and the position of a rubble and breccia zone that was formed as the salt was removed

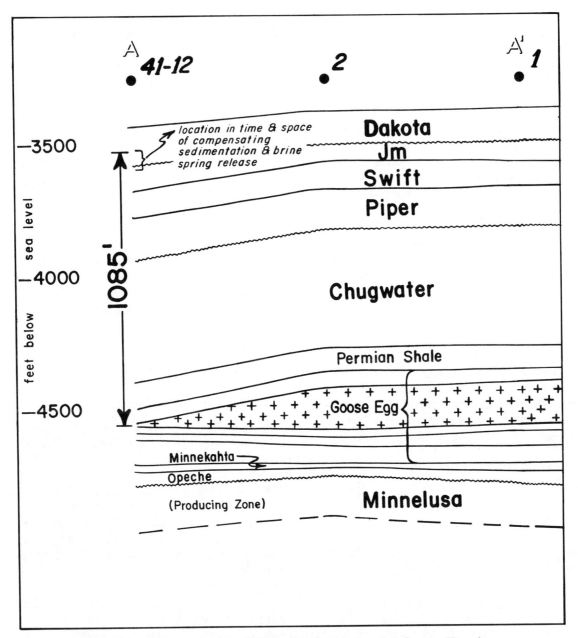

Fig. 2.—Structural cross section, Dillinger Ranch field, Campbell County, Wyoming.

and the beds above collapsed into the space which the salt had occupied. The well on the left contains 160 ft (48 m) of salt; the well on the right contains no salt. Fig. 6, a stratigraphic cross section through the same wells (datum: top of the Charles Formation), indicates that the salt removal first occurred during the time of deposition of the Tilston and Frobisher-Alida (below the Midale marker). A structural cross section of the Hurd area (Fig. 7), with two contour maps, shows the differing structural attitudes of several units caused by salt removal and compensation. The structure of the top of the Winnipegosis shows southwest dip. Strata from the Midale marker upward dip southwest. A structural contour map using the top of the Birdbear (Devonian) as datum shows a salt-collapse anticline. The brine from the Prairie evaporite solution rose approximately 2,244 ft (673 m) as shown in Figure 7.

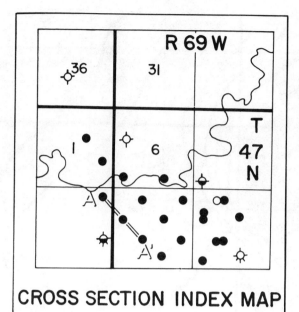

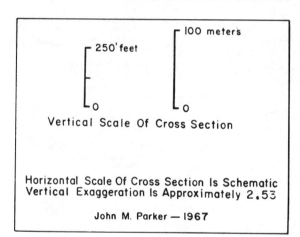

Fig. 2 (*Continued*)

North Dakota Permian and Jurassic(?) Salt Solution

Another area of salt solution and compensation, the Fryburg field area, Billings County, North Dakota, was described by Anderson (1966). He pointed out that the removal of salt by solution from the Permian in southwestern North Dakota could create collapse structures in the Jurassic and Cretaceous and petroleum reservoirs are possible in these latter rocks. He also included a map showing the known limits of the various Opeche, Goose Egg, and Jurassic salts in southwestern North Dakota.

Fig. 8 is a stratigraphic cross section in the

Fryburg oil field, datum being the top of the Amsden. This cross section shows the presence of variable thicknesses of salt in the Opeche Shale, the Goose Egg Formation, and the Piper Formation (the Dunham salt may be an upper member of the Triassic Spearfish Formation; however, it is considered herein to be a part of the Jurassic Piper Formation). Each of these salt beds may thicken or thin independent of the other, or all may thin in the same area. Fig. 9 is a stratigraphic cross section through the same three wells in the Fryburg field, datum being the top of the Dakota Formation. This figure shows the added sedimentation in the upper Swift which compensates for the removal of salt in the Dunham interval in the middle well and the Goose Egg interval in the right-hand well on the cross section.

The structural cross section (Fig. 10) shows a gentle arch at the Dakota level, northeast dip at the Piper level (which is one of the good seismic reflectors in this area), southwest dip at the Minnekahta level (which is another reflector), and northeast dip at the Amsden and lower units. With the extremely small structural relief in this area shown on the contour map (Fig. 11), it is easy to see how differential salt thinning can affect structure and thus the configuration of seismic reflections at various levels. The oil pays found to date in this field are below the salt in the Tyler and Mission Canyon Formations.

Montana Devonian Salt Solution

An excellent paper by Wilson *et al.* (1963) indicates the time and extent of salt removal from the Prairie Formation in south-central Saskatchewan and northeastern Montana. The effect of salt solution on petroleum accumulation and their interrelationship with the stratigraphy of potential petroleum reservoirs are also discussed.

The Outlook field in Montana is a good example of a salt-collapse structure and of Late Devonian migration of a pre-existing Middle Devonian oil pool. Fig. 12, a stratigraphic cross section in the field area, with the top of the Interlake Formation as datum, shows the removal of salt from the Prairie Formation and the subsequent thickening of the upper Duperow and Three Forks Formations, which compensates completely for the salt removal. Fig. 13 shows structural maps of the Outlook field with datums on top of the Silurian and the Souris River (De-

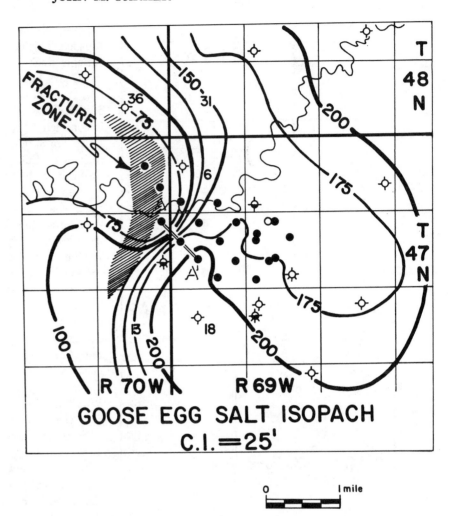

FIG. 3.—Goose Egg salt isopachous map C.I. = 25 ft (7.5 m); and Chugwater

vonian), and an isopach of the Prairie evaporite. The structural contours of the top of the Souris River reflect structure of the Duperow pay zone, because all Duperow compensation is in the upper part of that formation. The minimum thickness of the salt on the Prairie evaporite isopachous map is 50 ft (15 m). The regional or normal undissolved salt thickness in this area is 225 ft (67.5 m).

Fig. 14 is a structural cross section through the same three wells in the Outlook field (Fig. 12, A-A'). The Duperow oil pool was formed in and occupied a position above the Winnipegosis-Interlake pool before salt removal and collapse. Dur-

ing collapse this oil pool migrated southeastward to its present position. A drill-stem test and tests through perforations of oil-stained Duperow dolomites in the center well in Figure 14 recovered sulfurous salt water with small shows of oil and gas. These tests were made in the same stratigraphic zones that produce oil in the right-hand well in Figure 14. Therefore, oil and gas were present at one time in the Duperow dolomites in the position of the present Duperow depression. The above postulation of an oil field that has moved 2,000–3,000 ft (600–900 m) in Devonian time requires acceptance of the following: (1) this oil field in particular, and Williston

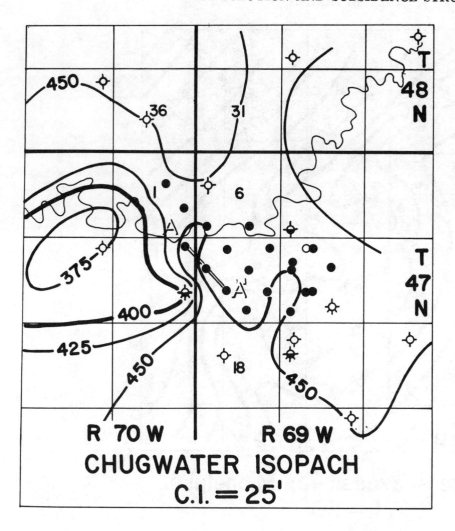

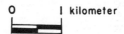

isopachous map, C.I. = 25 ft (7.5 m) ; Dillinger Ranch field, Campbell County, Wyoming.

basin oil fields in general, had a history of early folding and early oil migration into old highs; and (2) the pools in this field, as well as in the Williston basin, did not migrate long distances to their present positions since the beginning of the Laramide orogeny.

SALT SOLUTION AND SALT ESCAPE

The means of salt solution and the subsequent formation of collapse structures require two things: (1) the presence below the salt beds of an aquifer with sufficient hydrostatic head to dissolve salt and carry it upward through the sedimentary section to the sea floor; and (2) the ex-

istence of tectonic activity to cause fracturing or faulting sufficient to allow entry of water from the aquifer upward to the salt and then upward to the ocean floor. Faulting has been postulated by others as the type of tectonic activity which most probably produced channels for the ascending brines. However, the subsurface control obtained during the last 10 years in the Powder River basin shows that there is no faulting in the Minnelusa trend area; thus, fracture systems must be postulated as the escape channels for the ascending brines in that area.

From the available knowledge of Jurassic paleogeography, it is reasonable to assume that hilly

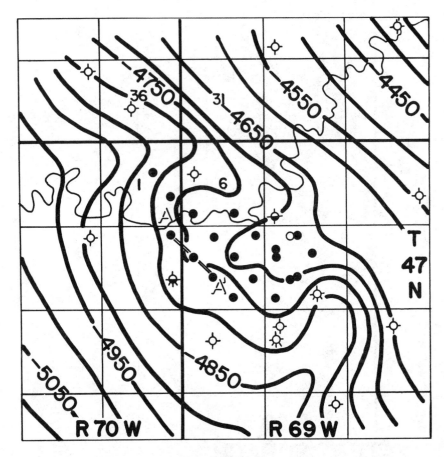

Structure — Eroded Top Minnelusa
C.I. = 50'

Fig. 4.—Structural map. Datum is eroded top of Minnelusa, C.I. = 50 ft (15 m), Dillinger Ranch field, Campbell County, Wyoming.

terrane existed during this period in the eastern Nevada-western Idaho area and that low, emergent plains were present in the Minnesota-Iowa-eastern Dakotas region. Either, or both, the eastern or western boundaries of the Jurassic basin might have provided (1) the necessary fresh-water intake outcrops of Minnelusa-equivalent rocks (or other rocks hydrodynamically communicating with the Minnelusa), and (2) the elevation differential required to produce the necessary amount of hydrostatic head. The source of the water and the hydrostatic head that existed during Devonian and Mississippian time to achieve the solution and removal of the Prairie salts are

speculative, but both a water source and a proper hydrostatic head must have existed.

RELATION OF SALT SOLUTION TO PETROLEUM ACCUMULATION

Two general types of situations exist in which salt solution may be related to petroleum accumulation. One situation is that in which salt removal and the subsequent formation of differential structure have occurred above the potential petroleum reservoir zone. In this case the relation of the present petroleum accumulation to the salt collapse structure is indirect and is tied to intervening geologic events which are them-

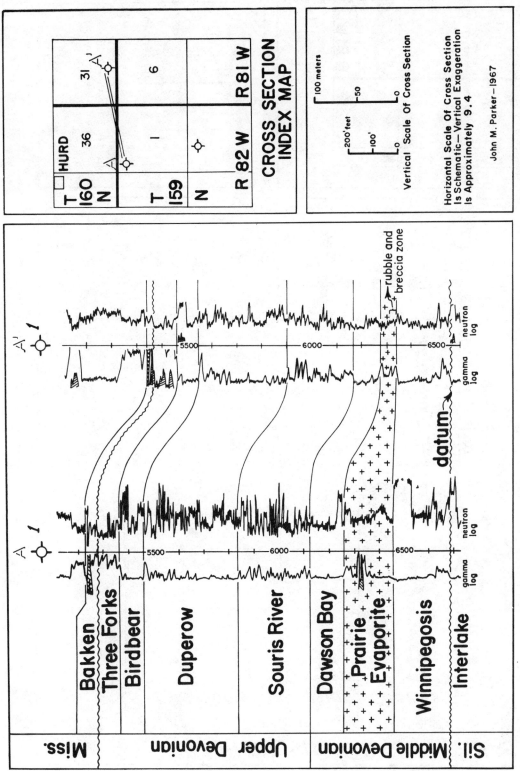

Fig. 5.—Stratigraphic cross section. Datum is top of Interlake, Hurd area, Bottineau County, North Dakota. Cross pattern = salt.

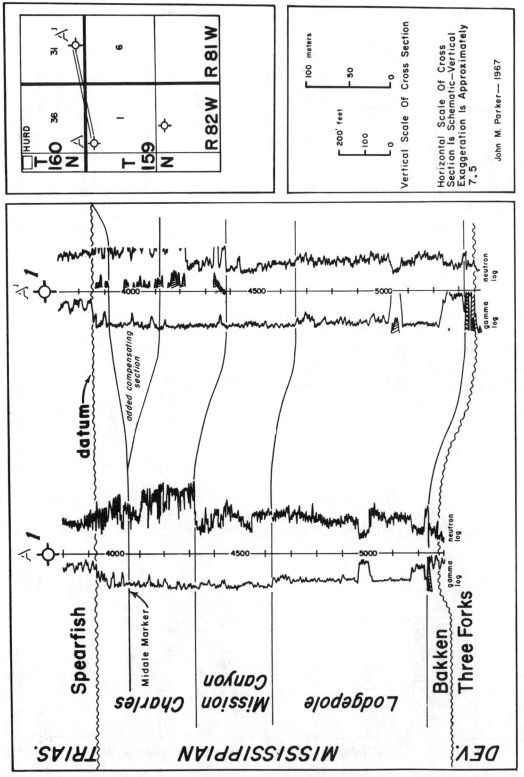

FIG. 6.—Stratigraphic cross section. Datum is top of Charles, Hurd area, Bottineau County, North Dakota.

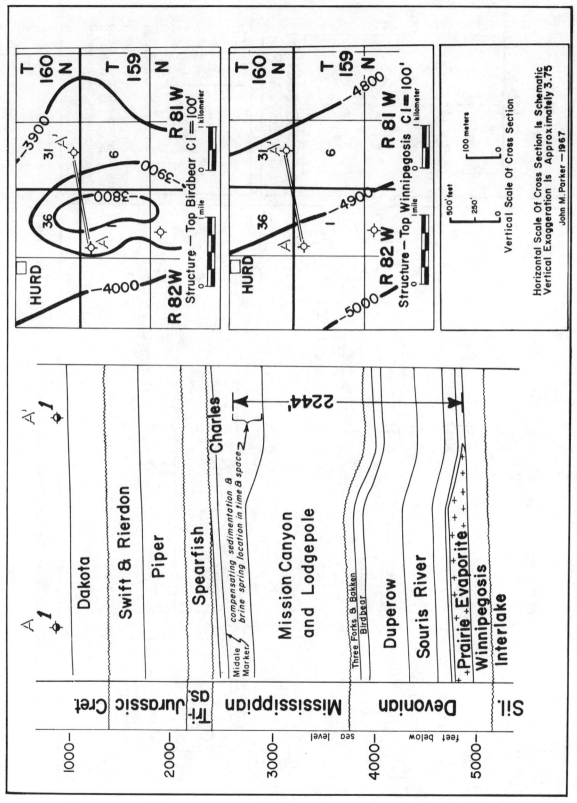

FIG. 7.—Structural cross section, and structural maps. Datum is top of Birdbear, C.I. = 100 ft (30 m); and structure map, in which datum is top of Winnipegosis, C.I. = 100 ft (30 m), Hurd area, Bottineau County, North Dakota. Cross pattern = salt.

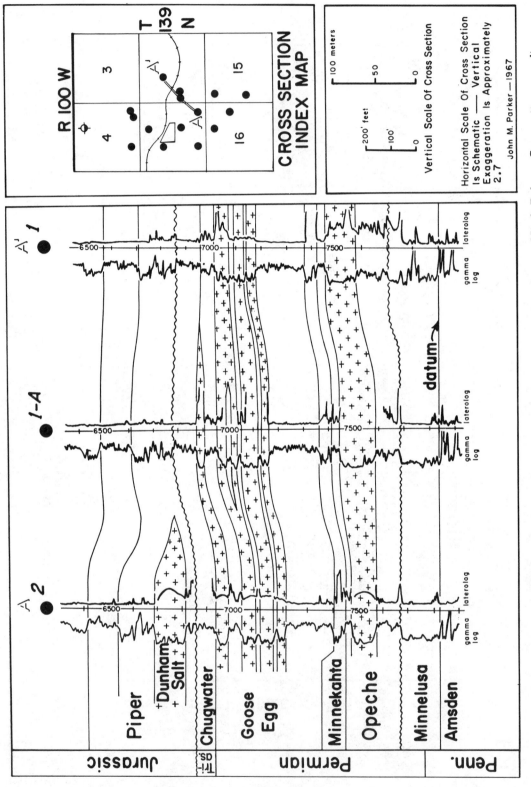

Fig. 8.—Stratigraphic cross section. Datum is top of Amsden, Fryburg field, Billings County, North Dakota. Cross pattern = salt.

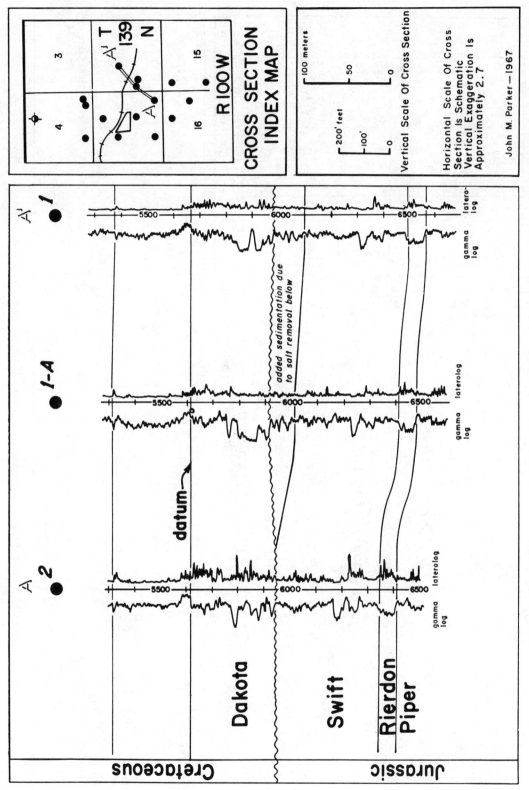

FIG. 9.—Stratigraphic cross section. Datum is top of Dakota, Fryburg field, Billings County, North Dakota.

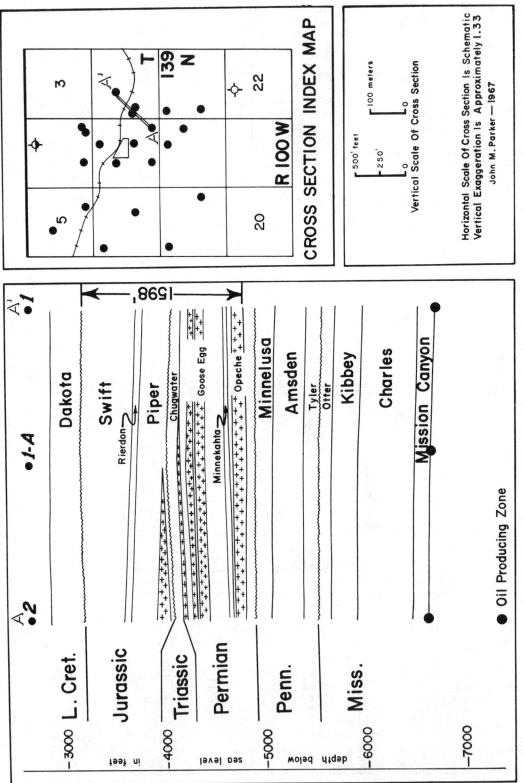

Fig. 10.—Structural cross section, Fryburg field, Billings County, North Dakota. Cross pattern = salt.

selves related to the formation of the salt-collapse structure, as shown in the Dillinger Ranch field example.

The other situation is that in which the salt removal occurred below the level of potential petroleum reservoirs and subsidence structures formed in the oil and gas reservoirs; and in this case the formation of a salt-subsidence structure directly accomplishes the migration of a petroleum accumulation into a new position on the crest of the salt-solution structure, as is shown in the Outlook field example.

The relation of the time of salt solution to the time of petroleum accumulation discussed in this

paper raises one corollary premise, *i.e.*, that oil can be formed and can migrate with a thin sedimentary cover. Specifically, in the Dillinger Ranch field example, it was stated that in the time of early to middle Chugwater deposition, migration of oil occurred in the Minnelusa Formation. This time of migration assumes that the burial depth of 1,000–2,500 ft (300–750 m), which was present during the oil-generation period in the Opeche and Minnekahta sediments and during the subsequent oil-migration period, provided sufficient overburden pressure to effect this generation and migration. This depth of burial is small compared with the depths that some geolo-

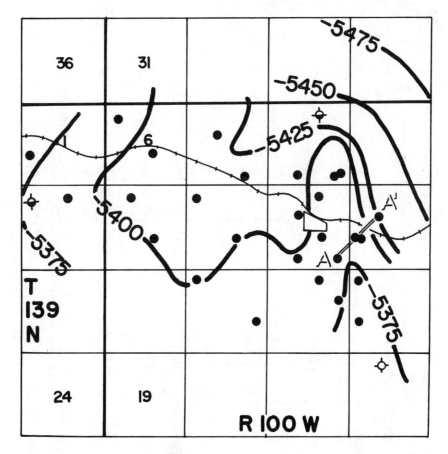

Structure – Top Lower Amsden (Tyler)
C.I. = 25'

FIG. 11.—Structural map. Datum is top of lower Amsden (Tyler), C.I. = 25 ft (7.5 m), Fryburg field Billings County, North Dakota.

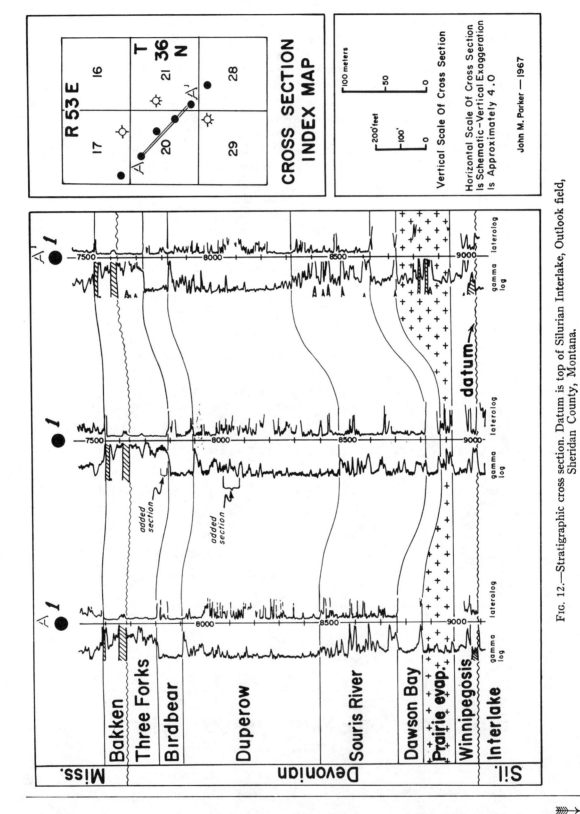

CROSS SECTION INDEX MAP

Horizontal Scale Of Cross Section
Is Schematic–Vertical Exaggeration
Is Approximately 4.0

Vertical Scale Of Cross Section

John M. Parker — 1967

Fɪɢ. 12.—Stratigraphic cross section. Datum is top of Silurian Interlake, Outlook field, Sheridan County, Montana.

Fɪɢ. 13.—Structural map: datum is top of Silurian; C.I. = 50 ft (15 m). Isopachous map: Prairie evaporite; C.I. = 25 ft (7.5 m). Structural map: datum is top of Souris River; C.I. = 50 ft (15 m); Outlook field, Sheridan County, Montana.

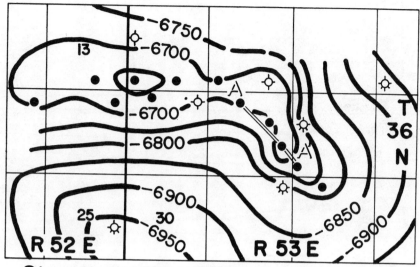

Structure—Top Silurian, C.I.=50'

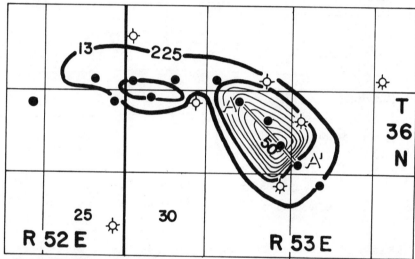

Isopach—Prairie Evaporite, C.I.=25'

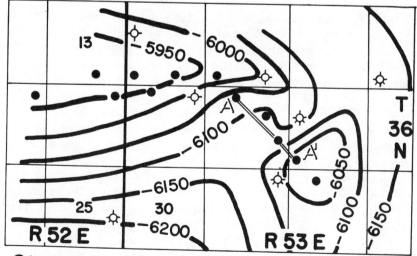

Structure—Top Souris River, C.I.=50'

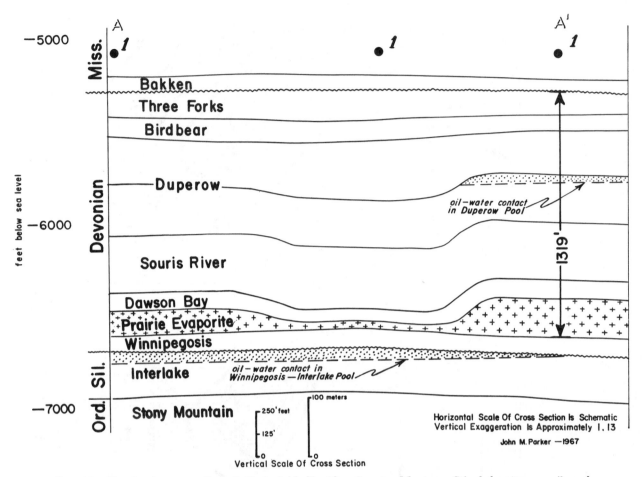

Fig. 14.—Structural cross section, Outlook field, Sheridan County, Montana. Stippled pattern = oil pool.

gists have indicated to be necessary for transformation of organic matter into petroleum and subsequent migration. However, the time span between the end of deposition of the Minnekahta and the deposition of the middle Chugwater is approximately 10–30 million years. Therefore, ample time transpired for the necessary temperature, pressure, and chemical conversions that are required to change oil-source materials to oil.

CONCLUSIONS

Looking for oil accumulations associated with salt solution and subsidence structures in low-dip areas requires precise stratigraphic correlations and detailed mapping. All of the structural contour and isopachous maps presented in this paper have contour intervals of 25 ft (7.5 m), 50 ft (15 m), or 100 ft (30 m). In this large area, the correct mapping of local structure with datums on the tops of prospective pay zones in areas

where salt removal has occurred requires the ability to look beyond the presence or absence of salt in a borehole. Geologists have postulated the presence of regional lineaments along and above which salt has been removed. In actuality, the local salt-removal areas may be irregular in size, shape, and distribution, although as a group they may occur in a regional band. In areas where salt removal has occurred, seismic and subsurface interpretation must include specific postulates regarding possible salt-thickness changes. These postulates can be derived by careful mapping of the thickening and thinning of the post-salt compensating units.

Oil fields, other than those discussed here, that have differential structure resulting from salt solution and removal include (1) numerous fields in the Powder River basin, (2) the Repeat field in Carter County, Montana, (3) the Birdbear fields in Roosevelt County, Montana, (4) the Flat

Lake field in Sheridan County, Montana, and adjoining parts of North Dakota and Saskatchewan, (5) the Hummingbird field of south-central Saskatchewan, and (6) the Medora-Scoria fields of Billings County, North Dakota. Still others remain to be found.

SELECTED BIBLIOGRAPHY

Anderson, S. B., 1966, A look at the petroleum potential of southwestern North Dakota: N. Dak. Geol. Survey Rept. Inv. 42, 3 sheets.
———— and John B. Hunt, 1964, Devonian salt solution in north central North Dakota, *in* Third International Williston basin symposium: Billings Geol. Soc., N. Dak. Geol. Soc., and Sask. Geol. Soc., p. 93–118.
Baillie, Andrew D., 1955, Devonian System of Williston basin: Am. Assoc. Petroleum Geologists Bull., v. 39, p. 575–629.

Bishop, Robert A., 1954, Saskatchewan exploratory progress and problems, *in* Western Canada sedimentary basin: Am. Assoc. Petroleum Geologists, p. 474–485.
de Mille, G., J. R. Shouldice, and H. W. Nelson, 1964, Collapse structures related to evaporites of the Prairie Formation, Saskatchewan: Geol. Soc. America Bull., v. 75, p. 307–316.
North Dakota Geological Society, 1961, Stratigraphy of the Williston basin–Devonian System: N. Dak. Geol. Soc., 47 p.
Roberts, Albert E., 1966, Stratigraphy of Madison Group near Livingston, Montana, and discussion of karst and solution-breccia features: U.S. Geol. Survey Prof. Paper 526-B, 23 p.
Stanton, Robert J., Jr., 1966, The solution brecciation process: Geol. Soc. America Bull., v. 77, p. 843–848.
Willson, W., D. L. Surjik, and H. B. Sawatzky, 1963, Hydrocarbon potential of the south Regina area, Saskatchewan: Dept. Min. Resources, Prov. Saskatchewan, Rept. 76.

Jour. Research U.S. Geol. Survey
Vol. 4, No. 4, July–Aug. 1976, p. 379–386

THIN-SKINNED TECTONICS AND POTENTIAL HYDROCARBON TRAPS—ILLUSTRATED BY A SEISMIC PROFILE IN THE VALLEY AND RIDGE PROVINCE OF TENNESSEE

By LEONARD D. HARRIS, Knoxville, Tenn.

Abstract.—Seismic data, although limited to a small part of the western half of the Valley and Ridge province in Tennessee, confirm that the structural style is thin-skinned and that there is a fundamental change from west to east in both the total section preserved and the structural complexities that exist in the subsurface. Cambrian to Upper Ordovician rocks in the Valley and Ridge province have been thrust westward over a 4-mile (6.4-km) projection of Cambrian through Mississippian rocks of the Appalachian Plateau. Structure within the plateau projection is characterized by splay anticlines, whereas structure in the Valley and Ridge is dominated by a series of imbricate thrust sheets containing isolated fault-bound masses of Cambrian sandstone and shale. Both splay anticlines and large fault-bound masses of rocks preserved in the subsurface in east Tennessee may be favorable fracture-porosity traps for the accumulation of hydrocarbons.

Field studies of the Pine Mountain thrust sheet in Kentucky, Virginia, and Tennessee, from 1920 to the late 1940's, led to the concept that thin-skinned tectonics was the dominant style of deformation in the complexly thrust-faulted southern Valley and Ridge province (Wentworth, 1921; Butts, 1927; Rich, 1934; Rodgers, 1949; Miller and Fuller, 1954). Advocates of this concept suggested that the Paleozoic sedimentary prism had been stripped from basement and moved many miles northwestward by a series of great thrust faults. These faults, taking advantage of lithologic contrasts in rocks, formed as nearly flat-lying bedding-plane thrusts in incompetent rocks and ramped steeply upward across competent rocks. Movement of thrust sheets up and over ramp zones produced major rootless folds by simple duplication of beds. Subsurface data from oil and gas test holes in and adjacent to the southern Valley and Ridge (Harris, 1970; Miller, 1973) have partly confirmed the general characteristics of thin-skinned tectonics as outlined by the earlier workers. A more positive confirmation was realized in 1974 when Geophysical Services, Inc., released to the public domain a 20-mi (32-km) segment of a migrated vibroseis seismic profile in the Valley and Ridge between the

towns of Kingston and Dixie Lee Junction, Tenn. (fig. 1). This important profile clearly demonstrates for the first time in the southern Valley and Ridge the characteristic "sledrunner" thrust-fault signature of thin-skinned tectonics (fig. 2). Major reflectors in the Rome Formation (Lower Cambrian) can be seen to dip steeply southeastward from the outcrop and to decrease gradually in dip in the subsurface to where they merge with a nearly flat-lying décollement zone.

My personal experience with seismic interpretation is limited, but because of my familiarity with the details of the geology of the area, I thought that it might be of interest to illustrate a geologist's point of view in an interpretation of a seismic profile in the southern Valley and Ridge. Because several geophysical variables were not considered in my interpretation, the structure as depicted should not be accepted in detail. Instead, the structural interpretation presented herein is intended to serve only as a model that perhaps furnishes insight into the expectable general subsurface condition in the southern Valley and Ridge province.

INTERPRETATION

At the surface, the seismic profile released by Geophysical Services, Inc., transects a stratigraphic sequence limited to three contrasting lithologic units of Cambrian to Upper Ordovician age (figs. 1 and 3). Vertically, the sequence includes, from the base up: the Rome Formation (a Lower Cambrian sandstone and shale unit), the Conasauga Shale (a Middle and Upper Cambrian shale and a thin limestone unit), and the Knox Group and Chickamauga Limestone, undivided (an Upper Cambrian to Upper Ordovician carbonate unit). All or parts of this sequence are repeated five times on the surface because of thrust faulting. A careful analysis of the repeated sections indicates that each contrasting lithologic unit of Cam-

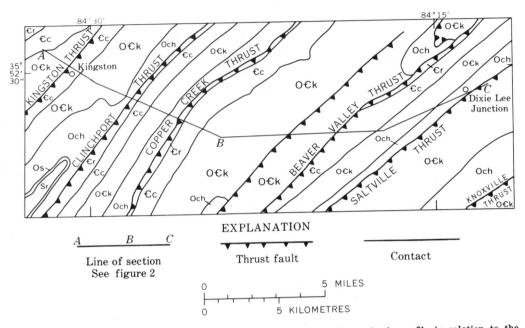

EXPLANATION

A   B   C

Line of section
See figure 2

▼▼▼▼
Thrust fault

Contact

0   5 MILES

0   5 KILOMETRES

FIGURE 1.—Index map showing the trend of the Geophysical Services, Inc., seismic profile in relation to the general structural grain of the Valley and Ridge province in part of east Tennessee. Segment *A–B* is normal to the structural grain; segment *B–C* trends obliquely to the grain; €r, Rome Formation; €c, Conasauga Shale; O€k, Knox Group; Och, Chickamauga Limestone; Os, Sequatchie Formation; Sr, Rockwood Formation. Outline of index map shown by dashed line on the generalized geologic map (fig. 4).

brian to Upper Ordovician age consistently produces a distinct recognizable seismic pattern. In addition, two other distinct seismic patterns were found to be restricted to the subsurface. The five major seismic signatures recognized include (fig. 3):

1. A basement pattern, which is characterized by a lack of through-going readily distinguishable reflector horizons.
2. Rome Formation pattern, which is characterized by a series of strong, nearly continuous heavy lines.
3. Conasauga Shale pattern, which tends to resemble a series of discontinuous ripple marks.
4. Knox Group and Chickamauga Limestone sequence undivided pattern, which resembles a discontinuous standard sandstone symbol.
5. An Upper Ordovician through Mississippian pattern, which resembles a coarse sandstone symbol arranged in a series of definite layers.

In utilizing the five patterns to interpret structures in the seismic profile, it became apparent that the major Cambrian through Upper Ordovician units selected appeared to increase in thickness toward the east (fig. 2). Surface measurements show that the thickness of this sequence ranges from about 7,000 ft (2,100 m) on the west to about 8,600 ft (2,580 m) on the east. In

addition, an apparent thickening occurs in the eastern two-thirds of the profile, segment *B–C*, because the trend of that part of the seismic line is oblique to the structural grain of the Valley and Ridge (fig. 1). Although superficially, seismic profiles oriented at an angle to the structural grain may seem to be of minimum value, this is not necessarily true. Instead, this orientation has the added advantage of allowing the geologist to view different perspectives of the subsurface in a single seismic profile. Segment *A–B* of the profile illustrates a normal or near normal cross-sectional view; segment *B–C* shows the longitudinal characteristics of thrust sheets and demonstrates their lateral continuity (fig. 2).

In the past, because of limited subsurface control, most structural interpretations in the Valley and Ridge of east Tennesee have been based on the extrapolation of surface features into the subsurface (summarized by Rodgers, 1970). The structural style shown in these sections is based on the characteristics commonly attributed to thin-skinned tectonics. Thus, deformation is interpreted to be confined to the sedimentary sequence above basement and to consist of a series of major thrust faults that join a master décollement at or near the sedimentary rock-basement inter-

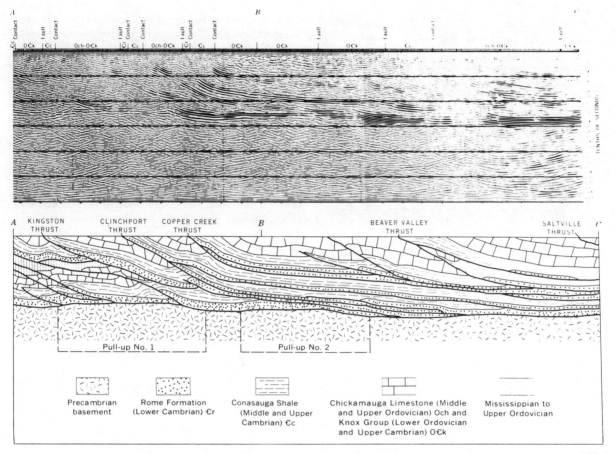

FIGURE 2.—Seismic profile (upper) and interpretive structure section (lower). Rectangular areas designated by dashed line and labeled "pull-up No. 1" and "pull-up No. 2" are probably acoustically produced warps in the basement surface related to velocity contrasts. The location of line *A–B–C* is shown in **figure 1.**

face. The geometry and space relation shown in the seismic profile and my interpreted structure section (fig. 2) confirm that the style of deformation in east Tennessee is thin-skinned. In addition, these same data point out that because of the lack of subsurface control, most previous interpretations are simply generalizations of the complexities that exist in the subsurface. The availability of subsurface control affords the opportunity to scrutinize rigorously some of the basic assumptions of thin-skinned tectonics. Accordingly, three of the major characteristics commonly attributed to thin-skinned deformation are considered in the following:

1. Deformation confined to the sedimentary sequence and little or no basement involvement

The interpreted contact between basement and the overlying sedimentary sequence in the seismic profile

appears to be warped into a series of broad folds (fig. 2). Because the seismic profile is a time section, these folds, which seem to involve basement, are probably acoustically produced pull-ups related to the contrasts in relative velocities of energy impulses in different rock units. Estimated velocities for the Cambrian to Upper Ordovician rock units range from about 14,000 ft (4,200 m) per second for the Conasauga Shale, about 17,000 ft (5,100 m) per second for the Rome Formation, to about 20,000 ft (6,000 m) per second for the carbonates of the Knox Group and Middle Ordovician. Consequently, the profile tends to expand vertically over areas in the subsurface where structural conditions have overthickened units of lower velocity, simply because in those areas a greater time interval is necessary for energy to travel down and back to the surface. Conversely, the vertical section tends to thin over areas where the higher velocity carbonate unit is

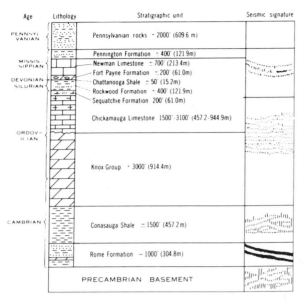

FIGURE 3.—Generalized stratigraphic section in the area traversed by the seismic profile.

structurally overthickened, because in those areas a shorter time interval is required for energy to travel down and back to the surface. As a result of major subsurface duplication, the contact between basement and the overlying sedimentary sequence appears to be warped into two broad pull-ups. Pull-up No. 1, in the western fourth of the profile, occurs where there is major duplication, both on the surface and in the subsurface, of high-velocity carbonate. This pull-up is accentuated by tectonic thickening of the lower velocity Rome Formation on either limb. Pull-up No. 2, a smaller feature near the center of the profile, appears to be related mainly to the pile-up of lower velocity Rome on both limbs and minor duplication of carbonate in the upper part of the profile.

Other than pull-ups, which are adequately explained as velocity contrasts accentuated by tectonic overthickening of individual lithologies, no structural features present in the sedimentary sequence are observed to affect basement. Apparently basement within this profile was not involved during deformation.

2. Presence of a master décollement

The seismic profile and the interpreted structure section show that the major reflector (the Rome Formation) dips southeast from the surface and joins a master décollement zone at or near the sedimentary rock-basement interface. The seismic character of the Rome

Formation immediately above the basal décollement suggests that the Rome is progressively overthickened tectonically from west to east. In the western half of the profile, the Rome forms a nearly continuous unit within which overthickening is a local phenomenon accomplished by splay thrusting. In contrast, the Rome interval in the middle and eastern part of the profile is a series of duplicated slabs. Apparently the development of local obstructions within the décollement zone during movement resulted in a minor deflection of the thrust fault upward to the top of the Rome, where the fault again becomes a bedding-plane thrust in the basal part of the Conasauga Shale. Continuing movement resulted in subsurface duplication and abandonment of large masses of the Rome and some Conasauga Shale. This same process appears to have been operative on most major thrust faults, even above the décollment.

Subsurface stacking and abandonment of large masses of the Rome Formation have had a profound effect on the distribution and thickness of the Rome on the surface in Tennessee. In the Valley and Ridge, displacement along most major thrust faults is scissorlike: it increases along strike toward the southwest. This increase is accompanied by a gradual decrease in the outcrop width of the Rome directly above major thrust faults (fig. 4). For example, the Rome is present above the Beaver Valley thrust, just north of Knoxville where the displacement begins. As displacement on the thrust increases toward the southwest, surface exposure of the Rome gradually tectonically thins to extinction. The same distribution pattern of the Rome can be seen from northwest to southeast normal to the structural grain of the Valley and Ridge. In general, the westernmost thrust sheets have the most continuous and widest outcrop pattern for the Rome. The outcrop pattern thins progressively eastward so that in the central and easternmost thrust sheets, the outcropping Rome is discontinuous or completely missing. This west-to-east tectonic thinning of the Rome tends to confirm Miller (1973) and Milici's (1975) suggestion that the orogenic cycle in the Valley and Ridge began with the formation of a single master plate which later imbricated from west to east. Milici has based his interpreted west-to-east sequential development of thrusting on the fact that throughout the southern Appalachians, many major thrust faults have been overridden and buried from the east by the next succeeding thrust fault. This type of structural relationship is evident in the area shown in the northeastern quarter of figure 4 where the Wallen Valley thrust has been overridden and buried from the east by the Clinchport thrust. For this kind of burial process to be operative

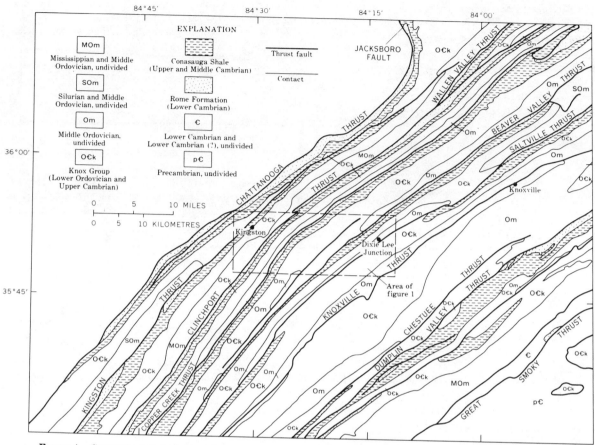

FIGURE 4.—Generalized geologic map showing the distribution of the Rome Formation and the Conasauga Shale.

throughout the southern Appalachians, movement must have ceased on all overridden thrust faults in a west-to-east sequence. Because movement ceased on the westernmost faults early in the sequential development of structure, these thrust faults have the least displacement and the greatest amount of preserved Rome. In contrast, the central and easternmost thrust faults, because of the accumulative effect of displacement on each succeeding thrust fault, have the greater displacement and the least amount of preserved Rome.

3. Formation of thrust faults

Regional structural analysis (Milici, 1975) and the relationship between the distribution of the Rome Formation and the amount of accumulative displacement on thrust faults, discussed earlier in this report, suggest that the orogenic cycle in the Valley and Ridge of east Tennessee began with the regional formation of a single master plate moving above a basal décollement (fig. 5). Apparently, minor splay thrusting distorted

the décollement zone near the west edge of the Valley and Ridge, where perhaps low confining pressures associated with relatively shallow depth of burial resulted in the deforming forces exceeding the shearing strength of the rocks. Continuing development of these subsurface obstructions caused the master plate to begin imbricating from west to east. Movement along the first imbricate thrust fault produced a large rootless anticline simply by duplication of beds. The geometry and space relations of this first imbricate thrust and its associated fold are identical with relationships predicted by earlier workers (Rodgers, 1949) for thrust faults that form in relatively undeformed rocks early in the orogenic cycle. In contrast, the geometry and space relations of the later developing imbricate thrusts are strongly influenced by the fact that the faults originate in deformed rocks at points of maximum stress where space problems exist. Space problems (deSitter, 1956; Gwinn, 1964) are created in the moving plate by drag where the edges of formations abut against the thrust surface, thereby tightening the developing rootless fold, and in areas where the moving plate is warped over the diagonal ramp zone in the autochthonous plate. Once the imbrication process is set in motion, the same factors that contribute to space problems tend to be repeated, so that as each space

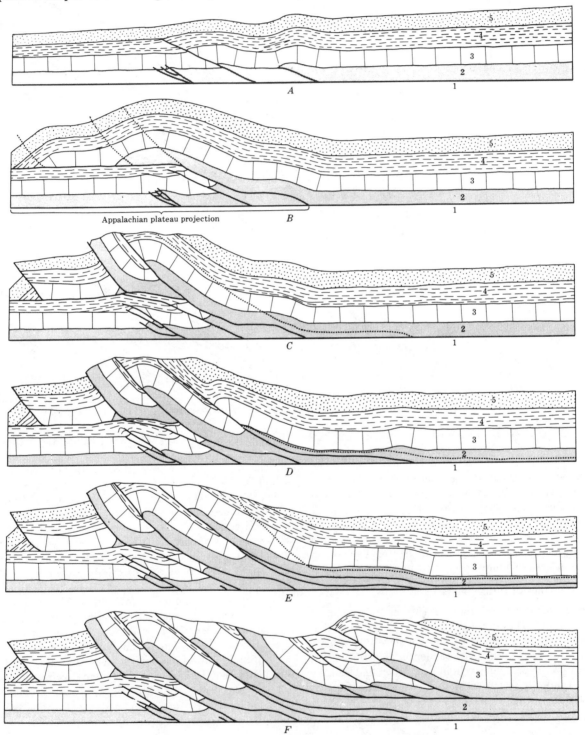

problem is resolved by thrusting, the process of imbrication shifts from west to east.

POSSIBLE HYDROCARBON TRAPS IN THE SUBSURFACE OF EAST TENNESSEE

Factors suggesting that the Valley and Ridge province of east Tennessee should be considered as a prime target for oil and gas exploration include the following: (1) the commercial quantities of oil that have been produced from the Trenton Limestone along the west edge of the Valley and Ridge a few miles north of the Tennessee State line at Rose Hill, Va. (Miller, 1973), (2) shows of oil and gas in shallow water wells in the area, (3) oil found in vugs and in fracture porosity in roadcut exposures and isolated outcrops of the lower part of the Knox Group and in parts of the Middle and Upper Ordovician limestone sequence, (4) a thick stratigraphic section in the Valley and Ridge that shows lateral facies changes, (5) the widespread occurrence of a thick shale sequence (Conasauga Shale) that under the proper structural conditions may act as a seal, and (6) an abundance of fracture porosity.

Although many of the above factors have been evident for years, the Valley and Ridge of east Tennessee remains as one of the last almost completely untested large areas in the Eastern United States. This lack of exploration is apparently based on a rigid adherence to the carbon ratio theory, the fact that the area is a little-understood complex structural province, and, in the past, the widespread exploitation of shallow production from structurally uncomplicated reservoirs in the adjacent Appalachian Plateau. Now that development of reservoirs adjacent to the Valley and Ridge has reached a mature stage, future exploration efforts should be directed eastward toward untested areas, like

FIGURE 5.—Sequential development of structure. Lithologic units are: (1) basement, (2) Rome Formation and Conasauga Shale, (3) Knox Group and Chickamauga Limestone, (4) Upper Ordovician through Mississippian, and (5) Pennsylvanian. A, Orogenic cycle initiated with the regional development of a basal décollement. Minor local obstructions within the décollement zone cause the master plate to begin to imbricate. B, Continuing growth of subsurface obstruction results in the formation of the first major imbricate thrust fault. Movement of the overriding plate thrusts Valley and Ridge rocks over a projection of the Appalachian Plateau and produces a large rootless anticline by duplication of beds. Dotted lines in section B through E mark the general position where later imbricate thrust faults form. C, Continuing movement and tightening of the rootless anticline create space problems at points of maximum stress. Formation of major or minor imbricate thrust faults resolves' the space problems. D, E, and F, Once imbrication begins, space problems continue to develop and are resolved by a west-to-east progression of imbrication.

the Valley and Ridge, especially if the reasons for past rejection are mainly demonstrably invalid.

The carbon ratio theory, which suggests that a relationship exists between the percent of fixed carbon in coals and the distribution and abundance of oil, evolved in the early 1900's (White, 1915). White's concept suggested that few or no commercial quantities of oil would be found in areas where the fixed carbon in coal exceeded 70 percent. Fixed carbon ratios in Virginia, West Virginia, and Pennsylvania were found to increase progressively eastward, and the 70-percent isocarb nearly coincides with the western edge of the Valley and Ridge. Although coal beds do not occur in the Valley and Ridge from Pennsylvania to Alabama, the 70-percent isocarb occurring near the western edge of the Valley and Ridge in West Virginia and Pennsylvania was extrapolated southward to rule out the entire province as a target for oil exploration.

Recent studies (Epstein, Epstein, and Harris, 1974) have shown that color changes in conodonts tend to follow trends on the fixed carbon of coal beds. Thus, in areas where coal beds do not occur, color changes of conodonts can be used to measure the percent of fixed carbon. Preliminary data from east Tennessee indicate that the fixed carbon percent tends to decrease westward from about 70 percent at the Saltville fault to less than 60 percent in Plateau rocks west of the Valley and Ridge, suggesting the possibility that commercial quantities of both oil and gas may occur in parts of the Valley and Ridge of Tennessee.

The fact that the Valley and Ridge of east Tennessee is a structurally complex province, within which thrust faulting and varying degrees of folding are associated, has in the past and continues in the present to deter petroleum exploration. Although restricted to a small part of the west half of the Valley and Ridge in Tennessee, the recently released seismic data furnish fresh insight into the kinds of subsurface structural traps present and their possible distribution. Analysis of the seismic data focuses attention on a fundamental change from west to east both in the complexities of subsurface structure and the total stratigraphic section involved (fig. 2). East of the Clinchport fault, the total stratigraphic interval is limited to a sequence that includes only Cambrian to Upper Ordovician. Subsurface structure is characterized by an abundance of large fault-bound masses of Rome Formation and Conasauga Shale and by minor splay thrusting. Data concerning the lateral limits of fault-bound masses are not definitive. However, segment B–C of the profile was shot at an angle to strike and individual masses are seen to persist; these facts suggest that at least some fault-bound masses are plates having considerable lateral extent. Where sealed by the Conasauga Shale, fault-bound plates appear to be structurally situated so that fracture-porosity zones within the Rome part of these plates may have acted as traps for hydrocarbons migrating updip.

West of the Clinchport fault, the subsurface includes a sequence of Cambrian through Mississippian strata buried beneath a thrust sheet containing Cambrian to Upper Orlovician rocks. Reconstruction of the development of structure in the area (fig. 5) indicates that the Cambrian through Mississippian sequence is a subsurface projection of the Appalachian Plateau buried beneath the Chattanooga and associated faults of the Valley and Ridge. Because the Chattanooga fault is a continuous structural element from the Jacksboro fault to the southern border of Tennessee, it seems reasonable to suggest that a Cambrian through Mississippian projection of the Appalachian Plateau may be present in the subsurface beneath the entire length of the Chattanooga thrust sheet and perhaps beneath part of the associated Kingston thrust sheet. Thus, a nearly complete buried Paleozoic section about 4 mi (6.5 km) wide and 100 mi (160 km) long may exist in the subsurface at the edge of the Valley and Ridge. As delineated by the seismic data (fig. 2), subsurface structure within the Appalachian Plateau projection is dominated by the occurrence of a large splay anticline produced by imbricate thrusting. The indication that splay anticlines occur in the subsurface is important for the following reasons: (1) splay anticlines on a worldwide basis have been found to be prime fracture-porosity traps, (2) this structure affords a favorable setting in which to test an almost complete Paleozoic section, and (3) similar kinds of structures associated with thin-skinned tectonics have produced quantities of gas, both along the west edge of the Valley and Ridge and in the adjacent plateau in Pennsylvania, Maryland, and West Virginia (Gwinn, 1964; Jacobeen and Kanes, 1974). There seems to be little doubt that the Appalachian Plateau projection beneath the Valley and Ridge of east Tennessee deserves careful consideration in any future oil and gas exploration program.

REFERENCES CITED

Butts, Charles, 1927, Fensters in the Cumberland overthrust block in southwestern Virginia: Virginia Geol. Survey Bull. 28, 12 p.

deSitter, L. U., 1956, Structural geology: New York, McGraw-Hill Book Co., 552 p.

Epstein, A. G., Epstein, J. B., and Harris, L. D., 1974, Incipient metamorphism, structural anomalies, and oil and gas potential in the Appalachian Basin determined from conodont color: Geol. Soc. America Abs. with Programs, v. 6, no. 7, p. 723–724.

Gwinn, V. E., 1964, Thin-skinned tectonics in the Plateau and northwestern Valley and Ridge provinces of the Central Appalachians: Geol. Soc. America Bull., v. 75, no. 9, p. 863–900.

Harris, L. D., 1970, Details of thin-skinned tectonics in parts of the Valley and Ridge and Cumberland Plateau provinces of the Southern Appalachians, chap. 10 in Fisher, G. W., Pettijohn, F. J., Reed, J. C., Jr., and Weaver, K. N., eds., Studies of Appalachian geology—central and southern: New York, Interscience Publishers, p. 161–173.

Jacobeen, Frank, Jr., and Kanes, W. H., 1974, Structure of Broadtop Synclinorium and its implications for Appalachian structural style: Am. Assoc. Petroleum Geologists Bull., v. 58, no. 3, p. 362–375.

Milici, R. C., 1975, Structural patterns in the Southern Appalachians—evidence for a gravity slide mechanism for Alleghenian deformation: Geol. Soc. American Bull., v. 86, no. 9, p. 1316–1320.

Miller, R. L., 1973, Structural setting of hydrocarbon accumulations in folded southern Appalachians: Am. Assoc. Petroleum Geologists Bull., v. 57, no. 12, p. 2419–2427.

Miller, R. L., and Fuller, J. O., 1954, Geology and oil resources of the Rose Hill district—the fenster area of the Cumberland overthrust block—Lee County, Virginia: Virginia Geol. Survey Bull. 71, 383 p.

Rich, J. L., 1934, Mechanics of low-angle overthrust faulting as illustrated by Cumberland thrust block, Virginia, Kentucky, and Tennessee: Am. Assoc. Petroleum Geologists Bull., v. 18, no. 12, p. 1584–1596.

Rodgers, John, 1949, Evolution of thought on structure of middle and southern Appalachians: Am. Assoc. Petroleum Geologists Bull., v. 33, no. 10, p. 1643–1654.

—— 1970, The tectonics of the Appalachians: New York, Interscience Publishers, 271 p.

Wentworth, C. K., 1921, Russell Fork fault, in The geology and coal resources of Dickenson County, Virginia: Virginia Geol. Survey Bull. 21, p. 53–67.

White, C. D., 1915, Some relations in origin between coal and petroleum: Washington Acad. Sci. Jour., v. 5, p. 189–212.

THE AMERICAN ASSOCIATION OF PETROLEUM GEOLOGISTS BULLETIN
V. 51. NO. 10 (OCTOBER, 1967), P. 2056-2114, 30 FIGS., 11 TABLES

THEORY OF PALEOZOIC OIL AND GAS ACCUMULATION IN BIG HORN BASIN, WYOMING[1]

D. S. STONE[2]

Denver, Colorado

ABSTRACT

A theory of oil and gas accumulation has been developed to account for the relations among stratigraphy, structure, and fluid distribution in the oil fields that produce from Paleozoic reservoirs in the Big Horn basin of Wyoming. This proposed theory relates the common oil-water contacts observed in the normal multi-zoned Paleozoic anticlinal fields to height of the oil column, formational thicknesses in the crestal area, and number of Paleozoic formations productive of hydrocarbons in each field. The similar chemical composition of the Paleozoic crude oils and of the associated formation waters, the vertical density stratification of fluids in the multi-zoned fields with large oil columns, and some unusual reservoir-pressure relations are cited in support of the concept of a "common-pool state."

The major conclusion of this study is that essentially all of the hydrocarbons in Paleozoic and Triassic reservoir rocks in the Big Horn basin were generated from the euxinic, dark-colored, organic-rich and phosphatic, fine-grained sediments of the marine facies of the Permian Phosphoria Formation. Primary migration probably was completed by Early Jurassic time when these hydrocarbons accumulated within regional stratigraphic traps created primarily by (1) updip facies change, pinchout, and truncation of the reservoir carbonates of the Phosphoria Formation, but also by (2) uneven Phosphoria-Goose Egg truncation of the underlying Tensleep Sandstone, both locally within the Big Horn basin and farther east beyond the area covered by marine Phosphoria rocks.

The hydrocarbons in these Phosphoria and Tensleep stratigraphic traps were later released as a consequence of fracturing and faulting associated with Laramide folding, and migrated into older Paleozoic reservoir rocks until fully adjusted to anticlinal structure in common pools. Vertical segregation of an original common pool into several separate pools was accomplished in some exceptional fields by (1) selective hydrodynamic tilting within the Tensleep zone, (2) leakage or redistribution of fluids through fault zones, or (3) escape of hydrocarbons to the surface and inspissation resulting from breaching of the original Triassic cap rocks.

INTRODUCTION

The Big Horn basin is in northwestern Wyoming and south-central Montana (Fig. 1). It is a topographic and structural basin surrounded by

[1] Manuscript received, April 21, 1966; accepted, August 25, 1966.

[2] Chevron Oil Co. Special thanks are due E. G. Bowman, Chief Geologist, Chevron Oil Co., Denver, for critically reading the early manuscript and for his interest and encouragement. The writer is indebted also to Roger Hoeger and Gilman Hill of Petroleum Research Corp. for reviewing the sections on formation-water chemistry and hydrodynamic modifications, and to S. R. Silverman of Chevron Research Co. for reviewing the sections on crude oil chemistry and alteration. Discussions with Donald I. Foster, James G. Crawford of Chemical & Geological Laboratories, Stanley McEnroy of Pan American Petroleum Corp., W. J. Wenger of U.S. Bureau of Mines in Laramie, Wyo., and members of the Denver Study Group are gratefully acknowledged. The critical comments of R. A. Rea, Sherman A. Wengerd, and R. B. Powers, reviewers for the *Bulletin*, were valuable in final revision of the manuscript. The original of Figure 8 was kindly loaned to the author by R. P. Sheldon of U.S. Geological Survey.

The writer is indebted to Chevron Oil Co., especially to Margie Rowell and Nina Prigodich, for the final typing of the manuscript, and to Donald C. Bartlett for the drafting. The ready assistance of B. K. Stieger

granitic mountains that attained their present form during the early Tertiary and were sculptured by late Tertiary and Quaternary erosion. On the east are the Pryor-Big Horn Mountains; on the south, the Owl Creek Mountains; and on the west, the Yellowstone-Absaroka volcanic plateau and the Beartooth Mountains (Fig. 2). On the north, the basin is separated from the Crazy Mountain syncline of south-central Montana by the Nye-Bowler left-lateral wrench-fault zone.

The Big Horn basin is the most prolific pro-

is especially acknowledged. Permission to publish this paper, including much of the previously unpublished chemical data, has been granted by Standard Oil Co. of California, the parent company, and Chevron Research Co. However, the original manuscript was written entirely at home, and special thanks are due my wife, Charity, for many hours of difficult typing in the early draft stage.

Original concepts presented and supported in this paper were discussed freely with W. N. Barbat in 1965–1966 while critically reviewing the first draft of his paper, "Crude-Oil Correlations and Their Role in Exploration," which appeared in the July 1967 *Bulletin* (v. 51, no. 7, p. 1255–1292). The final draft of that paper was subsequently expanded to include discussions of many of these concepts.

ducing basin in Wyoming and in the central Rocky Mountain region. Some of the first oil fields in the Rocky Mountains were discovered in large surface anticlines in the Big Horn basin, and these early fields still dominate the production and reserves statistics in this area. Significantly, nearly 40 percent of Wyoming's total 140 million bbl produced in 1965 came from 9 Big Horn basin fields in the 25 largest fields listed for the state; 7 of these are ranked in the top 10 Wyoming fields by cumulative production (Table I). These giant fields produce mostly from Paleozoic reservoirs, and in this paper, at least 95 percent of the total proved reserves of the Big Horn basin are attributed to a Paleozoic source (Fig. 3), with only 5 percent attributed to Mesozoic and Tertiary sources. When the facts are considered (1) that Wyoming in 1964 produced only slightly less than the combined total of Colorado, Montana, North and South Dakota, Utah, and Nebraska, and (2) that it has produced more than 56 percent of the total cumulative reserves in the Rocky Mountain region (*i.e.*, 2,500,000,000 bbl), the significance of Paleozoic oil in the Big Horn basin becomes apparent.

The impressive Paleozoic oil reserves of the Big Horn basin inspired this study of the Paleozoic oil fields. An understanding of the geologic conditions which account for the concentration of reserves in the Paleozoic of the basin seems requisite and compelling. Although reconstruction of the geologic history leading to the formation of so many giant oil fields in terms of source, migration, and final accumulation of the contained hydrocarbons was the immediate objective of this investigation, the writer hoped also that geologic and economic generalizations could be made that

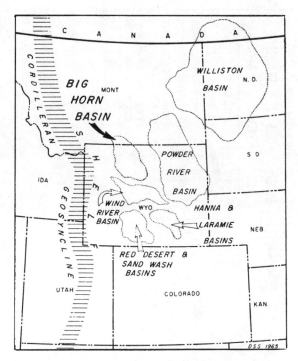

FIG. 1.—Index map, showing location of Big Horn basin, Wyoming and Montana.

might be applied to future exploration within the basin.

METHOD OF INVESTIGATION

The geology of every oil field in the Big Horn basin was reviewed; published information was used as a basis for further detailed analysis. Specifically, publications by Wyoming Geological Association (1957 a, b), Wyoming Geological Association and Billings Geological Society (1963), and Biggs and Espach (1960) provided the basic data including structural contour maps of the oil fields. Since publication of these maps, however, considerable development activity occurred and much up-dating was necessary. Commercial well-data cards, supplemented by key well logs, and current U.S. Geological Survey federal unit maps, supplied by the U. S. Geological Survey drafting department in Casper, Wyoming, provided the control for these revisions. Many published studies of specific oil fields also were consulted. In addition, drill-stem test and perforation data were analyzed to determine fluid conditions within each oil field so that the geometry of each accumulation could be outlined and estimates of reserves and recoveries made. Photogeologic maps,

TABLE I. SEVEN LARGEST BIG HORN BASIN FIELDS RANKED ACCORDING TO CUMULATIVE PRODUCTION*

Field	Year of Disc.	Rank in State	1965 Prod. (bbls.)	Cum. Prod. to 1-1-66
Elk Basin	1915	2	18,462,210	246,881,407
Oregon Basin	1912	3	7,930,762	130,470,217
Hamilton Dome	1918	4	7,807,900	117,156,042
Grass Creek	1914	5	5,363,537	101,082,775
Garland	1906	7	3,913,478	85,545,370
Frannie	1928	8	1,637,523	60,891,307
Byron	1918	10	2,732,740	59,021,908

* Data from "Production report for the twenty-five largest fields in Wyoming," published by the Oil and Gas Conservation Commission, Wyoming, August 1, 1966.

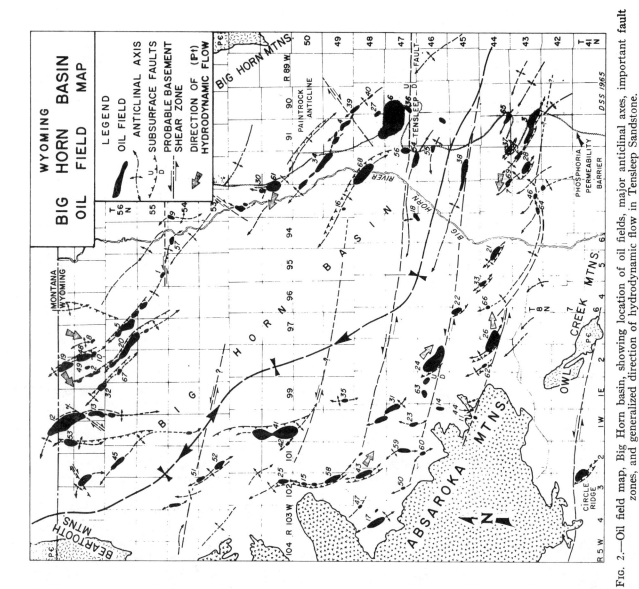

FIG. 2.—Oil field map, Big Horn basin, showing location of oil fields, major anticlinal axes, important fault zones, and generalized direction of hydrodynamic flow in Tensleep Sandstone.

NO.	FIELDS	LOCATION	PALEOZOIC GRAVITIES (API°)	PALEOZOIC RESERVES: GROUP*
1	ALKALI ANTICLINE	T55N-R95W	18-23	A
2	BIG POLECAT	T57N-R98W	27-28	B
3	BLACK MOUNTAIN	T43N-R91W	23-26	D
4	BONANZA	T49N-R91W	36	D
5	BYRON	T56N-R97W	18-24	E
6	COTTONWOOD CREEK	T47N-R91W	28-30	E
7	CORLEY	T43N-R93W	24	A
8	COWLEY	T57N-R97W	19	A
9	CRYSTAL CREEK	T54N-R93W	24	A
10	DEAVER, NORTH	T57N-R97W	22-25	B
11	DICKIE	T45N-R101W	20	A
12	ELK BASIN	T58N-R100W	23-31	G
13	ELK BASIN, SOUTH	T57N-R100W	28	C
14	ENOS CREEK	T46N-R100W	27-28	B
15	FERGUSON RANCH	T50N-R102W	14	A
16	FIVE MILE	T49N-R93W	52-55	A
17	FOURBEAR	T48N-R103W	12-15	A
18	FOURTEEN MILE	T46N-R94W	45-46	E
19	FRANNIE	T58N-R98W	17-28	F
20	GARLAND	T56N-R97W	19-23	F
21	GEBO	T44N-R95W	29	D
22	GOLDEN EAGLE	T45N-R97W	39	C
23	GOOSEBERRY	T47N-R100W	21-23	B
24	GRASS CREEK	T46N-R98W	22-25	F
25	HALF MOON	T51N-R102W	15	B
26	HAMILTON DOME	T44N-R98W	15-25	B
27	HIDDEN DOME	T48N-R94W	20	B
28	KIRBY CREEK	T43N-R92W	21-23	A
29	LAKE CREEK	T43N-R91W	30-34	C
30	LAMB	T51N-R92W	19-25	B
31	LITTLE BUFFALO BASIN	T47N-R100W	20	A
32	LITTLE POLECAT	T57N-R98W	27	A
33	LITTLE SAND DRAW	T44N-R96W	34-36	B
34	MANDERSON	T49N-R92W	37-43	A
35	MEETEETSE	T49N-R99W	31	A
36	MEYER GULCH	T47N-R90W	24	D
37	MURPHY DOME	T44N-R92W	30-34	D
38	NEIBER DOME	T45N-R92W	33-42	A
39	NOWOOD and SOUTHEAST	T48N-R90W	28-30	A
40	NOWOOD, MIDDLE DOME	T48N-R90W	27	F
41	OREGON BASIN	T51N-R100W	18-23	F
42	OREGON BASIN, WEST	T51N-R101W	18-20	A
43	PITCHFORK	T48N-R102W	14-19	C
44	PROSPECT CREEK	T45N-R100W	23	A
45	RALSTON	T56N0R101W	46	A
46	RED SPRINGS	T43N-R93W	11	A
47	ROSE CREEK	T48N-R103W	15	A
48	SAGE CREEK	T54N-R94W	19	C
49	SAGE CREEK, WEST	T57N-R98W	22-25	A
50	SHEEP POINT	T47N-R102W	14-17	A
51	SHOSHONE, NORTH	T53N-R102W	19	A
52	SHOSHONE, SOUTH	T53N-R101W	21	A
53	SILVER TIP	T58N-R100W	36-45	B
54	SLICK CREEK	T46N-R92W	23-35	A
55	SOUTH FORK	T43N-R93W	23	A
56	SOUTH FRISBY	T47N-R92W	30-32	A
57	SPENCE DOME	T54N-R94W	19	A
58	SPRING CREEK	T49N-R102W	12-15	C
59	SUNSHINE, NORTH	T47N-R101W	18-19	B
60	SUNSHINE, SOUTH	T46N-R101W	18-19	A
61	TORCHLIGHT	T51N-R93W	23-35	B
62	WAGONHOUND	T44N-R98W	26	A
63	WALKER DOME	T46N-R99W	26	A
64	WARM SPRINGS	T43N-R94W	18-23	B
65	WATER CREEK	T44N-R91W	21	A
66	WAUGH DOME	T44N-R97W	27	B
67	WHISTLE CREEK	T56N-R98W	28	B
68	WORLAND	T48N-R93W	20-45	D
69	ZIMMERMAN BUTTE	T44N-R93W	24	B

*RESERVE GROUPS
A= less than 1 MM bbls.: B= 1 to 10 MM bbls.: C=10 to 25 MM bbls.:
D= 25 to 50 MM bbls.: E= 50 to 100 MM bbls.: F= 100 to 250 MM
bbls.: G = 250 to 500 MM bbls.

published U. S. Geological Survey field studies, and many papers on stratigraphy, structure, and chemistry of the crude oils and waters also were consulted.

GEOLOGIC HISTORY

Figure 4 is a schematic history of the stratigraphy, sedimentation, and tectonic activity in the Wyoming shelf area. The vertical scale in this schematic diagram is approximate geologic time and the emphasis is on depositional cycles and periods of erosion.

The Big Horn basin was part of the stable shelf region that was east of the Cordilleran geosyncline during Paleozoic and much of Mesozoic times. Deposition in the neighboring Cordilleran geosyncline began in Precambrian time and, although interrupted by periods of erosion, continued until near the end of Cretaceous time, when the Tertiary basins of Wyoming began to form. More than 26,000 ft of Paleozoic miogeosynclinal sediments was deposited in west-central Utah (Roberts *et al.*, 1965), whereas only 2,000–3,500 ft of Paleozoic shelf sediments is preserved in the Big Horn region.

Nearly all of the Paleozoic systems in the Big Horn basin are represented by a transgressive-regressive cycle. They are separated by important erosional intervals. Isopachous maps of the Paleozoic rocks in the Big Horn region indicate that northerly tilting, deposition, emergence, and erosion occurred repeatedly, resulting in truncation of the Ordovician, Devonian, and Mississippian sediments from north to south (Thomas, 1965). No Silurian rocks are preserved. Except for the Cambrian clastic wedge, all pre-Pennsylvanian formations are predominantly carbonate rocks, but commonly have basal transgressive sandstone beds.

Toward the end of the Pennsylvanian, probably beginning in Missourian time (Agatston, 1954, p. 567), the northern Big Horn region was uplifted, and erosion of the upper Tensleep Sandstone beds began. Uplift continued into early Permian time and the seas retreated southward. Middle Permian transgression then deposited marine Phosphoria rocks over the partly truncated Tensleep Sandstone beds in most of the Big Horn basin, while the equivalent Goose Egg and Opeche redbeds and evaporites were deposited in eastern Wyoming. Except for the Dinwoody and

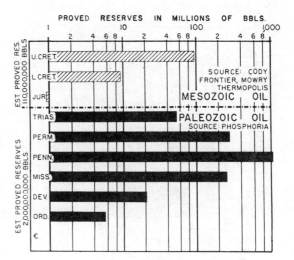

Fig. 3.—Stratigraphic distribution of proved reserves, Big Horn basin, Wyoming.

Thaynes-Alcova-"Curtis" cycles, a general westerly regression occurred in Triassic time in the Wyoming shelf area. No important erosional breaks can be detected from late Permian through Triassic time in Wyoming (Oriel, *in* McKee *et al.*, 1959).

The first post-Permian uplift and erosion occurred during the Early Jurassic in the Wyoming shelf area; no rocks of Early Jurassic age have been identified definitely in the Big Horn basin (Downs, 1952; Burk, 1956). Progressive northward truncation at the top of the Triassic is evident (Downs, 1952, p. 26) in the Big Horn basin, and the Triassic Chugwater Formation is overlain unconformably by the Middle Jurassic Gypsum Springs Formation, followed by the younger, shallow-water Sundance, and bright-colored, nonmarine Morrison sediments.

Jurassic and Lower Cretaceous formations thicken from south to north in the Big Horn basin area, but on a regional scale, they thicken westerly into the miogeosyncline. Cretaceous sedimentation began with the deposition of thin, discontinuous, basal transgressive sandstone on a low-relief Jurassic erosion surface. The very shallow, Early Cretaceous sea first spread westward, and later withdrew; a thick wedge of fine clastics was deposited. This wedge is characterized by gray-black marine shale. After several relatively rapid transgressions and regressions, resulting in the deposition of several Frontier Sandstone wedges that pinch out toward the northeast, the sea began to withdraw eastward in response to

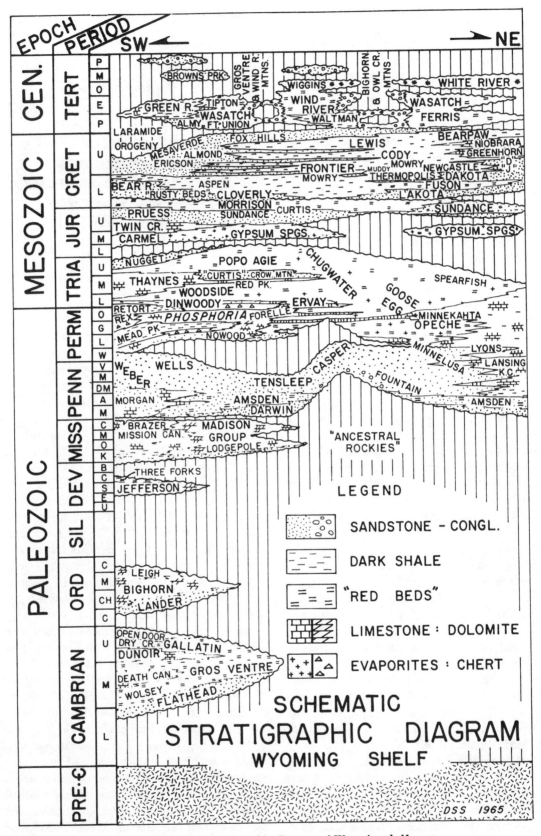

Fig. 4.—Schematic stratigraphic diagram of Wyoming shelf area.

the first pulses of the Laramide orogeny. This withdrawal is reflected in the predominantly regressive Mesaverde clastic wedge.

Laramide orogenesis intensified considerably at the beginning of the time of Fort Union deposition (Paleocene) and continued into the Eocene, accompanied by important folding and faulting, and followed by deposition and partial erosion of Absaroka volcanics on the western margin of the basin. The uplift of the peripheral Big Horn, Beartooth, Owl Creek, and Absaroka Mountains was accompanied by deposition of conglomerate along their margins; multiple unconformities formed in the section. Most of the folding which produced the present oil-field anticlines in the Big Horn basin probably occurred during the Paleocene-Eocene pulses of the Laramide orogeny, but continued compressional stress regionally tilted these structures during Tertiary time. The writer believes that deep basement, left-lateral, wrench-fault movements have been important in deter-

mining the location of some of the larger anticlines (see Chamberlin, 1940).

STRATIGRAPHY

The lithologic and producing characteristics of the Paleozoic and Triassic rocks (Fig. 5) are described under three subheadings: "reservoir rocks," "source rocks," and "cap rocks." These headings make it possible to consider the formations in ascending order, from Cambrian to Triassic.

RESERVOIR ROCKS

Cambrian.—At the base of the Cambrian section, lying on Precambrian granitic or metamorphic basement, is the nearshore, transgressive Flathead Sandstone. This thin basal unit is best developed in the western part of the basin, as at Oregon Basin, where it is 170 ft thick (Allen, 1954, p. B-49); however, it is either absent or poorly developed along the east flank, as at Gar-

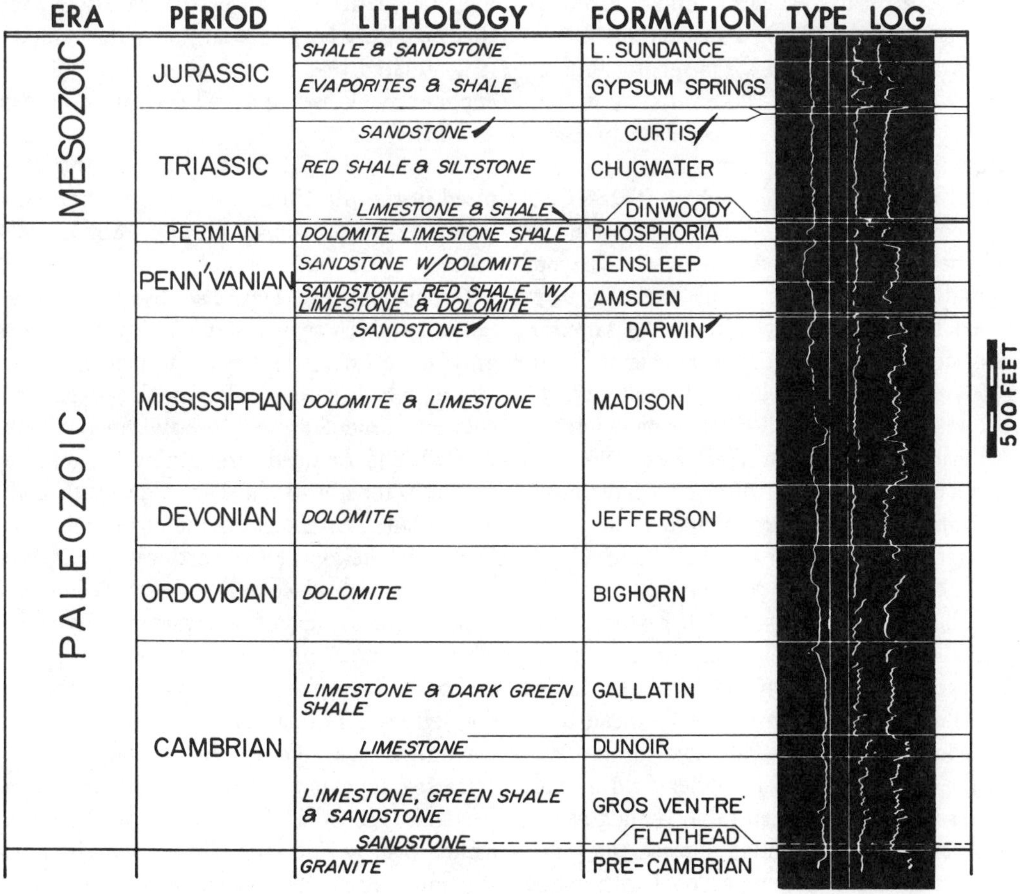

ERA	PERIOD	LITHOLOGY	FORMATION	TYPE LOG
MESOZOIC	JURASSIC	SHALE & SANDSTONE	L. SUNDANCE	
		EVAPORITES & SHALE	GYPSUM SPRINGS	
	TRIASSIC	SANDSTONE	CURTIS	
		RED SHALE & SILTSTONE	CHUGWATER	
		LIMESTONE & SHALE	DINWOODY	
PALEOZOIC	PERMIAN	DOLOMITE LIMESTONE SHALE	PHOSPHORIA	
	PENN'VANIAN	SANDSTONE W/DOLOMITE	TENSLEEP	
		SANDSTONE RED SHALE W/ LIMESTONE & DOLOMITE	AMSDEN	
	MISSISSIPPIAN	SANDSTONE	DARWIN	
		DOLOMITE & LIMESTONE	MADISON	
	DEVONIAN	DOLOMITE	JEFFERSON	
	ORDOVICIAN	DOLOMITE	BIGHORN	
	CAMBRIAN	LIMESTONE & DARK GREEN SHALE	GALLATIN	
		LIMESTONE	DUNOIR	
		LIMESTONE, GREEN SHALE & SANDSTONE	GROS VENTRE	
		SANDSTONE	FLATHEAD	
		GRANITE	PRE-CAMBRIAN	

500 FEET

FIG. 5.—Geologic column with type electric log from Byron field, Big Horn basin.

land-Byron. The Flathead generally has coarse, angular grains in a finer sand matrix. In some places the Flathead is conglomeratic with feldspar and quartz phenoclasts.

The Gros Ventre Formation, which overlaps the Flathead Sandstone, is a clastic unit composed of glauconitic and sandy limestone and thin sandstone beds in a dark grayish-green shale. Thickness of this unit ranges from 400 to 600 ft in the Big Horn basin.

The upper 400 to 500 ft of the Cambrian comprises the Gallatin Formation which is glauconitic (and locally pebbly) sandstone, limestone, and shale with thin sandstone interbeds. At the base of the Gallatin Formation, a thin limestone member, called the Du Noir Member, is recognized readily on mechanical logs.

Hydrocarbons have been recovered from the Flathead Sandstone at Oregon Basin (Fig. 13) and Elk Basin (Fig. 9) fields, and from the upper part of the Gallatin at Elk Basin (Figs. 9, 10), in the Big Horn basin. The Elk Basin Gallatin oil was a typical asphaltic Paleozoic crude held in a common pool with overlying Paleozoic oil and is believed to have come from a Phosphoria source. However, the unusual condensate found in the Flathead at Oregon Basin is believed to have come from Tertiary sources across the major east flank fault zone (D. P. McGookey, Texas Co., personal commun., 1966). This conclusion is based on nearly identical gas analyses of Oregon Basin Flathead (hanging wall) and nearby McCulloch Peaks Fort Union (footwall) samples (analyses supplied by R. B. Powers, Tenneco Oil Co., 1966). Similarly, the low-sulfur oil (W. J. Wenger, written commun., 1967) recovered from the Flathead at Elk Basin (see Rea, 1962, p. 118; also, supplied sample, 1967) may have been derived from Mesozoic sources across the fault which cuts the east flank of this field (Fig. 9).

The Flathead Sandstone provides an excellent reservoir, but the thickness of the geologic section from the top of the Phosphoria to the top of the Flathead is too great in the Big Horn basin fields to allow the inclusion of the Flathead in a common pool with younger Paleozoic formations. Only in the well-known Lost Soldier field of the Red Desert basin in south-central Wyoming has a relatively thin Paleozoic section allowed the Flathead to be included in a common pool with overlying Paleozoic zones (Krampert, 1949).

Ordovician.—The Ordovician is represented in the Big Horn basin by the Bighorn Dolomite. White, tan, and pink, finely crystalline to microsaccharoidal dolomite and cherty dolomite are the most common rock types, suggesting an aerated shelf environment of deposition. The formation thins (probably by erosional truncation at the top) from 450 ft on the north to zero at the outcrop in the extreme southeastern corner of the basin. A basal transgressive sandstone unit, the Lander Sandstone Member, is present in many places near the Cambrian-Ordovician unconformity.

The Bighorn Dolomite contains commercial oil only at Hamilton Dome, Garland, and Elk Basin. However, the oil in the Bighorn zone of each of these fields is similar chemically to that produced from all younger Paleozoic zones and is believed to be exotic and held in a common pool with them. Also, 5,000 bbl of 22° API gravity, black, high-sulfur (2.38 percent), asphaltic oil was produced from the Bighorn Dolomite in the Torchlight field. Although all of the Paleozoic zones in this field now have separate oil-water contacts, it is postulated that they were once part of a single common pool, having an oil column of more than 1,300 ft.

Silurian.—Silurian rocks have not been recognized in the Big Horn basin; they are absent as a result of pre-Devonian erosion (Shaw, 1954, p. 36).

Devonian.—The Devonian System in the Big Horn basin is represented primarily by a dark gray or brown crystalline, in places platy, dolomite, which is interbedded with pinkish or buff-colored carbonates and detrital rocks. Isolated, frosted, and rounded sand grains are reported to be characteristic of the Devonian south and east of Garland. Although in the upper part of the Devonian section of the northwestern basin area, some interbedded dolomite and varicolored shale appear which may be equivalent to part of the Three Forks Formation of Montana, in most of the basin Devonian rocks have been assigned to the Jefferson Formation of probable middle Late Devonian age. Only the carbonate of the Jefferson appears to serve as reservoir rock in the Devonian section. The use of the alternative name, "Darby Formation," for Devonian rocks in Wyoming has been recommended by the Wyoming Geological Association (1957a). The term

"Darby" does not appear to have been adopted by operators in the Big Horn basin who apparently prefer the name Jefferson.

The Devonian section thins from 300 ft on the northwest to zero on the southeast flank of the basin (Benson, 1966). Much section is missing both at the top, by erosion, and at the bottom, by transgressive onlap above important unconformities. Despite the darker color of some of the Devonian carbonate section, the general aspect of interbedded detrital mudstone, siltstone, and sandstone, and nondetrital light-colored carbonate suggests an aerated shelf environment in most of the Big Horn basin. Some thin, very fossiliferous, dark gray carbonaceous shale of Late Devonian to earliest Mississippian age is in the northern Big Horn basin (Sandberg, 1963; Benson, 1966, Fig. 16). As a possible source sediment, however, either the carbonaceous character of the shale should be considered negatively, or the volume of dark shale must have been too small to have supplied any significant amount of indigenous hydro-carbons to Devonian reservoirs in the Big Horn basin.

Commercial Devonian production has been established at Elk Basin, Garland, and Hamilton Dome. However, petroleum entrapped in the Devonian of the Big Horn basin, like that found in other sub-Phosphoria Paleozoic rocks, is held in common pools with overlying Paleozoic oil, and probably is derived from Phosphoria sources.

Mississippian.—The well-known Madison Group represents the Mississippian System in the Big Horn basin. Throughout Wyoming, the Madison is predominantly dolomite with characteristic limestone interbeds (Fig. 6). The entire Wyoming Mississippian section probably includes only equivalents of the Lodgepole and Mission Canyon Formations of the Williston basin (Andrichuk, 1958); it is bounded both above and below by regional unconformities (Levorsen, 1960, Figs. 6–14 and 6–15, p. 133). These rocks thin by onlap at the bottom and erosion at the top from 800 ft in the northern Big Horn basin to less

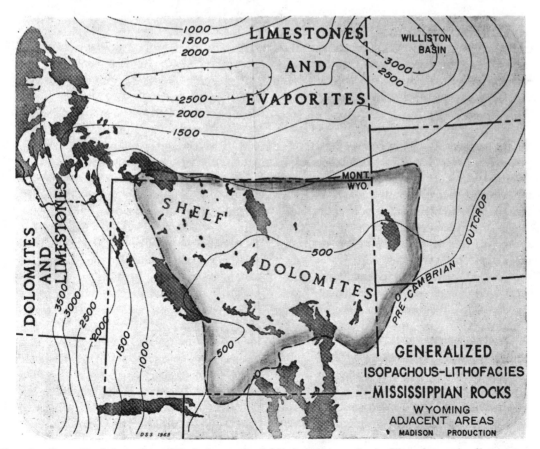

FIG. 6.—Generalized isopachous-lithofacies map of Mississippian rocks in Wyoming and adjacent areas.

TABLE II. COMPARISON OF AMOUNT OF VERTICAL CLOSURE WITH CRESTAL THICKNESSES OF INTERVAL FROM TOP OF PHOSPHORIA TO TOP OF MADISON FOR FIELDS PRODUCING FROM MADISON RESERVOIR

Fields	Approx. Vert. Anticlinal Closure (Feet)	Crestal Thickness from Top of Phosphoria to Top of Madison (Feet)
1. Black Mountain	1,000	970
2. Byron	2,200	400
3. Elk Basin	5,000+	425
4. Fourbear	1,250	600
5. Frannie	2,000	400
6. Garland	3,000	390
7. Grass Creeek	2,700[1]	860
8. Hamilton Dome	2,000[1]	680
9. Lamb-Torchlight	1,300(?)	550
10. Oregon Basin	1,500+	560
11. Pitchfork	2,000	780
12. Red Springs	1,300	1,000(?)
13. Sage Creek	500[2]	360
14. Silver Tip	600[1]	500
15. Spence Dome	700±	520
16. Spring Creek	1,100+	940

[1] With aid of fault
[2] or 1,000 ft with Frannie

than 500 ft in the southern basin fields. The relatively thin Wyoming section is in contrast to the more than 2,500 ft of Mississippian strata in the Williston basin of northeastern Montana and northwestern North Dakota (Pye, 1958, Fig. 17, p. 203; Carlson and Anderson, 1965, Fig. 24, p. 1841).

Madison reservoir dolomite and limestone generally are massive and highly porous, and are notorious as a troublesome lost-circulation zone. They are gray to buff or brown, finely crystalline or saccharoidal, and commonly contain chert. At the top of the section a nonreservoir, thin-bedded limestone and dolomite unit, with "tight" sandstone and varicolored shale, is present in many places.

The Madison ranks third in importance, after Tensleep and Phosphoria, as a producing reservoir in the Big Horn basin, accounting for more than 13 percent of the proved reserves of the basin (approximately 265,000,000 bbl). The Madison Formation produces in 16 Big Horn basin fields. However, *Madison production is obtained only in structural traps that also contain oil in overlying Paleozoic formations.* As shown in Table II, vertical anticlinal closure in excess of the crestal thickness from the top of the Phosphoria to the top of the Madison is required for

Madison production. *"No Madison stratigraphic accumulations are yet known in this basin"* (Thomas, 1965, p. 1870). Furthermore, from the evidence accumulated during this investigation, it is concluded (1) that the Madison Formation was incapable of generating and preserving indigenous hydrocarbons in commercial quantities in Wyoming, and (2) that Madison reserves are exotic hydrocarbons derived from Phosphoria source rocks.

Pennsylvanian.—The Pennsylvanian Amsden Formation and Tensleep Sandstone of the Big Horn basin were deposited in a very shallow-marine, shelf environment and contain more quartz clastics than older rocks. The Amsden may represent a complete transgressive-regressive cycle with many minor oscillations as interpreted from the percent ranges of "clastic quartz and clasticity" (Gorman, 1965). The basal transgressive Darwin Sandstone Member fills an uneven Mississippian erosion surface which has 50 ft or more of relief, and is typically cut by crevices and channels filled with basal Amsden clastic sediment, reportedly as deep as 200 ft below the top of the Madison (Agatston, 1954, p. 514). The Darwin is overlain by a carbonate section, which is overlain by a quartz-rich, regressive, carbonate and sandstone sequence. Amsden strata include gray to pink, cherty, microcrystalline dolomite in the upper part, and much sandstone and varicolored shale with some anhydrite beds in the lower part. Red and pink colors are prominent, indicating an oxidizing environment.

The Amsden section has a poor production record because of a lack of reservoir porosity. Considerable fracturing has been noted in the Amsden (Harris *et al.*, 1960) and may be necessary for commercial production. Fracturing is required for the inferred communication between Tensleep and Madison zones. Inasmuch as production obtained from the Amsden generally is reported together with that from the Tensleep, accurate recovery factors for this reservoir are difficult or impossible to calculate.

The Tensleep Sandstone of Middle Pennsylvanian (Desmoinesian) age is the most prolific producer in the Big Horn basin, accounting for more than 1,000,000,000 bbl or 50 percent of the basin's proved ultimate reserves (Fig. 3). It produces in 53 Big Horn basin fields. Todd (1964, p. 1068) described the Tensleep Sandstone as "white (N8), weathering dun (10 YR 8/2), hard-friable, massive to cross-bedded (3–7 feet), quartz- and carbonate-cemented, moderately to well-sorted, fine- to very fine-grained (2.7–3.3Φ) orthoquartzite." Interbeds of dolomite are common, particularly in the lower part. The occurrence of pink carbonate lentils below, which usually are expressed by higher, characteristically variable resistivity values on electric logs, generally serves as the basis for selecting the Tensleep-Amsden contact in the subsurface.

The Tensleep is a very shallow-water, predominantly regressive sandstone unit, deposited during the withdrawal of Pennsylvanian seas from the Wyoming shelf area prior to an important Late Pennsylvanian-early Permian period of epeirogenic uplift and erosion in northwestern Wyoming and southern Montana. Tensleep rocks preserved in the Big Horn basin range from 87 to 422 ft in thickness (Todd, 1964, p. 1068). At the erosional upper surface of the Tensleep, several cuestas and monadnocks have been preserved which, where surrounded and overlain by nonreservoir Phosphoria rocks and sealed by Tensleep carbonate interbeds below, provided early erosional truncation traps. According to Lawson and Smith (1966, p. 2216 and Fig. 24), Tensleep Sandstone porosity values range up to 27 percent above a depth of 8,000 ft but generally decrease to much less than 10 percent below this depth. Accompanying permeability values range from a few to more than 800 md. Strong water drive in the shallower fields generally permits substantial oil recovery from the Tensleep zone.

The widespread sandstone lithofacies of the Tensleep-Amsden has never been considered seriously by geologists as a source for the oil in these Pennsylvanian sandstone reservoirs. The source of the oil in the Tensleep has been assigned by most geologists to the overlying Phosphoria sediments (*e.g.*, Campbell, 1962, p. 500; Curtis *et al.*, 1958, p. 283; Hunt, 1953, p. 1864; Partridge, 1958, p. 304; Walton, 1947, p. 215). Communication between Permian and Pennsylvanian reservoirs also is a well-established concept (*Elk Basin*, Stewart *et al.*, 1955, p. 49; *Oregon Basin*, Walton, 1947, p. 220; *Silver Tip*, Thompson, 1954, p. 123).

Permian.—The Phosphoria (Park City) Formation of middle Permian age is unconformable on the Tensleep erosion surface throughout the Big Horn basin. It fills in the topographic lows in the Tensleep surface so that thicknesses of the two formations are approximately complementary (Agatston, 1954). The formation ranges in thickness from 200 ft in the north end of the basin to slightly more than 350 ft in the southwest part of the basin.

As the isopachous-lithofacies map (Fig. 7) illustrates, Permian rocks in the Big Horn basin may be divided into two distinct facies: a marine facies (including both fine-grained, dark-colored source rocks and lighter colored, carbonate-reservoir rocks) which covers most of the basin; and a red shale and evaporite nonreservoir facies, called the Goose Egg Formation, which is on the eastern side of the basin and covers all of eastern Wyoming. Only the reservoir carbonate is discussed in this section.

Although pre-Jurassic erosion has cut into rocks older than the Permian in central Montana, the Phosphoria wedges out northward in southern Montana by onlap of the Pennsylvanian erosion surface and overlap by Triassic redbeds (Alpha and Fanshawe, 1954, cross section; Campbell, 1962, p. 481; McKelvey *et al.*, 1953; Munyan, 1962; and Sheldon, 1963).

Phosphoria reservoir rocks include tan, brown, and gray dolomite and limestone, interbedded with dark-colored, phosphatic, and organic-rich shale and cherty carbonate. Although some of the Phosphoria carbonate probably was deposited in an oxidizing environment, certainly most of the darker colored marine Phosphoria rocks are the product of a generally reducing environment.

The Phosphoria Formation is productive in more fields (56) than any other formation in the Big Horn basin. This reservoir holds approximately 300,000,000 bbl of oil and oil equivalents or about 15 percent of the total Big Horn basin proved reserves (Fig. 3). Production from this formation is unique within the Paleozoic of this area in that it is not only found in structural traps but also in stratigraphic traps. In at least 16 Big Horn basin Phosphoria fields, stratigraphic variation contributes greatly to the formation of the Phosphoria trap and is essential in at least three fields (Cottonwood Creek, Manderson, and Water Creek).

The distribution of reservoir porosity within the Phosphoria carbonate facies is the result of both its depositional environment and superim-

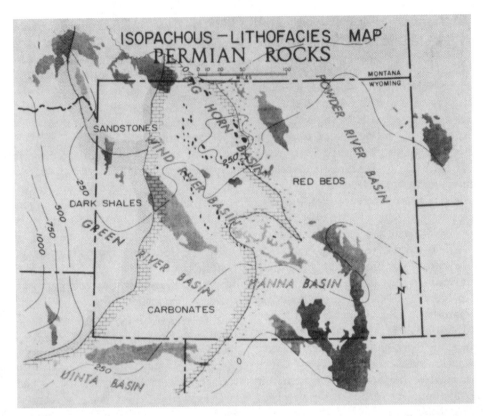

FIG. 7.—Isopachous-lithofacies map of Permian rocks in Wyoming and adjacent areas.

posed diagenetic changes. The most important diagenetic process affecting these rocks was dolomitization of original limestone, believed to have taken place nearly contemporaneously with deposition (Campbell, 1962, p. 495). Only a small part of the Phosphoria, usually near the top (Ervay Member), generally has attractive reservoir characteristics. Net pay in the Big Horn basin ranges from 10 to 75 ft, but probably averages 20 ft. Porosity values of less than 10 percent and permeability values of less than 10 md, with resulting recoveries less than 150 bbls per acre-foot, are typical of Phosphoria carbonate reservoirs. Fractures are important, and commonly required, for commercial production (at least in the lower part of the formation). Probably the best Phosphoria production in Wyoming comes from the well-known Cottonwood Creek field of the southeastern Big Horn basin (Fig. 2). In this major reentrant-stratigraphic accumulation, reservoir porosity is unusually well developed in an algal dolomite facies of the Ervay Member (Pedry, 1957; Willingham and McCaleb, 1966).

Triassic.—The Phosphoria Formation is overlain in western Wyoming by a thin sequence of compact Lower Triassic sandy and silty limestone, dolomite, and green mudstone known as the Dinwoody Formation. Because of its lithologic similarity to the underlying Phosphoria it commonly has been included with the Phosphoria as the Embar Formation. Inasmuch as little or no reservoir porosity exists in the Dinwoody, little production has been reported from this formation; most of the production probably has resulted from upward movement of Phosphoria hydrocarbons through fractures in the crestal parts of anticlines.

The remaining 400–1,200 ft of Triassic beds is assigned to the Chugwater Formation, a red shale and siltstone sequence. Despite the red colors and absence of fossils, however, the Chugwater is a shallow, quiet-water, aerated marine deposit, and thin electric-log correlation markers can be carried through this formation for 100 mi or more.

In the upper few hundred feet of the Chugwater, particularly in the southwestern part of the basin, several 10–15-ft sandstone tongues or lenses are present. These have been called Crow Mountain Sandstone Member (*cf.*, Tohill and Picard, 1966) or "Curtis" Sandstone, and are productive in several southwestern Big Horn basin fields. At least 50,000,000 bbl of proved reserves is attributed to the "Curtis" at Grass Creek field. However, the aromatic-naphthenic, sour crudes in the "Curtis" reservoir(s) are believed to be foreign to the Triassic, and derived from the underlying Phosphoria Formation (Collier, 1920, p. 69; Hunt, 1953, p. 1839). According to Hunt (p. 1870), "When crude-oil assay data on the Chugwater and Phosphoria oils are compared, only negligible differences are apparent." It is significant that "Curtis" production is found only in association with faulted anticlinal structures. "Curtis" Sandstone recoveries depend primarily on the mobility of the unusually viscous, low-gravity asphaltic oil which is commonly trapped at relatively shallow depths.

SOURCE ROCKS (PHOSPHORIA)

The concept that entrapped hydrocarbons generally are indigenous and derived from lateral stratigraphic equivalents of the reservoir rocks has been applied carelessly by many geologists without thorough investigation of all data bearing on the problem. For rocks such as the pre-Permian Paleozoic formations of western Wyoming, this is a hazardous assumption if considered together with their oxidizing or aerated depositional environments, their subjection to long and repeated periods of interformational and intersystemic erosion, the apparent absence of important primary stratigraphic traps within these formations, and the generally unfavorable structural "timing."

For example, Partridge (1958, p. 303) has stated that, "The oil in the Mississippian rocks of the Big Horn Basin is certainly the result of migration of the oil from the source facies of the deposition basin in the silled Williston basin to the east, and from the geosynclinal trough to the west. . . ." The fundamental difference in the chemical composition of the Madison oils in the Big Horn and Williston basins, however, strongly refutes this assumption, as do other factors. Moreover, Hunt and Forsman (1957, p. 108), from a study of the relation of crude-oil composition to stratigraphy in the contiguous and geologically analogous Wind River basin of Wyoming,

believed that, "The absence of typical source rocks in the Madison suggests that Madison oils in this area (typical of Madison oils throughout the Big Horn basin) may have originated in other sections." In the Wind River basin, as in the Big Horn basin, it would be difficult to relate the only two Madison accumulations at Circle Ridge and Beaver Creek to Madison facies (Fig. 6) as has been suggested by Partridge (1958). Data recorded by the Wyoming Geological Association (1957b) and Biggs and Espach (1960, p. 62), however, show that these two fields are unique in the Wind River basin in possessing vertical anticlinal closure in Paleozoic rocks exceeding 1,000 ft. In both fields this closure exceeds the stratigraphic thickness from the top of the Phosphoria to the top of the Madison reservoir. Similarly, in the Big Horn basin, the same correlation between stratigraphic thickness and vertical closure applies to all sub-Permian Paleozoic accumulations (Table II). For this reason, and other reasons presented later, it has been concluded that the oil found in pre-Permian rocks of the Big Horn basin, and probably also in the other parts of Wyoming, is not indigenous.

The writer believes that all commercial hydrocarbons in Paleozoic and Triassic reservoirs of the Big Horn basin were derived from a Phosphoria source. This conclusion is based on the following cumulative evidence:

1. The abundant dark-colored, phosphatic, organic, and calcareous mudstone and shale interbeds of the Phosphoria Formation in the Big Horn basin and in the Cordilleran geosyncline on the west indicate a reducing environment for much of the marine facies of the Phosphoria Formation. The scarcity of similar dark-colored, "source-type," organic shale in the sub-Phosphoria Paleozoic section provides a striking contrast, and implies a predominantly oxidizing environment for pre-Permian rocks in the Big Horn basin. Thus the dark-colored, organic-rich Phosphoria sediments provide not only the best, but the only logical choice of source for the petroleum in the Paleozoic reservoirs of the basin.

2. Nearly every well drilled into the marine facies of the Phosphoria Formation in the Big Horn basin and elsewhere on the Wyoming shelf has recorded "live" oil shows (Partridge, 1958, p. 304). As much as 24 gal of oil per ton of shale has been retorted and nearly 10 percent oil by

West

East

Southeast Idaho

Central Wyoming

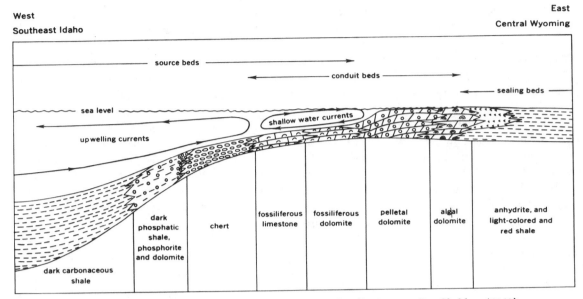

FIG. 8.—Sketch showing idealized sedimentation in Phosphoria sea, after Sheldon (1963).

weight of sample has been distilled from the Retort Phosphatic Shale Member in Beaverhead County, Montana (McKelvey *et al.*, 1959, p. 29). Asphaltic residue in fractured or porous zones is common in hand specimens of Phosphoria rocks (Stipp, 1947, p. 124).

3. Extracts of indigenous organic matter from Phosphoria cores ("carbonaceous dolomite and gray calcareous shale") are reported to have yielded relatively high concentrations of hydrocarbons and 14 bbl per acre-foot asphalt (Hunt and Jamieson, 1958). "In the Big Horn basin samples, the concentration of hydrocarbons in the Phosphoria is greater than that in the Frontier..." (p. 745). Kerogen content of the Phosphoria samples analyzed also was relatively high, but varied considerably.

4. All oils produced commercially from Paleozoic and Triassic reservoirs in the Big Horn basin and the rest of western Wyoming have strikingly similar chemical composition. They are different chemically from pre-Permian oils in other parts of the Rocky Mountain area outside of Wyoming. Therefore, if the oil in Big Horn basin Phosphoria reservoirs is indigenous (as most geologists believe), then a Phosphoria source for the oil in the sub-Phosphoria reservoirs is strongly suggested also by the chemical data.

5. In a study of the chemical relation of crude oils to source rocks by Brenneman and Smith (1958), a good agreement in both aromatic-type

analyses and infrared measurements on molecular distillation fractions was found between hydrocarbon extracts from a Phosphoria core and oil produced from the Tensleep in a Wind River basin well (Hudson No. 1 Shoshone, Sec. 30, T. 2 S., R. 2 E.). These results are pertinent to the Phosphoria source-rock question in the Big Horn basin, because the geological characteristics of both the Wind River and the Big Horn basins are closely comparable. Also, Sheldon (1967, p. 56 and Fig. 3) indicated that most of the same trace elements have been found in both the carbonaceous and phosphatic shales of the Phosphoria Formation and in crude oils in the Permian and, thus, probably all Paleozoic rocks in Wyoming (Hayden, 1961).

6. An ideal reentrant-stratigraphic trap of regional proportions was formed within the marine facies of the Phosphoria in the Big Horn basin and across most of western Wyoming and northwestern Colorado by the combination of (1) eastward and southward facies change from reservoir carbonate to nonreservoir redbeds and evaporites, and (2) northward wedgeout in southern Montana (Alpha and Fanshawe, 1954, cross section; Campbell, 1962, p. 481) and/or facies change (Munyan, 1962) augmented by erosional truncation and overlap by Triassic and Jurassic sediments (Fig. 7). Regional dip in the area was west into the Cordilleran geosyncline, at least through Middle Jurassic time, thus providing the

structural element of the trap (Sheldon, 1967, Fig. 8; Thomas, 1965, Fig. 14). This early trap is believed to have provided the means of storage and preservation of many billions of barrels of hydrocarbons through the long period between primary flush migration, which was completed probably by early Mesozoic (Early Jurassic) time, and final accumulation in anticlinal traps formed during the early Tertiary of the Laramide orogeny. Additional Phosphoria hydrocarbons probably were stored also in the upper part of the Tensleep Sandstone in irregularities beneath the Phosphoria-Tensleep unconformity. No other Paleozoic formation in the Big Horn basin has comparable intraformational or intrasystemic stratigraphic entrapment which could have preserved significant quantities of early-migrating indigenous hydrocarbons (Thomas, 1965). Moreover, throughout the Rocky Mountain area, all Paleozoic periods are separated by long intervals of erosion (Levorsen, 1960, p. 133–136) which presumably would have provided the escape of most, if not all, indigenous hydrocarbons that might have formed and migrated in pre-Permian time.

Figure 8 (from Sheldon, 1963, p. 159, Fig. 86) illustrates the environmental-stratigraphic setting which is believed to have generated and preserved so much oil in the Phosphoria. Sheldon (1963, p. 159, Fig. 86) stated that "these stratigraphic relations between the source rocks and the sealing beds make nearly ideal conditions for petroleum accumulation. The source beds are a basinward facies of the reservoir beds which in turn are a basinward facies of the sealing beds. Transgressions and regressions of these facies give rise to an interfingering of source, reservoir, and sealing beds."

CAP ROCKS (TRIASSIC)

The very effective cap rocks of the normal anticlinal Paleozoic fields in the Big Horn basin include the calcareous shale, siltstone, and tight carbonate of the Dinwoody Formation, the red shale and evaporite of the Triassic Chugwater Formation, and, in the area of Permian redbeds, the Goose Egg Formation. These are the fine-grained "impervious" and "infrangible" rocks which typify cap rocks of most producing areas. They have become ineffective as cap rocks only where intersected by important faults, or where surface erosion has cut so deeply that the Phosphoria Formation is exposed and escape or inspissation of contained hydrocarbons is possible.

Under special conditions, the impervious redbeds and evaporite of the lower Amsden Formation and other "tight" sections also have become cap rocks. This is apparent only in fields where segregation of an original common pool into separate zones has resulted from any of the several types of modifying mechanisms discussed later.

COMMON-POOL STATE

COMMON OIL-WATER CONTACTS

The geometric relation between hydrocarbon distribution and structure in the many large, multizoned, Paleozoic anticlinal fields of the Big Horn basin provides an important basis for the proposed theory of Paleozoic oil and gas accumulation. In studying this relation, several hundred drill-stem and production tests were reviewed in field areas to determine the type of fluid recovery and structural elevation of each test interval. The compiled data indicated that, for a majority of the anticlinal fields of the basin, structural elevations of oil-water contacts in the different Paleozoic zones or formations within any particular field differ insignificantly —generally less than 10 percent of the length of the oil column. It was therefore concluded that, *in the majority of the anticlinal fields of the Big Horn basin, all Paleozoic producing zones have, or originally had, a common oil-water contact, and are, or were, originally part of a common pool.*

The apparent minor variations in the level of the oil-water contacts observed in some of the fields can be explained as the result of areal and zonal differences in such factors as: (1) lithology and accompanying reservoir porosity; (2) relative permeability to oil, oil viscosity, and length of transition zones[3]; (3) pressure gradients in the water phase of the different zones; and (4) level of water encroachment in the different zones, which reflects the uneven field develop-

[3] Aufricht (1965, p. 2) has indicated that, for a Tensleep reservoir containing 14° API oil, 96 ft of oil column, above the 80 percent water-saturation level, is required to generate 1 psi pressure differential between the oil and water. This example illustrates the reason for the long transition zones observed in association with low-gravity Paleozoic oils.

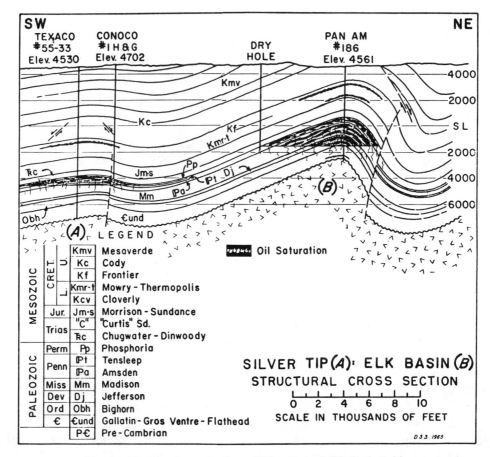

FIG. 9.—Structural cross section of Silvertip and Elk Basin fields.

ment of producing zones both in time and space, and probably also the relative strength of individual zonal water drives. Incomplete or erroneous information on fluid recoveries and test or completion intervals, particularly of the older fields and wells, may also be a contributing factor (Zapp, 1956; Arps, 1964).

The giant Elk Basin field, in the northernmost part of the basin in Wyoming, provides the most conspicuous example of the "common-pool state." This field produces from rocks of all Paleozoic systems except the missing Silurian and noncommercial Cambrian, and contains estimated ultimate Paleozoic reserves in the 500-million-bbl range. Figure 9B, a northeast-southwest structural cross section of the field, shows that an essentially level oil-water contact cuts across all Paleozoic zones in the field from Permian to Cambrian, without regard to lithology or age, encompassing them in a common pool having a vertical oil column measuring more than 2,300 ft. Figure 10 is

an illustration of the areal distribution of the oil in this common pool; the lines of intersection of the top of each producing Paleozoic formation with the common oil-water contact are shown.

Another excellent example of the common-pool state is the fields of the Garland-Byron structural complex, which are credited with an estimated total ultimate Paleozoic reserve of more than 225 million bbl of oil and oil equivalent. Figure 11 illustrates the relative vertical structural relationship between these two giant fields and includes a projected cross section of the more basinward Whistle Creek field. Figure 12 illustrates the areal pattern of the intersection of the top of the Paleozoic producing zones with the common oil-water contact at Garland.

The cross section (Fig. 11) reveals interesting and significant associations. The oil-water contact in both the Garland and Byron fields is essentially level, and common to all Paleozoic zones. However, at Garland, this contact is 1,000 ft higher than at Byron, though the former field is basinward of the latter. A large gas cap, also apparently common to all Paleozoic zones, is present at Garland, and absent at Byron (where the oil is probably undersaturated). Furthermore, the total vertical hydrocarbon column (gas and oil) at Garland is 1,500 ft *versus* 600 ft at Byron. Although there is less vertical closure at Byron (1,500 ft) than at Garland (3,500 ft), only a fraction of the available trap in both structures is filled with hydrocarbons. The writer suggests, therefore, that Garland may have had considerable vertical uplift relative to both Byron and Whistle Creek, subsequent to earlier structural entrapment of continuous phase, possibly undersaturated, oil. As a result of the reduced pressure environment thus created, free gas presumably came out of solution from the oil and expanded, causing downward displacement of the oil phase

to produce the large total column of hydrocarbons. D. I. Foster (personal commun., 1966), however, has suggested a correlation between the unusual gas cap at Garland and a buried Precambrian high area, revealed by Cambrian isopachous contours and by truncation at the top of the Bighorn Dolomite over the Garland anticline (Thomas, 1965, p. 1869).

In the huge bi-domed Oregon Basin field (Fig. 13B), in the western Big Horn basin, the Phosphoria, Tensleep, Amsden, Madison, and probably the very upper part of the Jefferson (of mostly nonreservoir lithology) have a common oil-water contact (Walton, 1947, p. 220) which appears to be slightly more irregular than in many other fields of the basin. Calculations from test data in the different zones give results which range up to 150 ft or 10 percent, of the 1,500-ft oil column in the south dome of the field. However, in-field

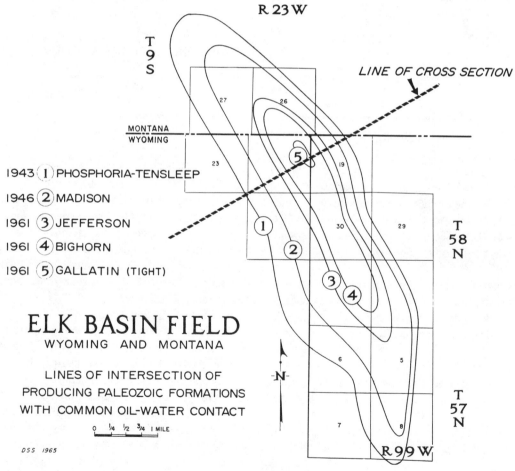

Fig. 10.—Map showing lines of intersection of producing Paleozoic formations with common oil-water contact in Elk Basin field.

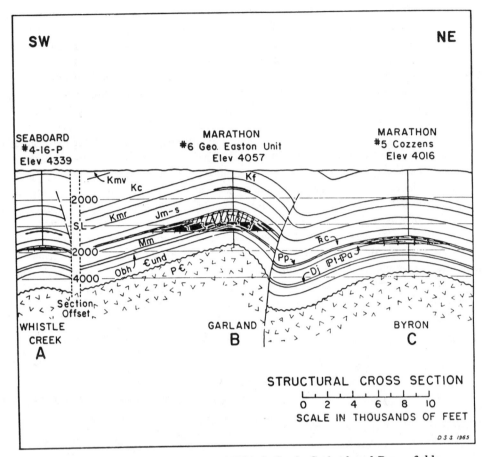

Fig. 11.—Structural cross section of Whistle Creek, Garland, and Byron fields.

minor faulting and particularly uneven field development, together with those variables listed previously, seem to explain adequately these minor variations in the common-pool state. Reserves of this Paleozoic accumulation are estimated to be considerably more than 200 million bbl of oil.

A small amount of gas and some oil are produced from the Triassic "Curtis" sandstones at Oregon Basin field. As no chemical analyses of these hydrocarbons are available, a genetic evaluation of this production cannot be made, but the predominance of free gas suggests either a Tertiary source across the east-flank fault zone or selective leakage of solution gas from a deeper Paleozoic reservoir. If this is Tertiary gas, it is the only exception to the observation that hydrocarbons in the Triassic are everywhere of the Paleozoic type.

The West Oregon Basin field is on a structural closure west of the south dome of Oregon Basin.

The small vertical closure, *i.e.*, approximately 100 ft, limits the accumulation to the Phosphoria, which has an oil-water contact little different from that at Oregon Basin (Fig. 13). This condition suggests original phase continuity between the two producing areas.

Two other examples of simple anticlinal accumulation, in which all Paleozoic zones comprise a common pool, follow.

(1.) In the Spring Creek field (Fig. 14A) of the western Big Horn basin, the northward-trending anticlinal structure is sharply folded and cut by a longitudinal, steeply east-dipping, thrust-fault zone, along the west flank. As shown in the structural cross section (Fig. 14A), an upper splay of this fault does not effect a seal on the accumulation; production is obtained from both sides of this splay, with a level oil-water interface cutting across all Paleozoic formations from Phosphoria through Madison, hanging wall and footwall, and enclosing all production in a common pool.

(2.) In the deep Worland field (Fig. 14B), in the eastern Big Horn basin, is another unusual situation. Within the Paleozoic, only the Phosphoria has yielded hydrocarbons in commercial quantities. These hydrocarbons include a sour gas

with an oil ring having variable gravities ranging from 45° API near the crest to 20° API near the oil-water contact (Morris, 1951, p. B-78). The significance of Worland field is that the effective vertical closure in the Phosphoria is more than

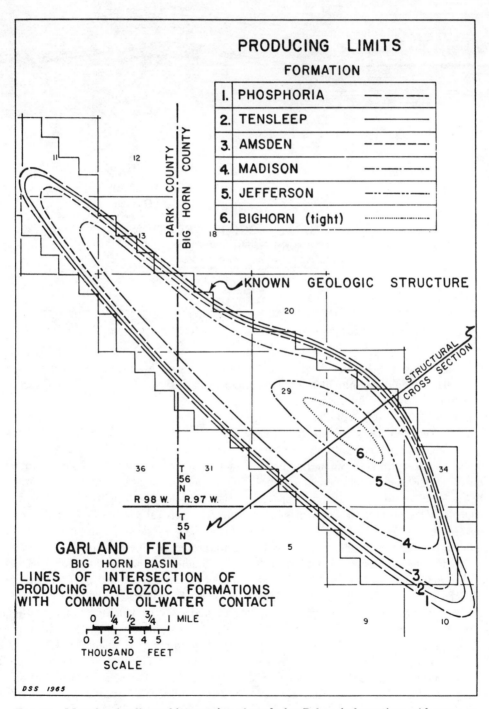

FIG. 12.—Map showing lines of intersection of producing Paleozoic formations with common oil-water contact in Garland field.

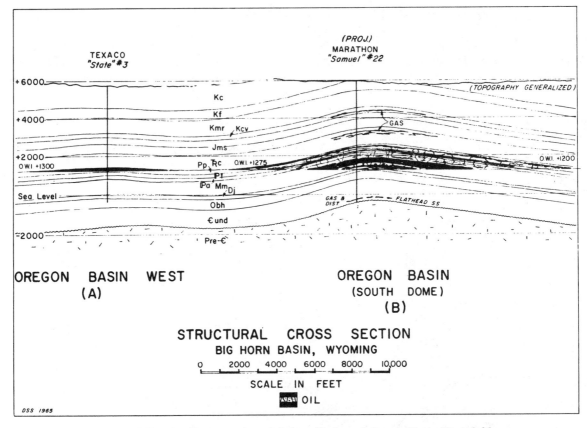

FIG. 13.—Structural cross section of Oregon Basin and Oregon Basin West fields.

750 ft and, according to the theory presented here, the Tensleep should contain hydrocarbons. The Tensleep does contain hydrocarbons; free oil, ranging from 22° to 26° API, has been recovered in all Tensleep tests within the area defined by an assumed common oil-water contact based on the apparent water level in the Phosphoria zone. The absence of commercial production from the Tensleep of the field thus appears to be explained by the absence of sufficient porosity and permeability.

The common-pool state probably also exists in other areas of the central Rocky Mountain province. The best-known example is the Lost Soldier field of the Red Desert basin, Wyoming, in which even some of the fractured Precambrian is included in the more than 2,000 ft of Paleozoic oil column (Krampert, 1949). In the Ashley Valley field of eastern Utah, the Phosphoria and Weber (Tensleep equivalent) are considered to comprise a common pool (Peterson, 1961) and contain identical Phosphoria-type crudes (Wenger and Ball, 1963). Many other examples of the common-pool state are available in the geologic literature (e.g., Gibson, 1965, p. 97; Hetherington and Horan, 1960).

SIGNIFICANCE OF EFFECTIVE VERTICAL CLOSURE

If all of the Paleozoic zones comprise a common pool with a common oil-water contact within an anticlinal closure, then a correlation should exist between the height of the oil column or effective vertical closure in the Phosphoria and the number of deeper Paleozoic formations that contain hydrocarbons on any particular structure. The number of these deeper sub-Phosphoria formations involved in the pool is determined by (1) the formational thicknesses in the crestal area and (2) the volume of hydrocarbons available for distribution. Whether oil can be produced in commercial quantities from any part of the Paleozoic section above the common oil-water interface, however, depends primarily on whether sufficient reservoir porosity and permeability, including fractures, are present.

Figure 15 was constructed to illustrate these

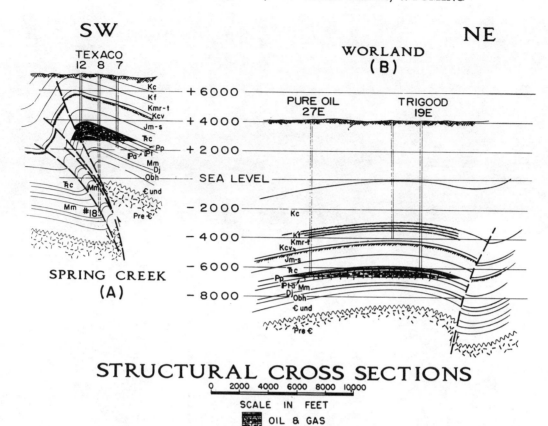

FIG. 14.—Structural cross sections of Spring Creek and Worland fields.

relations. Data on 39 fields are shown on this graph which includes only the "normal" anticlinal fields of the Big Horn basin, *i.e.*, those in the area of Phosphoria reservoir development and exhibiting essentially level, common oil-water contacts in the Paleozoic. Phosphoria stratigraphic accumulations and fields which show definite hydrodynamic modifications, which have been breached by surface erosion, or which are in the area of nonreservoir redbed Phosphoria facies,[4] are not included; these exceptions are discussed under separate subheadings. On this graph, the bars show the height of the Phosphoria oil column for each anticlinal field as measured downward from the top-of-Phosphoria base line (abscissa). At the same vertical scale, and measured in the same downward direction from the top of the Phosphoria, the crestal thicknesses of the productive Paleozoic formations in each field also are shown. Considerable thickness variation from field to field can be observed on the graph in

[4] These could have been included if the oil column were measured downward from the top-of-Tensleep rather than top-of-Phosphoria datum.

each of the Paleozoic formations. The fields have been arranged into five groups, according to the number of formations that are productive in each field, with "Phosphoria saturation only" on the right and "Phosphoria, Tensleep, Amsden, Madison, Jefferson, Bighorn (including Gallatin) saturation" on the left. Figure 15 illustrates that in the area of marine Phosphoria reservoir rocks, the number of productive Paleozoic formations or zones in any normal anticlinal oil field in the Big Horn basin is dependent upon the height of the Phosphoria oil column and the crestal thicknesses of the Paleozoic formations involved.

IMPORTANCE OF FRACTURES

Vertical communication between the Paleozoic producing zones in the Big Horn basin Paleozoic fields is believed to be primarily through fractures and faults. Prominence of fractures in the Paleozoic carbonate and calcareous sandstone is shown by core information from all productive formations. Fractures, usually "bleeding oil," are described from cores of producing zones in every Big Horn basin Paleozoic field; below the oil-

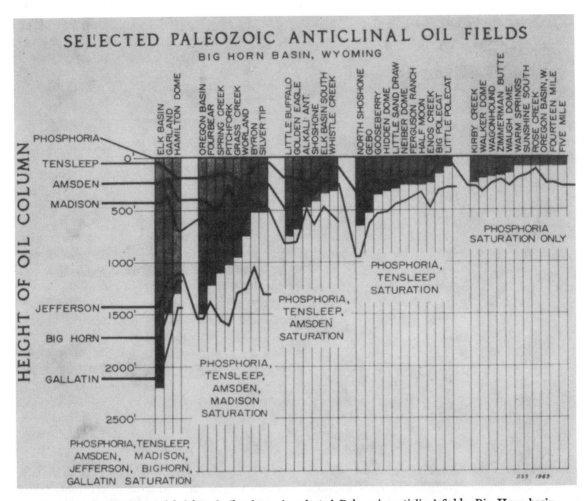

Fig. 15.—Correlation and height of oil column in selected Paleozoic anticlinal fields, Big Horn basin.

water interface, "dead oil" or "asphalt" stains in the fractures commonly are described.

Fractures formed because of the brittle nature of these predominantly carbonate reservoir rocks. Even the sandstones of the Tensleep and Amsden Formations act as brittle materials, largely because of the competent control of the surrounding massive, interbedded dolomites and limestones, and also as a consequence of the highly indurated or cemented nature of the sandstones. Drillers and operators commonly have remarked on the importance of fractures in the sandstone and, in many Tensleep pools drilled near the water level, they carefully have avoided penetrating too far into the Tensleep reservoir for fear of drawing bottom water into the well through fractures. Fracture communication between the Phosphoria ("Embar") and Tensleep zones generally was assumed by operators as shown by their common earlier practice of com-

pleting wells with both formations open to the well bore below casing (Walton, 1947, p. 220; McCoy et al., 1951, p. 1013).

In the Mississippian, Devonian, and Ordovician carbonates as well as in the Amsden and Phosphoria, fractures may provide a large part of the reservoir capacity (Stipp, 1947, p. 124, 126; Walton, 1947, p. 213; Willingham and McCaleb, 1966).

The Paleozoic fractures of Big Horn basin anticlines are primarily of both regional and local deformational origin and related to Laramide orogenesis. Harris et al. (1960), from a surface study of two typical Wyoming anticlines, concluded that, "The trend and concentration of fractures are controlled by the compressional structure configuration" (p. 1853). Although many Big Horn basin anticlines are asymmetric and faulted along their basin-edge flank, they are typically broad and simple, and may be classified as concentric folds (de Sitter, 1956). Most crestal fractures, therefore, are believed to be of tensional origin, produced as a consequence of uplift and arching of these folds. Along the steep flanks associated with longitudinal faulting, however, fracturing and minor secondary faulting are more likely the result of compressional or shear phenomena. Fractures produced by either of these mechanisms must have been formed by Laramide stresses which became active in Late Cretaceous time and continued through Paleocene and into Eocene time (Love, 1960, p. 205).

In addition to tectonic mechanisms, some fracturing may be of the gravity type, or may have formed as a result of diagenetic changes such as recrystallization or dolomitization.

Fracture communication in the axial regions of anticlinal accumulations is observed not only in the Rocky Mountains, but also in many producing areas of the world, with notable examples among the giant Middle East oil fields (Boaden and Masterson, 1952; Baker and Henson, 1952; Daniel, 1954; Dunnington, 1958).

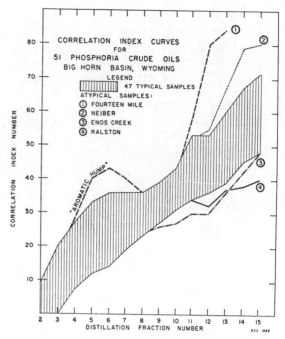

Fig. 16.—Correlation Index curves for 51 Phosphoria crudes, Big Horn basin.

GEOCHEMISTRY OF FORMATION FLUIDS
CRUDE-OIL CHEMISTRY

The natural division of pre-Tertiary Wyoming crude oils into two major groups, Mesozoic and Paleozoic, based on their chemical composition, has been well established (Hunt, 1953; McIver, 1962; and many others). Mesozoic crudes have a paraffin-naphthene base, low-sulfur, low-residuum, and high-gasoline content and are chemically "mature" according to Barton's rule (1934). These oils are associated with dark-colored, marine shale source rocks which enclose the shallow-marine sandstone reservoir rocks. In contrast, the Wyoming Paleozoic oils, with some notable exceptions, have been described as high-sulfur, high-residuum, low-gasoline, aromatic-naphthenic (asphaltic)-base crudes; they are relatively "immature" (H. M. Smith et al., 1959, p. 441).

In the Big Horn basin, chemical differences are insignificant between the crude oils from different Paleozoic formations in a single field, or from field to field in localized basin areas. Wenger and Lanum (1954a) have demonstrated this condition in their article on the Big Horn basin crude oils. However, minor differences occur in the chemical makeup of the Paleozoic oils of some fields, particularly in the deeper fields; it is proposed here

that these differences can be attributed, for the most part, to differences in the physical environments of the reservoir rocks.

Correlation index (CI) curves.—Figures 16, 17, and 18 illustrate the range of compositional variation for 51 Phosphoria, 40 Tensleep, and 11 Madison crude oils, respectively, in the Big Horn basin, utilizing the familiar U.S. Bureau of Mines correlation index (CI)[5] curves (Smith, 1940). Moreover, specific CI curves for Paleozoic crudes from three of the large multi-pay Big Horn basin fields have been plotted in Figures 19 through 21. The uniformity of the three curves from the Oregon Basin field (Fig. 19) is striking, and the curves for Elk Basin crude-oil samples (Fig. 20) show as much variation between 9 Tensleep samples (Espach and Fry, 1951) as do the curves for the oils from the Phosphoria, Madison (Biggs and Espach, 1960), Jefferson, and Bighorn (U.S. Bureau of Mines open file, 1964) zones.

[5] An empirical characterization number developed by the U.S. Bureau of Mines from routine Hempel distillation data. The magnitude of the correlation index number is related to the type of hydrocarbons that comprise the crude-oil fraction. The correlation index number is derived by use of the formula: $CI = 48640/K$ $(473.7G - 456.8)$, where K is the average boiling point of the fraction in degrees Kelvin, and G is the specific gravity of the fraction at 60° F.

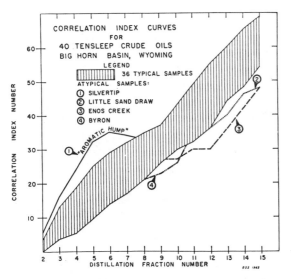

FIG. 17.—Correlation Index curves for 40 Tensleep crudes, Big Horn basin.

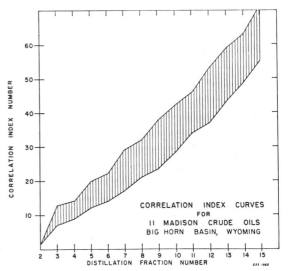

FIG. 18.—Correlation Index curves for 11 Madison crudes, Big Horn basin.

For those fields in which faulting and/or hydrodynamic effects such as cross-formational flow are apparent, a greater variation may occur in the chemical composition of contained crudes as expressed by CI curves. In the Hamilton Dome field, some differences can be seen between oils in the two deepest Paleozoic zones (more aromatic) and the three shallowest zones, including the "Curtis" Sandstone (More paraffinic); but the curves have similar slopes, and all five zones are believed to comprise a common pool with a common oil-water contact (Fig. 21 and Fig. 27C). A similar separation of the curves is observed in samples from the hydrodynamically modified Grass Creek and Torchlight fields. However, in comparison with the major differences in the CI curves (and basic chemical composition) between the crude oils from the Big Horn basin Paleozoic formations and from other Rocky Mountain producing formations and areas, these minor variations are not believed to have fundamental genetic significance, but are thought to be the result of

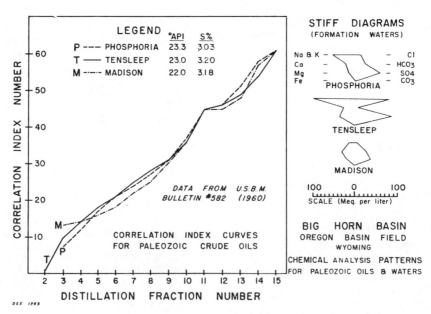

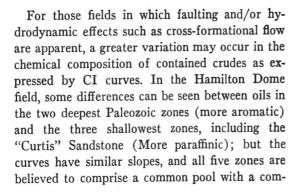

FIG. 19.—Chemical analysis patterns for Paleozoic oils and waters from Oregon Basin field, Big Horn basin.

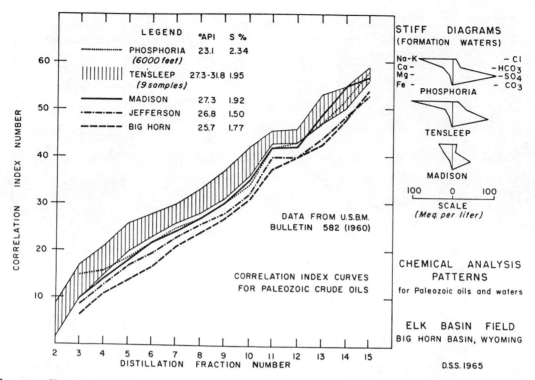

Fig. 20.—Chemical analysis patterns for Paleozoic oils and waters from Elk Basin field, Big Horn basin.

post-accumulation alteration of the original crude in response to physical and chemical environmental factors.

The similarity of all Paleozoic and Triassic CI curves agrees with the concept of a single Phosphoria source. In support of this statement, Figure 22 is a comparative graph of CI curves of three typical Madison crudes from the Big Horn basin and three from the Williston basin of North Dakota and Montana. These curves show that the Williston basin Madison crudes are unlike those from Wyoming. The Williston basin crudes have a paraffinic base with high gasoline and low-sulfur content and are more paraffinic in the higher boiling ranges, whereas "oils from Permian, Pennsylvanian, and Mississippian formations in these other areas [Big Horn basin] are generally more paraffinic in the gasoline range and contain more asphalt. Their CI values in few places indicate aromatic concentrations in the gasoline, and they are usually higher in the lubricating-oil range. CI curves of the Williston basin oils from Mississippian formations show more resemblance to those of Cretaceous oils produced in other areas outside of the basin" (Wenger and Lanum, 1954b, p. C-46, C-47).

The Williston basin crude oils furthermore are from fields in which stratigraphic control of the Madison reservoirs has been demonstrated (Fish and Kinard, 1959, p. 119; G. W. Smith et al., 1958), and are thus considered to be indigenous to the Madison. The geographic and geologic locations of these fields within this mildly deformed basin rule out exotic sources for many of the more centrally located fields. Thus the difference in basic chemical composition of the crude oils from the Madison in the two basin areas can be taken as an indication of a difference in source environments for the two crudes. This reasoning can also be applied to sub-Madison crudes in the same two areas (cf., Porter and Fuller, 1959).

Refractive index-specific gravity patterns and infrared-absorption spectra.—Hunt (1953, p. 1849, 1850) demonstrated that basic chemical similarities in the crude oils from the Paleozoic producing zones in Wyoming are also in refractive index-specific gravity and infrared absorption spectra patterns (Fig. 23). The crude-oil samples analyzed by Hunt include one from the Phosphoria in the Winkleman Dome field of the Wind River basin, which is clearly comparable with

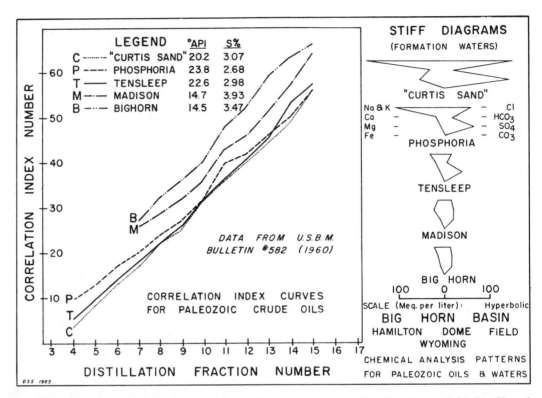

FIG. 21.—Chemical analysis patterns for Paleozoic oils and waters from Hamilton Dome field, Big Horn basin.

those from the Phosphoria, Tensleep, and Madison zones of the Kirby Creek, Grass Creek, and Elk Basin fields, respectively, in the Big Horn basin. The difference between these Paleozoic oils and Cretaceous oils is shown in Hunt's Figure 5 (1953, p. 1851). These same Paleozoic crudes

have about twice as many heavy aromatics, oxygenated compounds, and asphaltenes as Wyoming Cretaceous crudes (Hunt, 1953, p. 1853, Table V).

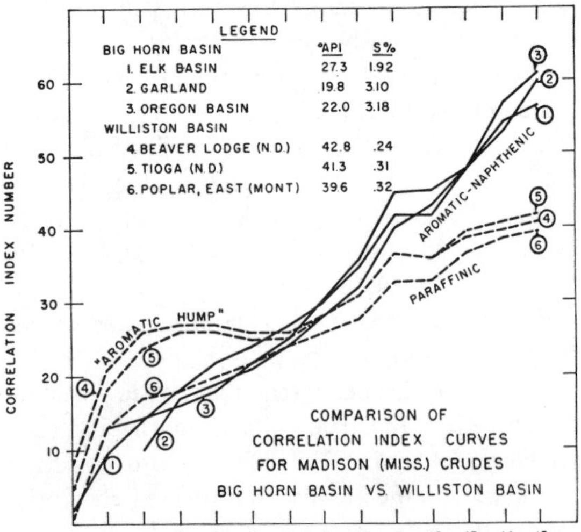

Similar results have been obtained in analyses by the Chevron Research Co. This work reveals that concentration of oxygenated compounds in Big Horn basin Paleozoic oils is somewhat higher than in the Cretaceous oils, but not unusually high. In addition, infrared data (i.e., intensity of absorption peak at 14.85ν) show that benzene is absent from most of these Paleozoic crudes, a condition confirmed also by Hempel distillation analyses data. These data are pertinent to the evaluation of the alteration effects discussed in a later section.

Sulfur, nitrogen, and trace metals.—In the literature, considerable importance has been attached to the relative concentrations of sulfur, nitrogen, and trace metals (held predominantly in porphyrin complexes) in the two types of Wyoming crudes. Paleozoic crudes are characterized by a high sulfur content, averaging 2.4 weight percent, whereas Cretaceous oils average 0.2 weight percent sulfur (Hunt, 1953, p. 1845; McIver, 1962, p. 248). Furthermore, in the Big Horn basin, the average nitrogen content in the Paleo-

FIG. 22.—Comparison of correlation index curves for Madison crudes from Big Horn and Williston basins.

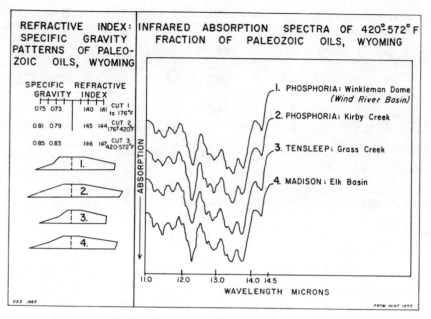

FIG. 23.—Refractive index-specific gravity patterns and infrared absorption spectra of the 420-572°F fraction of Paleozoic oils, Big Horn basin, after Hunt (1953).

zoic and Cretaceous crudes is calculated to be 0.22 percent and 0.04 percent, respectively. Finally, large amounts of vanadium-type porphyrins are in all heavier Paleozoic crudes (Bell, 1960; Hayden, 1961), whereas porphyrins are present only in small amounts or are absent from the lighter Cretaceous crudes. The nickel content generally is greater than the vanadium content in the Cretaceous oils according to Hayden (1961, p. 71).

Sulfur, nitrogen, and important trace metals are associated largely with the heavy asphaltic part of crude oils (Ball, 1962; Bell, 1960; Dunning *et al.*, 1953; Erickson *et al.*, 1954; Hayden, 1961; Kartsev *et al.*, 1959, p. 6; Rall *et al.*, 1962; H. M. Smith, 1962; H. M. Smith *et al.*, 1959). Thus, the high concentration of these nonhydrocarbons in the Paleozoic crude oils of the Big Horn basin is a predictable condition related to the high asphaltic content of these oils. However, the condition may not add much support to the established basic chemical differences between the aromatic-naphthenic Big Horn basin Paleozoic crudes, and the paraffinic Big Horn basin Cretaceous crudes or the Williston basin Paleozoic crudes. According to Sugihara (1962, p. 31), a positive correlation between V/Ni ratios and asphaltic content also exists, which may partly account for the high V/Ni ratios (generally great-

er than 4) found in Wyoming Paleozoic crude oils, and low V/Ni values found in Wyoming Cretaceous and Williston basin Paleozoic oils (from 1.5 to 2). Vanadium-to-nickel and other chemical ratios nevertheless can be valuable as characterization parameters.

The diverse array of minor trace elements in Paleozoic crudes of the Big Horn basin (Hayden, 1961) can be related also to the asphaltic character of these oils. However, Sheldon (1963, p. 9, Fig. 3) has pointed out that the identical elements occur in the Phosphoria carbonaceous and phosphatic shales; he has shown a rough correlation between those elements in the rock and the oil.

Wenger (personal commun., 1965) has found that nitrogen-to-carbon residue ratios generally are lower in Mississippian crudes of the Williston basin than in Mississippian crudes in the Big Horn basin. Similarly, a plot of the analyses of both Big Horn basin (Biggs and Espach, 1960) and Williston basin (Lindsey, 1954) Madison crudes suggests a lower sulfur-to-carbon residue ratio for the former oils (0.22) than for the latter oils (0.33). These facts support the concept of different source environments for these different crude oils.

Compilations of chemical data, such as those discussed above, also give strong support to the proposed theory of Paleozoic oil and gas ac-

209

TABLE III. SOME CHEMICAL CHARACTERISTICS OF PALEOZOIC CRUDE OILS FROM HAMILTON DOME FIELD

Formation	Est. Mean Elev. of Sample	API Grav.	Isotope Ratio C13/C12 (0/00)	Wt. % in Crude S	Wt. % in Crude N	Oxygenated Compounds Infrared* Absorption at 5.85 v	Trace Metals Ni	Trace Metals V	Trace Metals V/Ni	Percent Gasoline and Naphtha	Hempel Distillation Percent Carbon Residue in Crude	Hempel Distillation Correlation Index: Cut 5th	Correlation Index: Cut 10th	Correlation Index: Cut 15th
Permian, Phosphoria	+3,100	23.5	+0.1	2.65	.25	.037	7.6	46	6.0	14.4	7.8	13	32	56
Pennsylvanian, Tensleep	+3,000	22.4	+0.0	2.80	.35	.035	NA	H**	—	11.2	7.5	14	32	57
Triassic, Curtis	+2,925	19	+0.2	3.00	.38	.058	6.4	28	4.4	9.5	10.3	8.5	32	56
Mississippian, Madison	+2,400	14.7	+0.3	3.40	.38	.063	7.0	28	4.0	1.7	11.4	—	36	64
Ordovician, Bighorn	+2,100	14.5	+0.2	3.45	.45	.063	NA	H	—	2.2	10.6	—	40	65

Sources: Hayden (1961); Biggs and Espach (1960); Chevron Research Co.

* In optical density units.
** H; relatively high.
NA: not available.
Note: No benzene is present in any of these oils.

cumulation in the Big Horn basin. For example, the small variation in the V/Ni ratio in the crudes from the three main Paleozoic producing zones in the Hamilton Dome field of the Big Horn basin (Table III) and also from four zones in the Circle Ridge field (Table IV) of the neighboring Wind River basin agrees with the concept of common source and common reservoir. The Circle Ridge field is one of only two fields in the Wind River basin which has an effective vertical closure greater than the crestal thickness of the section from the top of the Phosphoria to the top of the Madison reservoir, i.e., 1,000 ft (Biggs and Espach, 1960, p. 62).

Carbon and sulfur isotopes.—Carbon C^{13}/C^{12} and sulfur S^{34}/S^{32} isotope ratios for petroleums have been investigated with mass spectrometric methods by several workers. Although carbon-isotope ratios of petroleums have been found to differ by only 1 percent, compared with 9 percent for all organic compounds (Silverman, written commun., 1966), the differences observed in carbon-isotope ratios of crude oils from different-age rocks generally are sufficiently uniform and distinctive to be useful as a characterization parameter in the differentiation of petroleums.

The average C^{13}/C^{12} ratio of +0.7 per mil for Paleozoic crudes of the Big Horn basin is in contrast to the somewhat higher average of about +2.1 for Wyoming Cretaceous crudes and lower average of −0.95 for Mississippian crudes from the Williston basin (data from Chevron Research Co., unpub.). Therefore, because the overall differences in C^{13}/C^{12} ratios between aromatic-enriched and aromatic-free fractions of petroleums

differ by only 1 per mil (Silverman and Epstein, 1958, p. 1006), it is probable that the differences in the average values for these three groups of crudes reflect differences in their source environments. In contrast, although significant variations occur within each group, it is important to the present theory of Paleozoic oil and gas accumulation in the Big Horn basin to note the small variation in C^{13}/C^{12} values between the different Paleozoic zones within the same Big Horn basin field (Tables III and V).

Thode *et al.* (1958) concluded similarly that sulfur-isotope ratios for geologically related petroleums appear to give information concerning the isotopic content of source sulfur at the time of petroleum formation. The evidence indicates that little or no fractionation of sulfur isotopes occurs during the maturation of oil in which sulfur is lost. Sulfur-isotope ratios also are independent of the sulfur content of the crude oil. The fact that hydrogen sulfide and associated petroleums have been found to have similar isotope contents is further evidence that little or no fractionation of sulfur isotopes takes place during mat-

TABLE IV. VANADIUM AND NICKEL CONTENTS OF PALEOZOIC CRUDE OILS FROM CIRCLE RIDGE FIELD, WIND RIVER BASIN, WYOMING (AFTER McIVER, 1962)

	V (ppm)	Ni (ppm)	V/Ni
Phosphoria	88	20	4.4
Tensleep	91	20	4.6
Amsden	91	20	4.6
Madison	93	20	4.7

TABLE V. SOME CHEMICAL CHARACTERISTICS OF CRUDE OILS FROM GARLAND FIELD

Formation	API Gravity	Benzene 14.85 ν	Carbon Isotopes C^{13}/C^{12} (0/00)	Wt. Percent in Crude		Oxygenated Compounds Infrared Absorption at 5.85 ν	Trace Metals			Percent Gasoline and Naphtha	Hempel Distillation			
				S	N		Ni	V	V/Ni		Percent Carbon Residue in Crude	Correlation Index: Cut		
												5th	10th	15th
Phosphoria	24.4	.03	+1.7	2.45	.15	.037	NA	H	NA	—	—	—	—	—
Tensleep	23.0	.00	+0.7	2.50	.28	.020	20	30	1.5	13.4	8.3	15	35	61
Amsden	22.0	NA	NA	2.88	.27	NA	NA	NA	NA	13.3	8.9	18	32	61
Madison	20.8	.00	+1.0	2.81	.28	.027	—	M	NA	9.8	9.6	17	32	60

Sources: Biggs and Espach (1960); Hayden (1961); Chevron Research Co.

uration (Thode *et al.*, 1958; Feely and Kulp, 1957).

From six unidentified samples of Wyoming Paleozoic crude oil, Thode *et al.* (1958, p. 2631, Table IX) found the average S^{34}/S^{32} ratio was −3.8 per mil, whereas in the same number of unidentified Wyoming Cretaceous samples the average ratio was −1.95 per mil, indicating a depletion in S^{34} for both groups. Also, average sulfur-isotope ratios in Alberta and Saskatchewan for the Mississippian Madison Group, Charles Formation and Mission Canyon Limestone, crudes were reported to be +1.8 and +5 per mil, respectively; other northern Rocky Mountain Paleozoic samples showed higher positive values. Unfortunately, the variation and narrow range in the values recorded and the small number of samples analyzed in Thode's studies impose limitations on the usefulness of these data in the evaluation of genetic differences between these petroleums. It is regrettable that no discrete sulfur-isotope data are presently available for the Paleozoic crudes from some of the larger fields of the Big Horn basin, such as Elk Basin, because such data could provide an additional test of the writer's concept of the common-pool state.

CRUDE-OIL ALTERATION

Petroleum composition is determined presumably not only by source material, depositional, generative, and transformational environments, and migrational processes, but also by any chemical changes within the reservoir after accumulation. Todd (1963) has proposed that alteration or "degradation" processes active during or immediately after Laramide orogenesis explain most of the differences not only among Tensleep crude oils but also among Paleozoic and Cretaceous crude oils (p. 611) in the Big Horn basin. The writer cannot agree with this viewpoint. In this section, several important alterational mecha-

nisms are considered in the light of available geochemical data, and the relative importance of these mechanisms in determining the chemical composition of the Paleozoic oils of the Big Horn basin is evaluated.

Water washing.—The idea that crude oils might be degraded through the washing action of moving artesian waters has been mentioned by many geologists and geochemists (K. E. Chave, personal commun., 1954; Cook, 1925; Hunt, 1953, p. 1870; Jones and Smith, 1965; McIver, 1962, p. 251; Nutting, 1934; Todd, 1963), but few analytical data on the subject have been published. The widely distributed Tensleep Sandstone of the Big Horn basin with its asphaltic crude, artesian water circulation, and obvious hydrodynamic associations provides a good model for this problem.

Most investigators of crude-oil degradation by moving water have been concerned primarily with the relative solubility of the various liquid hydrocarbons in water, although considerable work also has been done in the related field of gaseous hydrocarbon solution (Buckley *et al.*, 1958). The molecular weight is the prime factor influencing solubility of hydrocarbons in water (Erdman, 1962, p. 33); *i.e.*, light hydrocarbons are more soluble in water than heavier hydrocarbons but aromatics are more soluble than paraffins or naphthenes of the same molecular weight. Thus, small amounts of both light aromatic and light paraffin liquid hydrocarbons theoretically can be removed from a crude oil through solution in moving water. However, *light aromatics are removed preferentially*, unless, as inferred by Todd (1963), oxidation precedes solution, converting light paraffins to acids, ketones, aldehydes, alcohols, *etc.*, which are more soluble than light aromatics (McAuliffe, 1966). These oxidized paraffin compounds, however, are scarce in natural crude oils and normally not found in subsurface waters.

211

The simplest member of the aromatic series, benzene, is the most water-soluble pure liquid hydrocarbon found in natural petroleum (Seidell, 1941, p. 368; Baker, 1959, p. 873; Zarella et al., 1963). Toluene, the next compound in the aromatic series, is also the next in water solubility (Baker, 1959). The importance of this fact has been revealed recently by the development of the "Benzene analysis" method (Gulf Research and Development Co., Patent No. 2,767,320) which has been called "a new exploration tool." Benzene analysis is designed to take advantage of the discovery that subsurface waters in contact with an oil accumulation generally contain detectable quantities of dissolved benzene and other light aromatics, with this content decreasing laterally outward from the accumulation (Zarella et al., 1963).

The majority of Paleozoic oils in the Big Horn basin contain little or no benzene, as shown by both infrared absorption (viz. at 14.85ν) and by the U.S. Bureau of Mines routine Hempel analyses data. However, the absence or very low concentration of benzene in these oils, although suggestive, does not prove water washing per se, inasmuch as low concentrations of benzene could be an original condition related to the chemical "immaturity" of these crudes.

To evaluate further the water-washing hypothesis, the variations in the concentration of other light aromatics held in the lower boiling-point fractions of the Big Horn basin Paleozoic crudes were investigated utilizing the available data of the U.S. Bureau of Mines. At the beginning of this investigation, it was reasoned that, if all Paleozoic crudes in the Big Horn basin fields were derived from a single source (the Phosphoria Formation, as here proposed), and, if these crudes originally were of approximately identical chemical composition, any differences in light aromatic content observed between the oils entrapped in different zones within the same field or limited area of the basin, excepting the mature oils in the deep basin fields, should be attributed probably to the degradation effects of water washing. The relative importance of water washing in the different fields and zones, furthermore, should be related causally to such factors as the velocity of water flow through any particular zone, hydrodynamic gradients, and the extent of exposure of the oil to flowing water at the oil-water contact as determined, for example, by the relation between the height of the oil column and the areal size of the pool, and by the temperature and pressure in the reservoir (Levorsen, 1954, p. 531).

Under the similar hydrodynamic environment which is likely to exist within the Paleozoic reservoirs in any field area, the relative transmissibility (permeability × thickness) characteristics of these zones are probably the most important determinants of the rate of flow. Therefore, because the Madison, Tensleep, and Phosphoria zones in the Big Horn basin can be rated, in that order, on a scale of decreasing transmissibility (Roger Hoeger, Petroleum Research Corp., personal commun., 1964), the light aromatic concentration in associated crude oils should be expected to decrease proportionately from the younger to older formations where hydrodynamic flow is indicated. Wenger and Lanum (1954a, p. A-60) have noted that Madison oils have "no aromatic concentration" in the gasoline range; and that "one difference between Phosphoria and Tensleep oils is that fewer Tensleep oils show aromatic concentration in this range" (cf., Figs. 16, 17 and 18). The greater exposure of Madison oils to water washing at the oil-water interface resulting from the lower stratigraphic position and usual central-fold location of the Madison zone in the common-pool fields is also an important factor contributing to the absence of aromatics within this unit.

To place quantitative values on the importance of water washing of Paleozoic oils in terms of their relative light aromatic content, a comparison has been made of the distillation data on the fifth cut or fraction (302°F)[6] of Phosphoria, Tensleep,

---

[6] The aromatic concentration of the 5th fraction was chosen because (1) data are more abundant for this fraction than for lighter fractions below five, and (2) on the basis of the boiling points of aromatic compounds, it is expected that the most soluble light aromatics after benzene, and probably through propylbenzene (boiling point, 320°F), should be reflected in the composition of this fraction. Gooding et al. (1946, p. 4) observed that aromatic compounds distill through a fairly broad range of temperatures, from below to above their boiling points (e.g., toluene, 203°F to 251°F; boiling point, 231°F) because of the effects of mixture with other hydrocarbons. It has been pointed out also (Smith et al., 1950) that crudes with high CI values in the vicinity of the 4th and 5th fractions followed by decreases between the 5th and 7th fractions contain useful quantities of

TABLE VI. CORRELATION INDEX RATIOS AND RELATIVE AROMATIC CONTENT AS MEASURES OF IMPORTANCE OF WATER-WASHING OF PALEOZOIC CRUDES. (KEY: P=PHOSPHORIA; T=TENSLEEP; M=MADISON)

Field	API Gravity°			Correlation Index Fifth Cut			Correlation Index Ratios Fifth Cut			Estimated Percent Aromatic Content[1] Fifth Cut			Important Hydrodynamic Modification
	P	T	M	P	T	M	P:T	P:M	T:M	P	T	M	
Black Mountain	26.1	23.3	20.3	20	13	—	1.54	1.7	1.08	14.2	3.0	2.1	X
Byron	18.2	24.3	—	15	17	—	0.88	—	—	6.5	7.0	—	—
Elk Basin	23.1[2]	27.1	27.3	19	21	18	0.90	1.06	1.11	12.0	11.0	11.0	—
Enos Creek	27.1	27.1	—	13	13	—	1.00	—	—	1.0	1.0	—	?
Frannie	26.8[3]	27.5	17.1	18	19	(14)[4]	0.95	(1.28)	(1.36)[4]	13.0	14.0	(1.0)[4]	X
Garland	—	23.0	19.8	—	15	17	—	—	1.88	—	—	7.5	—
Golden Eagle	48.7	41.7[5]	—	23	18	—	1.28	—	—	19.1	10.1	—	?
Gooseberry	23.5	22.6	—	15	14	—	1.07	—	—	8.0	6.0	—	—
Gooseberry, North	19.0	20.8	—	20	14	—	1.43	—	—	11.8	6.9	—	?
Grass Creek	24.5	24.2	—	18	15	—	1.20	—	—	21.0	7.0	—	X
Half Moon	14.1	15.3	—	—	15	—	1.20	—	—	2.9	3.7	—	—
Hamilton Dome	23.8	22.6	14.7	13	9.9	None	1.31	—	—	7.0	4.0	None	X
Lake Creek	25.2	30.0	—	22	14	—	1.57	—	—	16.0	10.5	—	X
Lake Creek, Northwest	32.6	30.6	—	26	16	—	1.63	—	—	21.0	7.0	—	X
Little Buffalo Basin	20.7	20.7	—	15	14	—	1.07	—	—	—	8.0	—	—
Little Sand Draw	36.1	35.0	—	18	18	—	1.00	—	—	6.8	7.1	—	—
Manderson (& SW)	33.4	36.4	—	28	25	—	1.12	—	—	29.0	20.0	—	—
Neiber (& SE)	40.6	33.2	—	29	20	—	1.45	—	—	—	15.0	—	?
Oregon Basin (North)	23.3	23.0	22.0	17	18	16	1.06	1.07	1.00	8.0	7.0	—	—
Silver Tip	43.2	43.0	—	31	32	—	0.97	—	—	30.0	29.0	—	—
Torchlight	31.3	35.0	20.5	20	14	12	1.80	2.17	1.17	26.0	6.0	2.0	X

1. From specific dispersion graph in McKinney and Garton (1957, Fig. 1) based on method described in Holliman et al. (1950).
2. Sample is structurally the lowest from Elk Basin field (see Fig. 24).
3. Data from U. S. Bureau of Mines open file (Montana sample).
4. Estimated from Sage Creek data—not recorded on Frannie sample.
5. Tensleep data, taken from Hempel analysis in Biggs and Espach (1960), is of doubtful accuracy.

and Madison crude oils in the same fields. Table VI shows two types of measurements: (1) ratios of the correlation index values for Phosphoria-Tensleep, Phosphoria-Madison, and Tensleep-Madison crude samples; and (2) a determination of the approximate volume percent aromatic concentration for crudes from the three formations. Aromatic percentages were taken from the specific gravity *versus* specific dispersion graph by McKinney and Garton (1957, p. 28, Fig. 1) which was constructed from relations reported by Holliman et al. (1950).

Inasmuch as the CI number for a particular fraction increases with the aromatic content of the fraction, the ratio of the Phosphoria and Tensleep CI numbers, for example, is an indication of the relative difference in light aromatic content of the crudes from these two formations within the fields listed. Significantly, a positive correlation exists between the higher CI ratios for the fifth fraction (*i.e.*, generally greater than 1.25) and the presence of important hydrody-

toluene, provided the index value is 30 or higher. Considering Baker's (1959) observation that toluene is about four times more abundant than benzene in nearly all crude oils, and because benzene content is nil in a large number of the Big Horn basin Paleozoic crudes, it is clear that measurement of the concentration of toluene and higher aromatics should be more meaningful in the evaluation of the importance of water washing than measurement of benzene content.

namic modification or tilt on the oil-water contact in the Tensleep reservoir. Extensive areal exposure of the oil to water flow also is indicated for these fields. For example, Black Mountain, Lake Creek, Northwest Lake Creek, and Torchlight fields all have values for the ratio CI Phosphoria: CI Tensleep (fifth fraction) greater than 1.5; these same fields exhibit strong hydrodynamic effects in the Tensleep reservoir and have small vertical oil columns compared with the productive area.

The Phosphoria-Tensleep data for Frannie field do not seem to indicate removal of aromatics from the Tensleep crude even though this field is in an area of active hydrodynamic flow (Figs. 26, 27B). The fact that the Phosphoria sample was obtained 650 ft lower on the structure than the Tensleep sample, and that the Phosphoria is very thin and "tight" in this field and seldom, if ever, produced as a single segregated zone, may provide a partial explanation for this inverse relation; a similar condition at Elk Basin (item 3, Table VI) is associated clearly with a much lower structural location for the Phosphoria sample than for other crude-oil samples, as shown in Figure 24. The greater vertical oil column in the Frannie field compared with other hydrodynamically modified fields, or a postulated tar mat at the oil-water contact (Fanshawe, 1960, p. 15; Lawson and Smith, 1966; Todd, 1963), could

213

have protected the oil from extensive water-washing effects.

In further support of these data as a measure of water washing, a plot of aromatic content (CI values or percent aromatics) of the fifth fraction against depth (or temperature) did not show any well-defined pattern, indicating that depth is not an important controlling factor.

Table VI indicates that *there is no distinctive correlation between crude-oil gravity and intensity of water washing* as proposed by Todd (1963). Furthermore, an *increase* in aromatic content relative to paraffin in Tensleep petroleums has not occurred through water washing in the Big Horn basin, as claimed by Todd (1963, p. 611). On the contrary, at the same molecular weight, aromatics are more soluble than paraffins, and light aromatics, not paraffins, seem to have been removed preferentially from the Tensleep crude in some fields, thus causing a proportional increase in the concentration of equivalent molecular-weight paraffins in the lighter part of the oil. If some light liquid paraffins also were removed from this same part of the oil, available chemical data (infrared-absorption and Hempel analyses data) do not seem to indicate that this mechanism was important. It is possible, however, that the common undersaturated condition of Paleozoic oil pools may be a partial reflection of the removal by circulating waters of some solution gas and "ethane plus" gaseous paraffin hydrocarbons (*cf.*, Buckley *et al.*, 1958). Although this mechanism might account for some API gravity reduction, it would not affect materially the fundamental composition of the parent liquid petroleum, nor would the preferential removal of light aromatics from crude oil cause an appreciable increase in heavy aromatics over light paraffins.

Oxidation by water washing.—Todd also has placed considerable emphasis on oxidation in the possible alteration of Paleozoic crudes in the Big Horn basin. Oxygenated compounds are present in these Paleozoic oils, but not in unusually high concentrations. Significantly, the highest concentration of oxygenated compounds in the residual fractions (about 572°F, shown by Hunt, 1953, p. 1853, Table V) is listed for a Phosphoria sample from the Kirby Creek field. However, the CI curve for this crude is not unusual, nor is the light aromatic content of 8.3 percent for the 5th fraction low compared with other Paleozoic

crudes in the basin. According to Kartsev *et al.* (1959, p. 6), on the other hand, the known oxygen compounds of definite structure in petroleum comprise the acids and phenols, but the total amount of acids in crude oils is only a few percent at the most.

The general subject of oxygen in crude oil is a difficult one. A review of the pertinent geochemical literature (Hill, 1953; Lochte, 1962; and others) has led the writer to the conclusion that, at the present state of knowledge, it is not possible to evaluate critically through the study of the oxygen content of petroleums the importance of petroleum oxidation by reaction with oxygenated water.

On this general subject, however, Weeks (1958, p. 50) has offered the following pertinent comments:

Chemists who have critically studied the problem consider that the oxygen in asphaltic oils is an "impurity" remaining from the original organic matter, as a result of incomplete reduction of the organic matter, and that it was not added after the evolution and accumulation of the oil. There are cases of paraffin-oil accumulations exposed at outcrops with no sign of "oxidation" to asphaltic or to any other type of oil. The relatively high content of porphyrins in asphaltic oils seems to negate the common assumption that asphalts are oils that have been oxidized by contact with air or oxygen-bearing waters. The writer believes that important oxidation of hydrocarbons might only occur through the agency of bacteria or other catalyst.

The present writer would add that, in many Rocky Mountain fields, accumulations of chemically mature, paraffinic, Lower Cretaceous oil, showing no evidence of oxidation to asphaltic oil, are associated with fresh water under hydrodynamic control.

In view of the above considerations, Todd's (1963) proposal that dilute meteoric waters carrying free oxygen available for reaction have played an important role in oxidizing the Paleozoic oils seems highly questionable (Welte, 1965, p. 2266).

Oxidation by sulfate reduction.—Crawford (1949, p. 265; 1964) has indicated that in several deeper Big Horn basin Paleozoic fields, a high H_2S gas content is accompanied by high concentrations of carbonate and bicarbonate ions (from 10,000 to 40,000 ppm) in the associated formation waters, and that these, together with organic acids, are the natural end products of the sulfate-reducing reaction begun or aided by anaerob-

TABLE VII. COMPILATION OF DATA ON FIELDS WITH HIGH HYDROGEN SULFIDE AND BICARBONATE CONTENTS

Field	Formation	Depth (Feet)	°A.P.I. Gravity	Percent Gasoline & Naphtha	Carbon Residue in Crude (wt. %)	Correlation Index: Cut		S (wt. %)	H_2S (%)	Total Solids (PPM)	HCO_3 (PPM)	SO_4 (PPM)
						5th	15th					
Elk Basin	Tensleep	4,800	22.0–33.0	26.8	3.9	24	59	1.95	5–18.00	10,000±	3,000	4,200
Five Mile	Phosphoria	11,650	52.3–54.6	81.0	0.0	33	—	0.30	13.85	25,000	12,000	—
Fourteen Mile	Phosphoria	13,000	46.0	72.5	0.0	40	—	0.93	18.96	NA	NA	NA
Golden Eagle*	Tensleep	9,350	48.0	(est. 60.0)	(0.5)	(23)	72	0.92	40.00 (13)	6,780	4,700	1,100– 1,500
Manderson	Phosphoria	7,000	33.4	30.5	1.5	28	60	1.75	17.09+	75,000	32,500	5,145
Neiber	Phosphoria	10,040	40.6	54.6	0.8	29	80	1.88	37.2	37,753	24,300	3,778
Silver Tip	Phosphoria- Tensleep	8,800	43.0	55.9	0.4	32	65	1.13	34.0	NA	NA	NA
Slick Creek	Phosphoria	10,600	33.2	29.7	3.4	28	55	2.00	20–39.00	40,000	10,000	9,386
Worland	Phosphoria	10,450	37.0	38.5	1.2	27	54	1.55	30–63.00	44,355	16,000	6,900

* Data are either from Hunt (1953, p. 1867, Table VIII) or from Phosphoria analysis in Biggs and Espach (1960, p. 389), as the Tensleep distillation analysis in Biggs and Espach (p. 390) may represent a mislabeled sample, some other inadvertent error, or possibly an exotic Cretaceous oil (Wenger and Lanum, 1954, p. A-60).

ic bacteria.[7] The equation below (Feely and Kulp, 1957, p. 1808) illustrates the reduction of sulfate by petroleum (in this example, cyclohexane) through the enzyme catalytic action of anaerobic bacteria in which carbon dioxide and ultimately bicarbonate and H_2S are produced and petroleum is oxidized.

$$2C_6H_{12} + 9SO_4^- + 18H^+$$
$$\xrightarrow[\longleftarrow]{\text{anaerobic bacteria}} 9H_2S + 12CO_2 + 12H_2O.$$

$$Ca^{++} + 2OH^- + CO_2 \rightarrow CaCO_3 + H_2O.$$

The released carbon dioxide is dissolved in the formation waters (as HCO_3 ion) and may be precipitated later as carbonate cement below the oil-water contact. According to Ginter (1930, p. 152), "The inanimate organic matter represents the source of energy and nitrogen; the sulphate radical, the source of oxygen; and the carbon dioxide is a product of respiration." The reaction may be self-limiting because high H_2S concentrations are believed to inhibit bacterial activity (ZoBell et al., 1943, p. 1183).

Table VII is a listing of nine Big Horn basin fields reported by Biggs and Espach (1960) to contain a higher-than-normal percentage of hydrogen sulfide in the solution gas produced from Paleozoic zones. Corresponding data are shown regarding producing depths, API gravity, and

chemical characteristics of the oils and associated formation waters. It should be cautioned that the data in this table may be subject to significant quantitative errors because of unknown conditions and methods of measurement. Also considerable variation in the concentrations of these chemical constituents is evident in several fields, such as Elk Basin (Espach and Fry, 1951).[8] Some of the fields not listed may contain H_2S in important but unreported concentrations.

Several interesting and significant relations are evident from the data shown in Table VII. For example, with the exception of Elk Basin, all fields listed produce from below 6,900 ft, six of them from below 9,300 ft. This observation would seem to suggest that, as reported by Ginter (1934), Feely and Kulp (1957), and others, higher temperatures and pressures are required before important sulfate reduction can occur. However, because sulfate reduction by direct reaction with crude oil does not appear to occur except at temperatures well above those found today, or probably existent in the past, in any of the deep Big Horn basin fields (Bastin, 1926; Feely and Kulp, 1957; Toland, 1960), it is likely that bacterial catalysis is involved. Furthermore, the poor quantitative correlation between H_2S content and depth (temperature) for the nine fields shown in the table might be explained as

[7] These bacteria are now classified under a single genus, Desulfovibria (Feely and Kulp, 1957, p. 1809).

[8] A gradation in H_2S content of the gas from Phosphoria to Madison has been noted at Garland by Biggs and Espach (1960, p. 108).

due to variable bacterial reduction which is controlled by multiple environmental factors.

Laboratory experiments have indicated that bacterial reduction increases with increasing temperatures to at least 150°F. Rogers (1919) and ZoBell (1944–45, p. 74) have reported finding thermophilic bacteria flourishing at temperatures as high as 190°F, which is approximately the reservoir temperature expected at 10,000 ft in the Big Horn basin on the basis of the projection of the Elk Basin thermal gradient (Espach and Fry, 1951, p. 11 and Fig. 13). These observations are in agreement with the conditions associated with the high H_2S concentration in the Big Horn basin.

Todd (1963, p. 611) has questioned the ability of bacteria to "flourish at the depths of most Big Horn basin fields," but they reportedly have been found in oil field waters of other basins at similar depths (Bastin, 1926; Bastin and Greer, 1930; Ginter, 1930; Miller, 1949). Stone and ZoBell (1952) report the isolation of bacteria from a sample taken at 9,000 ft and these bacteria were found active at pressures of 5,000 psi. Furthermore, in a thorough discussion of the problem, Beerstecher (1954) has shown that neither the high temperature nor the pressure found in deep oil-field environments can bar microörganisms from participation in the process of petroleogenesis. Therefore, the absence of any substantive data on the existence of bacteria in the 10,000-ft oil-field environment is not as damaging to the proposed bacterial sulfate reduction in the deep Big Horn basin fields as the question of how the bacteria arrived there. Have they survived the long interval from the time of deposition, or were they emplaced later through water circulation? Beerstecher (1954, p. 103) stated that "for subterranean microorganisms, life would seem practical as long as the vast stores of oxidizable materials (hydrocarbons, hydrogen, iron, sulfur) remain available as a basic energy source."

The formation waters from those fields listed in Table VII are more saline than is normal for the basin, the most concentrated water being at Manderson with 75,000 ppm total solids. This gross correlation within the Big Horn basin of higher salinity with higher H_2S content is in agreement with Ginter's observation that bacterial activity increases with sodium chloride concentration up to 60,000 ppm (Ginter, 1934, p. 914).

More striking, however, are the high bicarbonate (HCO_3) concentrations and low sulfate concentrations, both predictable consequences of sulfate reduction.

According to ZoBell (1944–45), Feely and Kulp (1957), and Kartsev et al. (1959, p. 18), crude oils which have been oxidized by bacterial action should be depleted in heavy paraffins and correspondingly enriched in heavy naphthenes and aromatics. However, the CI numbers of the 15th fraction listed in Table VII lend only limited support to this concept. For example, the field with the highest H_2S content (i.e., Worland with as much as 63 percent H_2S) exhibits the lowest aromaticity for the 15th cut (CI 54). Whereas at Neiber Dome, a very high aromaticity (CI 80) is associated with 37 percent H_2S and above average HCO_3 concentration. This high CI value may be questionable because the Tensleep crude in the same field has a more normal CI of 54 for the same fraction. In general, however, the CI values for the 15th and other high boiling-point fractions of the crudes listed in the table are not significantly different from the average for all Paleozoic crudes in the basin (i.e., 59 for the 15th fraction). This observation suggests that oxidation of the heavy fractions of the listed crudes by bacterial reduction of sulfate has not been selective in the type of hydrocarbons attacked.

Maturation.—The above average gasoline and high light aromatic content of many deeper oils, indicated by the characteristic "aromatic hump," suggests that cracking and/or dehydrogenation, possibly by sulfur (Jones and Smith, 1965, p. 161 and Fig. 37), has occurred with the aid of higher temperatures and/or bacterial enzyme calyctic action (Barton, 1934; Brooks, 1952; Hunt, 1953, p. 1863; Kartsev et al., 1959, p. 14; Levorsen, 1954, p. 503, 504; McNabb et al., 1952; Silverman, 1963, p. 101; Welte, 1965; ZoBell, 1944–45, p. 71). This process commonly is referred to as "maturation" and appears to be a widely accepted explanation for the clear relation between crude-oil gravity and depth of burial in most basins of the world. Some of the geochemical evidence for petroleum maturation is discussed by Silverman (1963, p. 54, 55). The generally low-carbon residue, sulfur, and nitrogen, and higher light aromatic content in the deeper oils listed in Table VII are believed to have resulted from a maturation process, probably involving thermal catalytic cracking and dehydrogenation of the more typical, lower gravity Paleozoic crude oil found in shallower fields of the Big Horn basin.

Although evidence for limited alteration by water washing, oxidation through sulfate-reducing bacterial activity, and thermal cracking or maturation, possibly aided by bacterial enzyme catalysts, can be found for several Paleozoic oils of the Big Horn basin, there is no indication from these data that a paraffin-base crude of the Big Horn basin Cretaceous type or the Williston basin Paleozoic type could ever have provided the protopetroleum for the distinctive aromatic-naphthenic Paleozoic crudes of the Big Horn basin, as suggested by Todd (1963, p. 611).

FORMATION-WATER CHEMISTRY

The chemical classification of Big Horn basin crude oils into two major groups, Mesozoic and Paleozoic, is paralleled in the two-fold division of formation waters in the basin, except for the distinctive, highly concentrated waters of the Triassic (Fig. 21). The Triassic has been characterized as "a universal dividing bed between the relatively soft primary type waters of post-Triassic rocks and the hard, secondary saline waters of the pre-Triassic rocks" (Crawford, 1964, p. 55).

Beginning at the top of the Paleozoic column, Crawford (1964) characterized Phosphoria waters as dilute (2,000 ppm) to concentrated (40,000 ppm) and averaging 7,000 ppm. Phosphoria waters generally are characterized by a sulfate-chloride ratio greater than 1, calcium and magnesium concentrations being prominent. The higher concentrations of total solids usually are associated with high H_2S content in the solution gas.

Tensleep waters, although they resemble Phosphoria waters in composition, are more dilute, averaging 3,000 ppm. The trend is toward higher concentrations and salinity basinward, or greater dilution flankward, reflecting hydrodynamic conditions and the high transmissibility of the Tensleep aquifer. "Some of these waters, even associated with oil, are so dilute that they can be considered potable, and others are under artesian head and flow. Secondary characteristics, usually secondary salinity, are an integral part of the chemical system and often govern" (Crawford,

1964, p. 59). Amsden and Darwin waters resemble all other pre-Triassic waters, and are dilute, ranging from 300 to 2,900 ppm.

Very dilute to moderately dilute waters, usually well below 5,000 ppm total solids, are found in the Madison carbonates; a normal concentration averages 2,000 ppm. These "waters can often be distinguished from overlying Tensleep waters by a higher calcium-magnesium chloride ratio" (Crawford, 1964, p. 63). The low concentration of solids in Madison waters of the basin is interpreted also to be a reflection of a uniformly high transmissibility and dilution by meteoric waters.

Only three fields produce oil in commercial quantities below the Mississippian Madison Group in the Big Horn basin—Hamilton Dome, Garland, and Elk Basin. Thus, there is a paucity of water analyses from the Devonian (Jefferson), Ordovician (Bighorn), and Cambrian (Gallatin) zones. Crawford (1964, p. 68) reports that waters from these units at Hamilton Dome "have a general resemblance to each other but differ in minor points; these waters would undoubtedly be the most difficult to correlate" (Crawford, 1964, p. 66, Table 10). At Grass Creek, "the waters that were swabbed from the Devonian, Bighorn and Gallatin, were exactly the same—3,000 ppm total solids and secondary saline in character." This is good evidence for fluid communication through the large fault zone that cuts the southwestern flank of the field.

In Figures 19, 20, and 21, Stiff diagrams (Stiff, 1951) are shown for Phosphoria, Tensleep, and Madison waters from the Oregon Basin, Elk Basin, and Hamilton Dome fields, based on data from Biggs and Espach (1960). "Curtis" Sandstone and Bighorn Dolomite water samples from Hamilton Dome field also are included. The Tensleep sample from Oregon Basin is not typical, having well-above-average concentration. Crawford (1964, p. 64) stated that "the important Madison production at Oregon Basin is associated with water that averages close to 4,000 ppm total solids and in many ways resembles Tensleep water. . . ." and that "difficulty might be encountered in distinguishing Phosphoria and Tensleep waters . . . particularly in certain portions of the field" (Crawford, 1964, p. 68).

The basic similarity of the three waters from the Elk Basin field is apparent from the Stiff diagrams (Fig. 20); they differ only in the amount

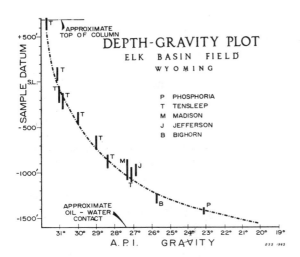

FIG. 24.—Depth-gravity plot showing density stratification of crude oils in Elk Basin field.

of dilution. Furthermore, the similar patterns of the Tensleep, Madison, and Bighorn samples from Hamilton Dome are obvious from the Stiff diagrams in Figure 21. However, a somewhat higher concentration is indicated for the Phosphoria sample, and the "Curtis" water is distinctive, having a much higher concentration than that of the deeper waters. The high concentration of Triassic waters may be a reflection of the presence of evaporites in this part of the section.

Thus, the formation waters produced from the Paleozoic zones in the fields of the Big Horn basin are chemically very similar; they are of moderately low concentration (i.e., from a few hundred to a few tens of thousands of parts per million total solids), and secondary saline in character, with noticeably high sulfate concentrations. Any differences observed in waters from the Big Horn basin Paleozoic reservoir rocks can be attributed to degree of dilution by meteoric waters entering at the outcrop (Crawford, 1964) and to basic petrographic differences in the rocks with which they are associated (Crawford, 1949, p. 285). This is not to say that differences in ionic concentrations do not occur in the different Paleozoic zones, but simply that a general chemical correlation between all Paleozoic waters of the Big Horn basin is evident. This correlation is in agreement with the concept of a common-pool state for the Paleozoic zones of the major anticlinal fields of the basin and suggests that considerable cross-formational communication exists.

The fact is also important that, not only are the paraffinic Madison and sub-Madison crude

oils of the Williston basin area different from those produced from equivalent rocks in the Big Horn basin, but also the Paleozoic formation waters co-produced with these oils are of much higher concentrations in the Williston basin (e.g., 50,000 to more than 300,000 ppm; Porter and Fuller, 1959, p. 181; Murray, 1959, p. 58; Crawford, 1964, p. 35) than in the Big Horn basin, probably reflecting their common association with evaporites. This relation is in accord with the concept that the sub-Permian Williston basin Paleozoic oils accumulated in an entirely different geological environment than the sub-Permian Big Horn basin Paleozoic oils.

DENSITY STRATIFICATION OF FLUIDS

A marked vertical density stratification of fluids can be observed in most large Paleozoic fields of the Big Horn basin. In the fields exhibiting a common-pool state the gravity of the oil in each formation or zone is approximately the same for the same datum elevation (cf., Walton, 1947, p. 220). Hence, the API gravity of any Paleozoic crude-oil sample from a particular field can be correlated with its vertical position within the oil column; the lower gravity values are associated with the lower structural elevations and the higher gravity values with the higher structural elevations. Overall gradients range from an average decrease of 1° API for each 30 ft below the top of the oil column at Worland field to an average decrease of 1° API for each 275 ft below the top of the column at Spring Creek field. Gradients between ½° and 1° per 100 ft are most common, and the larger gradients appear to be associated with the deeper, higher gravity fields. Plots of API gravity against sample depth, however, reveal a smaller change within the upper part of the oil columns than in the lower part near the oil-water interface. Thus, in fields with long oil columns, if the API or specific gravity values of several oil samples from different places within the oil column are plotted (abscissa) against their structural elevations (ordinate), a "J-shaped" curve results (cf., Hetherington and Horan, 1960).

The Elk Basin field of the Big Horn basin is an excellent example of fluid-density stratification within a common Paleozoic reservoir. A depth-gravity plot of available samples from all Paleozoic zones in this field is shown in Fig. 24. A smooth depth-gravity curve has been drawn through the sample points ranging from 32° API in the Tensleep at the crest (datum about +650 ft) to 23° API in a Phosphoria sample low on the flank (datum about −1,400 ft); by extension of this curve, a gravity value below 20° API is suggested for the oil-water contact. Significantly, this gradient is in the crude oil samples from all Paleozoic producing units, *i.e.*, Phosphoria (1 sample), Tensleep (9), Madison (1), Jefferson (1), and Bighorn (1). Accompanying the vertical stratification of crude oil gravity values at Elk Basin, a radical change in the saturation pressure, the gas in solution, and the H_2S content of the solution gas has been reported in the Tensleep (Espach and Fry, 1951). These physical characteristics range from 1,250 psi, 490 cu ft/bbl, and 18 percent H_2S within the crestal area of the anticline, to 530 psi, 134 cu ft/bbl, and 5 percent H_2S low on the flanks of the structure.

Data from Stewart *et al.* (1955) suggest that a gradient in interstitial-water saturation also may be present. Water saturation appears to grade from only a few percent at the top of the pool to more than 80 percent at the oil-water contact. Some workers have come to the conclusion that these low, crestal water saturations indicate that the Tensleep reservoir may be preferentially oil-wet.

Low water saturations also have been observed in several other Big Horn basin Paleozoic fields. Todd (1963, p. 614) has suggested that this condition may be the result of *in situ* oxidation of petroleum, resulting in precipitation of solid petroleum around the sand grains. The very low interstitial water saturations generally recorded in the structurally highest parts of anticlinal traps, however, seem to indicate that neither water washing at the oil-water contact nor petroleum oxidation through bacterial reduction of sulfates in the associated formation waters can be the cause of this phenomenon.

A stronger density stratification exists in the deep Worland field in the eastern part of the basin. The crude-oil gravity values measured in the Phosphoria reservoir of this field range from about 45° API at the crest to about 20° API at the oil-water contact; this vertical gravity stratification also occurs in the Tensleep Sandstone (uneconomic reservoir), as indicated by free-oil recoveries from drill-stem tests in several wells. As at Elk Basin,

a radical change in solution-gas content accompanies this density stratification; gas-oil ratios are reported to range from 18,000 to 5,000 cu ft/bbl between −5,900 to −6,200 ft, and from 5,000 to 1,500 cu ft/bbl from −6,200 to −6,600 ft within the Worland structure (Morris, 1951, p. B-78; Wold, 1952). The gas in this field is sour, containing from 30 to 63 percent H_2S (Biggs and Espach, 1960); these percentage concentrations probably also reflect position within the structure.

A less striking density stratification is in the large Paleozoic fields associated with well-defined hydrodynamic gradients. Table III shows a gradation in gravity values from 23.5° API for a Phosphoria sample near the top of the oil column, 19° API for a "Curtis" sample near the middle, and 14.5° API for a Bighorn Dolomite sample near the oil-water contact at Hamilton Dome (see also Krampert, 1947, p. 233). The known range of gravity values for other Paleozoic fields is shown in Figure 2.

Hedberg (1964, p. 1797) has noted that density stratification is common and has urged that this phenomenon be investigated further. Vertical density stratification of fluids within an oil column, particularly in large, fractured anticlinal accumulations, has been reported in fields in many parts of the world (Kartsev *et al.*, 1959, p. 21). Some examples are the Rangely field in Colorado (Cupps *et al.*, 1951), the Hawkins field in Texas (Wendlandt *et al.*, 1946), and the Burgan field of Kuwait (Hetherington and Horan, 1960; Boaden and Masterson, 1952).

Two questions concerning such fluid relations are: (1) what is the significance of the observed crude-oil density stratification in the Big Horn basin; and (2) how did this density stratification form? If the vertical stratification of crude-oil gravity discussed above is common to all Paleozoic zones in most larger anticlinal fields of the Big Horn basin as the available data strongly suggest, then another important parameter can be presented in support of the common-pool state interpretation, and further verification of the proposed theory of Paleozoic accumulation in the Big Horn basin is indicated.

The "how" of the observed density stratification is a more difficult problem, and additional research is needed. Hetherington and Horan (1960, p. 109) have indicated, however, that "since the hydrocarbons of oil and gas reser-

TABLE VIII. API GRAVITIES FOR PHOSPHORIA, TENSLEEP, AND MADISON CRUDE OILS

Formation	No. of Samples	Aver. API Grav.	Range API Grav.
Phosphoria	52	27°	12°–55°
Tensleep	39	26°	12°–43°
Madison	17	19°	11°–37°

voirs are mixtures of large numbers of chemical compounds of varying density and exist in gravitational fields, their compositions and properties should vary from top to bottom," and ". . . this must be true of all oil bodies, and for that matter of bodies of any fluids, particularly those which are mixtures of solutions not subject to constant agitation." Judging from the common occurrence of the density stratification phenomenon indicated by the present investigation, the writer agrees that gravitational segregation within a petroleum reservoir probably is the rule, not the exception. The scarcity of accurate sample data, particularly in fields with small oil columns, probably has been the prime reason for the apparent lack of recognition of this condition.

In opposition to this view, several authors favor degradation or removal of light fractions from crude oils by moving water as a mechanism for reducing API gravities, particularly near oil-water contacts. Specifically, Todd (1963, p. 615) has stated that, "Cross-formational migration of artesian water accounts for subnormal API gravity of the petroleum produced" in the Big Horn basin. This proposed mechanism, however, is not satisfactory for several reasons. The apparent selective removal of the more soluble light aromatics from the crudes has not been found to lower the gravity of several oils. Furthermore, some of the largest density gradients have been recorded in deeper, central-basin fields such as Elk Basin and Worland, where water drive is absent, salinity values are higher than normal, and hydrodynamic flow is unimportant.

The lower crude-oil gravity values generally associated with Madison crudes are exactly as expected considering the density stratification described. Madison gravity values generally are lower than Phosphoria and Tensleep gravity values simply because Madison production usually is located within the lower part of the original common-oil column (i.e., near the oil-water contact) because of its stratigraphic position below the

Phosphoria and Tensleep. The lower value of API gravity listed for each of the three Paleozoic zones shown in Table VIII is approximately the same, as expected according to the present theory. The low, 19° API average gravity listed for the Madison reflects the fact that all Madison production, except at Silver Tip field, has so far been found at shallower depths. Furthermore, the field gravity ranges in Figure 2 indicate that Paleozoic crude-oil gravity values change not only with depth but also—and probably more important—with location in the basin.

Although it has been concluded that water washing does not adequately explain density stratification in the Big Horn basin Paleozoic fields, it is possible that cross-formational flow at the oil-water contact of some large fields has played a secondary role. Many heavier compounds associated with the asphaltic parts of crude oils, including the oxygen-, sulfur-, and nitrogen-bearing organic compounds, have polar properties and are attracted to water (Silverman, 1965, p. 56). Porphyrins, for example, are concentrated commonly near oil-water contacts (Dunning et al., 1953). Other trace-metal constituents in crude oils exhibit vertical density stratification in oil tanks, and Hayden (1961, p. 21) suggests that similar stratification may occur in oil reservoirs.

Erdman (1965, p. 48) has stated that ". . . if the principal driving force for aromatization is dissolved oxygen perhaps from meteoric waters or sulfur produced by micro-organisms at the oil-water interface, an accumulation of asphaltic material near the interface would be expected." Inasmuch as petroleum as an energy source is oxidized in the process of bacterial reduction of sulfates, it is probable that API gravity would be lowered if any gases and light liquid hydrocarbons also produced became stratified in the oil column according to their density. The greatest supply of sulfate for bacterial reduction is near the oil-water transition zone (Stone and ZoBell, 1952) and this is where the most active petroleum oxidation should occur. The so-called tar-mat phenomenon thus might be explained in the deeper fields. However, most tar mats have been proposed for the shallower, basin-edge fields (Fanshawe, 1960; Lawson and Smith, 1966; Todd, 1963) where bacterial-sulfate reduction, as reflected in high H_2S gas content, appears to be unimportant. Could an explanation for the reported tar mat at the oil-water contact in the Frannie field be provided by the fact that uranium, probably derived from the nearby Pryor Mountain Madison uranium deposits and transported by artesian-water circulation, was found to be highly concentrated in a Madison crude sample from this field (Bell, 1960), and "petroleum, if subject to radiation from uranium in the crust of the earth for a sufficient period of time, should be altered to a plastic or solid form" (Hayden, 1961, p. 78)?

RESERVOIR PRESSURE COMPARISONS

Reservoir pressure information was sought on the larger, multizoned fields of the Big Horn basin with the expectation that additional evidence for intercommunication between Paleozoic producing zones might be discovered. However, this inquiry was frustrated from the start for several reasons, some of which are listed below:

1. Most large anticlinal fields were discovered in the early part of the century and virgin pressures were not usually recorded.

2. Most fields are operated by major oil companies or are held under multi-operator federal units; thus, the meager pressure data recorded are restricted and cannot be released without the consent of all field operators.

3. The intermittent and irregular discovery and development histories of the different Paleozoic zones in most of the common-pool fields probably have seriously upset any original state of pressure equilibrium which existed between zones before discovery, because it is unlikely that fluid communication is sufficiently perfect to allow equilibrium conditions to be maintained between zones after significant fluid withdrawal. Also variations in the hydrodynamic gradient, strength of water drive, and/or level of water encroachment in the different zones have produced apparent zonal segregation and probable secondary pressure differentials in many places.

The Elk Basin field provides an example of the kind of problems found in analyzing the limited pressure data from Paleozoic reservoirs of the larger Big Horn basin fields. The Paleozoic reservoirs in this field were discovered and developed during relatively modern times (Tensleep oil was discovered in 1942; Madison, 1946; Devonian and Ordovician, 1961) compared with other giant fields of the basin. An original pressure-depth plot for the Tensleep reservoir at Elk Basin has been constructed by Espach and Fry (1951, Fig. 12) from subsurface oil-sample data; this curve gives an oleostatic gradient of approximately 0.35 psi/ft for the Tensleep oil through the large

2,300-ft oil column. The initial Madison reservoir pressure in the year of discovery, 5 years after the Tensleep discovery, was 2,264 psi at a datum of −700 ft, somewhat less than the initial Tensleep pressure of about 2,330 psi at this datum, but considerably more than the 1,330 psi Tensleep datum pressure at this same date (Harvey and Krebill, 1954, p. A-42; Stewart et al., 1955). This relation shows partial drawdown of Madison pressures as a result of the few years of production from the Tensleep before discovery of the Madison producing zone.

In 1961, when pre-Madison production first was established, pressures of 2,132 psi at a −900-ft datum and 2,237 psi at a −1,150-ft datum for the Jefferson and Bighorn dolomites, respectively, were reported by the operator. As anticipated, these pressures are considerably lower than original Tensleep or Madison pressures for the same datum points. These initial pressures in the Jefferson and Bighorn were considerably higher than the Tensleep pressure of about 1,325 psi at a datum of −400 ft reported for 1954 (Stewart et al., 1955), 5 years after pressure-maintenance operations were started in the Tensleep. In one drill-stem test of the Jefferson Formation in the discovery well, however, a pressure of 1,400 psi at a datum of −788 ft was recorded, and the test was considered reliable (Roger Hoeger, Petroleum Research Corp., personal commun., 1964). A possible explanation for this low pressure might be found in the assumption of highly efficient, local fracture communication with the Tensleep, which should have registered a comparable pressure at this datum in 1961.

The reason that little or no water movement has occurred in the Tensleep at Elk Basin ". . . to compensate for extensive petroleum withdrawal" (Todd, 1963, p. 607) is that fluid, which has been migrating into the former productive area, is mostly oil (Stewart et al., 1955, p. 50), and not that a tar mat is present at the Tensleep oil-water contact (Fanshawe, 1960, p. 17). The oil is supplied through very efficient *gravity drainage* within the 2,300-ft Tensleep oil column, augmented by a pressure-maintenance program of nitrogen injection into the gas-cap area. This program was begun in 1949 (Harvey and Krebill, 1954). The originally higher pressured, underdeveloped sub-Tensleep units probably also have contributed oil to the Tensleep. These sub-Ten-

TABLE IX. COMPILATION OF DATA ON TENSLEEP FIELDS SHOWING HYDRODYNAMIC MODIFICATIONS

Fig.	Field	Tilt on Oil-Water Contact (Ft Per Mi) <200	>200	>500	Flow Direction	API Gravity	Aver. Topogr. Elev. (Ft)	Aver. Potentiometric Surf. Elev. (Ft)	Faulting Communication	Diversion	References*
—	Black Mountain	X			NW	23.1	5,700	+4,800	X?	X	Dobbin (1947, p. 804)
27; 28	Frannie			X	SSE	27.5	4,400	+4,200		?	L. B. Curtis (1954); Fanshawe (1960); Hubbert (1953); Lawson and Smith (1966)
29	Grass Creek	X			ESE	24.2	5,675	+4,700	X		McCanne (1947); Hill et al. (1961)
28	Hamilton Dome	X			ESE	22.6	5,600	+4,700	X		Collier (1920); Krampert (1947) Anonymous (1954)
26	Lake Creek, Northwest	X			NW	30.6	4,765	+4,600		X	Miller and Stiteler (1952); Green and Ziemer (1953)
—	Lake Creek	X			NW	30.0	5,075	+4,700		X	Summerford (1952); Green and Ziemer (1953)
26; 28	Murphy Dome	X			NW	34.0	4,700	+4,500		X	Partridge (1958); Lawson and Smith (1966)
27	North Deaver			X	SE	22.0	4,200	+3,900		X	Résumé, Petroleum Information (1960)
27	Sage Creek			X	SSE	23.5	4,150	+4,100			Elmer (1959); Sharkey et al. (1946); Lawson and Smith (1966)
30	Torchlight	X			NW	35.0	4,002	+4,800	X		Stewart et al. (1955); Willingham and Howald (1965)

* Wyoming Geol. Assoc. (1957a, b) and Biggs and Espach (1960) are useful references for all fields listed.

sleep units, such as the Madison, are in communication with the Tensleep and probably under active water drive.

Limited information for some of the other large Paleozoic fields in the Big Horn basin can be interpreted similarly. However, at Grass Creek field, initial pressures of 1,272 psi at +2,000 ft in the Tensleep, 1,340 psi at +1,500 ft in the "Curtis," and 1,512 psi at +1,303 ft in the Darwin Sandstone Member above the Madison (Donald Paape, Pan American Petroleum Corp., personal commun., 1965; Stanley McEnroy, Marathon Oil Co., personal commun., 1965) may reflect a pressure-depth relationship which is controlled by hydrodynamic flow through the fault zone.

It is unfortunate that only general qualitative conclusions can be made at this time from formation-pressure relations. However, the limited data available do appear to support the concept of zonal intercommunication in some larger multizoned Paleozoic fields. Inasmuch as some unit operators in the area probably have access to more complete data than are available to the writer, it is hoped that their information and the data here compiled in support of the common-pool state will stimulate further investigation and publication of pressure relations in the Big Horn basin Paleozoic fields.

EXCEPTIONS TO COMMON-POOL STATE

GENERAL

Not all Paleozoic fields in the Big Horn basin exhibit the common-pool state, but instead, show vertical segregation into several separate pools. These "exceptional" fields are believed to be the result of secondary modification of an original common pool.

HYDRODYNAMIC MODIFICATION AND SEGREGATION

Several fields along the edges of the Big Horn basin are under the influence of hydrodynamic control. This control is observed best in the Tensleep (Partridge, 1958; Summerford, 1952; Zapp, 1956). More obvious examples of hydrodynamic tilt in the Tensleep are listed in Table IX together with the approximate elevation of the Tensleep potentiometric surface, general direction of flow, and pertinent reference articles. Several other fields may exhibit lesser tilting of the oil-water contacts in the Tensleep, including Little Buffalo Basin (south plunge), Pitchfork (southeast plunge), and Garland (north plunge).

Hydrodynamic controversy.—Support of the importance of hydrodynamic modification of many Tensleep pools along the Big Horn basin edges is abundant. Some geologists, however, have explained the tilts as due to stratigraphic changes in the Tensleep Sandstone, multiple step faulting, geologic tilt of an original level pool sealed by a tar mat, or a combination of these factors (Fanshawe, 1960; Lawson and Smith, 1966; Todd, 1963). Whereas cross faulting and stratigraphic variations exist in some fields and affect the shape of Tensleep pools, the available evidence that these mechanisms are the sole causes of the observed tilted oil-water contacts is not very convincing. Although the crude-oil density stratification previously described probably has resulted in the formation of a lower gravity, more viscous layer near some oil-water contacts, the writer has been unable to find any published proof that the proposed impermeable tar mats or seals exist in the Big Horn basin.

Probably the most convincing evidence favoring the hydrodynamic concept is the general agreement between hydrodynamic gradients in the Tensleep, as defined by mapping of the Tensleep potentiometric surface, and the observed direction and approximate amount of tilt on Tensleep oil-water contacts in the associated fields (Roger Hoeger, Petroleum Research Corp., personal commun., 1964; John D. Haun, personal commun., 1966; Hubbert, 1966, p. 2515).

However, some geologists probably would take exception to the above statement. Fanshawe (1960, p. 18), followed by Todd (1963), has stated that "No fields have been found in the northern Rockies in which a tilted water table is caused by existing hydrodynamic gradients," and presumably referring to fault "barriers" and "congealing" of oil at the interface, he concluded ". . . that interference factors to hydrodynamics are more critical than the existing gradients."

Todd (1963) also has proposed that secondary cement, both silica and carbonate, localized below oil-water contacts, has held oil pools in place after regional orogenic tilt. However, Gilman Hill (personal commun., 1965), though acknowledging the secondary-cementation mechanism, stated that the locally "tight" reservoir rocks thus created below the original oil-water contacts should cause local "crowding" of potentiometric-surface contours in these areas of reduced transmissibility, because a lower hydrodynamic pressure gradient will exist in the more highly transmissible parts of an aquifer, and correspondingly higher pressure gradients will exist in those parts with lower transmissibility. The described conditions help explain the present highly tilted oil-water contacts in such fields as Frannie and Sage Creek, where cementation may exist below the oil-water contact. However, these fields still can be considered to be in equilibrium with their present hydrodynamic environments because a return to static pressure conditions should allow the oil to seek a normal, level adjustment to structure in time, except for possible minor capillarity anomalies or small stratigraphic accumulations left behind by permeability variations.

The usefulness of generalized regional potentiometric maps (Todd, 1963, Figure 5) is questionable because they are inadequate for local analysis of oil-field tilt factors, particularly where local faulting or secondary cementation is suspected. Moreover, it commonly is impossible to obtain sufficient pressure information to define accurately the local potentiometric contour configurations. The tendency toward lower crude-oil gravity values at the oil-water interface resulting from density stratification, possibly progressing to the extreme of the so-called tar mat, suggests a further complication, because the tilt factor is inversely proportional to API gravity (Hubbert, 1953). The writer believes, therefore, that Fanshawe's all-inclusive "paleohydrodynamic" explanation is not supported by the available evidence. Hubbert (1966, p. 2516) has expressed surprise at Lawson's and Smith's (1966) stratigraphic-tar mat hypothesis ". . . in view of the fact that Tensleep fields in the Big Horn basin are among the most conspicuously hydrodynamically tilted fields in the country. . . ."

Todd's (1963, p. 611 and Fig. 9) correlation, based on "a scatter plot," of "high head values with low-gravity oil, and low-head values with high-gravity oil" also is questionable. If such a correlation exists, which has not been proved, presumably the reason is that, in basins bordered by high mountain outcrops, artesian water flows naturally from the areas of higher intake at the outcrop toward areas of lower potential, generally nearer the center of the basin. Coincidentally many shallow fields, especially on the west side of the basin, contain lower gravity oils, whereas deeper fields contain more mature, higher gravity oils.

Selective Tensleep tilting.—Although the transmissibility of the Phosphoria Formation is poor because of its generally low permeability, the opposite is true of the more continuous, "blanket-type" Tensleep sandstone aquifer. Consequently, important tilted oil-water contacts have not been recognized and probably do not exist in the Phosphoria, but are common in the Tensleep. In the Madison aquifer, however, transmissibility is so uniformly high across the basin that a lower hydrodynamic gradient exists in the Madison and local hydrodynamic variations are small. Thus measurable tilts are lacking also in Madison pools (Roger Hoeger, Petroleum Research Corp., personal commun., 1965). This interpretation may gain support from the fact that pressures measured in the water phase of the Madison from many parts of the basin plot as a fairly straight pressure-depth line (Stanley McEnroy, Pan American Corp., personal commun., 1965). Madison waters are correspondingly dilute; they are well below 5,000 ppm total solids (Crawford,1964, p. 63), reflecting easy entry of meteoric waters at the outcrop, and subsurface dilution of formation waters. However, salinity is not related directly to the hydrodynamic gradient, and the amount of dilution therefore is not

a reliable measure of the importance of hydrodynamic flow. In the three fields which are commercially productive from Paleozoic reservoirs below the Madison, inclusion with overlying Paleozoic units in a common reservoir can be demonstrated (Fig. 15), and zonal segregation caused by hydrodynamic modification has not occurred.

Thus the Tensleep Sandstone is the only Paleozoic formation in the Big Horn basin in which hydrodynamic flow has seriously modified the geometry of the hydrocarbon accumulation from that normally expected in a static environment. Where significant off-structure tilting has occurred in the Tensleep, segregation of an original common oil pool has resulted. This is illustrated by the Lake Creek and Lake Creek Northwest fields (Fig. 25) which are shown in Figure 1 of Green and Ziemer (1953) and also reproduced by Levorsen (1954, p. 547, Fig. 12–6). At Frannie (L. B. Curtis, 1954; Hubbert, 1953, p. 2016–2019) and Sage Creek (Elmer, 1959) fields, in the northeastern Big Horn basin, the Tensleep pool has been tilted off the basinward, or southwest, side of the structure (Figs. 26 and

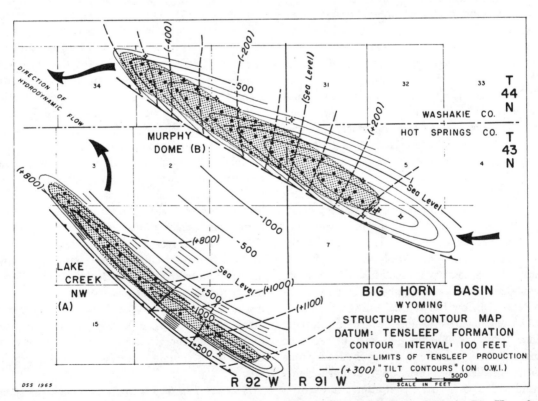

FIG. 25.—Structural contour map of Murphy Dome and Northwest Lake Creek fields in Big Horn basin showing hydrodynamic "tilt contours" on oil-water contacts.

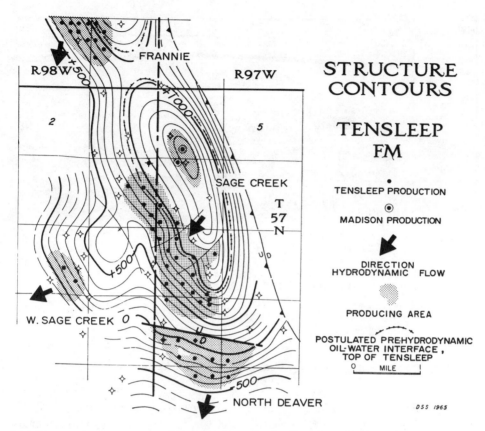

FIG. 26.—Structural contour map of Frannie, Sage Creek, West Sage Creek, and North Deaver fields, Big Horn basin, showing general direction of hydrodynamic flow in Tensleep Sandstone.

27B), but the Phosphoria, which is thin and "tight" here, and Madison pools have remained behind in "normal" adjustment to the anticlinal closure. It is of interest that the Tensleep discovery well at Sage Creek was located on the basis of hydrodynamic analysis.

As a test of this concept, a comparison was made of the estimated volume of recoverable oil, *i.e.*, 75,000,000 bbl in the present hydrodynamically controlled Tensleep pool at Frannie, with the theoretical volume which would have been originally held under normal static adjustment to structure in a common pool with the Madison. The computed reserves were essentially the same. Similar comparisons at Sage Creek (Fig. 26) cannot be made because much of the Tensleep oil has apparently been swept out of the original structure as indicated by the existence of several smaller hydrodynamically controlled accumulations outside of the closure (*e.g.*, North Deaver).

Cross-formation flow along faults.—Other types of hydrodynamic modifications of postulated original common anticlinal pools are represented also in the basin. At Murphy Dome (Fig. 25B, Fig. 27A) for example, the Phosphoria has no appreciable reservoir porosity, and some original Tensleep oil has escaped through the fault zone along the west flank and moved up into the lower pressured Triassic "Curtis" Sandstone. Dilute water is produced from the Tensleep at both Murphy and Lake Creek (Crawford, 1964, p. 60). At Black Mountain, multiple cross faulting complicates the interpretation, but approximately 1,100 ft of vertical anticlinal closure (Wyoming Geological Association, 1957a) has preserved Madison production on the crest of the structure below a segregated and strongly tilted Tensleep pool.

In Table IX, which is an incomplete list of fields exhibiting hydrodynamic effects, the Grass Creek (Fig. 28) and Hamilton Dome (Fig. 27C) fields are shown. In both of these giant fields, fluid communication exists not only among the various Paleozoic productive zones, but also includes the overlying Triassic "Curtis" Sandstone in a common reservoir. The flanking left-lateral

225

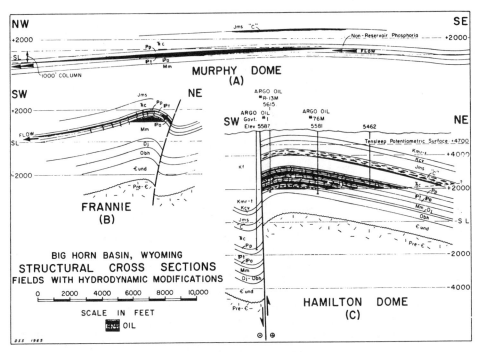

Fig. 27.—Structural cross sections of Murphy Dome, Frannie, and Hamilton Dome fields showing effects of hydrodynamic modification.

wrench (?)-fault zones (Fig. 2), associated with both fields (with locally 3,000–4,000 ft of vertical separation), and the numerous smaller faults that cut the Paleozoic at Hamilton Dome, must provide the avenue for fluid communication between the Paleozoic and Triassic "Curtis" in these fields. At Hamilton Dome, not only are the oil-water contacts of all of Paleozoic producing zones nearly at the same level, but also the waters are chemically similar (Fig. 21). Moreover, Crawford (1964, p. 61, 62, 65) states that at Grass Creek, ". . . the waters that were swabbed from the Devonian, Bighorn, and Gallatin were exactly the same—3,000 ppm total solids and secondary saline in character."

The oil-water interface on the "Curtis" zone at Grass Creek exhibits considerable southeast tilt, and Paleozoic zone(s) may have some tilt also (Hill et al., 1961, p. 62), although it is much less. This agrees with the regional hydrodynamic environment. On the structural cross sections of these two fields, and in Table IX, the approximate elevation of the Tensleep potentiometric surface lies well below topographic elevations. This relation helps to explain why these fields are filled with oil beyond the level of true anticlinal closure (i.e., below the elevation of the contact of the top

of the reservoir with the fault; Hill et al., 1961) and why the "Curtis" reservoir contains Paleozoic (Phosphoria) oil. Because the oil can rise in the fault zone no higher than the level of the Tensleep potentiometric surface, and thus cannot escape to the ground surface, an apparent fault seal has been created on these fields. However, a lower datum pressure in the "Curtis" zone relative to underlying Phosphoria, Tensleep, and Madison, and overlying Cretaceous sandstones, has resulted in vertical migration of Paleozoic oil into the "Curtis" reservoir, and a much longer "Curtis" oil column. Oil has not migrated southwest across the fault because no reservoirs are available (i.e., entry pressures are too high) in the footwall Cretaceous shale section adjacent to the hanging-wall Paleozoic and Triassic zones.

On the other hand, where the potentiometric surface lies above the ground surface, fault planes can provide conduits for escape of hydrocarbons, and oil will not be held in Paleozoic anticlinal traps beyond the highest level of contact of the top of the reservoir with the fault zone. An example is the "exceptional" Torchlight-Lamb field complex along the eastern rim of the Big Horn basin (Fig. 29A). Segregation of Paleozoic reservoirs is obvious in this structural complex,

226

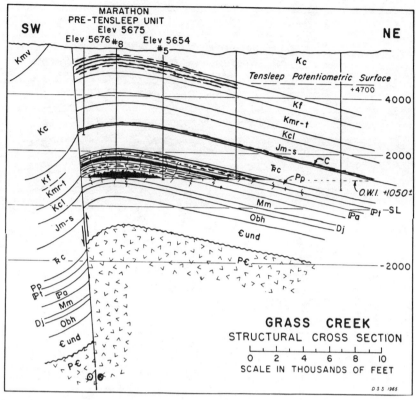

FIG. 28.—Structural cross section of Grass Creek field, Big Horn basin.

because the Phosphoria, Tensleep, probably several Madison, and Bighorn zones have separate oil-water contacts. The Phosphoria pool primarily contains gas and is characterized by extremely poor and irregular porosity development. The Tensleep pool has been segregated from the Phosphoria by an off-structure hydrodynamic tilt, between 160 and 170 ft per mile in a northwesterly direction (Stewart *et al.*, 1954; Willingham and Howald, 1965). Madison pools are segregated from the overlying pools with the "tight" Amsden redbeds serving as an apparent secondary cap rock, although probably oil-saturated. Only a small accumulation of lower gravity oil was present in the Bighorn Dolomite at the top of the structure.[9]

A fault is drawn between the Torchlight and Lamb structures in the map accompanying Biggs and Espach's discussion (1960, Fig. 163, p. 271) and limited seismic data indicate that this interpretation may be correct. If such a fault is present, an explanation for this zonal segregation within the Paleozoic may be provided. Because the Tensleep potentiometric surface in this area is approximately 600 ft above the surrounding ground level (Table IX), this fault zone should have been an escape route for any hydrocarbons below the level of independent anticlinal closure on each segregated zone. In this example, any interzonal tight layers presumably would provide secondary cap rocks under the modified subsurface conditions of oil viscosity and limiting entry pressure, temperature, and relative permeability to oil.

At Torchlight and at Lamb, according to this interpretation, some independent closure above the fault contact has permitted retention of the several small pools described above. However, in Paintrock anticline, farther south along the east flank of the basin (Fig. 2), a longitudinal fault zone at the crest has provided an escape route for all hydrocarbons originally trapped, because no independent anticlinal closure remains which is not cut and drained by the fault zone (Hill *et al.*, 1961).

[9] A drill-stem test sample of this oil has generously been given to the writer by B. D. Rea of Barlow and Haun, Inc., and is being chemically analyzed.

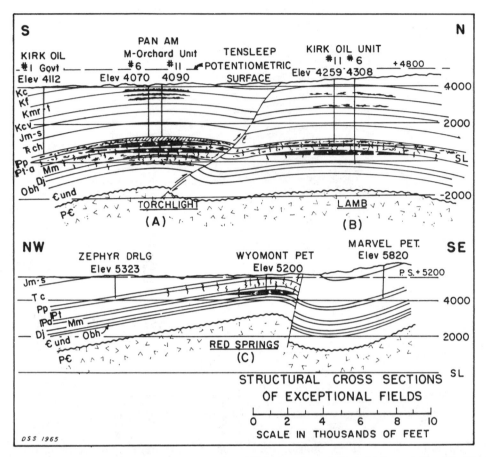

Fig. 29.—Structural cross sections of exceptional Torchlight-Lamb and Red Springs fields, Big Horn basin.

Hence, under conditions of a strong hydrodynamic gradient in zones of variable continuity and transmissibility, the ideal common-pool state can be disrupted and become segregated into several separate pools primarily as a consequence of off-structure tilt within the Tensleep Sandstone part of the original pool. However, faulting also may facilitate zonal segregation in those areas where a high potentiometric surface allows escape of hydrocarbons along faults from the part of the original common pool outside of the unfaulted, independent anticlinal closure.

 EROSIONAL TRUNCATION OF CAP ROCK

In a few shallow fields, where surface erosion has cut through the Triassic Chugwater cap rock and into the top of the Phosphoria Formation, oil commonly has been lost to the surface, or inspissated within the Phosphoria and, in some places, within the Tensleep reservoirs.

At Red Springs (Collier, 1920, Pl. IX; Dobbin, 1947, p. 805) near the southern margin of the

basin, the Phosphoria has been breached in this manner, and production is limited to the Madison zone (Fig. 29B). Amsden redbeds are the secondary cap rock. The producing depth is about 1,000 ft, whereas the oil has a gravity of about 11° API, and contains more than 4 percent sulfur (Table X). Red Springs is the only field in the Big Horn basin in which the Madison produces without reported Phosphoria and Tensleep saturation, but Table II shows that the structure has the required total closure in excess of the thickness of the Phosphoria-to-Madison interval at the crest. Thus, it is assumed that, before surface erosion reached the Phosphoria level, the Madison must have been included in a common pool with overlying Paleozoic rocks. The fault shown on the cross section (Fig. 29B) provides part of the closure, because the Madison potentiometric surface is below topographic elevations over the structure, and oil in the Madison cannot escape to the surface through the fault zone.

Oil of comparable gravity, sulfur content, and

TABLE X. COMPILATION OF DATA ON FIELDS MODIFIED BY EROSIONAL TRUNCATION OF CAP ROCK

Field	Producing Formation	Depth of Sample (Ft)	°API Gravity	Correlation Index: Cut		S (wt. %)	Carbon Residue in Crude (wt. %)	Oldest Rocks Exposed Over the Crest
				5th	15th			
1. Crystal Creek	Tensleep	964–997	18.6	20	59	3.95	9.0	Chugwater
2. Red Springs	Madison	902–1,082	10.7	—	67	4.08	13.0	Phosphoria
3. Spence Dome	Madison	434–470	18.5	—	59	3.13	9.4	Phosphoria
4. Warm Springs	Phosphoria	900–1,000	18.6	18	59	3.64	8.5	Chugwater
	Phosphoria	1,000–1,100	23.1	19	55	3.06	6.9	Chugwater
5. Wildhorse Butte	Phosphoria	1,120–1,202	12.1–14	—	67	4.41	13.0	Chugwater

carbon residue has been produced from the Phosphoria at a similar depth in the nearby Wildhorse Butte field (Table X), where structural closure appears to be absent and no other Paleozoic zones are productive. The Chugwater Formation crops out above this field.

Phosphoria beds also are exposed in the crestal region of the Spence Dome field along the northeastern flank of the basin. Although oil or gas saturation has been found in all Paleozoic zones in this field through the Madison, only Darwin and Madison production has been reported officially (Biggs and Espach, 1960, p. 252). The field has a structural closure of probably 700 ft and the thickness of the Phosphoria-to-Madison interval is estimated to be 520 ft, thus providing the basic requirement for oil in the Madison reservoir (Table II).

ACCUMULATION IN AREAS OF NONRESERVOIR PHOSPHORIA FACIES

Bonanza, Hidden Dome, and Nowood (including Nowood Southeast and Middle Dome) fields produce from the Tensleep or a basal Phosphoria sandstone reservoir within the area of the Goose Egg redbed facies of the Phosphoria along the eastern margin of the Big Horn basin (Fig. 2). These fields therefore are exceptions to the common-pool state involving Phosphoria marine carbonates and an explanation for the occurrence of Paleozoic oil in these fields should be provided.

Most geologists probably would agree that lateral migration through Tensleep Sandstone carrier beds must account for such accumulations. This could have occurred almost any time after deposition of the Phosphoria source sediments,

providing there were sufficient burial and basin-ward tilt, and that a trap was available to capture the oil. Studies of the unconformable relation between the Phosphoria and underlying Tensleep Sandstone have indicated that the thicknesses of the two formations are compensatory; i.e., the Phosphoria has filled in the irregularities in the Tensleep erosion surface (Agatston, 1954). Apparently in the Bonanza-Nowood area, an elongate, northwest-trending thick lens of Tensleep Sandstone has been preserved beneath the Phosphoria-Tensleep unconformity, providing a truncational subcrop trap in the upper Tensleep (D. I. Foster, personal commun., 1965). This accounts for 160 ft of a total 210 ft of vertical Tensleep oil column at Bonanza field (Lawson and Smith, 1966), which is an anticlinal accumulation filled to the spill point.

However, Middle Dome produces from a basal Phosphoria (reworked Tensleep) channel sandstone (Lawson and Smith, 1966), which appears to have a similar northwesterly trend and borders the eastern edge of the Bonanza upper Tensleep buildup (D. I. Foster, personal commun., 1965). It seems probable that easterly migrating Phosphoria oil was captured at the unconformity in the Bonanza-Nowood area before Laramide folding. It is also possible, however, that post-Laramide oil has been spilled from the down-plunge, Manderson Phosphoria stratigraphic accumulation to be caught in these well-defined Laramide anticlinal traps. The fact that the oil in Bonanza and Nowood is of unusually high gravity, higher than in the down-flank Hidden Dome field, could be explained by either mechanism. Whether the oil in these anticlinal fields remigrated from a down-dip Phosphoria stratigraphic trap or a Tensleep stratigraphic trap in the vicinity of the present fields does not alter the conclusion that the oil in these fields was derived from Phosphoria sources.

In the Frannie and Sage Creek fields of the northeastern Big Horn basin (Fig. 26), the Phosphoria is a nonreservoir facies. Some authors also have considered stratigraphic changes to be important for accumulation in these fields. Elmer (1959) described a fivefold zonation of the Tensleep separated by "tight" dolomitic layers at Sage Creek. Lawson and Smith (1966, p. 2215) proposed subcrop stratigraphic control for the Frannie Tensleep pool. (The indicated distribution of Tensleep oil in their Figure 24 is misleading.) However, the areal coincidence of all productive Tensleep porosity zones at Sage Creek and the hydrodynamic implications of this strongly tilted accumulation seem to indicate interzonal communication. Also, good water recoveries have been obtained in Tensleep drill-stem tests taken in many wells in the vicinity of these two fields and at elevations equivalent to those which are oil-productive inside the field areas. Therefore, the writer suggests that, although some of the oil in these fields could have been trapped early by combined lateral migration and Tensleep stratigraphic variations, much of the oil may have been emplaced later as a result of Laramide folding and fracturing of Phosphoria source and storage rocks in the vicinity of these fields. The association of marine Phosphoria facies and Madison production in the Frannie and Sage Creek fields is indicative of original fracture communication in a common pool.

Migration and Accumulation of Oil and Gas
general

Interpretation of the relation between geologic history and time of migration and accumulation of hydrocarbons in any suite of rocks and within any basin area is a difficult and speculative undertaking. Nevertheless, such an interpretation must be attempted if all pertinent geological factors and concepts are to be placed in perspective. Although much controversy continues in all phases of this subject, considerable progress has been made toward a more organized approach to the solution of specific time-of-migration and accumulation problems. Based largely on the earlier work of many geologists, Levorsen (1954) and Gussow (1955) proposed formally several analytical criteria which can place the otherwise subjective analysis of the time of migration and accumulation on a more scientific basis. The discussion presented below, therefore, is an attempt to evaluate these criteria as they relate to the history of migration and accumulation of Paleozoic oil in the Big Horn basin.

COMPACTION AND DEPTH OF BURIAL

It is now generally accepted that hydrocarbons can form in the source beds soon (geologically) after deposition (Hanson, 1959). However, before petroleum can be formed, expelled from the source sediments, and migrate into the reservoir, the source sediments must be buried and undergo compaction. Based primarily on Hedberg's (1936) compaction studies, and following Cheney's (Van Tuyl and Parker, 1941, p. 80) earlier conclusion, Gussow (1955, p. 554) proposed that probably 2,000 ft of overburden is required before sufficient hydrocarbons are forced from source sediments and "flush migration" can occur. However, Hedberg (1964, p. 1777) subsequently indicated that "There seems to be little reason in direct pressure effects alone for postulating any minimum depth of burial requirement," with certain exceptions. Kidwell and Hunt (1958), for example, have reported very low-grade migration and accumulation in Recent sediments buried less than 200 ft deep. Nevertheless, Gussow's proposed 2,000-ft overburden depth requirement appears to be qualitatively valid in many places (Welte, 1965).

Although this minimum overburden thickness would seem to indicate an Early Cretaceous time of migration for Paleozoic Phosphoria oil in the Big Horn basin (based on present stratigraphic thicknesses), other criteria discussed below do not generally support this possibility; flush migration in Early Jurassic time seems more likely. The present thickness of the Triassic section in the Big Horn basin ranges from about 1,200 ft near Thermopolis in the southern part of the basin to less than 500 ft at the Montana border (Thomas, 1965, p. 1862, Fig. 14). However, "the present thickness of the Chugwater Formation in the Big Horn basin is not the original thickness at the close of Triassic time" (Downs, 1952, p. 27). Pre-Middle Jurassic erosion removed an unknown thickness of Triassic and Lower Jurassic rocks, and caused one of the most widespread unconformities in the Rocky Mountain region. Moreover, the angle of progressive northward truncation of the top of the Triassic section suggests that the original Triassic section in the Big Horn basin area may have been as thick as 1,500 or 2,000 ft. This overburden thickness would meet the suggested requirement for expulsion of continuous-phase oil and the start of flush migration within the Phosphoria Formation and Tensleep Sandstone. Regardless, depth of burial of some Phosphoria source sediments farther west in the miogeosyncline probably was sufficiently great (Fig. 7), even before Early Jurassic time, to begin the expulsion and migration process.

REGIONAL TILT

It has been proposed also by many geologists that, before migration can occur on a regional scale, some minimum amount of tilt must be present along the migration path. Southwest tilt into the miogeosyncline was present even in Permian time, but this may not have been sufficient to initiate migration. The first important post-Phosphoria regional tilt in the Big Horn basin area was in a south-southwesterly direction and was created by Early Jurassic epeirogenic uplift. Isopachous contours of the Triassic (cf., Pye, 1958, p. 210, Fig. 20; McKee et al., 1959, p. 5; Sheldon, 1967, Fig. 8; Thomas, 1965, p. 1872, Fig. 14), if interpreted paleostructurally, show that westerly tilt into the Cordilleran geosyncline probably was great enough by Early Jurassic time to initiate primary migration in the underlying Phosphoria source rocks. Differential loading through time, with the most pressure applied over the main area of deposition of the fine-grained Phosphoria-source sediments, should have squeezed hydrocarbons out of these sediments and into the vertically and laterally adjacent carbonate and sandstone carrier beds, through which these hydrocarbons would have moved updip, eastward across the Wyoming shelf. According to Sheldon (1967, p. 62), "The migration of Permian oil probably began soon after deposition of the Triassic rocks, and continued until cementation impeded the paths of migration or tectonism broke them up." In terms of the first important regional tilt, then, primary "flush" migration of Paleozoic-Phosphoria oil on a regional scale probably was complete by Early Jurassic time, long before the Laramide orogeny.

CAPACITY OF TRAPS AND BUBBLE-POINT PRESSURE ANALYSIS

The earliest time of final accumulation can be estimated for some structural traps by study of the relation between gas saturation and the proportion of the trap which is filled with hydrocarbons (Gussow, 1955; Levorsen, 1954, p. 555). For example, if the oil in a certain trap is saturated with gas and a large free gas cap is present, and, if the trap is filled to the spill point, it usually can be assumed that the accumulation was not completed until its present elevation or depth of burial was reached. However, if the oil in the trap is undersaturated with gas and the trap is only partly filled, the accumulation probably was completed before the trap attained its present depth of burial and reservoir pressure.

In using the bubble-point pressure and trap-capacity criteria, an important uncertainty results in assuming that oil generally migrates at its saturation or bubble-point pressure when it has its greatest mobility and buoyancy, and its lowest viscosity; but this seems reasonable because, even if the oil were gas-deficient at the beginning of migration, the decreasing pressure along the updip migration path should increase gas saturation and eventually create some free gas (cf., Gussow's [1954] "differential entrapment" hypothesis). If the oil in a particular trap actually accumulated at its bubble point and the total volume of gas is the same today as it was at the time of accumulation, then the bubble-point pressure in undersaturated pools, converted to equivalent depth of burial, may be used to estimate the approximate overburden thickness at the time of accumulation. If hydrostatic conditions prevailed at the time of migration and accumulation, the depth of overburden in undersaturated pools is equal approximately to the bubble-point pressure × 2.3 ft/psi. However, pressures may have been above or below hydrostatic. In fields such as Garland, which has a free gas cap and is not filled to the spill point, it is theoretically possible to calculate the elevated pressure at which all gas would go back into solution in the oil, provided that reliable pressure-volume-temperature data are available, and thus arrive at an estimate of overburden thickness at the time of accumulation.

A general survey of the common-pool-state Paleozoic fields in the Big Horn basin indicates that the majority are filled to less than 50 per-

TABLE XI. ESTIMATE OF TIME OF MIGRATION FROM BUBBLE-POINT PRESSURE ANALYSIS

Field	Bubble Point Pressure (psi)	Depth* Equiv., (Ft)	Inferred Time of Migration	Source of Pressure Data
1. Cottonwood Creek	2,000–	4,600	Lower Cretaceous	Estimated from Standing's (1952) charts
2. Elk Basin	530–1,250	1,219–2,875	Lower Jurassic to Lower Cretaceous	Espach and Fry (1951)
3. Frannie	362	833	Triassic to Lower Jurassic	Curtis (1954)
4. Gebo	77–226	176–519	Triassic to Lower Jurassic	Biggs and Espach (1960)
5. Slick Creek	4,000–	9,200	Tertiary	Estimated from Standing's (1952) charts
6. Torchlight	59	136	Triassic	Willingham and Howald (1965)

* Bubble point pressure times 2.3 ft/psi.

cent of capacity on the basis of height of the oil column, which is equivalent to a much smaller percentage of the trap on a volumetric basis. However, a few fields are filled to, or nearly to, the spill point. Significantly, these fields are for the most part either very deep or very shallow. For example, the deep category includes Silvertip (Fig. 9) and some of the Phosphoria condensate fields of the east-central basin area. In the shallower category are Alkali and Bonanza anticlines on the east flank, and Oregon Basin and Fourbear fields on the west flank of the basin. The fields on the east flank might be considered as "terminal" traps (Gussow, 1955, p. 558).

The data further indicate that the deep Paleozoic fields generally are saturated with gas at the top of the structure, but undersaturated at the oil-water contact (e.g., Elk Basin and Worland). The shallow terminal traps, however, are greatly undersaturated and often are described as containing no solution gas. Some fields on the east side of the basin in this category are notable also for being in the area of nonreservoir Phosphoria rocks, thus implying lateral migration through the Tensleep.

The few data on bubble-point pressures available to the writer are shown in Table XI with corresponding theoretical overburden depths computed according to the hydrostatic pressure assumption. Clearly, the range of bubble-point pressures and converted depths in this table is greater than that which could have existed at the proposed Early Jurassic time of flush migration. In the deep fields, for example, these data suggest that accumulation was not completed until the present depth was reached. Is this condition as unlikely as it first appears?

Considerable evidence supports the hypothesis that the high temperatures associated with these deep fields have caused some type of thermal-catalytic cracking with the possible aid of anaerobic bacteria. Thus it may well be that a more normal immature oil "matured" to yield additional gas and other light ends in these deep fields by a modified process of separation-migration similar to that proposed by Silverman (1965). The deeper traps which are filled to a greater extent than most shallower traps, however, cannot be explained by fluid volumetric factors alone, because the high pressures associated with these greater depths would decrease the percentage of the trap occupied by hydrocarbons under constant reservoir volume conditions. Perhaps the answer is in the notable reduction of reservoir porosity (volume) in the Tensleep Sandstone at these greater depths (Lawson and Smith, 1966), apparently resulting from greater packing of the grains accompanied by secondary cementation (cf., Atwater and Miller, 1965). The Tensleep at Worland, although saturated with oil, is too "tight" to produce commercially. However, the unusual "tightness" of the Worland Tensleep Sandstone may be related also to the condition that the more porous upper Tensleep generally is missing at Worland because of truncation by the overlying Phosphoria, and that only the "tighter" lower Tensleep is available as a reservoir. It is possible, but unlikely, that secondary hydrocarbon generation occurred within the Phosphoria after early Laramide compression due to the higher temperatures in these deep fields (*cf.*, Welte, 1965).

In the shallow basin-edge fields, the low bubble-point pressures may suggest loss of gas from the original crude. This could have been accomplished by selective leakage from the trap, solution in circulating meteoric waters, or both. It is also possible in the east-side pools, such as Bonanza, that, if the oil in the Tensleep was derived from secondary spilling out of deeper pools, only greatly undersaturated crude normally found in the lower part of long oil-columns migrated updip to be trapped in the shallower pool.

The strongly undersaturated condition of the moderately deep Gebo field (Table XI) probably cannot be explained by water washing, because little hydrodynamic action can be demonstrated and the associated formation water is unusually saline. Secondary spilling cannot be the reason because this field is not updip from any known deeper pools. Some gas could have escaped through the associated flank fault, however, because the Paleozoic oil in the Triassic "Curtis" Sandstone in this field proves that the fault zone is as a carrier of fluids.

The bubble point-overburden computation for Frannie, a terminal trap in the northern basin area, however, agrees reasonably well with an Early Jurassic time of migration. Although this field has a strong hydrodynamic gradient, the Tensleep crude does not appear to have been degraded much as a result of water washing.

Thus, the use of the bubble-point method for estimation of time of accumulation of Paleozoic oil in the Big Horn basin may be "fraught with pitfalls" (Gussow, 1954, p. 557) and should be used with caution. Probably some post-entrapment modification of gas saturations in the Paleozoic fields of the basin has occurred. This modification can be attributed to the petroleum maturation process in the deep fields and, in the shallow fields, to the loss of gas through solution in moving water and selective leakage from the trap. The Phosphoria source and reservoir rocks apparently were never buried deeply except in the central part of the Big Horn basin. Hence the crude oils contained in the fields along the edge of the basin remained immature through the long time interval between primary flush migration

and final accumulation in Laramide structures. Only in those fields buried more than 6,000 ft in the central basin area can the Paleozoic oils be considered to be mature.

FORMATION OF TRAPS

The earliest time of trap formation has been used frequently for estimating time of migration and accumulation. A great deal has been said already about the Big Horn basin stratigraphic traps resulting from Phosphoria facies change and wedgeout and by subcrop irregularities at the Phosphoria-Tensleep unconformity. Traps were available almost from the time of deposition, even to the present (*e.g.*, Cottonwood Creek field). The time of trap formation thus is of little help in determining the time of primary flush migration of Phosphoria oil.

Lawson and Smith (1966) recently proposed that accumulation in many Tensleep pools in the Big Horn basin, although now associated with structural closure, is the result partly or wholly of stratigraphic variables in the Tensleep Sandstone. The variables envisioned include: (1) intraformational change in permeability and/or lithofacies; (2) incised channeling with later infilling of basal Phosphoria shale providing a truncational subcrop trap; and (3) a combination of (1) and (2), the later Laramide folding being superimposed. These stratigraphic conditions presumably exist in some places along the eastern side of the Big Horn basin (*e.g.*, Bonanza shows evidence of no. 3); they exist on the eastern side of the Powder River basin as revealed by many of the recently discovered Minnelusa oil fields. Their importance in determining the location of the present-day large anticlinal accumulations in the Big Horn basin, however, has not been proved, particularly because no wholly stratigraphic Tensleep accumulations of commercial significance have been found. Regardless, because of the coincidence of accumulation and structure, and because of the indication that fracturing is universally present in Big Horn basin Tensleep pools (*cf.*, Lawson and Smith, 1966), it is doubtful that stratigraphic changes in the Tensleep are of much importance to *present-day* accumulation geometry in the larger multizoned Paleozoic fields.

The implication, therefore, seems to be that

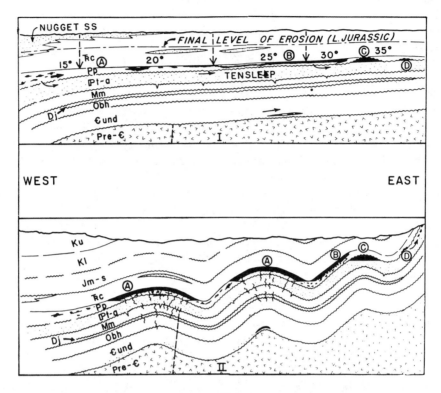

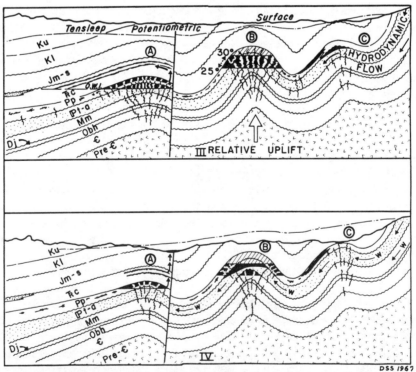

FIG. 30.—Conceptual sequence of geologic events leading to development and later modification of Paleozoic oil fields in Big Horn basin, Wyoming.

I. *End of Triassic:* epeirogenic uplift, tilt and erosion, formation of continuous phase hydrocarbons and completion of primary flush migration of Phosphoria oil: (a) compaction and diagenesis, (b) regional stratigraphic entrapment at Phosphoria carbonate-redbed facies boundary, (c) expulsion of Phosphoria oil into Tensleep

Phosphoria oil may have migrated long distances through Tensleep (and equivalent) carrier beds, from marine Phosphoria source rocks on the west, possibly as far as the eastern Powder River basin (Sheldon, 1967). If the very considerable volume of oil held in Minnelusa reservoirs is Phosphoria-derived, this might suggest that Tensleep stratigraphic interruptions in the Big Horn basin area were not important in terms of the volume of hydrocarbons entrapped. However, many geologists believe that Minnelusa oil probably had its source in the interbedded Minnelusa carbonate (B. F. Curtis *et al.*, 1958, p. 289; R. R. Berg, personal commun., 1965) or in the thin, lower Minnelusa black shale of the Hartville area (A. E. Allen, 1966).

Though most Big Horn basin anticlinal fields are less than half filled with oil and gas, the total volume of hydrocarbons now in these anticlines would seem to be much larger than in any earlier structural closures present at the suggested Early Jurassic time of flush migration. Thus, it must be concluded that final localization of accumulation in these fields could not have been completed until at least Late Cretaceous or Paleocene time (Love, 1960) and possibly not until the Eocene. This relation, then, indicates that Paleozoic oil in the large anticlinal pools of the Big Horn basin must have remigrated from earlier "storage" accumulations and adjusted to Laramide structure.

A conceptual sequence of the migration and accumulation process in the Paleozoic fields of the Big Horn basin is shown in Figure 30. This diagram illustrates that the hydrocarbons now found in the Paleozoic anticlinal fields originally migrated and accumulated within the Phosphoria carbonates and the upper Tensleep Sandstone,

probably during Triassic and Early Jurassic times. These hydrocarbons were stored in the regional Phosphoria and in more local Tensleep stratigraphic traps, until Laramide folding fractured the brittle Paleozoic rocks, including the "tight" lower Phosphoria, allowing the oil and gas to spread into previously oil-free reservoir space in underlying Paleozoic rocks, forming common pools which were fully adjusted to structure. Subsequent modifications of some original common pools by hydrodynamic flow, erosional breaching, and fault leakage have been outlined.

Some oil and gas which had been expelled into the Tensleep Sandstone must have undergone long-range lateral migration into areas beyond the eastern edge of Phosphoria marine rocks. Along the easterly migration path, several truncational subcrop and other stratigraphic traps probably captured much of this petroleum, possibly as far as several hundred miles from its original source (Sheldon, 1967).

The fact that available trap space in most common-pool Paleozoic fields is only partly filled with hydrocarbons provides another indication that spilling from the Phosphoria and/or Tensleep reservoirs has occurred. If the anticlines had been filled to the Phosphoria spill point with stratigraphically trapped oil before fracturing, later loss of much of this oil to older formations through faults and fractures would have resulted in the creation of an oil-water contact well above the original spill-point level (Fig. 30). The proposed theory of accumulation and adjustment to structure thus can meet the requirement of a partly filled trap.

Some authors have claimed or implied that Tensleep crude-oil gravity values in the Big Horn

---

Sandstone and stratigraphic entrapment in truncational unconformity trap, and (d) longer range lateral migration through Tensleep and equivalent carrier beds.

II. *End of Paleocene:* Laramide folding, crestal fracturing, and structural accumulation: (a) beginning of spilling of Phosphoria oil through fractures and faults into older Paleozoic formations, (b) basinward tilting of part of original Phosphoria primary stratigraphic trap, (c) adjustment of Tensleep stratigraphic oil to structure, and (d) spilling of oil out of tilted Tensleep stratigraphic trap.

III. *End of Eocene:* intensified folding, crestal fracturing, and faulting; differential uplift and erosion; and beginning of hydrodynamic flow; (a) adjustment of common-oil pool through fault communication in area of low-pressure potential, (b) completion of spilling process to produce a common-pool state, accompanied by relative vertical uplift, gas cap formation, and density stratification of hydrocarbons, and (c) surface-water invasion and start of hydrodynamic flow, resulting in tilting of Tensleep oil pool.

IV. *Recent:* development of present land surface and hydrodynamic environment: (a) deeper erosion and adjustment of oil to new hydrodynamic environment by leakage to surface through fault zone, (b) segregation into separate pools as result of severe hydrodynamic modification, and (c) loss of oil to surface and inspissation resulting from erosional truncation of cap rock.

basin grade uniformly from low to high values basinward or with depth in the Big Horn basin (Todd, 1963). This is not entirely correct. Figure 2, for example, shows that crude-oil gravity values on the west side of the basin generally are lower than those on the east side of the basin at similar depths. Furthermore, crude oil gravities in the fields of the southeastern arm of the basin are higher than normal for their depth, though many of these fields are under a strong hydrodynamic gradient and are associated with dilute formation waters. Thus, considering all local modifying factors previously discussed, including the density stratification within individual fields, crude-oil gravity values may be more directly related to location in the basin than to depth of burial, except for the mature oils in the deeper fields below 6,000 ft.

One possible explanation for this geographic crude-oil gravity variation is that it reflects original density stratification in the primary, west-dipping, Phosphoria accumulation, a stratification similar to that now found in the larger Paleozoic fields. D. B. Rea (written commun., 1966) has suggested that the lower gravity values on the west side of the Big Horn basin may be related to Absarokan igneous activity, but the writer suggests that higher gravity values are a more likely result of such igneous activity. However, some lighter crudes in the eastern Big Horn basin, such as those found at Bonanza and Nowood, could be explained by possible early entrapment along the Phosphoria-Tensleep unconformity.

SUMMARY AND CONCLUSIONS
GENERAL

From all data presented in this paper, there can be little doubt that the marine facies of the Permian Phosphoria (Park City) Formation is the source of most hydrocarbons now found in the Paleozoic and Triassic reservoirs of the Big Horn basin. Marine Phosphoria rocks are petroliferous and are the only Paleozoic rocks in this area that contain important dark-colored, organic-rich, fine-grained "source" rocks with well-defined characteristics of a reducing depositional environment; all other Paleozoic formations in the basin are composed mostly of light-colored "oxidized" rocks, and are bounded both above and below by important regional intersystemic unconformities representing long periods of erosion.

The Phosphoria marine facies must have provided not only the source but also the reservoir rocks in which hydrocarbons were trapped during the period of time preceding the Laramide orogeny by updip and lateral regional facies change and wedgeout. Early primary stratigraphic entrapment and storage within the Phosphoria carbonate facies, and locally along the Phosphoria-Tensleep unconformity, explain how so much oil was available for later structural accumulation in Laramide anticlinal traps.

The number of fields producing from the Paleozoic formations decreases with increasing geologic age of the reservoir rock, from 56 with Permian Phosphoria production, to 3 with Ordovician Bighorn production. This condition shows that hydrocarbons are normally not produced from any Paleozoic zone or formation in the Big Horn basin unless accompanied also by oil saturation in all overlying Paleozoic zones with reservoir porosity. The few exceptions include only those fields where secondary segregation has occurred through selective hydrodynamic displacement of the oil in the Tensleep reservoir, or where overlying Paleozoic reservoirs have lost their original hydrocarbons at the surface as a result of exposure through erosional truncation of the cap rock, or by leakage through fault zones in areas of high-pressure potential.

Thus the majority of the multizoned Paleozoic oil fields of the Big Horn basin are single oil pools with a single oil-water contact common to all productive Paleozoic formations.[10] Variation in the elevation of these common oil-water contacts, however, can range up to 10 percent of the total oil column as the result of differences in (1) relative permeability and oil viscosity, (2) pressure gradients in the water phase, (3) strength of water drive (or water encroachment), (4) withdrawal rates, or (5) development histories in the different formations or zones. Original migration and communcation between zones were through fractures, aided by faults in some fields. Density stratification (gravitational segregation) of fluids is present in the larger fields and apparently is common to all Paleozoic productive zones. Moreover, limited pressure information indicates noticeable drawdown effects in deeper Paleozoic

[10] Some of these same conclusions were reached independently by Lawson and Smith (1966).

zones discovered after development of shallower zones.

The petroleums in the different Paleozoic formations of the Big Horn basin are of the same fundamental chemical composition. Although there is evidence for selective alteration by water-washing, bacterial oxidation through sulfate reduction, or thermal maturation, the fundamental chemical character of the oil has not been changed materially by these processes. These crude oils have an intermediate aromatic-naphthenic base, are notably paraffinic in the gasoline range, and contain considerable asphalt. They are unlike Paleozoic oils from other Rocky Mountain basins. Mississippian Madison oils from the deep Williston basin, for example, are more aromatic in the gasoline range and contain more paraffins in the heavier fractions. Other chemical characterization parameters such as V/Ni, carbon C^{13}/C^{12}, and sulfur S^{34}/S^{32} ratios show a close similarity among samples of all Big Horn basin Paleozoic crude oils, but differ significantly from corresponding ratios for the Paleozoic crude oils of the Williston basin. Thus the Paleozoic oils from the two basin areas could not have been derived from the same source. The Mississippian oils of the Williston basin are associated with euxinic source shales and stratigraphic entrapment and are considered to be indigenous; Mississippian oils of the Big Horn basin are associated with "oxidized" rocks in "common-pool" anticlinal fields and are clearly "exotic."

Formation waters from the Paleozoic productive zones in the Big Horn basin also are chemically similar; they are of moderately low concentration and exhibit secondary salinity, usually with proportionately high sulfate concentrations except where sulfate-reducing bacteria apparently have been active. Differences in waters from the several Paleozoic formations in the Big Horn basin probably can be attributed to degree of dilution by circulating surface waters or to other environmental factors such as the petrography of the reservoir rock. These waters differ considerably from their counterparts in equivalent Williston basin Paleozoic reservoirs.

In the Big Horn basin, single-zone Paleozoic fields are the exception. Single-zone Phosphoria fields are either stratigraphically controlled or are associated with structural closures which are too small to allow spilling into older reservoirs. Commercial single-zone Tensleep fields exist only in areas of mostly nonreservoir Phosphoria facies, and although commonly associated with stratigraphic changes in the Tensleep, the fields are believed to be primarily controlled by structural closure and hydrodynamic flow. The oil in these Tensleep pools presumably accumulated by early lateral migration into truncational subcrop traps at the Phosphoria-Tensleep unconformity, followed by later adjustment to Laramide structure, or possibly by spilling of Phosphoria oil updip from more basinward pools. The only single-zone Madison field in the Big Horn basin is Red Springs (Fig. 29B), and in this field the overlying Paleozoic productive zones have lost most of their original hydrocarbons at the surface as a consequence of erosional breaching.

ECONOMIC SIGNIFICANCE OF THEORY

The economic significance of the concept of a single, Phosphoria source for all oil in the Paleozoic and Triassic reservoirs of the Big Horn basin seems self-evident. To date, no commercial stratigraphic accumulation has been found in any pre-Pennsylvanian formation in Wyoming and, according to the present theory, none is likely to be found. Moreover, although stratigraphic variations within the Tensleep Formation and at the Phosphoria-Tensleep unconformity probably contribute to some accumulations, the total reserves possibly attributable to stratigraphic entrapment are extremely small compared with structurally entrapped Paleozoic oil in the Big Horn basin. Stratigraphic variations within the Permian-Pennsylvanian section are much more important to accumulation in the neighboring Powder River basin (Wyoming Geological Association and Billings Geological Society, 1963).

The present theory states that, *in the area of Phosphoria reservoir carbonate deposition, sub-Phosphoria Paleozoic acccumulation normally can exist in the Big Horn basin only if vertical anticlinal closure is present in excess of the thickness of the Phosphoria Formation.* It is apparent, therefore, that on a Big Horn basin structural prospect with only 500 ft of anticipated vertical closure, a Madison test could not be justified if the thickness of the interval from the top of the Phosphoria carbonates to the reservoir porosity in the Madison were 700 ft in the area of the prospect. Probably many deep and expensive

Madison wildcat tests would not have been drilled had this condition been considered. For the same reasons, however, the presence of hydrocarbons in the Devonian and Ordovician rocks at Elk Basin should have been anticipated (and probably was) long before their discovery in 1961, considering the very large oil column in the Phosphoria-Tensleep zone of this field.

Although this discussion generally has been confined to conditions found within the area of the Big Horn basin, a cursory review of producing conditions in the Paleozoic fields of the other basins of the western Wyoming shelf area indicates that similar conditions probably exist. In the Wind River basin, only two fields produce from the Madison Formation, i.e., Circle Ridge and Beaver Creek, and both have vertical closures exceeding 1,000 ft—well in excess of the thickness of the Phosphoria-to-Madison section. However, it is significant that (1) no production has ever been found in the Madison of the Powder River basin where the Phosphoria is represented by the redbed, Goose Egg facies, (2) Paleozoic oil columns are generally small, and (3) Tensleep-Minnelusa accumulations generally are related to stratigraphic factors.

Sheldon (1967, p. 11) has estimated that approximately 400 cu mi of Phosphoria black shale source rock would have yielded about 100 billion bbl of oil assuming that only 5 percent of the original organic matter deposited with the sediments was converted to petroleum. If all of the aromatic-naphthenic base hydrocarbons so far discovered in the Paleozoic reservoirs of the Wyoming shelf area of Phosphoria carbonate deposition are attributed to Phosphoria sources, the Phosphoria has accounted for more than 3 billion bbl of proved recoverable oil reserves, plus many times this amount in unrecoverable oil. And if possible undiscovered Paleozoic hydrocarbons in the Wyoming shelf area, together with present and future Minnelusa reserves were included, at least another 1 billion bbl could be added to this recoverable total. With or without Minnelusa oil, the total is several times the amount of reserves attributable to all other formations combined in the same Rocky Mountain area.

References Cited

Agatston, R. S., 1954, Pennsylvanian and Lower Permian of northern and eastern Wyoming: Am. Assoc. Petroleum Geologists Bull., v. 38, no. 4, p. 508–583.

Allen, Albert E., Jr., 1966, Environments of deposition in the Minnelusa and their interpretation from gamma sonic logs: paper presented before Wyo. Geol. Assoc., Casper, Wyo., Aug. 31-Sept. 2, 1966.

Allen, B. W., 1954, Three zones produce in Oregon basin: Petroleum Engineer, v. 26, no. 2, p. B-43–B-50.

Alpha, Andrew G., and John R. Fanshawe, 1954, Tectonics of northern Bighorn basin area and adjacent south-central Montana, in Pryor Mountains-northern Bighorn basin, Montana: Billings Geol. Soc. Guidebook, p. 72–79.

Andrichuk, John M., 1958, Mississippian Madison stratigraphy and sedimentation in Wyoming and southern Montana, in Habitat of oil: Am. Assoc. Petroleum Geologists, p. 225–267.

Anonymous, 1954, Hamilton Dome field, Hot Springs County, Wyoming, in Southern Big Horn basin, Wyoming: Wyo. Geol. Assoc. Guidebook, p. 104–107.

Arps, J. J., 1964, Engineering concepts useful in oil finding: Am. Assoc. Petroleum Geologists Bull., v. 48, no. 2, p. 157–165.

Atwater, Gordon I., and E. E. Miller, 1965, The effect of decrease in porosity with depth of future development of oil and gas reserves in south Louisiana (abs.): Am. Assoc. Petroleum Geologists Bull., v. 49, no. 3 (pt. I), p. 334.

Aufricht, W. R., 1965, Distribution and production of oil and water in Tensleep-Minnelusa reservoirs: Soc. Petroleum Engineers Paper SPE 1162 (prepr.).

Baker, E. G., 1959, Origin and migration of oil: Science, v. 129, no. 3353, p. 871–874.

Baker, N. E., and F. R. S. Henson, 1952, Geological conditions of oil occurrence in Middle East fields: Am. Assoc. Petroleum Geologists Bull., v. 36, no. 10, p. 1885–1901.

Ball, John S., 1962, Nitrogen compounds in petroleum: Am. Petroleum Inst. Proc., v. 42, sec. 8, p. 27–30.

Barton, D. C., 1934, Natural history of the Gulf Coast crude oil, in Problems of petroleum geology: Am. Assoc. Petroleum Geologists, p. 109–155.

Bastin, E. S., 1926, The problem of the natural reduction of sulphates: Am. Assoc. Petroleum Geologists Bull., v. 10, no. 12, p. 1270–1299.

——— and F. E. Greer, 1930, Additional data on sulphate-reducing bacteria in soils and waters of Illinois oil fields: Am. Assoc. Petroleum Geologists Bull., v. 14, no. 2, p. 153–159.

Beerstecher, Ernest, Jr., 1954, Petroleum microbiology: New York, Elsevier Press, 375 p.

Bell, Kenneth G., 1960, Uranium and other trace elements in petroleums and rock asphalts: U. S. Geol. Survey Prof. Paper 356-B, p. 45–65.

Benson, Anthony L., 1966, Devonian stratigraphy of western Wyoming and adjacent areas: Am. Assoc. Petroleum Geologists Bull., v. 50, no. 12, p. 2566–2603.

Biggs, Paul, and Ralph H. Espach, 1960, Petroleum and natural gas fields in Wyoming: U. S. Bur. Mines Bull. 582, 538 p.

Boaden, E., and E. C. Masterson, 1952, Some aspects of field operations in Kuwait: London, Inst. Petroleum Jour., v. 38, p. 395–414.

Brenneman, M. C., and P. V. Smith, Jr., 1958, The chemical relationships between crude oils and their source rocks, *in* Habitat of oil: Am. Assoc. Petroleum Geologists, p. 818–849.

Brooks, Benjamin T., 1952, Evidence of catalytic action in petroleum formation: Ind. Eng. Chemistry, v. 44, no. 11, p. 2570–2577.

Buckley, Stuart E., C. R. Hocott, and M. S. Taggart, Jr., 1958, Distribution of dissolved hydrocarbons in subsurface waters, *in* Habitat of oil: Am. Assoc. Petroleum Geologists, p. 850–882.

Burk, C. A., 1956, Stratigraphic summary of the pre-Niobrara formations of Wyoming, *in* Wyoming stratigraphy: Wyo. Geol. Assoc., p. 91–96.

Campbell, Charles V., 1962, Depositional environments of Phosphoria Formation (Permian) in southeastern Bighorn Basin, Wyoming: Am. Assoc. Petroleum Geologists Bull., v. 46, no. 4, p. 478–503.

Carlson, C. G., and S. B. Anderson, 1965, Sedimentary and tectonic history of North Dakota part of Williston basin: Am. Assoc. Petroleum Geologists Bull., v. 49, no. 11, p. 1833–1846.

Chamberlin, R. T., 1940, Diastrophic behavior around the Bighorn basin: Jour. Geology, v. 48, p. 673–716.

Collier, A. J., 1920, Oil in the Warm Springs and Hamilton domes, near Thermopolis, Wyoming: U.S. Geol. Survey Bull. 711, p. 61–73.

Cook, C. W., 1925, Fractionation of petroleum during capillary migration: Econ. Geology, v. 20, p. 639–641.

Crawford, James G., 1949, Waters of producing fields in the Rocky Mountain region: Trans. A.I.M.M.E. Tech. Pub. 2383, v. 179, p. 264–285.

——— 1964, Rocky Mountain oil field waters: Casper, Wyo., Chem. and Geol. Labs., 68 p.

Cupps, Cecil Q., Philip H. Lipstate, Jr., and Joseph Fry, 1951, Variance in characteristics of the oil in the Weber sandstone reservoir, Rangely field, Colorado: U. S. Bur. Mines Rept. Inv. 4761, 68 p.

Curtis, Bruce F., John W. Strickland, and Robert C. Busby, 1958, Patterns of oil occurence in the Powder River basin, *in* Habitat of oil: Am. Assoc. Petroleum Geologists, p. 268–292.

Curtis, L. B., 1954, Geological and Tensleep reservoir summary of Frannie field, Park County, Wyoming, and Carbon County, Montana, *in* Pryor Mountains-Big Horn basin, Montana: Billings Geol. Soc. Guidebook, p. 126–129.

Daniel, E. J., 1954, Fractured reservoirs of Middle East: Am. Assoc. Petroleum Geologists Bull., v. 38, no. 5, p. 774–715.

Dobbin, C. E., 1947, Exceptional oil fields in Rocky Mountain region of United States: Am. Assoc. Petroleum Geologists Bull., v. 31, no. 5, p. 797–823.

Downs, George R., 1952, Summary of Mesozoic stratigraphy, Big Horn basin, Wyoming: Wyo. Geol. Assoc. Guidebook, p. 26–31.

Dunning, H. N., J. W. Moore, and M. O. Denekas, 1953, Interfacial activities and porphyrin contents of petroleum extracts: Ind. Eng. Chem., v. 45, p. 1759–1765.

Dunnington, H. V., 1958, Generation, migration, accumulation, and dissipation of oil in northern Iraq, *in* Habitat of oil: Am. Assoc. Petroleum Geologists, p. 1194–1251.

Elmer, N. C., 1959, Complex entrapment at Sage Creek field, Big Horn basin, Wyoming: Denver, Petroleum Information, Geol. Record, p. 97–100.

Erdman, J. Gordon, 1962, Oxygen, nitrogen, and sulfur in asphalts: Am. Petroleum Inst. Proc., v. 42, sec. 8, p. 33–41.

——— 1965, Petroleum—its origin in the earth, *in* Fluids in subsurface environments: Am. Assoc. Petroleum Geologists Mem. 4, p. 20–52.

Erickson, R. L., A. T. Myers, and C. A Horr, 1954, Association of uranium and other metals with crude oil, asphalt, and petroliferous rock: Am. Assoc. Petroleum Geologists Bull., v. 38, no. 10, p. 2200–2218.

Espach, Ralph H., and Joseph Fry, 1951, Variance characteristics of the oil in the Tensleep Sandstone reservoir, Elk Basin field, Wyoming and Montana: U. S. Bur. Mines Rept. Inv. 4768, 24 p.

Fanshawe, John R., 1960, Significance of interruptions to hydrodynamics in northern Rocky Mountain province, U. S. A.: Internatl. Geol. Congr., 21st Sess. Pt. 11, Copenhagen, p. 7–18.

Feely, H. W., and J. L. Kulp, 1957, Origin of Gulf Coast salt-dome sulphur deposits: Am. Assoc. Petroleum Geologists Bull., v. 41, no. 8, p. 1802–1853.

Fish, Andrew R., and John C. Kinard, 1959, Madison Group stratigraphy and nomenclature in the northern Williston basin: Denver, Petroleum Information, Geol. Record, p. 117–129.

Gibson, George R., 1965, Oil and gas in southwestern region—geologic framework, *in* Fluids in subsurface environments: Am. Assoc. Petroleum Geologists Mem. 4, p. 66–100.

Ginter, R. L., 1930, Causative agents of sulphate reduction in oil-well waters: Am. Assoc. Petroleum Geologists Bull., v. 14, no. 2, p. 139–152.

——— 1934, Sulphate reduction in deep subsurface waters, *in* Problems of petroleum geology: Am. Assoc. Petroleum Geologists, p. 907–925.

Gooding, R. M., N. G. Adams, and H. T. Rall, 1946, Determinations of aromatics, naphthenes and paraffins by refractometric methods: Ind. Eng. Chem., v. 18, p. 2–13.

Gorman, Donald R., 1965, The stratigraphy of the Amsden Formation of Wyoming (abs.): Am. Assoc. Petroleum Geologists Bull., v. 49, no. 3 (pt. I), p. 341.

Green, Thomas H., and C. W. Ziemer, 1953, Tilted water table at Northwest Lake Creek field, Wyoming: Oil and Gas Jour., July 13, p. 178.

Gussow, William Carruthers, 1954, Differential entrapment of oil and gas: a fundamental principle: Am. Assoc. Petroleum Geologists Bull., v. 38, no. 5, p. 816–853.

——— 1955, Time of migration of oil and gas: Am. Assoc. Petroleum Geologists Bull., v. 39, no. 5, p. 547–574.

Hanson, William E., 1959, Some chemical aspects of petroleum genesis, *in* Researches in geochemistry, P. H. Abelson, ed.: New York, John Wiley and Sons, p. 104–117.

——— 1960, Origin of petroleum, *in* Chemical technology of petroleum, by William A. Gruse and Donald R. Stevens: New York, McGraw-Hill, p. 228–254.

Harris, J. F., G. L. Taylor, and J. L. Walper, 1960, Relation of deformational fractures in sedimentary rocks to regional and local structures: Am. Assoc. Petroleum Geologists Bull., v. 44, no. 12, p. 1853–1873.

Harvey, R. J., and F. K. Krebill, 1954, Elk Basin—the Cinderella field: Petroleum Engineer, v. 26, no. 2, p. A-37-A-48.

Hayden, Harold J., 1961, Distribution of uranium and other metals in crude oils: U. S. Geol. Survey Bull. 1100-B, p. 17–99.

Hedberg, Hollis D., 1936, Gravitational compaction of clays and shales: Am. Jour. Science, 5th ser., v. 31, no. 184, p. 241–87.

———— 1964, Geologic aspects of origin of petroleum: Am. Assoc. Petroleum Geologists Bull., v. 48, no. 11, p. 1755–1803.

Hetherington, G., and A. J. Horan, 1960, Variations with elevation of Kuwait reservoir fluids: London, Inst. Petroleum Jour., v. 46, p. 109–114.

Hill, Gilman A., W. A. Colburn, and J. W. Knight, 1961, Reducing oil-finding costs by use of hydrodynamic evaluations, in Petroleum exploration, gambling game or business venture: Englewood, N. J., Prentice-Hall, Inc., p. 38–69.

Hill, J. Bennett, 1953, What is petroleum?: Ind. Chem. Eng., v. 45, no. 7, p. 1398–1401.

Holliman, W. C., et al., 1950, Composition of petroleum; properties of distillates to 100°F: U. S. Bur. Mines Tech. Paper 722, 55 p.

Hubbert, M. K., 1953, Entrapment of petroleum under hydrodynamic conditions: Am. Assoc. Petroleum Geologists Bull., v. 37, no. 8, p. 1954–2026.

———— 1966, History of petroleum exploration and its bearing upon present and future exploration: Am. Assoc. Petroleum Geologists Bull., v. 50, no. 12, p. 2504–2518.

Hunt, John M., 1953, Composition of crude oil and its relation to stratigraphy in Wyoming: Am. Assoc. Petroleum Geologists Bull., v. 37, no. 8, p. 1837–1872.

———— and J. P. Forsman, 1957, Relation of crude oil compositon to stratigraphy in the Wind River basin, in Southwestern Wind River basin: Wyo. Geol. Assoc. Guidebook, p. 105–112.

———— and George H. Jamieson, 1958, Oil and organic matter in source rocks of petroleum, in Habitat of oil: Am. Assoc. Petroleum Geologists, p. 735–746.

Jones, Theodore S., and Harold M. Smith, 1965, Relationships of oil composition and stratigraphy in the Permian basin of West Texas and New Mexico, in Fluids in subsurface environments: Am. Assoc. Petroleum Geologists Mem. 4, p. 101–224.

Kartsev, A. A., et al., 1954 (Eng. trans., 1959), Geochemical methods of prospecting and exploration for petroleum and natural gas: Berkeley, Univ. Calif. Press, 349 p.

Kidwell, Albert L., and John M. Hunt, 1958, Migration of oil in Recent sediments of Pedernales, Venezuela, in Habitat of oil: Am. Assoc. Petroleum Geologists, p. 790–817.

Krampert, E. W., 1947, Hamilton Dome, Hot Springs County, Wyoming, in Big Horn basin: Wyo. Geol. Assoc. Guidebook, p. 229–233.

———— 1949, Commercial oil in Cambrian beds, Lost Soldier field, Carbon and Sweetwater Counties, Wyoming: Am. Assoc. Petroleum Geologists Bull., v. 33, no. 12, p. 1998–2010.

Lawson, Donald E., and Jordan R. Smith, 1966, Pennsylvanian and Permian influence on Tensleep oil accumulation, Big Horn basin, Wyoming: Am. Assoc. Petroleum Geologists Bull., v. 50, no. 10, p. 2197–2220.

Levorsen, A. I., 1954, Geology of petroleum: San Francisco, W. H. Freeman and Co., 703 p.

———— 1960, Paleogeologic maps: San Francisco, W. H. Freeman and Co., 174 p.

Lindsey, K. B., 1954, Petroleum in the Williston basin, including parts of Montana, North and South Dakota, and Canada, as of July 1953: U. S. Bur. Mines Rept. Inv. 5055, 70 p.

Lochte, H. L., 1962, Oxygen compounds in petroleum: Am. Petroleum Inst. Proc., p. 42, sec. 8, p. 32–33.

Love, J. D., 1960, Cenozoic sedimentation and crustal movement in Wyoming: Am. Jour. Science, Bradley Volume, v. 258A, p. 204–214.

McAuliffe, Clayton, 1966, Solubility in water of paraffin, cycloparaffin, olefin, acetylene, cyclo-olefin, and aromatic hydrocarbons: Jour. Phys. Chem., v. 70, no. 4, p. 1267–1275. Paper presented before Div. Petroleum Chem., Am. Chem. Soc., Chicago, Ill., Aug. 30–Sept. 10, 32 p.

McCanne, Rolland W., 1947, Grass Creek oil field, in Big Horn basin: Wyo. Geol. Assoc. Guidebook, p. 223–228.

McCoy, Alex. W., III, et al., 1951, Types of oil and gas traps in Rocky Mountain region: Am. Assoc. Petroleum Geologists Bull., v. 35, no. 5, p. 1000–1037.

McIver, Richard D., 1962, The crude oils of Wyoming —product of depositional environment and alteration, in Symposium on Early Cretaceous rocks of Wyoming and adjacent areas: Wyo. Geol. Assoc. Guidebook, p. 248–251.

McKee, E. D., et al., 1959, Paleotectonic maps of the Triassic system: U.S. Geol. Survey, Misc. Geol. Inv. Map 1–300, 33 p.

McKelvey, V. E., R. W. Swanson, and R. P. Sheldon, 1953, Phosphoria Formation in southeastern Idaho and eastern Wyoming, in Guide to the geology of northern Utah and southeastern Idaho: Intermtn. Assoc. Petroleum Geologists Guidebook, p. 41–47.

———— et al., 1959, The Phosphoria, Park City, and Shedhorn formations in the western phosphate field: U. S. Geol. Survey Prof. Paper 313-A, 47 p.

McKinney, C. M., and E. L. Garton, 1957, Analyses of crude oils from 470 important oil fields in the United States: U. S. Bur. Mines Rept. Inv. 5376, 276 p.

McNabb, J. G., P. V. Smith, Jr., and R. L. Betts, 1952, The evolution of petroleum: Ind. Eng. Chemistry, v. 44, no. 11, p. 2557–2563.

Miller, John R., and Chester C. Stiteler, 1952, Northwest Lake Creek field, Hot Springs County, Wyoming, in Southern Big Horn basin: Wyo. Geol. Assoc. Guidebook, p. 108–109.

Miller, L. P., 1949, Rapid formation of high concentrations of hydrogen sulfide by sulfate-reducing bacteria: Contr. Boyce Thompson Inst., v. 15, p. 437–465.

Morris, J. I., 1951, The Worland, Wyoming field: Petrol. Eng., reference annual, p. B–77—B–85.

Munyan, Arthur C., 1962, A different stratigraphic interpretation of Pennsylvanian-Permian-Triassic sequence, southern Montana and adjacent areas: Mont. Bur. Mines and Geology Spec. Publ. 23, Strat. Paper 1, 12 p.

Murray, George H., Jr., 1959, Examples of hydrodynamics in the Williston Basin at Poplar and North Tioga fields: Denver, Petroleum Information, Geol. Record, p. 55–59.

Nutting, P. G., 1934, Some physical and chemical properties of reservoir rocks bearing on the accumulation and discharge of oil, in Problems of petroleum geology: Am. Assoc. Petroleum Geologists, p. 825–832.

Partridge, John F., Jr., 1958, Oil occurrence in Permian, Pennsylvanian, and Mississippian rocks, Big Horn basin, Wyoming, in Habitat of oil: Am. Assoc. Petroleum Geologists, p. 293–306.

Pedry, J. J., 1957, Cottonwood Creek field, Washakie County, Wyoming, carbonate stratigraphic trap: Am. Assoc. Petroleum Geologists Bull., v. 41, no. 5, p. 823–838.

Peterson, V. E., 1961, Ashley Valley oil field, Uintah County, Utah, in Symposium of the oil and gas fields of Utah: Intermtn. Assoc. Petroleum Geologists, no page nos.

Petroleum Information, 1960, Rocky Mountains oil and gas operations for 1959: Denver Petrol. Inf., 30th ann. resume, 159 p.

Porter, J. W., and J. G. C. M. Fuller, 1959, Lower Paleozoic rocks of northern Williston basin and adjacent areas: Am. Assoc. Petroleum Geologists Bull., v. 43, no. 1, p. 124–189.

Pye, Willard D., 1958, Habitat of oil in northern Great Plains and Rocky Mountains, in Habitat of oil: Am. Assoc. Petroleum Geologists, p. 178–224.

Rall, H. T., et al., 1962, Sulfur compounds in petroleum: Am. Petroleum Inst. Proc., v. 42, sec. 8, p. 19–27.

Rea, Bayard D., 1962, Pre-Mississippian geology of Elk Basin field, Park County, Wyoming, and Carbon County, Montana, in Three Forks-Belt Mountains area; symposium of Devonian System of Montana and adjacent areas: Billings Geol. Soc. Guidebook, p. 115–118.

Roberts, Ralph J., et al., 1965, Pennsylvanian and Permian basins in northwestern Utah, northeastern Nevada, and south-central Idaho: Am. Assoc. Petroleum Geologists Bull., v. 49, no. 11, p. 1926–1956.

Rogers, G. Sherburne, 1919, The Sunset-Midway oil field, California; Part II, Geochemical relations of the oil, gas, and water: U.S. Geol. Survey Prof. Paper 117, 103 p.

Sachanem, A. N., 1945, The chemical constituents of petroleum: New York, Reinhold Publishing Corp.

Sandberg, Charles A., 1963, Dark shale unit of Devonian and Mississippian age in northern Wyoming and southern Montana: U.S. Geol. Survey Prof. Paper, 475-C, Art. 64, p. C17–C20.

Seidell, A., 1941, Solubilities of organic compounds: New York, D. Van Nostrand, v. 2, 926 p.

Sharkey, H. H. R., et al., 1946 Geologic and structure-contour map of Sage Creek Dome, Fremont County, Wyoming: U.S. Geol. Survey Oil and Gas Inv. Prelim. Map 53.

Shaw, Alan B., 1954, The Cambrian and Ordovician of the Pryor Mountains, Montana, and northern Bighorn Mountains, Wyoming, in Pryor Mountains-northern Big Horn basin, Montana: Billings Geol. Soc. Guidebook, p. 32–37.

Sheldon, Richard P., 1963, Physical stratigraphy and mineral resources of Permian rocks in western Wyoming: U. S. Geol. Survey Prof. Paper 313-B, 273 p.

———— 1967, Long-distance migration of oil in Wyoming: The Mountain Geologist, v. 4, no. 2, p. 53–65.

Silverman, S. R., 1963, Investigations of petroleum origin and evaluation mechanisms by carbon isotope studies, in Isotopic and cosmic chemistry: Amsterdam, North-Holland Publishing Co., p. 92–102.

———— 1965, Migration and segregation of oil and gas, in Fluids in subsurface environments: Am. Assoc. Petroleum Geologists Mem. 4, p. 53–65.

———— and Samuel Epstein, 1958, Carbon isotopic compositions of petroleums and other sedimentary organic materials: Am. Assoc. Petroleum Geologists Bull., v. 42, no. 5, p. 998–1012.

Sitter, L. U. de, 1956, Structural geology: New York, McGraw-Hill Book Co., Inc., 552 p.

Smith, G. Wendell, et al., 1958, Mississippian oil reservoirs in Williston basin, in Habitat of oil: Am. Assoc. Petroleum Geologists, p. 149–177.

Smith, H. M., 1940, Correlation index to aid in interpreting crude-oil analyses: U. S. Bur. Mines Tech. Paper 610, 34 p.

———— 1962, The role of sulfur and nitrogen compounds in organic geochemistry: Am. Petroleum Inst. Proc., v. 42, sec. 8, p. 132–135.

———— et al., 1959, Keys to the mystery of crude oil: Am. Petroleum Inst. Proc., v. 39, p. 433–465.

Smith, N. A., H. M. Smith, and C. M. McKinney, 1950, Refining properties of new crudes; Part I—Significance and interpretations of Bureau of Mines routine crude oil analyses: Petroleum Processing, v. 5, p. 609–614.

Standing, M. B., 1952, Volumetric and phase behavior of oil field hydrocarbon systems: New York, Reinhold Publishing Corp. 123 p.

Stewart, F. M., F. H. Gallaway, and R. E. Gladfelter, 1954, Comparison of methods for analyzing a water drive field, Torchlight Tensleep reservoir, Wyoming: Jour. Petroleum Technology, v. 6, no. 9, p. 105–111.

————, D. L. Garthwaite, and F. K. Krebill, 1955, Pressure maintenance by inert gas injection in the high relief Elk Basin field: Trans. A.I.M.M.E. Tech. Note 4008, v. 204, p. 49–57.

Stiff, Henry A., Jr., 1951, The interpretation of chemical water analysis by means of patterns: Trans. A.I.M.M.E. Tech. Note 84, Petrol,. Technol., v. 192, p. 376–379.

Stipp, T. F., 1947, Paleozoic formations of the Bighorn basin, Wyoming, in Big Horn basin: Wyo. Geol. Assoc. Guidebook, p. 121–130.

Stone, Robert W., and Claude E. ZoBell, 1952, Bacterial aspects of the origin of petroleum: Ind. Eng. Chemistry, v. 44, no. 11, p. 2564–2567.

Sugihara, J. M., 1962, Trace metal constituents in petroleum: Am. Petroleum Inst. Proc., v. 42, sec. 8, p. 30–32.

Summerford, H. E., 1952, Inclined water levels, in Southern Big Horn basin, Wyoming: Wyo. Geol. Assoc. Guidebook, p. 98–103.

Thode, H. G., R. K. Wanless, and R. Wallouch, 1954, The origin of native sulfur deposits from isotope fractionation studies: Geochim. et Cosmochim. Acta, v. 5, p. 286–298.

———— Jan Monster, and H. B. Dunford, 1958, Sulphur isotope abundance in petroleum and associated materials: Am. Assoc. Petroleum Geologists Bull., v. 42, no. 11, p. 2619–2641.

Thomas, Leonard E., 1965, Sedimentation and structural development of Big Horn basin: Am. Assoc. Petroleum Geologists Bull., v. 49, no. 11, p. 1867–1877.

Thompson, J. C., 1954, Résumé of several fields in Wyoming and Montana south and west of the Elk Basin field, in Pryor Mountains-northern Bighorn basin, Montana: Billings Geol. Soc. Guidebook, p. 117–125.

Todd, Thomas W., 1963, Post-depositional history of Tensleep Sandstone (Pennsylvanian), Big Horn basin, Wyoming: Am. Assoc. Petroleum Geologists Bull., v. 47, no. 4, p. 599–616.

—— 1964, Petrology of Pennsylvanian rocks, Big Horn basin, Wyoming: Am. Assoc. Petroleum Geologists Bull., v. 48, no. 7, p. 1063–1090.

Tohill, Bruce, and M. Dane Picard, 1966, Stratigraphy and petrology of Crow Mountain Sandstone Member (Triassic), Chugwater Formation, northwestern Wyoming: Am. Assoc. Petroleum Geologists Bull., v. 50, no. 12, p. 2547–2565.

Toland, W. G., 1960, Oxidation of organic compounds with aqueous sulfate: Am. Chem. Soc. Jour., v. 82, p. 1911–1916.

Van Tuyl, F. M., and Ben H. Parker, Jr., 1941, The time of origin and accumulation of petroleum: Colorado School Mines Quart., v. 36, no. 2, 180 p.

Walton, Paul T., 1947, Oregon Basin oil and gas field, Park County, Wyoming, in Big Horn basin: Wyo. Geol. Assoc. Guidebook, p. 210–222; also: Am. Assoc. Petroleum Geologists Bull., v. 31, no. 8, p. 1431–1453.

Weeks, L. G., 1958, Habitat of oil and some factors that control it, in Habitat of oil: Am. Assoc. Petroleum Geologists, p. 1–61.

Welte, Dietrich H., 1965, Relation between petroleum and source rock: Am. Assoc. Petroleum Geologists Bull., v. 49, no. 12, p. 2246–2268.

Wendlandt, E. A., T. H. Shelby, Jr., and John S. Bell, 1946, Hawkins field, Wood County, Texas: Am. Assoc. Petroleum Geologists Bull., v. 30, no. 11, p. 1830–1856.

Wenger, W. J., and W. J. Lanum, 1954a, Characteristics of crude oil from Big Horn basin fields: Petroleum Engineer, v. 26, no. 2, p. A-52–A-60.

—— and W. J. Lanum, 1954b, Characteristics of crude oil from Williston basin fields: Petroleum Engineer, v. 26, no. 7, p. C-43, C-45–C-47.

—— and J. S. Ball, 1963, Characteristics of crude oils from Utah, Utah Geol. Mineralog. Survey Bull. 45, p. 597–607.

Willingham, Robert W., and C. D. Howald, 1965, Case history of the Tensleep reservoir, Torchlight field, Wyoming: Petroleum Engineers Paper SPE 1166 (preprint), 4 p.

—— and James A. McCaleb, 1966, Influence of geological heterogeneities on secondary recovery from Permian Phosphoria reservoir, Cottonwood Creek field, Wyoming (abs.): Am. Assoc. Petroleum Geologists Bull., v. 50, no. 9, p. 2030 (full text publ. in this volume of the Bulletin, p. 2122–2132).

Wold, John S., 1952, Report on Worland field, Wyoming, in Southern Big Horn basin: Wyo. Geol. Assoc. Guidebook, p. 117–119.

Wyoming Geological Association, 1957a (rev. 1959): Stratigraphic penetration chart, Wyoming.

—— 1957b, Symposium, Wyoming oil and gas fields Supplement, 1963: Denver, Petroleum Information, 484 p.

—— and Billings Geological Society, 1963, Northern Powder River basin guidebook: 204 p.

Wyoming Oil and Gas Conservation Commission, 1966, Production report for the twenty-five largest fields in Wyoming: 31 p.

Zapp, A. D., 1956, Structural contour map of the Tensleep Sandstone in the Big Horn basin, Wyoming and Montana: U. S. Geol. Survey Oil and Gas Inv. Map OM 182.

Zarella, William R., 1965, Significance of hydrocarbon disposition in petroleum exploration: Am. Assoc. Petroleum Geologists, distinguished lecture presented at Denver, Colorado, Oct. 21, 1965.

—— et al., 1963, Analysis and interpretation of hydrocarbons in subsurface brines (repr.): Am. Chem. Soc., Los Angeles mtg., p. A-7—A-16.

ZoBell, Claude E., 1944–45, Influence of bacterial activity on source sediments, in Research on occurrence and recovery of petroleum: Am. Petroleum Inst., p. 69–78.

—— Carroll W. Grant, and Herbert F. Haas, 1943, Marine microörganisms which oxidize petroleum hydrocarbons: Am. Assoc. Petroleum Geologists Bull., v. 27, no. 9, p. 1175–1193.

BULLETIN OF THE AMERICAN ASSOCIATION OF PETROLEUM GEOLOGISTS
VOL. 43, NO. 5 (MAY, 1959), PP. 992-1025, 11 FIGS.

STRUCTURE AND ACCUMULATION OF HYDROCARBONS
IN SOUTHERN FOOTHILLS, ALBERTA, CANADA[1]

F. G. FOX[2]

Calgary, Alberta, Canada

ABSTRACT

The southern foothills of Alberta are composed of much compressed, imbricately faulted Paleozoic and Mesozoic rocks in an elongate belt between the Great Plains and the Rocky Mountains. Within the belt structural shortening evidently has been achieved mainly by thrust faulting, accompanied by the development of drag folds, most of which are not large. The major thrust faults lie in well defined sliding zones, and they generally transect the bedding at relatively low angles. This is true for supposedly competent as well as incompetent beds, although there is apparently some refraction of fault planes between beds of differing competence. At and near the surface the faults commonly dip steeply, but this is a result of rotation and is not indicative of the dips the faults had when formed.

Structures that originated as folds and were later faulted may not exist. Turner Valley anticline, which is commonly regarded as a faulted fold developed from a protofold, might be an exception, but the evidence is by no means conclusive.

In this area oil and gas have been found in Cretaceous, Mississippian, and Devonian rocks. The reservoirs tapped to date are associated with thrust faults and it appears that, with the possible but unproved exception of the Turner Valley pool, all the traps are fault traps. Migration of hydrocarbons might have started very early, but was not completed before the close of the Laramide revolution. There is no reason to suppose that all hydrocarbons migrated at once, or in any one direction, or that the process of migration was uninterrupted.

INTRODUCTION

In Alberta the foothills belt lies in the western part of the southern half of the Province, adjacent to the Rocky Mountain front. It is about 250 miles long and 12–25 miles wide. It has three geographical subdivisions. The southern part extends from the International Boundary to the Bow River, the central part lies between the Bow and Athabaska rivers, and the northern part extends from the Athabaska to the British Columbia boundary (Fig. 1).

This structural belt, sometimes euphemistically referred to as the "disturbed belt," comprises an area of intensely faulted and folded sedimentary rocks. It strikes generally northwest, but there are three major strike segments. From the International Boundary to Crowsnest Pass the strike is about 135°, from Crowsnest Pass to the North Saskatchewan River it averages 170°, and from the North Saskatchewan to the British Columbia boundary it averages 135°. The boundaries are not everywhere clearly distinguished. On the east the boundary is usually drawn to include the most easterly definite and persistent dips. This *apparent* boundary very nearly coincides with the present western limit of Paleocene beds. On the west it is the frontal thrust of the Rocky Mountains. As this is not everywhere the same fault there are places where the western boundary is ill defined.

[1] Read before the Association at Los Angeles, March 13, 1958. Manuscript received, November 20, 1958.

[2] Triad Oil Co. Ltd., Calgary, Alberta.

Preparation of this paper has been encouraged by the Management of Triad Oil Co. Ltd., Calgary, and permission to publish it is gratefully acknowledged. For helpful discussion of many problems and reading of the manuscript thanks are due to J. C. Scott, R. Lakeman, P. E. Kent, A. N. Thomas, E. W. Best, F. A. McKinnon, W. J. Hennessey, J. C. Sproule, and W. R. S. Henderson.

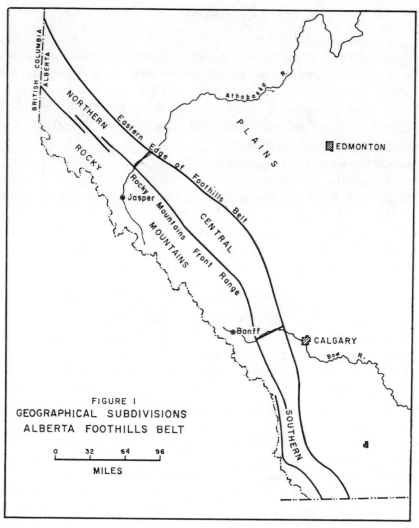

FIGURE I
GEOGRAPHICAL SUBDIVISIONS
ALBERTA FOOTHILLS BELT

Fig. 1

Between the boundaries the exposed stratigraphic succession comprises mainly Cretaceous rocks, but with important inliers of Carboniferous and Devonian rocks. The beds have been subjected to intense compression from the southwest, to which they have yielded mainly by northeasterly movement on thrust faults.

There are now four pools in the southern foothills producing or able to produce hydrocarbons. Besides Turner Valley, which is still the only oil pool, Jumping-pound and Pincher Creek produce gas, naphtha, and sulphur, and Savanna Creek is being developed as a gas pool. In recent months important discoveries of gas have been made in the Shell Waterton, Texaco Castle River, and Texaco Gladstone Creek wells. In the central foothills gas has been discovered in the Triad-B. A.-Stolberg and B. A.-Triad-Lovett River wells, which are 110 and 150 miles respectively northwest of Bow River.

STRATIGRAPHY

The formations that crop out in the southern foothills range in age from Mississippian to Paleocene, and from wells it is known that in the subsurface there are beds of Devonian and Cambrian age. These formations have been described in detail in other publications, and the stratigraphic succession has become firmly established. Some questions of correlation still exist, but these need not be discussed here. In Table I the names and thicknesses of the formations are given, with abridged lithological descriptions.

Generally the formations are thicker in the western foothills than they are in the east. Maximum aggregate thickness represented above the Proterozoic is about 30,000 feet. However, not all of these formations are present in any area, and nothing like this thickness of beds remains in one place. In the Crowsnest Pass area the total thickness of sediments deposited was probably of the order of 22,500 feet, and in the Bow River region 16,000 feet or more of beds were deposited. Rather less than these thicknesses remain to-day.

RELATION OF STRUCTURE TO STRATIGRAPHY

As the Paleozoic succession is composed mainly of carbonate rocks, of which some are notably massive, and the Mesozoic of shales, siltstones, and sandstones, it is generally assumed that the Paleozoic is the more competent. There can be little doubt of the general validity of this assumption, but it can not be accepted without certain reservations.

Several formations have evidently played a leading role in the structural evolution of the region by serving as sliding media for the major thrusts. An important sliding zone is suspected in the Cambrian in part of the region, for the same massive Cambrian dolomites are thrust onto Mesozoic beds to form the Front Range from Kananaskis River as far northwest as the Brazeau River, and perhaps the Athabaska, a distance of about 200 miles. In the Carboniferous sequence the Banff formation, comprising calcareous shale and silty and argillaceous limestones, is an important sliding zone. The Turner Valley sole thrust lies in the Banff formation for more than 20,000 feet—nearly 4 miles—across the strike (Fig. 4).

Perhaps the most striking zone of decollement in the foothills is the weak sequence of shale, siltstone, coal, and sandstone comprising the Fernie and Kootenay formations. At least four major thrusts (Mill Creek, Livingstone, Bear Creek, and Dyson Mountain faults) and many minor ones have reached the surface in these beds. Moreover, in numerous wells they have been found thrust over much younger Mesozoic beds, on faults that involve other formations at the surface.

The Blackstone formation ranges in thickness from 450 to about 1,000 feet, and comprises fissile shale with some siltstone and sandstone. Unfaulted sections are rarely found—indeed it is normal to find the formation cut by numerous thrusts, all at low angles to the bedding, and with important aggregate throw and displacement. (Throughout this paper "throw" refers to stratigraphic throw,

TABLE I

System	Formation	Thickness in Feet	Lithologic Description
Tertiary Paleocene	Paskapoo-Porcupine Hills	2,000	Sandstone, shale, mudstone, thin coal seams, cobble bed at base
		Disconformity	
	Willow Creek	350–2,700	Sandstone, shale, and mudstone
Cretaceous Upper	Edmonton—(Blood Reserve, St. Mary River, Lower Willow Creek)	1,000–1,500	Sandstone, shale, and coal, conglomeratic at base in places
	Bearpaw	0–600	Shale, siltstone, with thin sandstone beds
	Belly River	1,200–4,000	Sandstone, shale, some coal, basal sandstone very massive in places
	Wapiabi	1,100–1,800	Siltstone and shale
	Bighorn (Cardium)	30–450	Sandstone, arenaceous shale, lentils of chert conglomerate
	Blackstone	450–1,000	Siltstone and shale
	Crowsnest	0–1,800	Agglomerate, tuff, essentially confined to Crowsnest area
Lower	Blairmore	1,000–2,300	Sandstone, shale, some thin limestone, bentonitic, and tuffaceous beds
		Disconformity	
	Kootenay	50–700	Sandstone, carbonaceous shale, and coal
Jurassic	Fernie	100–900	Siltstone, shale, and fine-grained sandstone
		Disconformity	
Triassic	Spray River	0–50	Dolomite and limestone
		Disconformity?	
Permian?	Rocky Mountain	0–400	Arenaceous dolomite, limestone, quartzitic sandstone, siltstone, basal conglomerate in some places
		Disconformity	
Carboniferous Mississippian	Rundle group (Tunnel Mt.-Mt. Head-Turner Valley-Shunda-Pekisko)	900–2,000	Limestone, dolomite, some calcareous shale, commonly cherty, in places anhydritic
	Banff	500–900	Silty dark limestone, calcareous shale
	Exshaw	10–35	Black shale
		Disconformity?	
Devonian	Palliser	650–1,150	Limestone, dolomite, characteristically massive
	Alexo	100–200	Dolomite, silty to arenaceous
	Fairholme	1,350–1,750	Dolomite, limestone, argillaceous and silty limestone, anhydrite
	Ghost River(?)	270+	Variegated shale, dense dolomite, edgewise conglomerate
		Disconformity	
Cambrian		2,550±	Limestone, dolomite, argillaceous limestone, with sandstone or quartzite at base
		Unconformity	
Proterozoic		10,000+	Limestone, dolomite, argillite, quartzite, shale. Known only in Lewis overthrust sheet

and "displacement" to the distance between the faulted ends of beds as measured along the fault plane.) An excellent example of the response of the Blackstone beds to stress is provided by the Triad-B. A.-Stolberg well in the central foothills, 110 miles northwest of Bow River. In this well the Blackstone, which should be about 900 feet thick, was drilled through a vertical interval of 4,500 feet. This thickness must have been achieved either by great diapiric thickening or by repetition on numerous thrust faults. In either case, the incompetence of the Blackstone has permitted substantial structural shortening.

The Wapiabi formation is also composed of shale and commonly faulted, and in some places may be a major sliding zone. Its importance in this respect, however, is apparently much less than that of the Blackstone and Kootenay-Fernie formations.

Faults and their associated folds can be found in any formation in the foothills and some have important throw. It is generally true, however, that at the surface the major faults are found in the few formations discussed.

REGIONAL STRUCTURE

The southern foothills are bounded on the east by the Great Plains and on the west by the Front Ranges of the Rocky Mountains (Fig. 2). Both boundaries usually are represented on regional maps by long smooth sweeping curves. This is convenient but misleading.

On the east the thrust-faulted and closely compressed structures characteristic of the foothills seem to end just west of the west edge of Paskapoo exposures. However, there is no gradual diminution of intensity of compression from west to east—on the contrary the most easterly foothills structures are fully as complex as the most westerly ones. Moreover, in the Husky-Northern Pekisko wells, drilled 8 miles southeast of Turner Valley, faults were encountered in drilling that lie beneath unfaulted Paskapoo beds. The Consumer Co-op well, drilled 7 miles southeast of Turner Valley, penetrated at least one thrust fault that does not, as far as we know, cut the Paskapoo formation. Outcrops are relatively abundant and the Paskapoo beds dip uniformly east and at low angles; there is not the slightest suggestion that they are faulted (Fig. 3). Thus it seems obvious that the true eastern limit of foothills structure lies buried beneath the Paskapoo formation.

The western boundary is locally more distinct, but can not properly be delineated by a single unbroken line. South of the latitude of Savanna Creek it is marked by the base of the High Rock Range and the Clarke Range, which lie on the Lewis overthrust. However, from Savanna Creek northward the western edge lies farther east, at the base of the Highwood Range. This becomes part of the Fairholme Range in the vicinity of the Bow River Valley, and from the Highwood River north it lies on the McConnell fault. Thus the western boundary as envisaged by the writer must be shown by two lines *en echelon* as it is not everywhere marked by the same range or the same fault. It is a physiographic rather

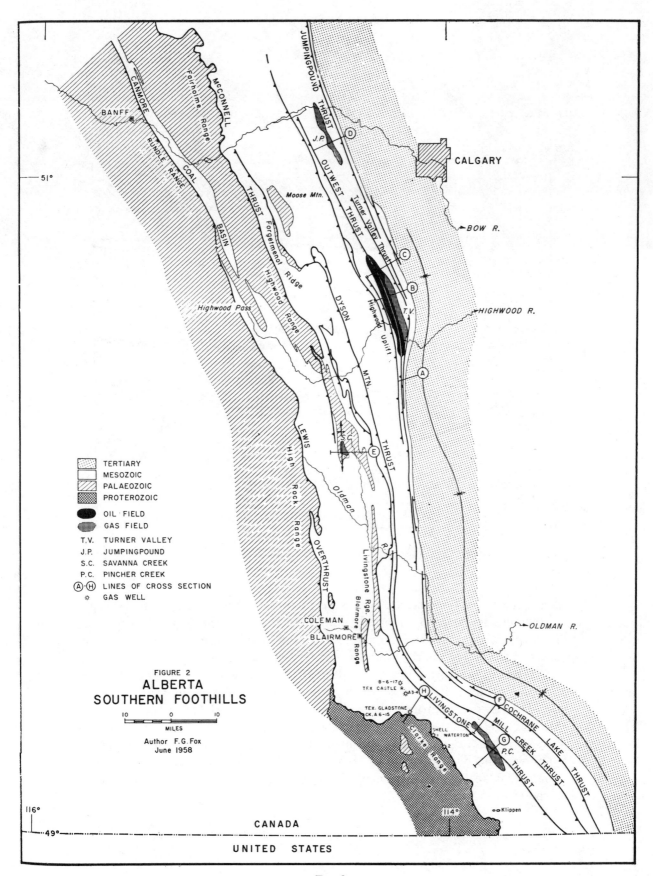

TERTIARY
MESOZOIC
PALAEOZOIC
PROTEROZOIC
OIL FIELD
GAS FIELD
T.V. TURNER VALLEY
J.P. JUMPINGPOUND
S.C. SAVANNA CREEK
P.C. PINCHER CREEK
A-H LINES OF CROSS SECTION
✿ GAS WELL

FIGURE 2
ALBERTA
SOUTHERN FOOTHILLS

10   0   10
MILES

Author F.G.Fox
June 1958

CANADA

UNITED STATES

FIG. 2

249

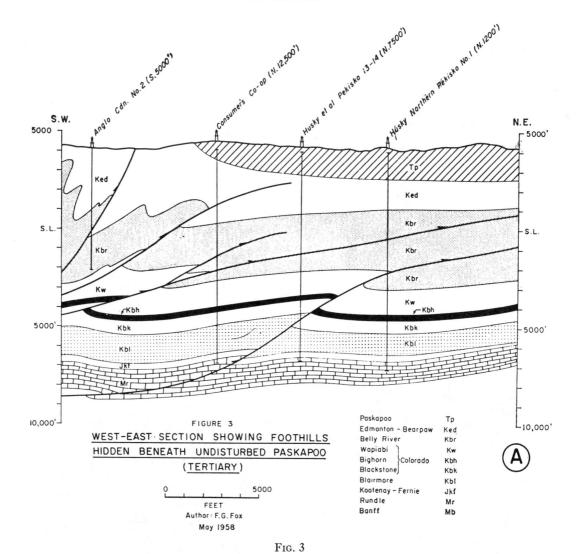

FIGURE 3
WEST-EAST SECTION SHOWING FOOTHILLS
HIDDEN BENEATH UNDISTURBED PASKAPOO
(TERTIARY)

Paskapoo	Tp
Edmonton – Bearpaw	Ked
Belly River	Kbr
Wapiabi	Kw
Bighorn } Colorado	Kbh
Blackstone	Kbk
Blairmore	Kbl
Kootenay – Fernie	Jkf
Rundle	Mr
Banff	Mb

0   5000
FEET
Author: F.G. Fox
May 1958

Fɪɢ. 3

than a structural boundary. There is no profound difference between the types of structure found in the foothills and those found in the Front Ranges; the difference is one of dimension only. Every kind of structure found in the mountains can also be seen in the foothills, and the reverse is true.

There are nine thrust faults of outstanding importance in the region (Fig. 2) and a multitude of lesser ones which, with a few minor exceptions, all dip westerly. The Lewis overthrust, with known length of at least 175 miles, maximum throw of 20,000–25,000 feet, and displacement of at least 25 miles, is the greatest of these. Most of the others have known lengths of 50–125 miles, throws of 5,000–8,000 feet, and displacements of 1½–8 miles. The McConnell fault has throw of about 13,000 feet and displacement of several miles in Bow Valley, and it has been traced at least 160 miles.

Near the eastern margin of the foothills easterly dipping thrust faults are not uncommon. Such faults are thought to be characteristic frontal structures. Throw and displacement are not great and they probably do not extend to any great depth. East-dipping faults in the central and inner foothills are uncommon, but are found in some places. Probably they were once frontal faults; all that are seen now are the eroded root zones.

The lesser faults of the region require no comment except that, although individually they may have relatively small throw and displacement, they have permitted in sum a very great amount of shortening. Moreover, several minor surface thrusts can unite at depth to produce a major displacement.

In this part of the foothills four important inliers of Paleozoic rocks are present, Moose Mountain, Forgetmenot Ridge, Livingstone Range, and Blairmore Range. The oldest beds exposed in any of these are of Mississippian age. The latter three are known to be thrust blocks, and the probability is that Moose Mountain is also underlain by a major sole thrust. If it is not, it is unique in the foothills.

The best known structures are those in which bore holes have been drilled, and even they are not fully known. Even where many wells have been drilled, as in Turner Valley, major problems remain unanswered and widely divergent structural interpretations are possible.

In the ensuing discussion reference is made to the dips of various fault planes, and to the angles at which they have cut across bedding planes. The dips are important, but the angles at which faults cut bedding planes are much more important and revealing.

TURNER VALLEY

Turner Valley oil and gas pool lies near the eastern edge of the foothills about 30 miles southwest of Calgary. The first well was drilled on the structure in 1913, near gas seepages along the south fork of Sheep Creek, and was completed as a gas well in the Blairmore formation. Eleven years elasped before the Paleozoic reservoir was discovered in 1924, and this important event was followed by 12 more years of sporadic development of the gas-cap area. Crude oil was not discovered until 1936, when a well drilled on the west flank encountered the Rundle at 6,396 feet and yielded an initial flow of 860 barrels per day of 45° API oil from a depth of about 6,500 feet. This well, Turner Valley Royalties No. 1, is generally regarded as the discovery well, although as Spratt and Taylor (1936, p. 715) have pointed out the Model well No. 1 provided a hint of the presence of crude oil in 1930. Model No. 1 was a small well, however, and the significance of its "discoloured" naphtha was not generally appreciated at the time.

After the successful completion of the Turner Valley Royalties well development of the west flank proceeded rapidly and reached its culmination with development of the Millarville sector, which is really part of the west flank, during the second great war. There are now about 300 oil wells and 96 gas wells capable of production from the Turner Valley formation of the Rundle group, which is of Mississippian age. No drilling is being done at present.

The structure as seen at the surface is an anticline with a core of Blackstone, Bighorn, and Wapiabi shales, and Belly River sandstone on both flanks. The Belly River can be traced without interruption around the south end of the structure, where there is simple southerly plunge, but the northwesterly closure of the Belly River beds is interrupted by several faults and minor folds. The shales exposed in the core are cut by numerous minor faults, and the whole anticline is underlain by the Turner Valley sole thrust. It is, in fact, a drag fold, and it owes its existence to movement on the sole thrust. Other faults of considerable importance are the Millarville and Outwest thrusts (Figs. 4 and 5).

The Turner Valley thrust is a major fault at depth, the throw being about 5,000 feet maximum and displacement at least 17,500 feet in the central part of the structure. There at depth beds of the Mississippian Banff formation have been thrust onto the Cretaceous Wapiabi shale. However, both throw and displacement decrease southeast and northwest from the central area. For example, at the north end of the producing area the total throw and displacement on the Turner Valley thrust, even with the added effect of several subsidiary thrusts, are only of the order of 3,500 and 2,000 feet, respectively. Moreover, there is a truly remarkable decrease in throw and displacement updip on the fault from the leading edge of the Rundle block. At the surface the fault involves only Belly River and Edmonton beds (Figs. 4 and 5); its throw ranges from about 1,000 to 2,500 feet and displacement ranges along strike from 1,000 to 2,000 feet.

The net result of these factors is that the sole thrust, which intersects the surface about 1¼ miles east of the axis of the fold, is arcuate and convex to the east in plan. It has been mapped northwest and southeast beyond the limits of the field.

Gallup (1954, p. 399) has described the sole fault as being concave to the east in surface plan, and as having "plunge" northwest and southeast from the central sector. However, the "plunge" of the fault plane is more apparent than real; it is due to the fact that the wells that penetrated the fault plane in the central sector of the field are updip from those that passed through it on the ends of the structure. The anticline or drag fold overlying the fault has plunge in two directions, but the fault plane itself has none.

Another fault of major importance, although modest proportions, is the Millarville thrust. This fault has its known southern extremity on the west limb of the structure near the center of the field, whence it strikes northerly across the northwest-plunging end of the main Turner Valley thrust block. On the northeast flank the strike swings northwest. On this fault the northwest flank has overridden the remainder of the plunging end of the anticline, forming the Millarville fault block. This is itself broken into segments by several subsidiary faults. Gallup (1954, p. 400) has described the Turner Valley and Millarville thrust blocks as separate structures, as the former is the reservoir in the southern two thirds,

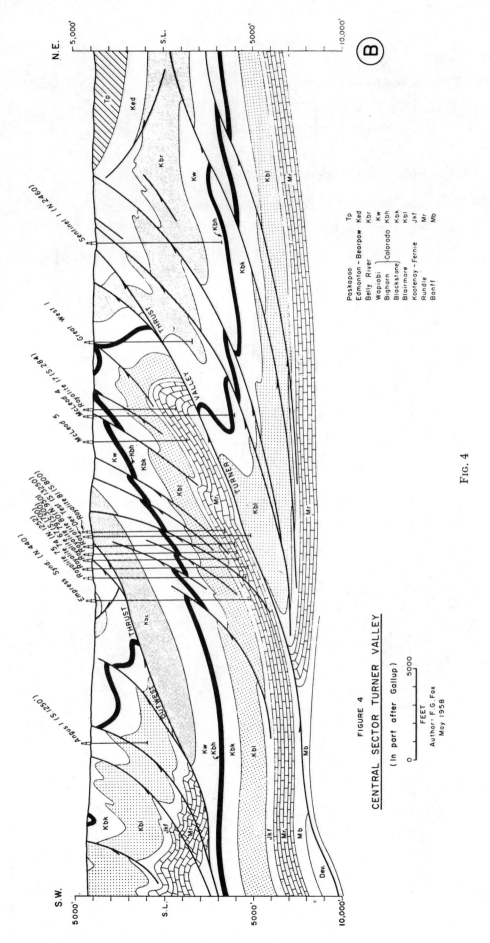

FIGURE 4

CENTRAL SECTOR TURNER VALLEY

(In part after Gallup)

FEET
0   5000

Author: F.G.Fox
May 1958

FIG. 4

Paskapoo   Tp
Edmonton – Bearpaw   Ked
Belly River   Kbr
Wapiabi   ⎫   Kw
Bighorn   ⎬ Colorado Kbh
Blackstone   ⎭   Kbk
Blairmore   Kbl
Kootenay – Fernie   Jkf
Rundle   Mr
Banff   Mb

253

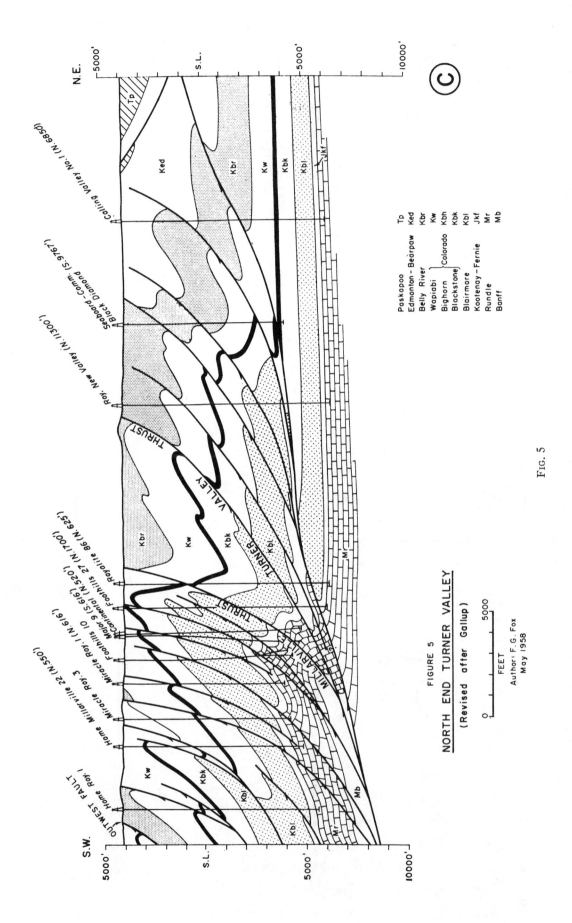

FIGURE 5

NORTH END TURNER VALLEY

(Revised after Gallup)

Author: F. G. Fox
May 1958

Paskapoo	Tp
Edmonton - Bearpaw	Ked
Belly River	Kbr
Wapiabi } Colorado	Kw
Bighorn	Kbh
Blackstone	Kbk
Blairmore	Kbl
Kootenay - Fernie	Jkf
Rundle	Mr
Banff	Mb

and the latter in the northern third of the field. However, it seems apparent that the Millarville fault is subsidiary to the Turner Valley thrust, and the Millarville block is part of the northwest-plunging end of the Turner Valley block; they are merely parts of the same major thrust sheet. Reservoir data support this view.

A third major fault, the Outwest thrust, lies on and partly conceals the west flank of the anticline. It is probably the sole thrust of the Highwood uplift, which is the next large structure west of Turner Valley. At the surface, where the throw is about 3,500 feet, Wapiabi and lower Belly River beds have been thrust onto higher Belly River. The throw increases downdip so that beneath the Highwood uplift it is about 6,000–7,000 feet, and displacement is of the order of 16,000 feet. The near-surface dip of the fault varies considerably. West of the central part of the field it has been penetrated by several wells and its dip is known to be about 35°. Shortly below the surface this is reduced to about 20°. Near the north end of the field the fault is thought to dip much more steeply near the surface—in Figure 5 it is shown dipping west at 60°.

In addition to these major thrusts numerous minor faults are present, particularly in the northern half of the field. Many of these are not recognized at the surface; their presence was revealed by drilling.

Gallup (1954), Link (1949), Hume (1957), and others regard Turner Valley as the descendant of a proto or ancestral fold, which had its beginnings at the close of Paleozoic time, and endured additional folding at the close of Jurassic time, when the accumulation of oil is supposed to have taken place. Gallup says (1954, p. 410), "The Laramide orogeny merely accentuated the folding and faulted the rocks of a pool already formed." This might be so but the evidence is inconclusive.

It is well known that at the close of the Paleozoic era there was an epeirogenic uplift of much of Alberta and eastern British Columbia, which resulted in erosion and eastward truncation of the Permo-Pennsylvanian and older beds. The Nevadian orogeny occurred in latest Jurassic or earliest Cretaceous time, and it culminated in the uplift of the Cordilleran geanticline on the site of the present Selkirk Mountains. East of this lay the Rocky Mountain geosyncline, its east shelf extending far beyond the present foothills area. It is not unreasonable to suppose, as Gallup does, that these orogenic movements caused the development of folds within the bordering geosyncline.

The following principal reasons for considering Turner Valley to represent such an early fold appear to be weighty, but on close examination they fail to carry conviction.

1. The absence of oil from, and the presence of water in, the Highwood uplift, which adjoins Turner Valley on the west. The Rundle in this uplift stands about 5,500 feet structurally higher than the lowest oil-bearing Rundle of Turner Valley. Where tested it has yielded only water with traces of oil and gas. This appears to pose an awkward problem. However, if a fold was formed at Turner Valley at the close of Paleozoic time and filled with oil as a result of Nevadian movements at the end of Jurassic time, and if the Highwood uplift is a Laramide

structure, the answer appears to be clear—the Highwood uplift was formed after the oil had migrated updip into the Turner Valley protofold. But it is not so clear as it seems. According to Gallup (1954, p. 404), "It is now recognized that the porous sections of Turner Valley are, for the most part, of organic origin," and "The porous zones of Turner Valley can not be correlated with porous zones of the Paleozoic uplifts only 1–2 miles west. . . . " These two statements raise a serious doubt in the writer's mind that the apparent absence of oil from the Highwood uplift must be attributed to development of a protofold at Turner Valley. Porous zones of organic origin are not noted for their lateral continuity, but rather for their lack of it. There may never have been any direct connection between the Turner Valley reservoir and the porosity in the Highwood uplift, and if there was not, the absence of oil from the uplift need not be explained by any special sequence of tectonic events. The evidence is far from conclusive.

2. The belief that differences in thickness of as much as 80 feet in the Kootenay formation (and presence of sandstones in the Millarville area that are not found in the central area), and 50 feet in the Fernie formation must be attributed to early uplift. These thickness variations seem to be impressive, occurring as they do in a relatively small area, and there is evidence of early uplift in the general area. However, considerable apparent variations in thickness of the incompetent Kootenay and Fernie sandstones and shales are to be expected wherever they are involved in an imbricate fault system, as they are at Turner Valley. Also, the uplift known to have occurred was regional, not local, and thin Fernie sections are found over a broad area, not only at Turner Valley. Finally, the Kootenay has a regional erosional disconformity at its top, the Fernie has one at its base and one within it, and these disconformities converge eastward. Thus substantial differences in the thicknesses of the two formations, even on a local scale, do not necessarily require the postulation of a protofold.

3. Hume (1957) says: "Drilling in Turner Valley confirmed the belief that accumulation had preceded faulting, because at the north end of the field the oil-water line in one fault block is much lower than in another, although presumably at the time of accumulation they were at the same level." It is true that the different oil-water levels suggest accumulation prior to development of the north end *subsidiary* thrusts, but it does not follow that it occurred prior to development of the sole thrust. The original trap might have been a thrust sheet, the north end of which was broken up later.

It seems to the writer, therefore, that although Turner Valley anticline apparently could have been developed from a protofold, it is not definite that it did. It could just as well be nothing more than a great drag fold, formed as a result of eastward movement on the Turner Valley thrust during the latest stages of the Laramide orogeny. If this is so, the oil is contained in a fault trap, into which it migrated very little, if any, earlier than the close of the Laramide orogeny, and certainly not before the formation of the Turner Valley thrust.

The problem is of more than academic importance. If there was a Nevadian protofold the Turner Valley oil might have migrated and been trapped as early as the end of Jurassic time, and other oil was either trapped in similar folds or migrated updip into the plains area. If there was not, then the migration of oil can not have been completed before the end, or nearly the end, of the Laramide revolution, when the Turner Valley thrust block was formed, and any similar thrust block might be found to contain oil or gas.

Whatever its earlier history Turner Valley is now a pool in a compound thrust sheet.

JUMPINGPOUND

Jumpingpound gas pool lies near the eastern edge of the foothills about 20 miles northwest of Turner Valley and 22 miles west of Calgary. It is about 15 miles long and 2 miles or less wide. The pool is in typical foothills country so far as its geology is concerned, but the hills are lower and less heavily timbered than most of the foothills.

Drilling began at Jumpingpound in 1914 when four wells were spudded. Three were abandoned in 1914: one when the rig blew down, one at 600 feet, and the third at 1,255 feet. The fourth struggled along for 7 years before it was finally abandoned in 1921. During those days the Blairmore was the only producing formation in Turner Valley (and the foothills) and it seems certain that these four wells were intended to test the Blairmore. However, the Rundle reservoir was discovered in Turner Valley in 1924 and it is safe to assume that all Jumping-pound holes drilled after that date had the Rundle as their principal objective. The old Northwest Company drilled four, one each in 1926 and 1928 and two in 1936. All were failures. The two earlier ones were abandoned when it became evident that the Rundle could not be expected at less than about 10,000 feet, which was too deep for cable tools. The two holes drilled in 1936, presumably also with cable tools, both penetrated faults and were abandoned at depths of 1,652 and 2,019 feet. Of these early wells at least five were located within the present productive area, and had the tools, money, and faith been available to drill them deep enough they would have been successful. In 1939 the Brown Consolidated well was drilled on the west flank of the structure, and was abandoned at 6,885 feet when it became evident that a major thrust had been penetrated at 6,570 feet. Following this, and after a seismic survey, Shell drilled a dry hole, Shell Norman, to 12,056 feet, about $4\frac{1}{2}$ miles west of the present field. This well reached the Rundle at 11,588 feet.

Finally, in 1944, Shell drilled the Shell 4-24-J well (later renamed Unit 1), which reached the Rundle at 9,618 (-5,598) feet and encountered a large flow of gas and distillate. It was completed at 9,947 feet. The Unit 2 well, drilled $\frac{3}{4}$ mile west of Unit 1, was about 1,000 feet structurally lower on the Rundle and found only water. Unit 3, drilled halfway between Nos. 1 and 2, found water, gas, and distillate, and established the water table at 6,436 feet subsea.

There are now 14 wells in the field. Of these 11 are gas wells, two are abandoned as marginal and non-commercial, and one is dry and abandoned. Two wells, Units 4 and 8, were drilled through a sole thrust and into the subthrust block, where Rundle, Banff, Devonian, and Cambrian (?) beds were penetrated. The subthrust beds are water-bearing.

Jumpingpound structure is generally anticlinal at the surface. The core of the fold exposes Blackstone shale in four small areas surrounded by Bighorn and Wapiabi beds. The Blackstone is contorted and broken by several faults. Belly River beds lie along the east flank and the northern half of the west flank, and 3 miles south of the Bow River, show some closure around the Wapiabi beds. The closure is interrupted by two west-dipping thrust faults but is reasonably distinct. Along the west flank the Belly River can be traced southeastward to about the center line of the fold (4 miles southeast of the line of section, Fig. 6), beyond which it disappears beneath a west-dipping thrust. Southwest of this fault Belly River, Wapiabi, Bighorn, and Blackstone beds appear repeatedly on a series of imbricate faults. Although the Bighorn in the crestal area of the fold is believed to close round the Blackstone, outcrops are non-existent in the critical localities and surface evidence for southeasterly closure of the structure as a whole is indeed tenuous.

The east flank is faulted, the Belly River having been thrust, on the Jumpingpound fault, onto Edmonton strata. As interpreted in Figure 6, the Jumpingpound thrust is the most easterly exposed fault. On Figure 7 it is considered to be the next known fault west.

Figure 6 was drawn through the Units 1 and 2 wells, and Unit 3 was projected about 800 feet to the line of section. It is a conservative though complicated interpretation, of which the dominating elements are the Jumpingpound thrust, the sole fault, and the intervening imbricate wedge of Mesozoic beds. The Jumpingpound fault is definitely folded. It is easily recognized in the wells; in Units 1 and 3 Blackstone beds overlie Belly River, and in Unit 2 farther west and downdip, Blairmore overlies Belly River as it should. In the vicinity of the wells there are six small flyer faults, and the main thrust emerges $1\frac{1}{2}$–2 miles east of the axis of the fold. In the field area the throw is about 6,000 feet and displacement is nearly 4 miles.

The sole fault is of more modest dimensions, its throw and displacement being 1,500 and 2,500 feet, respectively. It has been penetrated in two wells, Unit 8 north and Unit 4 south of the line of section, but it has not been seen at the surface. A flyer from this fault, penetrated in Units 1 and 3, divides the reservoir into three segments according to Martin (1956).

The imbricate wedge of Mesozoic beds, cut by at least six high-angle faults, is consistent with all available data, but it poses problems. None of the faults is known to exist above the Jumpingpound thrust, and it seems a fair assumption that they have their roots in the lower Mesozoic beds and extend upward to, but not across, the thrust. Such an imbricate system could have evolved in either of two ways. If the Jumpingpound fault formed before the sole fault, the interven-

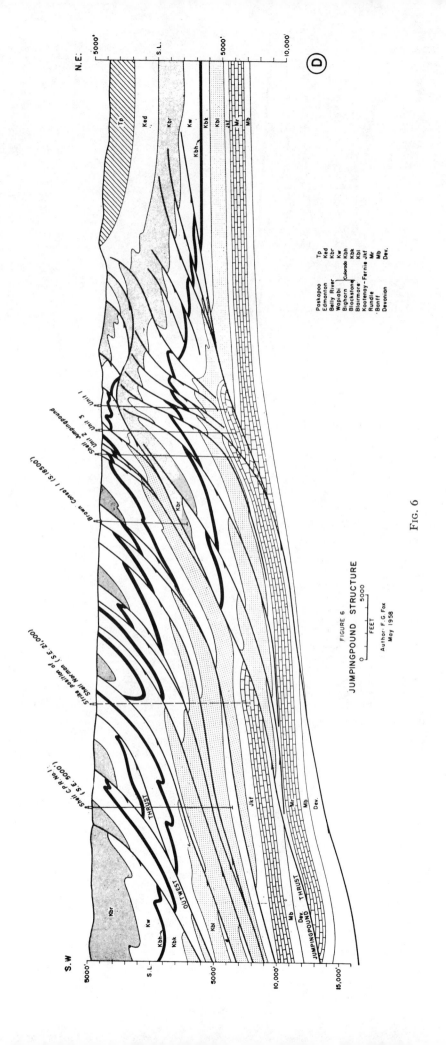

FIGURE 6

JUMPINGPOUND STRUCTURE

FEET
0   5000

Author: F. G Fox
May 1958

Paskapoo   Tp
Edmonton   Ked
Belly River   Kbr
Wapiabi   Kw
Bighorn   Kbh
Blackstone } Colorado Kbk
Blairmore   Kbl
Kootenay-Fernie Jkf
Rundle   Mr
Banff   Mb
Devonian   Dev.

Fig. 6

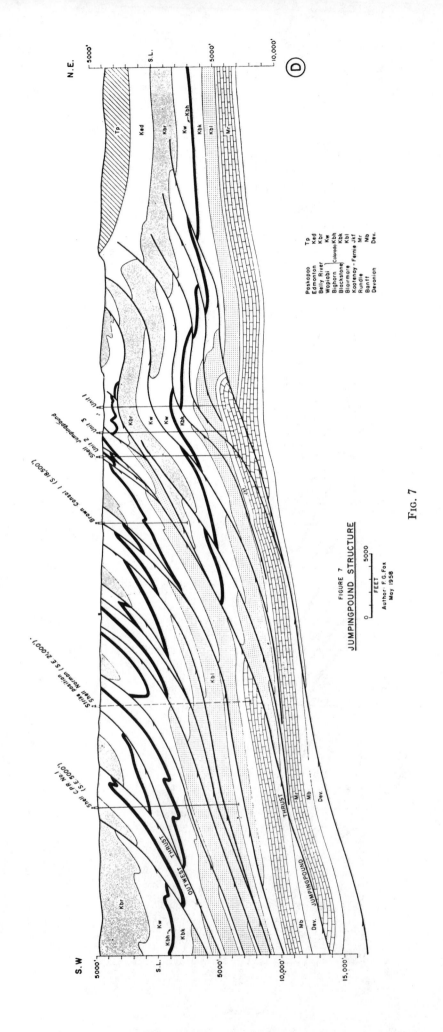

FIGURE 7

JUMPINGPOUND STRUCTURE

FEET
0   5000

Author: F.G.Fox
May 1958

Fig. 7

ing imbricate wedge itself probably would have been developed later during the folding of the former. Whether it caused the folding of the Jumpingpound thrust, or was caused by it, can not be told. Alternatively, if the sole fault formed first and the imbricate system next the Jumpingpound thrust later could have sheared off the near-surface extensions of the imbricate faults, and carried them north-eastward and upward some miles, where all trace of them has been lost due to erosion.

West of the field the Jumpingpound fault dips at 25° for about 3 miles, beyond which the dips are reduced to 10°–15°. The throw is reduced to about 3,000 feet, but displacement remains large, being about 3 miles. The fault is thought to underlie the Rundle found in the Shell Norman well, but was not penetrated.

Figure 7 illustrates a substantially different and simpler interpretation of the same data, and is preferred by the writer. Two of the main structural elements, the sole and Jumpingpound thrusts, are essentially unchanged, although the latter is shown to be less sharply folded and the former more distinctly folded than in Figure 6. The difference is to be seen in the intervening Mesozoic beds, where the excessively complicated imbricate structure has been replaced by a simple series of sub-parallel thrusts, which cut across the Mesozoic rocks at relatively small rather than large angles. They are all folded in harmony with the folding of the two main thrusts.

This structure probably developed in response to a single phase of compression —the final or nearly final pulse of the Laramide orogeny. The faults developed in succession from west to east, or from the top of the section down in non-folded beds, at low angles to the bedding planes. Formation of the Jumpingpound thrust was followed after some interval of steady pressure by development of the next fault below, and this by the next, and the process ended with the movement on the sole fault, which was accompanied by folding of the overlying beds and fault planes. The folding was caused partly by the rise of the Rundle on the sole fault and partly by the continuing compression. Compression, however, was diminish-ing and the whole process stopped when the Laramide orogeny ended—it simply ran down. Had compression continued long enough to move the Rundle eastward about 3 miles, and upward 3,000–4,000 feet, the resulting structure probably would have resembled the Turner Valley structure.

There is nothing in either interpretation to suggest that the Jumpingpound structure developed from a fold in the Rundle or above it. The leading edge of the Rundle block is presumed to have suffered some distortion due to drag, but the implication that the faults developed in unfolded beds is clear. Moreover, this is true of all the structures shown on Figures 6 and 7. The only folds to be seen were caused by movement on faults; that is, they are drag folds.

The gas and naphtha are, therefore, contained in a fault trap; if there were no sole fault there would be no trap and no Jumpingpound pool. Only the fault prevented the escape of the hydrocarbons updip into the plains.

Savanna Creek gas pool lies about 25 miles southwest of Turner Valley and 60 miles southwest of Calgary, in the Front Range of the Rocky Mountains. It is the most spectacular field in Canada, in both its scenic and geological aspects. Development is proceeding on and around Plateau Mountain—a somewhat flat-topped anticlinal mountain 9 miles long, 3 miles wide, and standing 8,200 feet above sea-level. The surface of the plateau is about 1,000 feet above the timber line. Local relief ranges from 500 to more than 1,500 feet.

The first well (Anglo-Canadian-Savanna Creek No. 1) was drilled in 1939. It was located near the crest of Plateau anticline and near the base of the Rundle group, and was intended to test the Devonian rocks. However, it passed from Rundle to Upper Cretaceous rocks at very shallow depth (less than 50 feet) and was abandoned in the lower part of the Blairmore formation at 3,372 feet (Fig. 8). Although the well was a failure it proved the presence near the surface of a thrust fault. The next well, Phillips-Husky-Northern Target (hereafter referred to as PHNT) Savanna Creek No. 1, was drilled in 1952. In this well the Rundle was penetrated at 4,439 feet and yielded some gas, but a second major thrust was encountered at 5,182 feet. The well was, therefore, suspended in the Kootenay formation at 5,628 feet. In 1954 drilling in the same well was resumed and the Rundle was penetrated a second time at 6,401 feet. In the lower part of this block in strata thought to represent the Turner Valley formation, and also in other parts of the Rundle group, large quantities of gas were discovered. The well was drilled through the Rundle and reached bottom in the Banff formation at 9,040 feet. Development has been relatively slow but there are now six wells ready to produce gas and one is being drilled. One well has been abandoned.

Savanna Creek area has been described in some detail by Scott, Hennessey, and Lamon (1957).

Plateau Mountain anticline is a broad deceptively simple fold in the Rundle formation. West-flank dips do not exceed 25°; dips on the east flank are 10° or less. The axis of the structure strikes nearly due north and south. It plunges south from the crest maximum at about $4\frac{1}{2}°$; the north plunge is less distinct and does not exceed 2°. It is oblique to the Highwood Range, which strikes north-westerly. Thus the north-plunging end of the fold passes from the Front Range into the foothills, and the south-plunging end merges with the west flank of the Livingstone Range. On the west it is flanked by a strongly asymmetrical syncline containing Blairmore and Blackstone beds; on the east by the Livingstone River syncline and Hailstone Butte.

The structure section (Fig. 8) was drawn through the discovery well. It is similar to a recent section by Scott *et al.* (1957), and, in its uppermost part, to a section by Webb and Spratt (1939, published by Link, 1949). Link (1949) has also published a section similar to this.

Certainly the most outstanding feature of this section is the presence of three major and numerous minor thrust faults folded in harmony. Although the folded

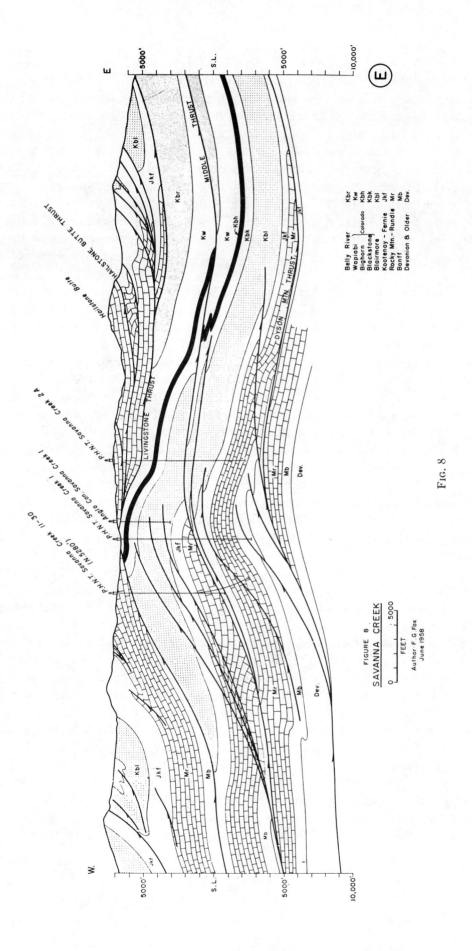

FIGURE 8
SAVANNA CREEK

FEET
0   5000

Author: F. G. Fox
June 1958

Fig. 8

Belly River   Kbr
Wapiabi ⎤   Kw
Bighorn ⎥Colorado Kbh
Blackstone ⎦   Kbk
Blairmore   Kbl
Kootenay – Fernie   Jkf
Rocky Mtn.– Rundle Mr
Bonff   Mb
Devonian & Older   Dev.

character of the Dyson Mountain and the Middle thrust is inferred, there is no question that the uppermost thrust is folded. As pointed out by Link (1949), the Anglo Canadian well not only proved the presence of a major thrust, but, in conjunction with outcrop data, demonstrated that the thrust is folded. Subsequent penetrations of the thrust plane by the PHNT Savanna Creek No. 1 and 2A wells not only confirmed these early conclusions of Webb, Spratt, and Link, but have revealed the presence of subsidiary bedding-plane thrusts. Less than 1,000 feet west of PHNT Savanna Creek No. 1 the fault dips west at 15° and the same distance east it dips east at 11°. Both dips increase somewhat down the flanks of the fold. The fault has cut across the Rundle at an angle to the bedding of 5° or less, and emerges in the foothills 6 miles east of the axis of the fold.

There is little doubt that this is the Livingstone thrust. Its stratigraphic throw is 6,500 feet or more; displacement is not less than 9 miles and may be as much as 12. The only single fault east of Plateau anticline with such dimensions is the Livingstone thrust. Furthermore, no group of thrusts has sufficient throw and displacement. For example, if the throw and displacement of the Hailstone Butte fault and the six thrusts next east of it be added together, the total throw is only about 4,200 feet and displacement about $2\frac{1}{4}$ miles. Moreover, the Hailstone Butte thrust, a low-angle fault that might be assumed to be the major thrust, can be eliminated on mechanical grounds. It has throw of only 1,500 feet, and on it at Hailstone Butte Rundle beds have been thrust over Fernie shales, whereas in the wells on the folded thrust Rundle has been thrust over Wapiabi. Thus for the folded fault in the wells to be the Hailstone Butte thrust would require that the fault cut down section as it approaches the surface. This might be possible, but it is unlikely. Finally, the Hailstone Butte thrust dies out completely a very short distance south of the line of section, whereas in wells south of the section the folded thrust remains strong.

The Middle thrust has stratigraphic throw of 1,500–2,100 feet, and displacement of 1–$2\frac{1}{2}$ miles. Where it emerges at the surface is not definitely known. On the ground the next obviously important fault east of the Livingstone thrust is the Dyson Mountain thrust, the trace of which lies 8 miles east of the axis of Plateau anticline. On this fault, Wapiabi, and perhaps Blackstone, beds have been thrust over Belly River, but northward both throw and displacement increase, so that south of Moose Mountain Kootenay has been thrust over Belly River. In Plateau anticline, on the Middle thrust, Kootenay and Fernie beds are cut off by the fault between the No. 1 and No. 2A wells, and the Blairmore and Blackstone formations are not thought to extend far east of No. 2A. As shown by Figure 8, the Middle thrust probably has Belly River beds both above and below it where it reaches the surface. This seems to preclude any updip connection between the Middle and Dyson Mountain faults. The Middle fault probably emerges at the surface in Belly River beds a short distance east of the Livingstone thrust and west of the Dyson Mountain thrust.

This conclusion is tentative only. A minor change in the interpretation shown in Figure 8 to suggest a greater eastward extension of Wapiabi, Blackstone, and Blairmore beds on the Middle thrust, and an assumption that throw and displacement on the fault increase northward, as they do on the Dyson Mountain fault, would permit the joining of the two faults. However, at the moment the writer is inclined to think that the Dyson Mountain fault lies deeper in Plateau anticline, as indicated by Figure 8.

The presence of the lowermost fault shown on the section is not merely inferred; it was penetrated by the PHNT Savanna Creek No. 3A well, nearly 2 miles north of the line of section. Stratigraphic throw on the fault is at least 2,500 feet. The displacement is unknown, but here shown to be about 3 miles. If this fault rises, with slightly increasing dips, east of the line of section, it should reach the surface about 8 miles east of Plateau anticline. Moreover, it can be inferred that at the surface the beds directly overlying the fault should belong to the Wapiabi formation. Both the assumption and inference are justified by surface data. There is an important thrust 8 miles east of Plateau anticline, and on it Wapiabi shale has been thrust onto Belly River beds; it is the Dyson Mountain fault.

The lowest Rundle penetrated to date, that below the Dyson Mountain fault, is probably the trailing edge of the Rundle of the Highwood uplift. At some depth below this Rundle there ought to be at least one, and perhaps two, more faults, the Outwest and Turner Valley thrusts. They may be merged or not. In either case the fault, or faults, should be present and should be nearly parallel with the bedding planes in Cambrian or, perhaps, Proterozoic beds.

The numerous subsidiary faults that cut the Rundle are also of great interest because of the uncommonly low angles of their intersection with it. This is especially true of the uppermost sheet. The Hailstone Butte fault has been mapped (Scott *et al.*, 1957) and was penetrated in the No. 2A well. There can be no doubt that it transects the bedding at an exceedingly small angle—less than 5°. A smaller fault, above the Hailstone Butte thrust, can be seen in the cirque wall east of the No. 3A well, 2 miles north of the line of section, cutting the beds at about 10° The subsidiary faults penetrated in the No. 1 and No. 11–30 wells are *inferred* to cut the Rundle at angles of 5°–15°. They are known to cut the Blairmore at angles of this order.

It might be argued that these faults should be shown to cut at greater angles across the Rundle and older Paleozoic than they do across Mesozoic beds. This would be consistent with the long held conviction that the Paleozoic is much more competent than the Mesozoic. It would also suggest application of the principle of step faulting as proposed for the Appalachians by Rich (1934) and persuasively applied to the southern foothills by Douglas (1950) and Hume (1957). However, it would not be consistent with what is seen in the uppermost thrust sheet, where the sole thrust (Livingstone thrust) and all the subsidiary thrusts are very low-angle (to the bedding) faults. Moreover, there is no suggestion whatever that any of the faults are step faults.

Some explanation for the folding of the faults is required. Douglas (1950) and Hume (1957) have suggested that some folded faults have been developed from step faults. Douglas (1950, pp. 84–85) says "Folding of the fault plane is, in part, only the result of the accentuation of its initial anticlinal and synclinal bends. An initial anticlinal bend is present in the thrust plane where it changes from a diagonal, crosscutting shear to a bedding plane thrust . . . , and an initial synclinal bend where it changes again to a diagonal shear. . . . " As applied to the Livingstone thrust farther south, in the vicinity of the Oldman River, Douglas' explanation appears to fit the observed facts very well. However, at Savanna Creek it does not apply (Scott *et al.*, 1957). There is no evidence that the faults cut across the Rundle at notably greater angles than they do across the Mesozoic formations. There is no suggestion that the present folded nature of the thrust planes is attributable to peculiarities in their original shapes. The Livingstone thrust, which may be a step fault at Oldman River, is a very low-angle thrust at Savanna Creek.

To the writer it seems likely that at Savanna Creek the faults cut at low angles across Paleozoic and Mesozoic beds alike, that folding did not begin until the faults were all formed, and that when folding began movement on the faults ceased. The thrust blocks then were folded together, perhaps as the result of drag over a deep-seated (Turner Valley-Outwest?) thrust, or perhaps as a result of their own inertia.

If there is a change in the character of the Livingstone thrust between Savanna Creek and Oldman River, it probably can be related in large part to the stratigraphy of the Rundle group. The presence locally of evaporites, shaly beds, or even thinly bedded rather than massive limestone and dolomite might greatly reduce its competence. For example, in the Triad-B. A.-Stolberg well the Rundle was penetrated three times. The upper two Rundle bodies are both thin slices lying on low-angle thrusts in evaporitic Rundle beds. They were simply peeled off the top of the formation (Thomas, 1958). Evaporites have not been observed in outcrop at Savanna Creek but they have been logged in the wells, and furthermore, there are several thinly bedded shaly zones.

It seems clear that the Paleozoic, and particularly the Rundle, in places is less competent than we have supposed, and is capable of forming structures similar in nearly every respect to structures in Mesozoic beds. Comparison of Figure 8 with Figure 7 reveals a remarkable general similarity between the Savanna Creek and Jumpingpound structures. Different beds are involved in the two structures, and the former is broader and higher, but they are otherwise much alike and they probably formed in the same way.

The entrapment of gas here was possible because the migration of hydrocarbons was not completed before the faults were formed. The trap was made by the development of these faults, and the folds we now see are merely distortions of the original fault trap. Once this was formed, folding of the thrust blocks localized the trapped hydrocarbons, but it was the faults that prevented their escape updip in the first place.

PINCHER CREEK

The Pincher Creek gas-condensate pool is 18 miles north of the International Boundary and 100 miles due south of Calgary, in about the middle of the foothills. It is not in typical foothills country (except structurally) as the characteristic elongate ridge and valley topography is missing. Most of the country is of low relief, and the plains extend to the mountain front.

The earliest drilling in Alberta was done in the general Pincher Creek area. Oil seeps along Cameron Brook in Waterton Park had been known to the Indians, presumably for generations, and by 1886 at least one white man, the famous John G. (Kootenai) Brown, had learned of them and put the oil to use. A survey of the plains on the east was made in 1884 by a banking firm, but drilling did not commence until 1891, when a well was started 2 miles southwest of the present field. It was abandoned at 190 feet and its successor, drilled 12 miles southwest in about 1893, was rather less successful.

In 1901 a well was started on Cameron Brook and drilled to 1,020 feet, where oil was found in the Cretaceous below the Lewis overthrust. This strike gave rise to many ill founded rumors of spectacular production, which were denied by its backers. It undoubtedly did produce some oil, although apparently the quantity was small, and it encouraged further drilling. At least 11 more wells were drilled nearby in the next 6 years, without success.

Drilling in the foothills had its real beginning in 1907 when a well was drilled to 1,800 feet near Pincher Creek. Exploration was slow, however, and during the next 40 years only 16 wells were drilled in the general area. All were dry and the deepest one was abandoned at 7,189 feet. Twelve of them were drilled after 1920. At least six of the early wells were drilled within 5 miles of the boundaries of the present field.

In 1942 Canadian Gulf Oil Company commenced geophysical work in the southern foothills. Gravity-meter surveys were made first and these were followed by extensive geophysical surveys during the next 4 years. In 1947 Canadian Gulf Pincher Creek No. 1 was drilled to 12,516 feet and completed in Rundle beds. On production tests it indicated an open-flow potential of 45 million cubic feet of gas and 1,670 barrels of condensate per day. Subsequent developments have been slow, mainly for economic reasons. The field, about 16 miles long and $2\frac{1}{2}$ miles wide, now has 10 wells testing or on production. Two others have been abandoned.

In 1929 and 1930 one company drilled 17 core holes in the vicinity, presumably seeking structural data. These ranged in depth from 225 to 1,754 feet.

The surface geology in the area reveals nothing of the presence of either a fold or fault slice at depth. The reservoir underlies a series of rather thin thrust sheets of Belly River and Wapiabi beds, and even the strike of the surface beds is at variance with the strike of the reservoir. There is no suggestion of anticlinal structure—on the contrary the distribution of the beds indicates that the series of thrust sheets might represent the severely faulted east limb of a southeasterly plunging syncline lying on the Harland Lake thrust. West of the north end of the field there is a suggestion that Belly River, Wapiabi, Bighorn, Blackstone, and Blairmore beds are closing westward around such a syncline. The west limb of the syncline, if it is a syncline, lies hidden beneath a fault.

West of the field area and the zone of imbricate faults there is a belt 6–9 miles wide, characterized by gentle folds and a few faults involving Colorado and Belly River beds, and extending to the mountain front. The Belly River is at the west against the mountains with the Colorado below it at the east.

Northeast from the field for 3 miles at the north end and 5 miles at the south end the exposures are mostly Belly River in thin steeply inclined fault sheets. Farther northeast to the foothills front younger beds are found with the Belly River in the fault slices. East of Cochrane Lake thrust in the frontal crumpled zone, uppermost Cretaceous and even Paleocene beds are found.

The area is one in which mapping and correlation of structures are perplexing as most of the outcrop is confined to stream channels. Also, the lack of good topographic expression of structure makes the tracing of beds and contacts most difficult.

Figures 9 and 10 are structure sections drawn through the northern and central parts of the field respectively (Fig. 2). Both illustrate the absence of anything resembling an anticline at the surface.

Both sections are characterized by the remarkably low angles at which the beds are presumed to lie at depth, and by the extremely low dips of the fault planes at depth. It is possible to make quite different sections, but there is strong evidence to support the interpretations offered here.

Near the western end of Figure 9, two wells, Alberta Gas and Fuel Yarrow, and Alberta Gas and Fuel Drywood, provide an important clue to the structure. Both wells penetrated a regular section (excepting a minor repetition of Blackstone beds at Drywood) until they were part way through the Kootenay formation. In the Kootenay both passed through a fault, the Drywood well to Wapiabi and the Yarrow well to Blackstone beds, and in each the fault has 2,000–2,500 feet of throw As there is no fault of this magnitude known to reach the surface between the two wells it is reasonable to suppose that both penetrated the same fault, the Twin Butte thrust, and further, that this fault reaches the surface east of the Drywood well. Thus, very low dip for at least one of the thrusts is fairly established.

Further evidence is provided by the attitude of the reservoir beds. Erdman *et al*. (1953, p. 156) point out that the Rundle beds dip southwest at 4°–8° in the field area. Also, the dips are known from geophysical data to remain at 10° or less for some distance west of the field (Erdman, pers. comm.). Moreover, the drilled thicknesses of the Blairmore and Blackstone beds directly above the Rundle indicate that they must be nearly flat.

With the establishment of the very low westerly dip of the reservoir block and of the Twin Butte thrust, it follows that the intervening structures also dip west

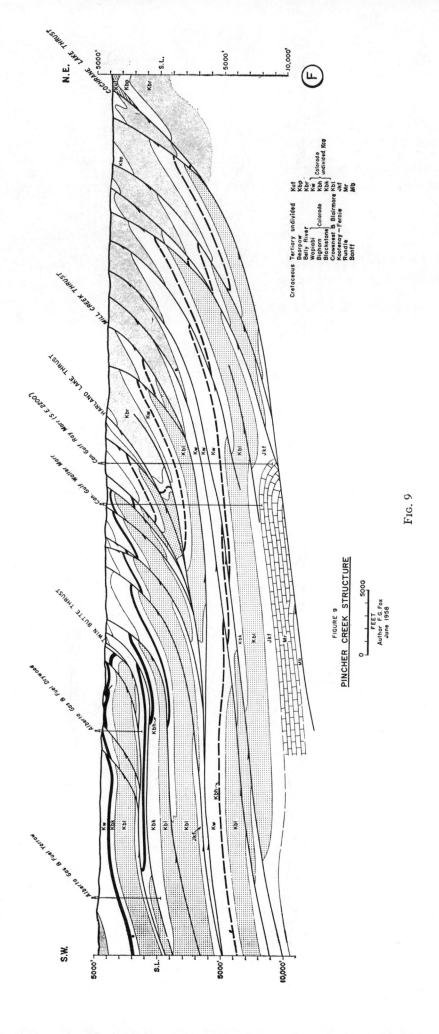

FIGURE 9

PINCHER CREEK STRUCTURE

0   5000
FEET

Author F. G. Fox
June 1958

FIG. 9

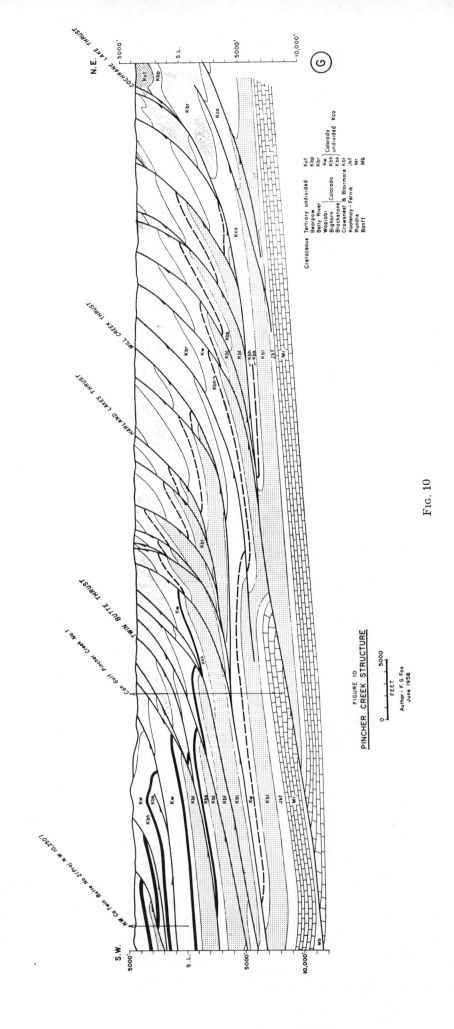

FIGURE 10

PINCHER CREEK STRUCTURE

FEET

0   5000

Author: F. G. Fox
June 1956

Cretaceous Tertiary undivided   Kut
Bearpaw   Kbp
Belly River   Kbr
Wapiabi   } Colorado   Kw
Bighorn   Kbh } undivided Kco
Blackstone   Kbl } Colorado
Crowsnest & Blairmore   Kbl
Kootenay – Fernie   Jkf
Rundle   Mr
Banff   Mb

FIG. 10

at very low angles—faults and bedding planes alike. Moreover, it also follows that the faults and strata east of the Gulf Marr wells (Fig. 9), however steep they may be at the surface, most probably are also nearly flat at depth. To assume that they are steep east of the Marr wells and nearly flat west of them would pose a difficult, though not insoluble, geometrical problem.

No well has yet been drilled through the Rundle and the sole thrust presumed to lie beneath. The presence of this fault is inferred from the fact that the eastern termination of the Rundle is abrupt, and the inference is supported by geophysical data and by drilling. A fold with a very steep northeast limb could produce an equally abrupt apparent termination, but experience in other areas and the presence of so many known faults in this area lead to the conclusion that the reservoir is not an anticline but a thrust sheet. The east dips at the leading edge of the Rundle block do not necessarily indicate that the structure is a faulted fold. They are mainly the result of drag against the fault, which probably formed in non-folded beds. The throw and displacement on the sole fault are unknown quantities. On Figure 10 the throw is shown to be about 3,000 feet and displacement 3 miles, but these values might well be far from the truth.

No single thrust of major proportions emerges at the surface east of the wells, and the place of emergence of the sole fault is, therefore, in doubt. The Mill Creek thrust is a considerable fault, but in the face of the evidence favoring low dips at depth it does not seem likely that it underlies the Rundle block. However, the Cochrane Lake thrust has been traced a considerable distance (Douglas, 1952) and is an important fault. It is, therefore, assumed to be the sole thrust. It is thought that it merges downward with several of the faults west of it (four in Figure 9 and seven in Figure 10), and that it is a major thrust at depth. An alternative possibility is that the sole thrust does not appear at the surface anywhere. Should this be so, the Cochrane Lake fault and its subsidiary faults might be even more nearly flat than they are shown to be.

Not far west of the field is the present eastern edge of the Lewis overthrust sheet. The eastward movement of this enormous thrust mass has evidently had a profound influence on the geology of the area. This is partly to be seen in Figures 9 and 10. The thrust sheet on which the Drywood and Yarrow wells (Fig. 9) were drilled, and the sheet on which the Twin Butte well (Fig. 10) was drilled are characterized not only by the very low dips of the thrust planes, but also by the remarkably gentle folds and low bedding dips associated with them from the surface down. These structural conditions persist westward to the mountain front. Farther east, as shown in both sections, the near-surface dips are steep. In the writer's opinion, the gentle structures on the west have low dips because they were once overlain by the Lewis overthrust sheet, the weight of which prevented any considerable folding, rotation, and steepening of the beds and fault planes below. At the latitude of Pincher Creek the thrust sheet probably extended about 9 miles east of its present front. Only two small klippen about 3 miles northeast of the mountains now remain.

The faults under the gentle westerly structures probably lie very near to the attitudes at which they were originally formed. Farther east, where they were free of excessive overload, they were rotated and steepened along with the bedding planes, especially near the surface.

The Lewis overthrust had other effects. At Pincher Creek the general strike at the surface is about N. 45° W., whereas the strike of the reservoir body (Rundle) is N. 30° W. For this reason the strike of the reservoir is commonly thought to be peculiar. The fact is there is nothing very peculiar about the strike of the Rundle; on the contrary it is reasonably consistent with the strike of the foothills as a whole. It is the strike of the surface beds that is peculiar. They have been bulged eastward in Pincher Creek area and south (Fig. 2) by the eastward movement of the Lewis thrust sheet, which apparently stripped the Colorado and younger beds from their foundations and pushed them eastward, accompanied by considerable piling and imbrication in front. The Rundle, being deeply buried and relatively remote from the thrust sheet, was less affected by its movement and, furthermore, was affected later than were the Cretaceous beds. This process accounts for the apparent great disparity in shortening between the Rundle and the Upper Cretaceous beds, and for their divergent strikes.

Erdman *et al.* (1953, p. 156) suggest that the reservoir block, with the Kootenay/Fernie and Blairmore beds immediately overlying it, is an early Laramide structure that was later overridden by the near-surface structures. This implies (as they point out) that the Lewis overthrust, which was the cause of the shallower structures, was formed later than the reservoir block and the Cochrane Lake thrust. The writer's view is that the Lewis overthrust is the earlier fault. As it pushed forward it moved a large part of the Cretaceous sequence with it, resulting in the formation below and in front of it of a series of low-angle faults. In front of the thrust sheet the subthrust Cretaceous beds were free to move, and rotation of beds and fault planes resulted. However, as rotation proceeded and the angles of dip increased, resistance to further rotation also increased. Ultimately the point was reached where relief from the stress was attained more readily by the formation of a new fault or series of faults farther in front of the thrust sheet, and these in turn suffered rotation. The end result was the formation of the Cochrane Lake thrust and rotation of its forward part, but this was only the end because the Laramide revolution was ending. Had it continued unabated more faults would have been formed east of the Cochrane Lake thrust. Thus Pincher Creek reservoir is in a very late, rather than an early, Laramide structure.

The Cochrane Lake thrust is not the only one involving the Rundle formation. The Mill Creek thrust and perhaps the Twin Butte thrust also involve the Rundle west of the line of section. The Waterton and Castle River blocks both lie structurally above and west of the Pincher Creek block. (This is not to imply that they lie on either of the faults last named).

A striking feature of both sections is the simplicity of the reservoir block compared with the structures that lie above it. Both also show how a major displacement at depth can be masked at the surface by a maze of subsidiary and relatively minor faults.

Pincher Creek is clearly an accumulation in a fault trap. There is no evidence that a fold ever existed there, and only the formation of the Cochrane Lake thrust and eastward dislocation of the Rundle prevented the escape of the hydrocarbons updip.

CASTLE RIVER-WATERTON

During the past 2 years important discoveries have been made in the inner foothills about 12 miles west of the north end of Pincher Creek field. To date, five wells have made wet-gas discoveries in Mississippian and Devonian rocks, and five more wells are being drilled.

The area being explored lies close in front of the Lewis overthrust, north of which for $3\frac{1}{2}$–$4\frac{1}{2}$ miles Belly River to Kootenay beds lie in regular succession, dipping generally southwest beneath the thrust. Farther north and east, several west-dipping faults cause repetition of parts of the Bighorn to Kootenay succession. Seven miles northeast of the mountain front is the Mill Creek anticline, on which two dry holes have been drilled. The structure in the area has exerted a strong influence on the topography, and the countryside exhibits typical foothills ridges and valleys.

This series of discoveries began with the completion of Shell Waterton No. 1 in early 1957, followed shortly by the Texaco Castle River A-3-4 well. Subsesequently, Texaco has completed two wells, Gladstone Creek A-6-15 and Castle River B-6-17, and Shell has completed Waterton No. 2 and is drilling two more wells. All have found gas and naphtha in the Rundle formation, and the Texaco Castle River B-6-17 well also yielded strong gas flows from Upper Devonian rocks.

At this stage in the development of the area firm conclusions regarding the subsurface structure are not possible, but a pattern seems to be emerging. It is tentatively suggested that the Castle River A-3-4 and B-6-17 wells reached the same Rundle thrust block, and that the Gladstone Creek A-6-15 and Waterton No. 1 and No. 2 wells reached a different, higher block. In the ensuing discussion these are alluded to as the Castle River and Waterton blocks respectively.

In the Waterton block, the Rundle was penetrated in Shell Waterton No. 1 at −5,239 feet; in Waterton No. 2 at −5,224 feet; and in the Gladstone Creek A-6-15 at −6,847 feet. Thus between Waterton 2 and 1 there is 15 feet of plunge in 4 miles and between Waterton 1 and Gladstone Creek there is 1,608 feet of plunge in $7\frac{1}{2}$ miles. This indicates that the crest maximum of the thrust block is southeast of Waterton 1 and that the block plunges northwesterly.[3] What its configuration is farther southeast will be learned only by further drilling.

---

[3] Data from a well newly completed show that the northwesterly plunge is not simple; the Waterton block is divided into two segments by a subsidiary thrust.

No conclusions can be drawn yet about the plunge of the Castle River block. The B-6-17 well encountered the Rundle at $-6{,}281$ feet and the A-3-4 well penetrated it 521 feet lower, at $-6{,}793$ feet. The latter well, however, is downdip from the former and this accounts for some, if not all, of the difference in elevation. Whether any of the difference is attributable to plunge can not yet be told.

Figure 11 presents an interpretation of the structure based on the belief that its evolution was generally similar to that of the Pincher Creek field. A close similarity between this section and the western parts of Figures 9 and 10 should be noted.

Although there is an abundance of outcrop data in the area as well as information from three wells, one close to and two on the line of section, the real key to this interpretation is the Crowsnest formation in outcrop and in the wells. At the surface it is found at A, B, and C. In the Gladstone Creek well it was penetrated at A', B', and C', and in Castle River it was found at C''. It is abundantly clear that in Gladstone Creek Crowsnest A' can be in continuity with A at the surface (admittedly interrupted by a minor fault), B' with B at the surface, and C' with C'' in Castle River A-3-4 and C at the surface. This is the simplest way to fit them together. Thus it is established that the dips of the beds and fault planes need not remain steep to any considerable depth.

An obvious and acceptable alternative interpretation can be made by connecting Crowsnest B' in Gladstone Creek with A at the surface, and C' with B. To do so requires a re-interpretation of the fault system and the acceptance of a more steeply inclined series of thrusts, at least between the wells. It also calls for a larger and higher third thrust mass of Paleozoic rocks west of Gladstone Creek. In Figure 11 this is an inconspicuous body. The interpretation offered here is preferred because it is simpler.

Several features of the structure as portrayed in Figure 11 should be noted. Most obvious is the simplicity of the structure of the reservoirs as compared with the overlying structures. No less important are the very low angles at which the faults are shown to transect the beds, and, therefore, the near parallelism of the bedding and fault planes. Further concerning the faults, it should be remarked that they are assumed to cut the Paleozoic at slightly less acute angles than those of their intersection with the Mesozoic beds. Finally, the nearly flat attitudes of beds and fault planes from the Gladstone Creek well west are directly attributable to the fact that the Lewis thrust sheet once extended at least 3 miles northeast of its present front in this locality.

The wells now being drilled should greatly advance the solution of the outstanding structural problems, although most of them are on strike extensions of the Waterton thrust block. More wells intended to probe the Castle River block, and wildcat wells at the northeast across the strike will ultimately permit a more firm interpretation of the structure. Meanwhile the relations of the Waterton and Castle River blocks to the Pincher Creek block are unclear.

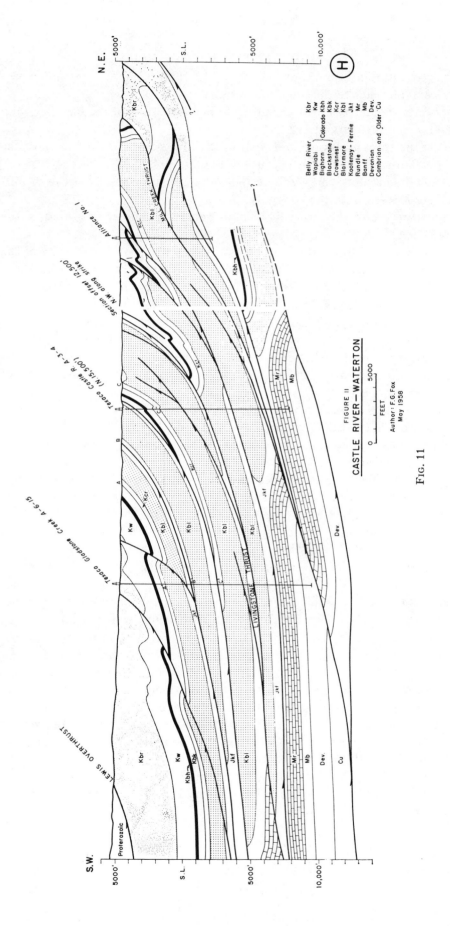

FIGURE II

CASTLE RIVER—WATERTON

Author: F.G.Fox
May 1958

0   5000
FEET

Belly River   Kbr
Wapiabi   Kw
Bighorn   Kbh
Blackstone   } Colorado   Kbk
Crowsnest   Kcr
Blairmore   Kbl
Kootenay - Fernie Jkf
Rundle   Mr
Banff   Mb
Devonian   Dev.
Cambrian and Older 'Cu

FIG. 11

REVIEW AND SYNTHESIS

In the structural interpretations (Figs. 3–11), there are many common characteristics and all point to the same conclusions.

1. In every productive area the Turner Valley formation of the Rundle group is the reservoir, or a reservoir.

2. All of the traps are associated with thrust faults and all, with the possible exception of Turner Valley, exist because of these faults.

3. The faults are not everywhere recognizable as major breaks at the surface (Figs. 4, 5, 6, 7, 9 and 10).

4. The structure of the reservoir block is generally simpler than that of the overlying beds.

5. On the major faults in the foothills, stratigraphic throw is generally of the order of 5,000–8,000 feet, and slip or displacement ranges from one to 8 miles. The most important zones permitting such displacements are in the Blackstone, Kootenay and Fernie, Banff, and unnamed Cambrian beds.

6. The thrust faults commonly have steep dip relative to a horizontal plane, but they transect the bedding planes at rather small angles in most places, and probably had very low dip when formed. Where they now dip steeply their attitudes are a result of rotation of beds and fault planes during the late stages of their evolution.

7. The refraction of fault planes between beds of differing apparent competence is less pronounced than is commonly supposed.

8. With the exception, perhaps, of Turner Valley, there is no evidence that local folding preceded faulting in any of these structures—in fact, the reverse is true. The sections illustrate a remarkable absence of folds, excepting drag folds, and suggest that the principal mechanism by which crustal shortening was accomplished in the foothills was low-angle thrust faulting.

9. The entrapment of hydrocarbons, in most places, could not have occurred before the formation of the sole thrust of any productive structure.

It is suggested that the present Front Range and Foothills region underwent epeirogenic uplift at the end of Paleozoic time, and later, near the end of Jurassic time, perhaps responded to the Nevadian orogenic movements by the formation of very broad regional folds. The Laramide revolution, which culminated no earlier than Paleocene time, caused the development of a series of major thrust faults, first in the west and then progressively farther east. These formed at low angles to the bedding planes and any horizontal plane, but their leading edges were generally steepened by rotation. The fundamental faults cut into the Paleozoic succession, and disrupted the continuity of the porous zones in the Devonian and younger beds, thus providing traps for hydrocarbons. Indigenous gas and oil, as they were forced to migrate by these movements, were caught in the fault traps. There is no reason to suppose that all the hydrocarbons migrated at one time, or in one direction, but at least some moved updip toward the east. During the final stages of the Laramide orogeny many of the low-angle faults were folded and some of the traps were distorted.

It is implicit in the whole of the foregoing argument that migration of gas and oil could have begun, but could not have been completed, before the formation of the most easterly fault that cuts the reservoir beds, which also appear to have been the carrier beds. No less clear is the implication that there ought to be oil-bearing structures other than Turner Valley, and this does not depend on the question of whether there were local protofolds formed at the end of Jurassic time. In either case it is difficult to believe that only one structure in the whole foothills belt caught and held any of the migrating oil.

REFERENCES

BEACH, F. K., AND IRWIN, J. L., 1940, *The History of Alberta Oil*, Dept. Lands and Mines, Gov. Alberta, Edmonton.

BEACH, H. H., 1943, "Moose Mountain and Morley Map Areas, Alberta," *Geol. Survey Canada Memoir 236.*

CLARK, L. M., 1954, "Geology of Rocky Mountain Front Ranges near Bow River, Alberta," *Western Canada Sedimentary Basin*, Amer. Assoc. Petrol. Geol., pp. 29–46.

DE WIT, R., 1953, Devonian Stratigraphy in the Rocky Mountains South of Bow River; *Alberta Soc. Petrol. Geol., 3d Ann. Field Conf. and Sympos.*, pp. 105–07.

DOUGLAS, R. J. W., 1950, "Callum Creek, Langford Creek, and Gap Map-Areas, Alberta," *Geol. Survey Canada Memoir 255.*

———, 1953, "Carboniferous Stratigraphy in the Southern Foothills of Alberta," *Alberta Soc. Petrol. Geol., 3d Ann. Field Conf. and Sympos.*, pp. 68–88.

ERDMAN, O. A., 1950, "Alexo and Saunders Map-Areas, Alberta," *Geol. Survey, Canada Memoir 254.*

———, BELOT, R. E., AND SLEMKO, W., 1953, "Pincher Creek Area, Alberta," *Alberta Soc. Petrol. Geol. 3d Ann. Field Conf. and Sympos.*, pp. 139–57.

GALLUP, W. B., 1954, "Geology of Turner Valley Oil and Gas Field, Alberta, Canada," *Western Canada Sedimentary Basin*, Amer. Assoc. Petrol. Geol., pp. 397–414.

HUME, G. S., 1941, "A Folded Fault in the Pekisko Area, Foothills of Alberta," *Trans. Roy. Soc. Canada*, Vol. 25, Sec. IV, pp. 87–92.

———, 1957, "Fault Structures in the Foothills and Eastern Rocky Mountains of Southern Alberta," *Bull. Geol. Soc. America*, Vol. 68, pp. 395–412.

HAGE, C. O., 1942, "Folded Thrust Faults in Alberta Foothills West of Turner Valley," *Trans. Roy. Soc. Canada*, Vol. 36, Sec. IV, pp. 67–78.

LINK, T. A., 1935, "Types of Foothills Structures of Alberta, Canada," *Bull. Amer. Assoc. Petrol. Geol.*, Vol. 19, No. 10, pp. 1427–71.

———, 1949, "Interpretations of Foothills Structures, Alberta, Canada," *ibid.*, Vol. 33, No. 9, pp. 1475–1501.

———, 1953, "History of Geological Interpretation of the Turner Valley Structure and Alberta Foothills," *Alberta Soc. Petrol. Geol., 3d Ann. Field Conf. and Sympos.*, pp. 117–33.

MARTIN, R., 1956, "The Jumpingpound Gas Field," *ibid., 6th Ann. Field Conf. Guide Book*, pp. 125–40.

McLAREN, D. J., 1953, "Summary of the Devonian Stratigraphy of the Alberta Rocky Mountains," *ibid., 3d Ann. Field Conf. and Sympos.*, pp. 89–104.

NORTH, F. K., 1953, "Cambrian and Ordovician of Southwestern Alberta," *ibid.*, pp. 108–16.

———, AND HENDERSON, G. G. L., 1954, "Summary of the Geology of the Southern Rocky Mountains of Canada," *ibid., 4th Ann. Field Conf. Guide Book*, pp. 15–81.

RICH, J. L., 1934, "Mechanics of Low-Angle Overthrust Faulting as Illustrated by Cumberland Thrust Block, Virginia, Kentucky, and Tennessee," *Bull. Amer. Assoc. Petrol. Geol.*, Vol. 18, pp. 1584–96.

RUTHERFORD, R. L., 1944, "Regional Structural Features of the Alberta Foothills and Adjacent Mountain Ranges," *Trans. Roy. Soc. Canada*, Vol. 38, Sec. 4, pp. 71–77.

SCOTT, J. C., 1951, "Folded Faults in Rocky Mountain Foothills of Alberta, Canada," *Bull. Amer. Assoc. Petrol. Geol.*, Vol. 35, No. 11, pp. 2316–47.

———, 1953, "Savanna Creek Structure," *Alberta Soc. Petrol. Geol., 3d Ann. Field Conf. and Sympos.*, pp. 134–39.

———, HENNESSEY, W. J., AND LAMON, R. S., 1957, "Savanna Creek Gas Field, Alberta," *ibid., 7th Ann. Field Conf. Guide Book*, pp. 113–31.

SPRATT, J. G., AND TAYLOR, V., 1936, "Oil Prospects along the West Flank of the Turner Valley Gas Field," *Bull. Canadian Inst. Min. and Met.*, Vol. 29, No. 10, pp. 713–22.

THOMAS, A. N., 1958, "Note on a Geological Cross Section through the Nordegg Area," *Alberta Soc. Petrol. Geol., 8th Ann. Field Conf. Guide Book*, pp. 121–27.

Reprinted by permission of the Society of Economic Paleontologists and Mineralogists, from *Strike-slip Deformation,
Basin Formation, and Sedimentation:* SEPM Special Publication No. 37, edited by K. T. Biddle and Nicholas H. Christie-
Blick, December 1985, pp. 51-77.

STRUCTURAL STYLES, PLATE-TECTONIC SETTINGS, AND HYDROCARBON TRAPS OF DIVERGENT (TRANSTENSIONAL) WRENCH FAULTS

T. P. HARDING AND R. C. VIERBUCHEN
Exxon Production Research Company
P. O. Box 2189
Houston, Texas 77252-2189;

AND

NICHOLAS CHRISTIE-BLICK
Department of Geological Sciences and
Lamont-Doherty Geological Observatory
of Columbia University
Palisades, New York 10964

ABSTRACT: A divergent (transtensional) wrench fault is one along which strike-slip deformation is accompanied by a component of
extension. Faulting dominates the structural style and can initiate significant subsidence and sedimentation. The divergent wrench fault
differs from other types of wrench faults by having mostly normal separation on successive profiles, negative flower structures, and a
different suite of associated structures. En echelon faults, most with normal separation, commonly flank the zone, and some exhibit
evidence of external rotation about vertical axes and have evidence of superimposed strike slip. Flexures associated with the wrench
fault are formed predominantly by vertical components of displacement, and most are drag and forced folds parallel to and adjacent
to the wrench. Hydrocarbon traps can occur in fault slices within the principal strike-slip zone, at culminations of forced folds, in the
flanking tilted fault blocks, and within less common en echelon folds oblique to the zone.

Divergent wrench faults occur at active plate boundaries, in extensional and contractional continental settings, and within plates far
from areas of pronounced regional deformation. Along transform margins and within wrench systems, divergent wrench styles tend to
develop where major strands or segments of strands bend or splay toward the orientation of associated normal faults (e.g., elements
of the San Andreas system in the Mecca Hills, California), and where major strands are regionally oblique to interplate slip lines (e.g.,
Dead Sea transform, Middle East). The style also develops at releasing fault oversteps and fault junctions (e.g., Ridge Basin, California),
and locally where crustal blocks rotate between bounding wrench faults. In extensional settings, divergent wrench faults may form
within graben doglegs and oversteps (e.g., between the Rhine and Bresse grabens, northern Europe), and they may separate regions
that experienced different magnitudes of extension (e.g., Andaman Sea area). Many oceanic fracture zones have divergent wrench
characteristics. The style has also been recognized in magmatic arcs (e.g., the Great Sumatran fault) and in both backarc and peripheral
foreland settings (e.g., Lake Basin fault zone, Montana) near convergent plate boundaries, and in intra-plate settings (e.g., Cottage
Grove fault, Illinois; Scipio-Albion trend, Michigan).

INTRODUCTION AND PREVIOUS WORK

Divergent wrench zones are defined as those strike-slip systems within which the principal displacement zone and, in many cases, the adjacent associated structures are dominated by extensional characteristics. This article illustrates the structural style of divergent wrench zones and summarizes their known tectonic habitats and associated hydrocarbon traps.

Past discussions of wrench-fault styles have emphasized the contractional structures that accompany-convergent strike slip (e.g., Wilcox et al., 1973; Sylvester and Smith, 1976). Several studies have shown individual structures attributable to extensional strike slip, but none has dealt with the full range of structural features that can be present. Nelson and Krausse (1981) presented the most thorough documentation, using an example from the Illinois basin where coal mining has provided unusually dense subsurface control. The results of that study are incorporated in our present work. The mechanics of oblique plate separation were discussed briefly by Harland (1971), and he first applied the term "transtension" to deformation in these zones. Wilcox et al. (1973) introduced the term "divergent wrench fault" and suggested that the structure along such faults is characterized by an increase in extensional block faulting and a decrease in en echelon folding. These authors described graben structures that were formed by divergent strike-slip

deformation of a clay-cake model. Courtillot et al. (1974) also reproduced extensional strike-slip structures with a clay-cake model of a transform fault between simulated spreading centers. Earlier, Bishop (1968) documented fault patterns in New Zealand, portions of which are now recognized as essential elements of the structural style. Harding (1983) has recently illustrated a divergent wrench fault on seismic reflection data from the Andaman Sea and has established criteria for differentiating these zones from other styles on seismic profiles (Harding, 1985). D'Onfro and Glagola (1983) have published seismic reflection profiles of another divergent wrench zone in southeast Asia.

Our overview of divergent wrench systems presented here is drawn from these previous studies, from new field work in the Mecca Hills of southern California, and from additional seismic reflection profiles.

EXTENSIONAL STRIKE-SLIP STYLES

Two important changes occur in the structures associated with wrench faults when strike slip is accompanied by significant divergence. First, lateral compression is decreased to the extent that en echelon folds and thrusts are present only diffusely or concentrated in localized settings. Second, regional extension enhances brittle deformation to the degree that the style is commonly dominated by faults.

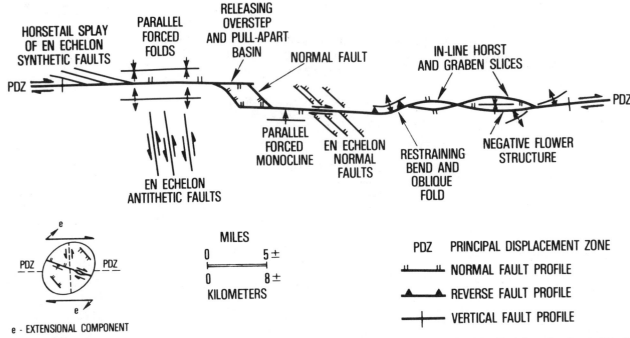

FIG. 1.—Assemblage of structures associated with divergent wrench faults and simplified strain ellipse for right-slip deformation (inset). Not all divergent wrench zones contain every feature and those structures present may be combined in different ways. The assemblage shown here is a composite of examples presented in this paper and others cited in the references.

Principal Strike-Slip Displacement Zone

A linear, throughgoing principal displacement zone is the fundamental element of the structural assemblage (PDZ in Fig. 1). This zone accounts for most of the lateral displacement between opposing blocks. Other faults in the assemblage tend to be shorter and to have smaller displacements. In some cases, the deformation consists solely of the principal displacement zone; in other instances, the zone is flanked on one or both sides by associated external structures.

Tchalenko and Ambraseys (1970) demonstrated that the principal displacement zone is composed mostly of synthetic strike-slip faults (i.e., Riedel shears; Riedel, 1929) and interconnecting P shears. Strike slip along P shears is in the same sense as along the synthetic faults, but P shears and synthetic faults step in opposite senses (Fig. 2). The prediction of structures and their orientations shown in Figure 2 is simplistic, and is most applicable to the early stages of a single tectonic event and to one wrench zone. Protracted or episodic deformation involving two or more major wrench faults can be considerably more complex. Orientations of structures in left-slip systems may be determined by viewing Figures 1 and 2 in reverse image.

The overall dip of the principal displacement zones is thought to be near-vertical at depth in many instances, and where there is definitive control, faults with normal separation dominate in transverse profiles (i.e., hanging-wall block apparently downthrown; Fig. 3). As with other types of wrench faults, the direction of fault dip may change along

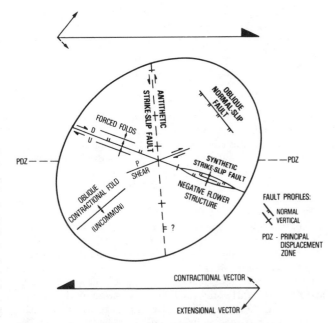

FIG. 2.—Strain ellipse with idealized orientations of main structures initiated along a right-lateral divergent wrench fault, compiled from clay-cake models and from geologic examples. Most of the structures and their orientations are present in other styles of wrench faults, but the comparative abundance of the various features differs significantly. Segments of the principal displacement zone (PDZ) that splay or bend to the right tend to have normal separation, but reverse faults can also be present depending on the way in which displacements are distributed along the PDZ.

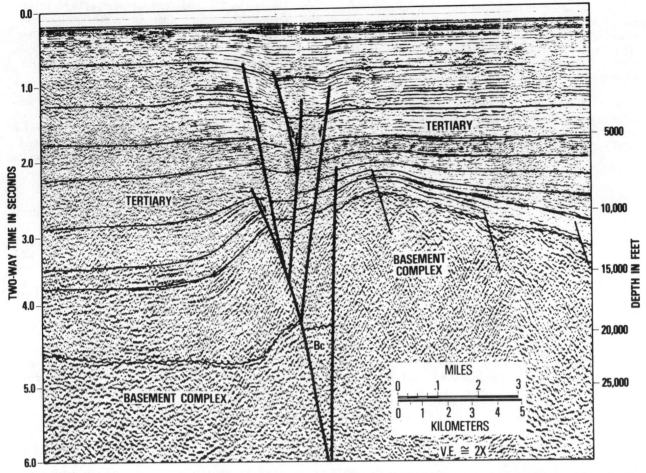

FIG. 3.—Seismic profile across a major fault zone in the Bering Sea. This area has undergone several deformations with different structural styles. Divergent wrench fault characteristics are most apparent in the younger Tertiary sediments above 2.0 seconds. Interpretation courtesy of W. A. Spindler (personal commun., 1984).

strike, and the apparently upthrown side can also change along strike or from one structural level to another. In some examples, the principal displacement zone appears to terminate abruptly at depth against a subhorizontal crustal detachment (Royden, 1985 this volume; Cheadle et al., in press).

Along some divergent wrench systems the principal displacement zone splays upward and outward at shallower levels (Emmons, 1969). The fault slices may form the core of a negative flower structure, which is defined as a shallow synform bounded by the upward and outward spreading strands of strike-slip faults that have mostly normal separation (above 2.0 seconds in Fig. 3; Harding and Lowell, 1979; D'Onfro and Glagola, 1983; Harding, 1983, 1985). The fold is termed a synform because at depth the flanks may have tilted independently and may never have been parts of an integrated flexure. Divergent wrench faults composed of a single strand may be more common, however.

Faults with reverse separation in transverse profile (i.e., hanging-wall block apparently upthrown) are present within some principal displacement zones. Contractional structures are throught to be most common where irregularities in the zone's trace cause opposing blocks to converge (e.g., restraining bend in Fig. 1; Crowell, 1974; Aydin and Nur, 1982). The contractional features commonly distinguish divergent wrench faults from normal-slip faults, which otherwise may have similar profile characteristics (Harding, 1985). However, successive intervals of normal slip and block rotation (in cross section) in areas of very great extension can over-steepen the initial normal faults so that they acquire reverse separation (Proffett, 1977). The presence of both normal and reverse separation then ceases to distinguish the two styles. Releasing bends (Crowell, 1974) and releasing oversteps (Christie-Blick and Biddle, 1985 this volume; see Fig. 1) may cause blocks to diverge and enhance the development of extensional strike-slip features. Several examples of these relationships are discussed below.

Most folds associated with the divergent wrench faults we have studied occur adjacent to and parallel with the principal displacement zone. In the absence of regional horizontal shortening, the more subtle vertical components of movement appear to be the major cause of fold formation.

The flexures resemble the forced folds that develop above normal fault blocks: anticlines or monoclinal knees adjacent to and parallel with the relatively higher side of the principal displacement zone; synclines or monoclinal ankle flexures adjacent to and parallel with the edge of the apparently down-dropped block. These flexures commonly bound negative flower structures at shallow levels (e.g., above 2.0 seconds in Fig. 3). Where the principal displacement zone consists of a single strand, a single monoclinal knee or ankle may face toward the relatively down-dropped block, or both high- and low-side flexures may be absent.

External Associated Structures

Deformation of the terrane flanking the principal displacement zone may be negligible. The regional dip in several examples continues unchanged across the zone, and external structures are diffuse or absent (see shallower reflections on Fig. 3). In other examples, faults oblique to the trace of the principal displacement zone are the most common type of associated structure.

External Faults.—The external faults, where present, occur in belts along one or both sides of the throughgoing wrench and are en echelon or oblique to the trend of the zone: left-stepping for right-lateral systems (Fig. 1) and right-stepping for left-lateral zones (Fig. 4). The strike of the external faults ranges from slightly oblique to the principal displacement zone (e.g., synthetic or Riedel strike-slip faults; Riedel, 1929) to nearly transverse (e.g., antithetic or conjugate Riedel faults; Fig. 2). The external fault sets commonly contain several fault orientations, but most external faults tend to be subparallel to the direction predicted for normal faults by our simple strain ellipse. In addition, where control is sufficient for determination, most of the external faults are characterized by normal separation and steep dips.

The fault set of the Lake Basin fault zone is an example of a distinct right-stepping en echelon pattern (Fig. 4). There is no principal displacement zone at the surface, but a left-lateral divergent wrench fault is believed to be present at depth (Chamberlain, 1919; Alpha and Fanshawe, 1954; Smith, 1965). A segment of the Dasht-e Baȳaz earthquake

system studied by Tchalenko and Ambraseys (1970) in central Iran similarly consists solely of en echelon fractures. The relationships there are unusually clear and these authors interpreted the pattern as representative of an initial stage in the development of the throughgoing principal displacement zone.

Individual dislocations along the Lake Basin fault zone strike at angles of 30° to 70° (most are between 45° and 60°) to the overall trend and have been interpreted as normal faults (Smith, 1965). Seismic control near the west end of the zone demonstrates steep fault dips with normal and, possibly, several reverse separations (Fig. 5). The variety of trends and separations probably results from the interaction of two factors: (1) pre-deformational anisotropies, and (2) external rotation and reactivation of the faults.

External rotation may significantly alter the structural pattern, and the style is, therefore, best expressed in the early stages of deformation. Attempts to identify it in later stages on the basis of observed or reconstructed fault patterns may be misleading. The effects of rotation have been demonstrated on a small scale by means of earthquake fractures in southern California (Terres and Sylvester, 1981; Fig. 6A in Christie-Blick and Biddle, 1985 this volume). There, clockwise external rotation of a planar anisotropy oriented oblique to the right-slip Imperial fault occurred during the earthquake of 15 October 1979 (see inset map in Fig. 6 for location of Imperial fault). The earthquake fractures approximate en echelon normal faults, and antithetic strike slip was superimposed on them by the rotation. Both reverse and normal separations are present. The sense of external rotation matched the sense of strike slip on the Imperial fault, and imposed the antithetic displacements in a mechanism roughly analogous to the slip generated by inclining a row of dominoes. Such a mechanism has been discussed by Freund (1970), and also by Luyendyk et al. (1980) to explain how paleomagnetically determined block rotations of 70° to 80° may be accommodated in the Miocene tectonics of southern California.

External Associated Folds.—Folds outside the principal displacement zone are generally sparse. Where developed, they are oblique or en echelon to the master fault: left-step-

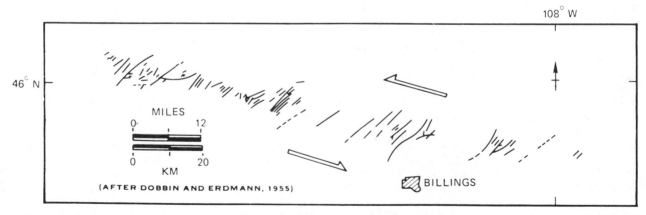

FIG. 4.—Surface map of Lake Basin fault zone, Montana. Orientations of most en echelon faults correspond with the idealized normal-fault direction in Figure 2 for left-lateral wrench faulting, but some faults are more nearly transverse and several are less oblique than predicted.

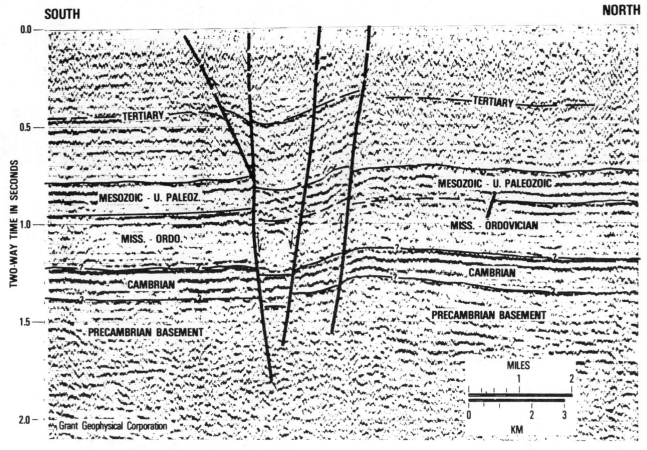

Grant Geophysical Corporation

FIG. 5.—Seismic profile near west end of the Lake Basin fault zone. Zone includes both faults with normal separation and faults with reverse separation. Most faults dip steeply and they may converge at greater depth into a single, throughgoing principal displacement zone. The deformation has been dated as post-Paleocene and possibly post-Eocene by Alpha and Fanshawe (1954). Profile courtesy of Grant Geophysical Corporation, 1983; interpretation courtesy of G. H. Weisser (personal commun., 1983).

ping for left-lateral faults and right-stepping (Fig. 1) for right-lateral faults. Where we have observed external folds, they result from localized convergent deformation, for example, at a restraining bend (Fig. 1). They may also form during an earlier, convergent phase of strike slip that had the same sense of offset as the subsequent divergent phase.

Other folds and faults may be inherited from earlier deformation unrelated to strike slip or may be superimposed by later tectonic events. Some of these structures could appear by coincidence to be congruent with the structural relationships of a wrench zone, and recognition of this disparity is a prerequisite for the correct analysis and identification of style.

DIVERGENT WRENCH FAULTS OF THE MECCA HILLS

Plate-Tectonic Setting of the Mecca Hills Faults

The Mecca Hills faults are three subparallel, right-slip faults that splay northward from the San Andreas wrench system along the northeast margin of the Salton Trough (Fig. 6). The divergent wrench style consists here almost solely of the principal displacement zone. The present tectonic framework originated around 4.5 Ma and is considered to be a northwestward continuation of the Gulf of California ridge-and-transform system (see inset map in Fig. 6; Crowell and Sylvester, 1979; Crowell, 1981). The subsurface axis of the trough lies within a releasing overstep (Fig. 1) between the San Andreas and Imperial faults. Their side-stepped pattern is indicated by earthquake epicenters and the surface traces of active faults (Rockwell and Sylvester, 1979). Away from the San Andreas and Imperial faults, the Salton Trough is flanked by high basement terranes that form the limbs of a broad arch (J. C. Crowell, personal commun., 1979).

On the northeast flank of the basin, convergent wrench faults dominate the structural style of the San Andreas and other major northwest-striking faults (Sylvester and Smith, 1976). The strike of the Mecca Hills splay faults is midway between that of the convergent strands and the north-south orientation predicted for normal faults associated with the San Andreas system (compare orientations with the strain ellipse of Fig. 2). Structural features typical of divergent wrench faults have been observed along three of the splays: the Painted Canyon, Eagle Canyon, and Hidden Springs

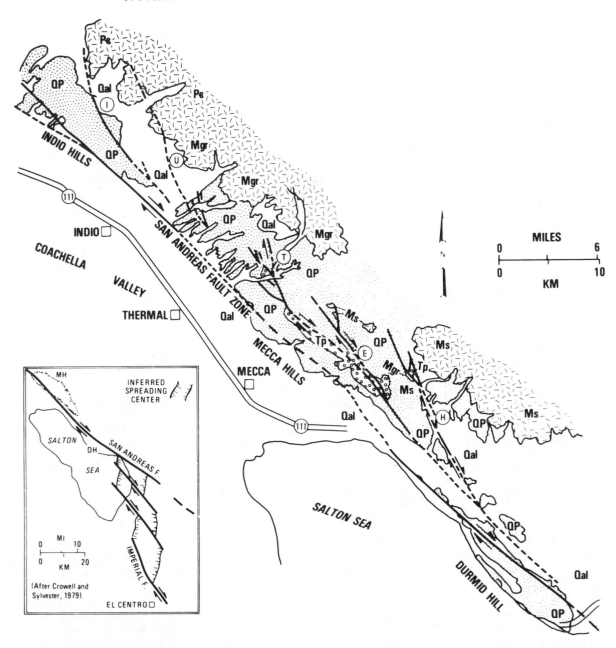

RECOGNIZED AND POTENTIAL DIVERGENT WRENCH FAULTS

(H) HIDDEN SPRINGS FAULT (FIG. 8)

(E) NORTHERN SEGMENT OF
EAGLE CANYON FAULT (FIG. 7)

(T) THERMAL CANYON SPLAY OF
PAINTED CANYON FAULT

(U) UNNAMED FAULT

(I) INDIO HILLS FAULT

Qal RECENT ALLUVIUM

QP PLIOCENE-PLEISTOCENE
NONMARINE SEDIMENTS

Tp PLIOCENE NONMARINE SEDIMENTS

Mgr MESOZOIC (?) GRANITIC ROCKS

Ms MESOZOIC OROCOPIA SCHIST

Ps PRECAMBRIAN COMPLEX

(GENERALIZED AFTER ROGERS, 1965, AND JENNINGS, 1967)

faults (Fig. 6). The investigation of two faults to the north with similar orientations is incomplete.

Evidence of Right Slip on the Mecca Hills Faults

Fine-grained Pliocene rocks adjacent to the Painted Canyon fault are offset approximately 3 km from their source (A. G. Sylvester, personal commun., 1979). Along another portion of this fault, Hays (1957) reported distributions of rock colors that suggest about 0.8 km of right slip. In addition, the en echelon pattern of folds associated with the southeastern segments of the Painted Canyon and Eagle Canyon faults suggests right slip.

The distribution of Pliocene sedimentary rocks along the Eagle Canyon fault indicates right slip of 1.6 km (Hays, 1957). Along the southern segment of this fault the northwestern boundary of an en echelon fold set lies approximately 1.3 km farther southeast than its position across the fault (Fig. 7). The folds are thought to have formed mainly during accumulation of the right slip (Sylvester and Smith, 1976), but they could have developed independently, which would negate their validity as offset evidence.

Along the Hidden Springs fault, the northeast limit of a distinctive basal breccia is apparently offset 1.2 km right-laterally, but according to Hays (1957), irregularities in the original depositional pattern could account for much of this displacement. Near Shaver's Well, displacement on the Hidden Springs fault dies out rapidly into horsetail splays (Fig. 8). Here, tongues of a schist breccia are offset as much as 45 m right-laterally on individual faults (Hays, 1957).

Structural Styles of the Mecca Hills Faults

Both the Painted Canyon and the Eagle Canyon faults have convergent strike-slip styles along their southern and central segments where they are closest to the San Andreas fault zone (Fig. 6). Along the central segment of the Eagle Canyon fault, beds are steeply dipping to overturned, the fault has a vertical or upthrust profile (segment with sawtooth symbols in Fig. 7), and drag folds are tight where present. Farther south, the limbs of tight, en echelon folds are offset across the fault, and it is difficult to differentiate these external structures from features of the principal displacement zone.

Deformation along the northern, divergent trace of the Eagle Canyon fault (segment with double tick symbol in Fig. 7) is much less intense and less complex than that to the south. A major normal fault, oriented north-south within the extensional sector predicted for the regional right-slip deformation, is the main external structure in this area (left of mid-point of Fig. 7). Beds dip 5° to 15° basinward across the area, with only minimal disruption. The principal displacement zone consists of a single strand with normal separation (Fig. 9). At the locality of Figure 9 (shown in Fig.

7), the zone's dip appears to flatten with depth, but this may be only a local irregularity. A gentle upturn of beds within the hanging-wall block locally reverses the basinward dip and forms a low-side syncline (Fig. 10). The syncline is thought to reflect the vertical component of the deformation. An accompanying high-side, forced anticline lies opposite the syncline only 0.5 km to the southeast (Fig. 7), and drag folds that demonstrate shortening are noticeably absent within the principal displacement zone. To the northwest, the Eagle Canyon fault terminates into a large horsetail structure composed of closely spaced, subvertical to vertical faults. There is little or no folding or other distortion of the regional dip of beds within the horsetail.

The Thermal Canyon fault (T in Fig. 6) is the north-northwest-striking splay of the Painted Canyon fault and is the segment of that system that has a divergent style. The contractional structures that dominate the southern portions of the Painted Canyon fault appear to be absent, and the principal displacement zone has normal separation. Horizontal slickenside striae are the only structural indication of strike slip.

The Hidden Springs fault (H in Fig. 6) is consistently oriented within the extensional sector of the regional right-slip deformation. The principal displacement zone of this fault generally dips steeply to vertically, and along much of its length has an extensional strike-slip style (Fig. 8). The zone terminates northward into a complex horsetail composed of several major fault strands. In a tributary canyon just south of Shaver's Well, the main faults bound narrow, steeply dipping slices. One strand splays upward and outward, forming a fault architecture similar to a negative flower structure (Fig. 11). Other subvertical faults within the canyon abut the throughgoing north-northwest-striking faults at high angles. Both the main strands and the external faults have abundant horizontal slickenside striae, but there is little or no evidence of horizontal shortening within the offset beds.

South-southeast of the horsetail, the strand of the Hidden Springs fault illustrated in Figure 11 turns eastward into the contractional sector of the deformation (i.e., turns subparallel to the orientation predicted for contractional structures associated with right slip; compare with Fig. 2), and forms a restraining bend (segment with sawtooth symbols in Fig. 8). The profile of the principal displacement zone opposite this bend is that of a downward-steepening reverse fault, and beds immediately adjacent to the fault are shortened by drag folding (Fig. 12).

COTTAGE GROVE FAULT SYSTEM

Plate-Tectonic Setting of the Cottage Grove Fault System

The structural style of the Cottage Grove fault system is dominated by external en echelon faults and has been elab-

---

Fig. 6.—Late Cenozoic tectonic framework of Salton Trough, California, in the vicinity of the Mecca Hills (inset), and location of divergent wrench faults identified along the northeast side of the San Andreas fault. Lack of continuous outcrops adjacent to the San Andreas fault makes it difficult to demonstrate a direct connection with the divergent wrench faults, but a splay configuration is inferred from the tectonic relationships. Abbreviations in inset map: MH, Mecca Hills; DH, Durmid Hill.

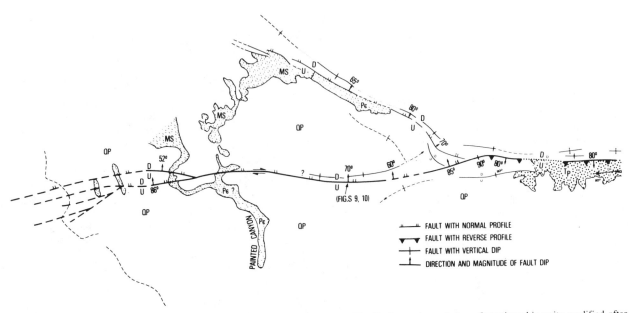

FIG. 7.—Generalized surface geologic map of the Eagle Canyon fault, Mecca Hills. Dating and correlation of stratigraphic units modified after

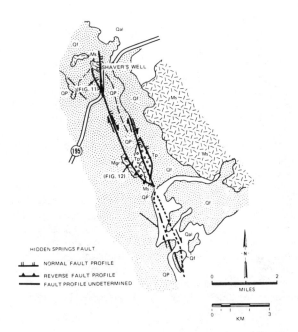

FIG. 8.—Generalized geologic map of northern segment of the Hidden Springs fault, Mecca Hills (after Jennings, 1967). Qf, Pleistocene fanglomerate; see Figure 6 for explanation of other symbols and for setting of fault.

FIG. 9.—Profile of Eagle Canyon fault. Outcrop is approximately 25 m high. Exposed segments of fault surface are indicated by arrows. Basal Pliocene-Pleistocene conglomerates on left are juxtaposed against stratigraphically higher, lighter colored buff sandstones and pebbly sandstones of the hanging-wall block on the right. View toward northwest. See Figure 7 for location and Figure 10 for associated fold.

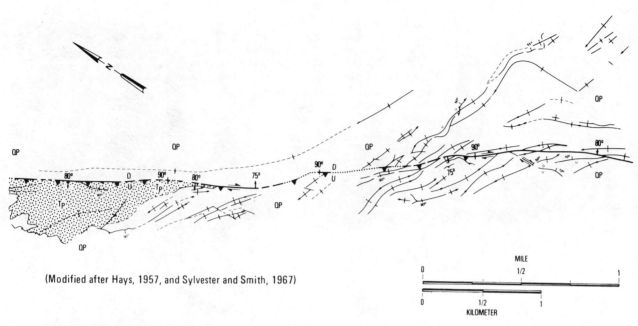

(Modified after Hays, 1957, and Sylvester and Smith, 1967)

Jennings (1967). See Figure 6 for explanation of symbols for rock units and setting of fault.

FIG. 10.—Drag syncline on eastern, relatively down-dropped side (foreground) of northern segment of the Eagle Canyon fault. Position of fault is indicated by arrow. Visible portion of cliff face is approximately 20 m high. Fault dips toward viewer and follows face of darker colored outcrop from left margin of photograph. View toward west-northwest. See Figure 7 for location and Figure 9 for profile of fault zone.

FIG. 11.—Portion of a divergent horsetail at the northwest termination of the Hidden Springs fault. Exposure of faults is 12 to 15 m high. Splays are closely spaced and steeply dipping; most have horizontal slickenside striae here or in adjacent outcrops. Displaced Pliocene-Pleistocene gravels exhibit little or no folding. Faults merge downward into single, subvertical zone at right base of photo. View toward south-southeast. See Figure 8 for location. T, displacement toward viewer; A, away from viewer.

orately documented by Nelson and Krausse (1981; see their illustrations for further details of the style). The system trends east-west for possibly 110 km across the southwest flank of the Illinois Basin (Fig. 13). The basin is a gentle, intracratonic downwarp. The broad arches and domes at the periphery of the downwarp were formed by mid-Ordovician time, except the Pascola arch, which is thought to have developed in the late Paleozoic (Buschbach and Atherton, 1979; Sloss, 1979). The basement warps and adjacent forelands (Appalachian and Arkoma Basins in Fig. 13) separated the Illinois Basin from orogenic belts that were active on the southeast (Appalachian thrust front) and southwest during the late Paleozoic.

The origin of the intracratonic sag is uncertain, but it may be related to thermal contraction of lithosphere, extended during the formation of an earlier graben system (Ervin and McGinnis, 1975; Sloss, 1979). Aeromagnetic and gravity surveys have delineated a deeply buried northeast-trending trough, the Reelfoot rift of Ervin and McGinnis (1975) or the Mississippi Valley graben of Kane et al. (1981). The

northeasternmost segment of this graben corresponds closely with the southward plunge of the Cambrian-Ordovician precursor of the Illinois Basin (Fig. 1 in Sloss, 1979). The Cottage Grove fault system may have developed along a segment of the older graben system (Heyl and Brock, 1961; Buschbach and Atherton, 1979).

Precise dating of the age of faulting is not possible because of the absence of stratigraphic section between the deformed Pennsylvanian beds and the surficial Pleistocene cover. Igneous dikes in the nearby fluorspar district (4 in Fig. 13) have been dated as 267 ± 20 Ma, or Early Permian (Zartman et al., 1967), and are similar compositionally to dikes intruded into extensional faults of the Cottage Grove system. Nelson and Krausse (1981) accept this as evidence of a late Paleozoic age for the wrench faulting.

Evidence of Right Slip on the
Cottage Grove Fault System

Clark and Royds (1948) were the first of a number of authors (e.g., Heyl and Brock, 1961; Wilcox et al., 1973) to suggest that the Cottage Grove system contains structures that resulted from strike-slip deformation. Nelson and Krausse (1981) have recently summarized this structural evidence: (1) The orientations of structural elements generally repeat those predicted for right slip by the strain-ellipse and clay-cake models for right slip (compare Figs. 2 and 14; Wilcox et al., 1973). (2) The fault sets resemble the fracture systems developed along seismically active faults that have explicit evidence of lateral displacements (e.g., Fig. 8 in Tchalenko and Ambraseys, 1970). (3) The fault pattern and elements of the fold pattern are also similar to zones that have geologic evidence of historic strike slip (e.g., Bishop, 1967).

The boundaries of a Pennsylvanian stream deposit suggest as much as 1.6 km of right slip, but other channels limit the offset to a smaller amount. Nelson and Krausse (1981) have concluded that the maximum lateral displacement is on the order of several hundred meters.

Characteristics of the Principal Strike-Slip Displacement Zone

The principal displacement zone includes the main faults of the Cottage Grove fault system and has structural char-acteristics typical of wrench zones. These are (1) subvertical faults, (2) inconsistent apparent upthrown side, (3) interchanging normal and reverse separations across the zone as a whole, and (4) narrow, steeply dipping, in-line horst and graben slices that form a braided swath and have normal and reverse separations on individual faults (Nelson and Krausse, 1981). Nelson and Krausse describe the master zone as typically tens to hundreds of meters wide, with vertical displacements as great as 60 m.

Discontinuities occur in the major strands of the principal displacement zone and are occupied by considerably shorter, northwest-striking en echelon faults (Fig. 15). These en echelon faults are steeply dipping, and most have normal separations. Some of the faults have horizontal slickenside striae and reverse separations that are interpreted by Nelson and Krausse (1981) as consequences of oblique slip. These authors, in an interpretation similar to ours and others (e.g., Tchalenko and Ambraseys, 1970), have interpreted the structure of the discontinuities as characteristic of an early stage in the development of the Cottage Grove system.

Folds associated with the known extent of the system lie immediately adjacent to the upthrown side of the principal displacement zone, and most trend subparallel with the zone (Fig. 16). The zone-parallel folds verge toward the appar-

FIG. 12.—Steep upthrust profile at restraining bend in trace of the Hidden Springs fault. Fault bounds a large block of Mesozoic(?) granite on the left. Exposed segment is approximately 6 m high. Pliocene-Pleistocene sandstones and gravels in footwall on right are dragged steeply upward in apparent response to a vertical, convergent component of displacement. View toward the northwest. See Figure 8 for location. T, displacement toward viewer; A, away from viewer.

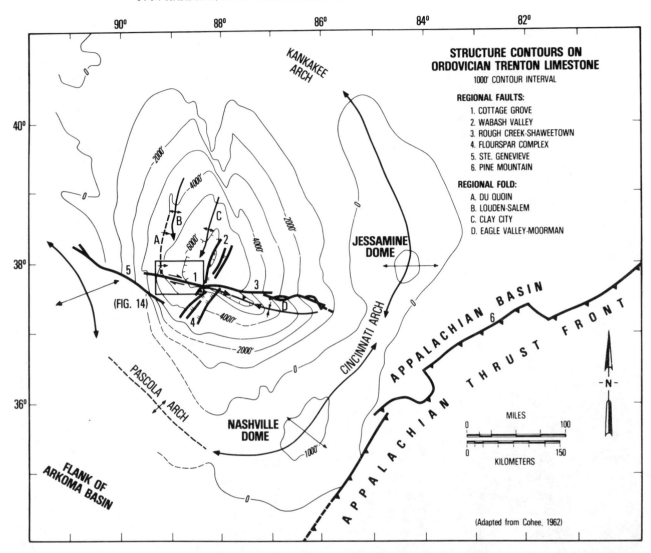

FIG. 13.—Intraplate tectonic setting of the Illinois Basin and Cottage Grove fault system (fault 1). Peripheral arches and flanks of foreland basins (Arkoma Basin, Appalachian Basin, and Black Warrior Basin south of Nashville Dome) separate the Illinois Basin from the convergent margin structures to the east (e.g., Appalachian Thrust Front) and south (south of Arkoma Basin).

ent down-dropped block (Figs. 15, 16). Nelson and Krausse (1981, p. 56) interpreted the structures as drape folds caused by vertical components of the deformation. Two small anticlines adjacent to the eastern portion of the principal displacement zone have a fundamentally different pattern. They lie oblique to the trend in a right-stepping en echelon pattern that is compatible with the right-slip deformation.

Characteristics of the External Structures

Left-stepping en echelon faults are present within a belt 5 to 16 km wide along both sides of the entire mapped trace of the principal displacement zone (Fig. 14). Individual faults are as long as 11 km and have a maximum vertical separation of 20 m (Nelson and Krausse, 1981). The orientation of these faults relative to the principal displacement zone varies from that predicted for en echelon normal faults to the more nearly transverse trend anticipated for antithetic

strike-slip faults (compare Figs. 2 and 14). Most dislocations are thought to be steeply dipping (60° to 90°) normal faults by Nelson and Krausse, and they consider faults on opposite sides of the principal displacement zone to have developed independently.

Nelson and Krausse (1981) also report that some of the external faults show evidence of oblique slip, such as oblique and horizontal slickenside striae, and changes in dip direction, separation sense (normal and reverse), and apparent upthrown side on individual dislocations. Strike slip may be dominant on a number of faults, and at several localities, reverse and normal separations occur on adjacent, parallel strands (Fig. 17). We suggest that, similar to the Lake Basin fault zone, external clockwise rotation of the blocks bounded by the en echelon faults occurred during wrench deformation and could have selectively rejuvenated some of the faults in a reverse- and strike-slip sense. In addition, other breaks may have originated as antithetic strike-slip faults.

Several anticlines are west of the principal displacement zone and trend oblique to its projection (Bremen anticline, Wine Hill dome, and Campbell Hill anticline in Fig. 16).

Their location and right-stepping en echelon map pattern suggest an extension of the zone of right-slip deformation. The folding may have been accentuated by shortening at a restraining bend formed where the system's inferred continuation trends more nearly due west.

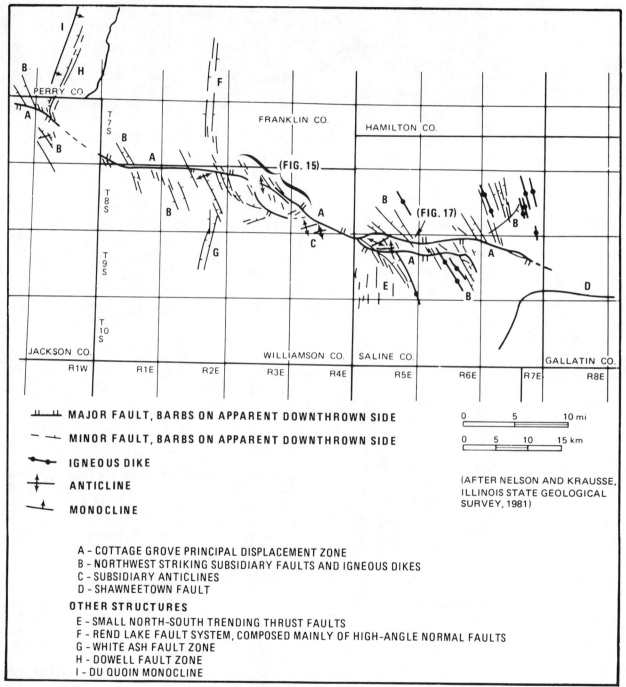

MAJOR FAULT, BARBS ON APPARENT DOWNTHROWN SIDE

MINOR FAULT, BARBS ON APPARENT DOWNTHROWN SIDE

IGNEOUS DIKE

ANTICLINE

MONOCLINE

(AFTER NELSON AND KRAUSSE, ILLINOIS STATE GEOLOGICAL SURVEY, 1981)

A - COTTAGE GROVE PRINCIPAL DISPLACEMENT ZONE
B - NORTHWEST STRIKING SUBSIDIARY FAULTS AND IGNEOUS DIKES
C - SUBSIDIARY ANTICLINES
D - SHAWNEETOWN FAULT

OTHER STRUCTURES

E - SMALL NORTH-SOUTH TRENDING THRUST FAULTS
F - REND LAKE FAULT SYSTEM, COMPOSED MAINLY OF HIGH-ANGLE NORMAL FAULTS
G - WHITE ASH FAULT ZONE
H - DOWELL FAULT ZONE
I - DU QUOIN MONOCLINE

Fig. 14.—Main structural elements of the Cottage Grove fault system (subsidiary anticlines are incompletely shown; see Fig. 16). Structures D through I are considered not part of the Cottage Grove fault system by Nelson and Krausse (1981), and several are known to be of a different age. See Figure 13 for regional setting.

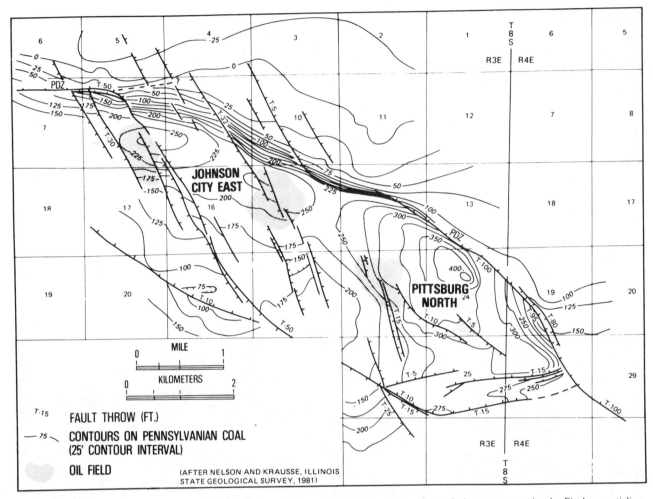

Fig. 15.—Detailed structure map of the principal displacement zone (PDZ) of the Cottage Grove fault system opposite the Pittsburg anticline (located in Figs. 14, 16). A gap in the PDZ occurs in Sections 8, 9, 10, 14, and 15 and is replaced by elements of an en echelon fault set. Other north-northwest-striking faults are external to the zone. The fold lies along the apparent high side of a projection of the PDZ and verges toward the relatively down-dropped, northeastern block.

ANDAMAN SEA FAULT

Plate-Tectonic Setting of the Andaman Sea Fault

The Andaman Sea fault illustrates the profile characteristics of divergent wrench faults over a much greater range of depth than the Mecca Hills and Cottage Grove examples and further demonstrates the manifestation of the style in seismic reflection data. The wrench fault lies within the Andaman Sea region, a marginal basin bounded on the west by the Andaman-Nicobar Ridge and subduction zone, and on the east by the magmatic arc terranes of the Malay Peninsula (Fig. 18; C.C.O.P., 1981). On the eastern margin of the sea, north-striking normal faults were active from the Oligocene to the early Miocene (Curray et al., 1979). In the central portion of the sea, magnetic anomalies in oceanic crust demonstrate that seafloor spreading has been occurring there since at least 13.5 Ma (Lawver and Curray, 1981). Geophysical data and the trend of the central rift valley indicate that segments of the spreading ridge trend east-northeast to north-northeast and are linked by right-slip transform faults that strike north-northwest (Fig. 18).

Evidence for Right Slip on the Andaman Sea Fault

The Andaman Sea fault trends north-northwest subparallel to right-slip faults of similar age elsewhere in the region (Fig. 19; Eguchi et al., 1979). The zone has a number of features characteristic of wrench faults: (1) a relatively straight, throughgoing trace; (2) changes in the apparent upthrown block that occur along strike (Fig. 19) and with depth (Fig. 20); (3) flanking folds that have a right-stepping en echelon pattern near the northern end of the fault; and (4) a possible restraining bend suggested by reverse-separation faults at the zone's southern end. Several of the en echelon folds appear to be offset 2.5 to 3 km right-laterally (folds A-A′ and B-B′ in Fig. 19).

Structural Style of the Andaman Sea Region

The structural assemblage of the Andaman Sea region repeats elements present in the above examples (compare Figs. 14, 16, and 19) and corroborates the observation that divergent strike slip has a unique and definable structural style. Deformation is characterized by a single major fault flanked by a set of discontinuous, oblique faults with normal separation (Fig. 19, and right side of Fig. 20).

The external fault set is asymmetric; dislocations are more numerous east of the principal displacement zone. Their profile characteristics are relatively simple and are repeated from one external fault to the next. The features suggestive of external rotation, cited for the Lake Basin and Cottage Grove zones, are absent. En echelon folds occur only near an apparent restraining bend formed where the wrench fault turns toward a more northwesterly course (north of profile 1 in Fig. 19). Elsewhere there are few if any structures

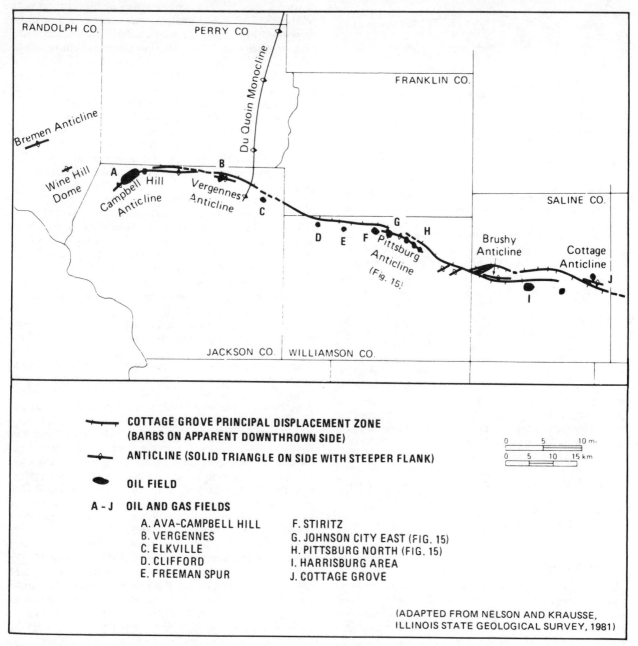

Fig. 16.—Map of the distribution of anticlines and oil and gas fields associated with the Cottage Grove fault system. Other oil and gas fields external to the system (most numerous to the north) are not shown. The inception of the Du Quoin monocline predates strike-slip deformation, and this structure is not considered to be part of the Cottage Grove system (Nelson and Krausse, 1981).

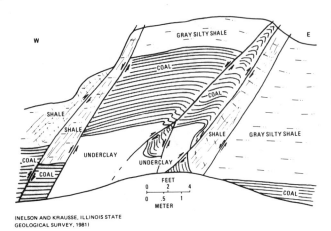

FIG. 17.—Detailed cross section of northwest-striking en echelon faults of the Cottage Grove fault system. Individual strands with reverse separation are tentatively interpreted as inverted normal faults that have undergone external rotation about a vertical axis. Other elements have retained at least part of their original normal separation. Beds within the fault zone are tightly folded. See Figure 14 for location.

indicative of regional shortening, and reflections from Neogene strata have consistent, low dips across the region (Figs. 20, 21).

The principal displacement zone is more complex and variable than the external normal faults (compare faults in Figs. 20 and 21). Its profile characteristics resemble those illustrated in Figures 3 and 5: (1) Faults splay upward and outward and most have normal separations; and (2) monoclinal flexures at the shoulders of the zone verge inward toward a central, down-dropped slice and bound a shallow synform (Fig. 21). Similar features characterize the seismic reflection profiles of another wrench fault in southeast Asia described by D'Onfro and Glagola (1983). A negative flower structure is present on profile 2 and includes a subsidiary fault with a reverse separation within Miocene strata on the zone's east side (Fig. 21). The reverse fault developed at a restraining bend within the zone's southern segment (Fig. 19). The shallow synform is not developed at profile 1, and a change in the dip separation at depth is the most distinctive expression of wrench faulting here (Fig. 20). The apparent upthrown block changes from the west side above 3.3 seconds to the east side below 3.5 seconds. The top of the basement in the western block is only approximated from regional control, but its displacement to deeper levels is corroborated by the presence of stratal reflections considerably below 3.5 seconds.

PLATE–TECTONIC SETTINGS FOR EXTENSIONAL STRIKE SLIP

Divergent wrench faults occur in a wide variety of plate-tectonic settings, including transform, divergent, and convergent plate boundaries; extensional and contractional continental settings; and within plates far from areas of pronounced regional deformation (Fig. 22). Evidence for extensional strike slip is least common, however, in regions dominated by crustal shortening, or where there is minimal

basement-involved deformation (e.g., passive continental margins). The zones and settings evolve through time (not shown in the instantaneous sketches of Fig. 22), and the faults may exhibit extensional strike-slip during only part of their history. Divergent segments may occur within systems that elsewhere have convergent styles (e.g., Fig. 22 a, b) or may dominate the entire length of the wrench zone (e.g., Fig. 22 d, e).

Transform Plate Boundary and Intra-Wrench Settings

Divergent wrench styles tend to develop within transform or wrench systems (1) where the principal displacement zone bends or splays into the extensional sector of regional deformation (Fig. 22a); (2) at releasing fault oversteps and fault junctions (Fig. 22b); and (3) locally where crustal blocks rotate between bounding wrench faults (Fig. 22c).

The geometrically simplest habitat of divergent wrench style is a releasing splay or bend in a single wrench fault. A releasing splay develops where splay faults, when viewed down the trace of a right-slip zone, diverge away from the viewer on the right side of the master fault or converge with the zone in a direction away from the viewer on the left side (Fig. 22a). The converse is true for left-slip zones. The Mecca Hills faults are examples of right-slip releasing splays (Fig. 6).

A similar pattern of orientation change defines a releasing bend. The western segment on the Bocono fault adjacent to the La González Basin, northwestern Venezuela, is an example (Fig. 23a). Quaternary normal separation is on the order of several kilometers, and right slip is indicated by displaced drainages and stratigraphy, offset moraines, and seismicity (see references in Schubert, 1980). Other examples of releasing bends have been summarized recently by Mann et al. (1983).

The entire wrench system may have a divergent wrench style where a straight fault is consistently oblique to the regional interplate slip lines. The principal displacement zones of the Dead Sea transform south of Lebanon, for example, are predominantly of divergent wrench style judging from descriptions of surface features by Freund et al. (1970), Schulman and Bartov (1978), Garfunkel (1981), Eyal and Reches (1983), and Manspeizer (1985 this volume). Restoration of the 105 km of left slip documented along this zone includes a clockwise rotation of the eastern block (Figs. 4b to 9b in Freund et al., 1970), suggesting an oblique divergent plate motion of several degrees (see also Garfunkel, 1981). Plate-tectonic analyses of the Red Sea – Gulf of Aden area also imply extensional strike-slip along the Dead Sea transform (e.g., Le Pichon and Francheteau, 1978; Cochran, 1983; Mann et al., 1983). The inception of the folds in Israel that are oriented oblique to the Dead Sea transform, and interpreted by Freund (1965), Vroman (1967), and Wilcox et al. (1973) to be the result of wrench-related deformation, significantly predates the initiation of strike slip (Eyal and Reches, 1983). In addition, Eyal and Reches (1983) have determined that the stress field related to the Dead Sea transform is different from the stress field involved in the development of the folds. The folds are therefore not considered an essential element of the divergent wrench style south of Lebanon.

Releasing oversteps and releasing fault junctions (Fig. 22b) also have patterns that result in the oblique separation of blocks, and they are similar to those defined for releasing splays and bends. Indeed, Mann et al. (1983) argued that oversteps generally evolve from bends in an earlier throughgoing fault, although other origins are possible (Aydin and Nur, 1985 this volume). A possible example of a releasing overstep characterized by divergent wrench style is the Hula graben of the northern Dead Sea rift (Fig. 23c;

Freund et al., 1968). Numerous other examples of pull-apart basins are cited by Aydin and Nur (1982) and by Mann et al. (1983), but not all such basins are associated with divergent wrench faults. The southern San Andreas fault opposite the Salton trough, for example, is dominated by a convergent wrench fault style (Sylvester and Smith, 1976).

Extension occurs at releasing fault junctions in the manner discussed by Kingma (1958), Lensen (1959), and Crowell (1974). At the junction that bounds the Ridge Basin (Fig.

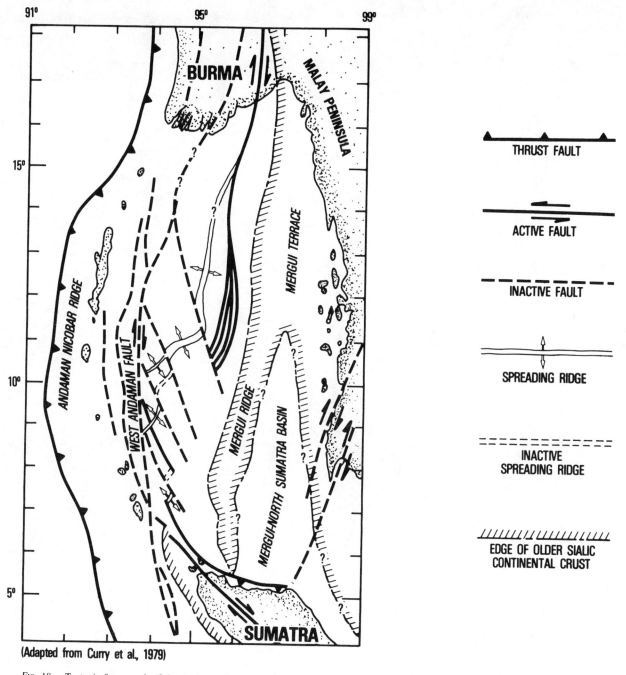

(Adapted from Curry et al., 1979)

FIG. 18.—Tectonic framework of the Andaman Sea. Arrowheads have been added to identify spreading centers and are diagrammatic.

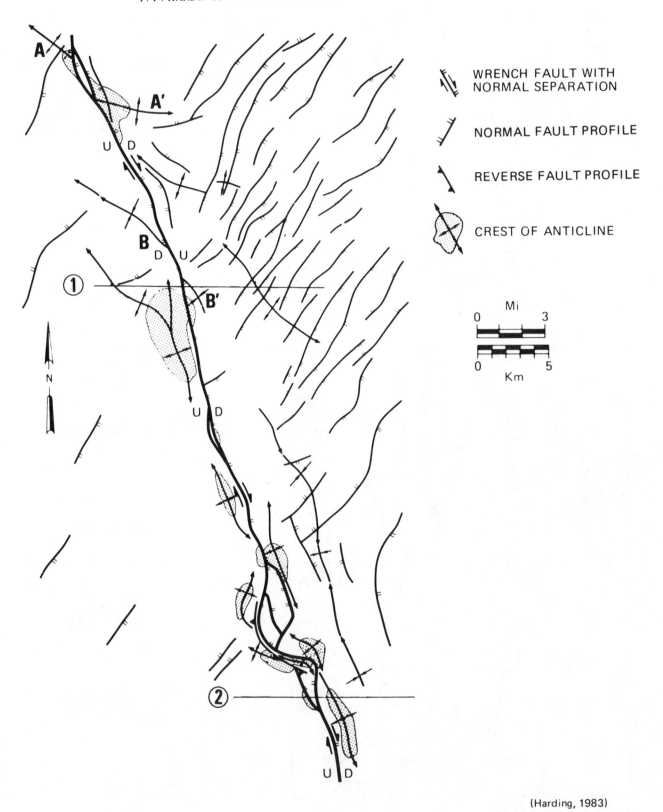

WRENCH FAULT WITH
NORMAL SEPARATION

NORMAL FAULT PROFILE

REVERSE FAULT PROFILE

CREST OF ANTICLINE

(Harding, 1983)

FIG. 19.—Tectonic map of structures mapped on a lower Miocene reflection and position of the Andaman Sea seismic profiles. A–A′ and B–B′ identify possibly correlative folds that appear to be offset in a right-lateral sense.

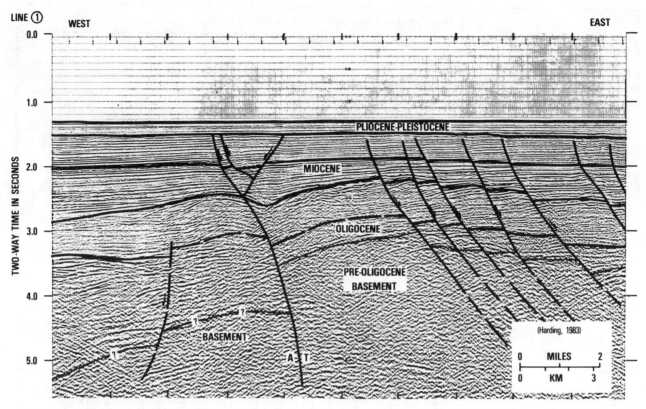

FIG. 20.—Interpreted seismic profile 1 across the Andaman Sea wrench fault. Deeper stratal reflections on west side of wrench fault (left center) indicate a change in the apparently down-dropped block at depth. Obliquely striking normal faults dip toward the right on eastern half of profile. See Figure 19 for structural setting of line. T, displacement toward viewer; A, displacement away from viewer.

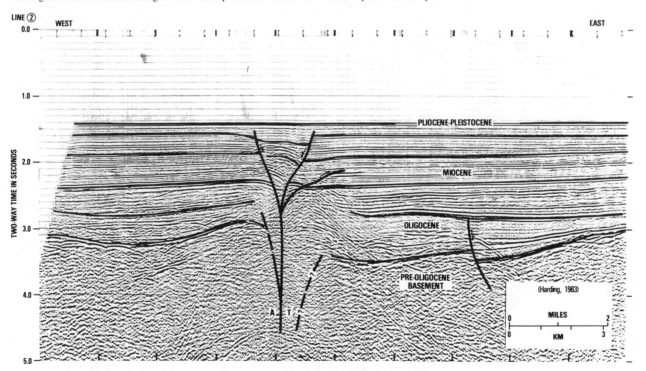

FIG. 21.—Interpreted seismic profile 2 across the Andaman Sea wrench fault. This profile illustrates the characteristics of a negative flower structure and changes in the development of the fault zone that occur along strike from profile 1 (compare with Fig. 20). See Figure 19 for structural setting of line. T, displacement toward viewer; A, displacement away from viewer.

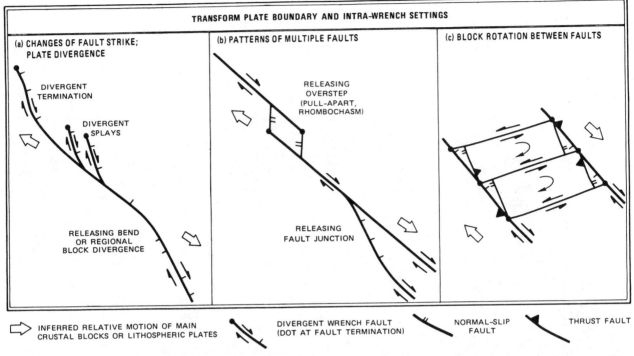

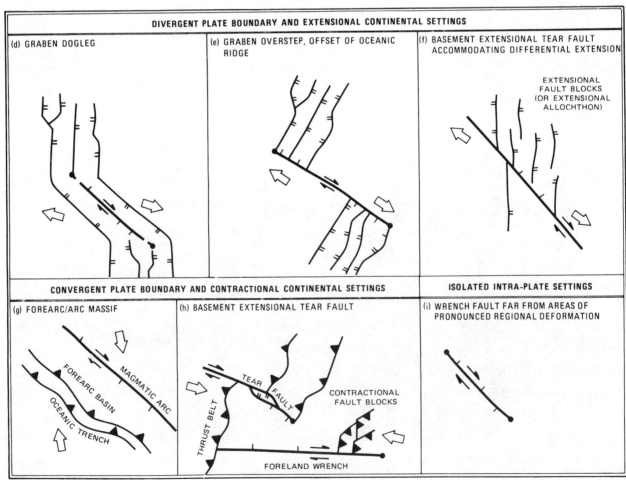

23b), the San Gabriel fault was active from about 12 Ma to 5 Ma and has predominantly normal separation (Crowell, 1982). However, the adjacent Liebre fault, active from 8 Ma to between 6 and 5 Ma and approximately parallel to the San Gabriel, is characterized by reverse separation. This difference in style illustrates the fallibility of using patterns predicted from Figures 1 and 2 in areas where two or more deformational events are superimposed or where more than one major strike-slip fault is involved. In addition, the fault patterns shown in Figure 23 do not always result in basin subsidence.

A possible scheme for the rotation of crustal blocks between wrench faults is shown in Figure 22c (scale is uncertain). Abundant paleomagnetic evidence demonstrates widespread, predominantly clockwise rotation of crustal blocks in western North America during Cenozoic time (Beck, 1980; Luyendyk et al., 1980). Any significant block rotations similar to those shown in Figure 22c should lead to the initiation of sedimentary basins and to alternations of convergent and divergent wrench style where block corners alternately rotate toward or away from the through-going fault (indicated with normal- and thrust-fault symbols). These structural complexities have been corroborated by earthquake hypocentral location and first-motion studies in the eastern Transverse Ranges, California (Nicholson et al., 1985). The structural patterns that result from such rotational deformation are not included in Figures 1 and 2.

Divergent Plate Boundary and Extensional Continental Settings

Well-known examples of divergent wrench faults within divergent plate boundaries and extensional continental settings are oriented oblique or transverse to the trend of normal fault systems. The most common are apparently localized in the vicinity of graben doglegs and oversteps (Fig. 22d; Harding, 1984). Doglegs are short, oblique graben segments connecting subparallel straightaways. Extensional strike slip can occur within the middle segment of the dogleg, particularly if the regional plate separation is oblique to that segment. Several transform faults in the northern Red Sea (see Fig. 1 in Coleman, 1974) may be interpreted in this manner. Freund (1982) has also observed that obliquely oriented strike-slip faults, not necessarily confined to doglegs, are an occasional part of graben structures. He has attributed the formation of these faults to rare instances of crustal attenuation parallel to the graben axis.

Graben oversteps occur where the ends of graben segments overlap and are not connected directly by normal faults (Fig. 22e). Continued extension ultimately causes strike slip along a line connecting the grabens (Courtillot et al., 1974). Zones of en echelon faults connect the overstepped ends of the Rhine and Bresse grabens and have been interpreted by Illies (see Fig. 10 in Illies, 1974) as left-lateral wrench zones or oblique-slip faults with left-slip components. The zones resemble the belts of external faults associated with diver-

gent wrench faults, and he has compared the deformation to an incipient ridge-ridge transform fault system.

A somewhat different class of divergent wrench fault occurs in broad areas of pervasive regional extension, such as the Basin and Range Province of southeastern California and adjacent Nevada (Figs. 22f, 23d). There, wrench faults act as major basement tear faults, separating terranes that in late Cenozoic time experienced different amounts of extension (Wernicke et al., 1982; Burchfiel et al., 1983). These faults, however, do not necessarily have a divergent style throughout their lengths. Specific examples are the Garlock fault (Davis and Burchfiel, 1973), the Las Vegas Valley shear zone (Guth, 1981), and the Furnace Creek fault zone (Wright and Troxel, 1970; Stewart, 1983; Cemen et al., 1985 this volume). The Andaman Sea fault (Fig. 19) may also be an extensional tear fault. In both the Basin and Range and Andaman Sea examples, the associated normal faults are oblique rather than en echelon because they are of regional extent, and not arranged solely along wrench zones.

Divergent wrench faults are also known from oceanic regimes. The western rift zone of southwestern Iceland, for example, is connected to the Reykjanes segment of the North Atlantic mid-oceanic spreading center by an obliquely oriented zone of right-stepping, en echelon fissure swarms (Tryggvason, 1982). The swarms are geometrically similar to the en echelon normal faults between the Rhine and Bresse grabens. Elsewhere, many oceanic transform zones consist of valleys and ridges within and parallel to the principal displacement zone and of external, oblique slip faults (Fox and Gallo, 1984). The structural style of such fracture zones is similar to that of some continental divergent wrench faults, but the style does not develop in the same way. The transform valley is thought to be a manifestation of restricted partial melting at the relatively cold ridge-transform intersection. This results in reduced segregation of basaltic melt and thinner oceanic crust. Oblique faulted topography adjacent to the transform arises from reorientation of the principal stress axes as a result of welding of the upper mantle across the transform (Fox and Gallo, 1984).

Convergent Plate Boundary and Contractional Continental Settings

We have recognized divergent wrench faults in three convergent margin and contractional continental settings thus far: arc massif, and both backarc and peripheral forelands (cf. Dickinson, 1974); additional settings are probable.

Regions of high heat flow and thin, weakened lithosphere, such as arc massifs, appear to have localized longitudinal wrench faults within convergent margin settings (Fig. 22g; Fitch, 1972). Along the axis of the central Barisan Mountains in Sumatra, the right-lateral Great Sumatran fault consists of a braided zone that includes in-line grabens, pull-apart basins, and linear topographic depressions. The structural depressions are bounded by major right-lateral faults and subordinate, but still important, normal faults

---

F<small>IG</small>. 22.—Cartoon of plate-tectonic settings and patterns of divergent strike-slip faulting. Examples are based on known occurrences (except style of fault in d, which is assumed), and are consistently drawn for a right-slip system to facilitate comparison. There is undoubtedly a greater variety of possible fault patterns and complexities than is shown. Scale is variable, and the patterns in one sketch may constitute local details of another.

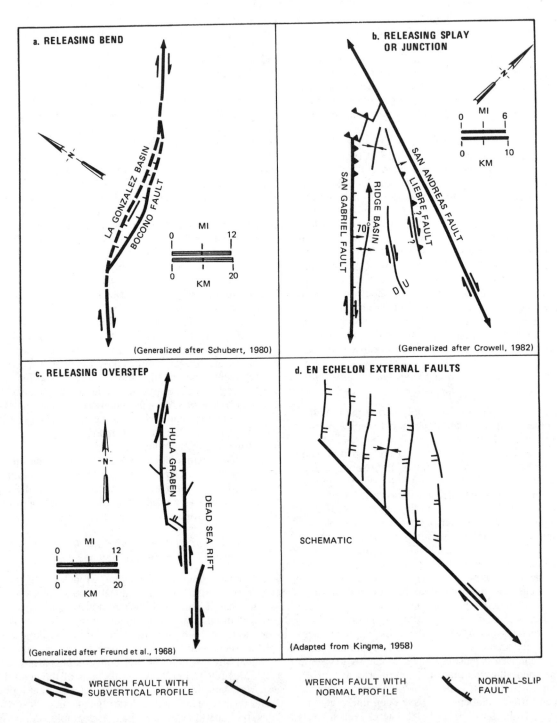

Fig. 23.—Four common patterns of strike-slip faults and secondary faults associated with strike slip that sometimes result in basin subsidence. Fault profile styles in (a) are from Giegengack (1977) and Schubert (1980). Fault profiles in (b) are generalized from descriptions in Crowell (1982); the reverse segment at northwest end of San Gabriel fault was superimposed on an earlier extensional profile. Fault profiles in (c) are from Horowitz (1973) and Schulman and Bartov (1978).

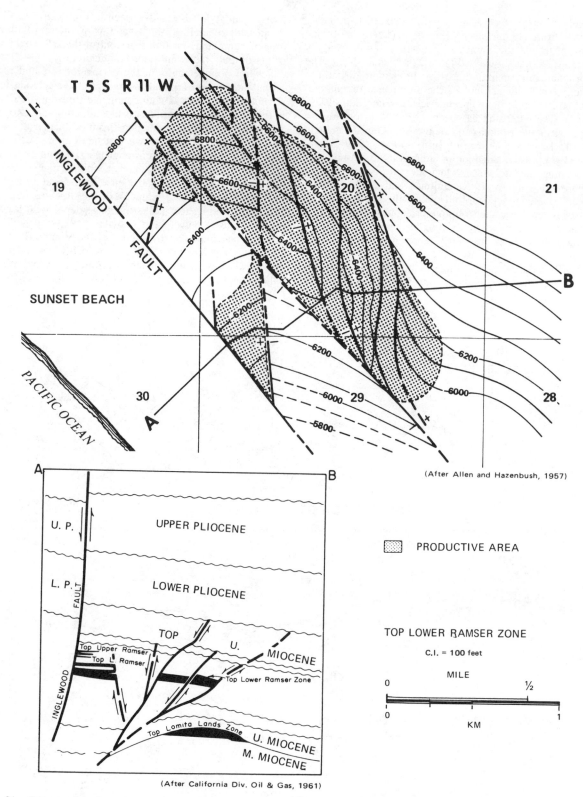

(After Allen and Hazenbush, 1957)

▓ PRODUCTIVE AREA

TOP LOWER RAMSER ZONE

C.I. = 100 feet

(After California Div. Oil & Gas, 1961)

FIG. 24.—Structure of the Sunset Beach oil field, Los Angeles Basin, California. Shallow, north-south, en echelon fault blocks are closed southward by throughgoing, northwest-striking faults, and westward by normal faults. The Inglewood fault is part of the regional Newport-Inglewood wrench zone, which has a maximum right slip of 760 m on structural markers near the top of the Miocene (Harding, 1973).

(see Fig. 9 in Posavec et al., 1973). Strike slip probably resulted from oblique encroachment between the Indian Ocean plate and the Sumatran plate in a setting roughly analogous to that shown in Figure 22g (Fitch, 1972). The extensional component of the strike slip is more difficult to explain, especially because of the dominance of contractional structures in the adjacent backarc (Hamilton, 1979, Plate I). The extension has been attributed to oceanward migration of the frontal parts of the Sumatran arc system by D. E. Karig and W. Jensky (in Hamilton, 1979). Dickinson (1974) related extension in arc massifs to crustal uplift and arching, and to volcano-tectonic subsidence. Recent unpublished work by T. R. Bultman and T. P. Harding indicates that contractional folds within the Sumatran backarc basins may owe their oblique orientation to the Great Sumatran fault more to pre-existing basement grain than to right-slip tectonics as previously reported by Wilcox et al. (1973).

Wrench faults oriented oblique to the trend of major tectonic zones within the region of convergence may also have a divergent style (foreland wrench in Fig. 22h). The Lake Basin fault zone (Figs. 4, 5) has this orientation within an orogenic foreland. Factors that control the strike-slip style of the oblique faults are poorly understood.

Several oblique faults in the peripheral foreland region of the outer West Carpathians appear to have acted as tears, separating eastern areas with continuing thrusting from areas in the west where thrusting was already complete (Royden, 1985 this volume). The Vienna Basin is located at a complex releasing overstep between at least three such faults (see pull-apart within extensional tear fault in Fig. 22h), each consisting of several branches with normal separation. Reflection seismic data, subsidence history, and heat-flow data suggest that the Vienna Basin is allochthonous, and that the bounding tear faults merge at depth with a gently dipping crustal detachment (Royden, 1985 this volume). Oblique tear faults, such as those of the Vienna Basin, are commonly late structures, and are invariably associated with belt-parallel extension, which is promoted by pronounced salients and re-entrants in the orogen (Dahlstrom, 1970).

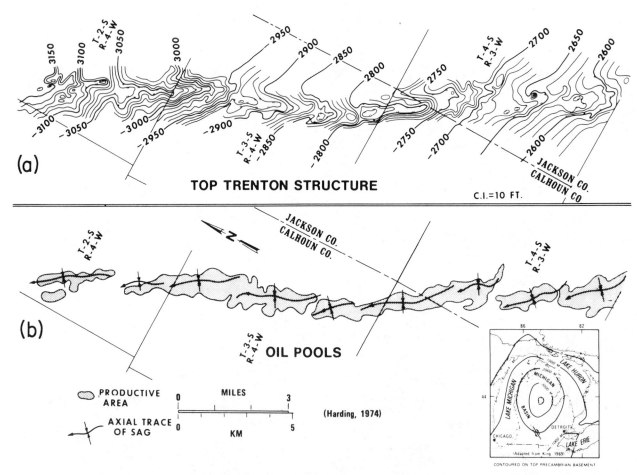

FIG. 25.—Structure map and setting (inset) of the Scipio-Albion trend, Michigan Basin, Michigan. (a) Map of central and northern parts of the field contoured on the top of the Ordovician Trenton Formation, which directly overlies the productive interval. (b) Pattern of sag axes determined from the structure contours.

Intraplate Occurrences

The Cottage Grove fault of the Illinois Basin illustrates the intracratonic setting of divergent wrench faults (Figs. 13, 22i), and the Scipio-Albion trend of the Michigan Basin provides another example (see the following discussion). Pre-existing zones of crustal weakness are recognized as a potentially important determinant of location and orientation. Convergent wrench faults are also known from mid-plate localities, but the cause of the mid-plate strike slip and controls on its style are unknown.

HYDROCARBON TRAPS ASSOCIATED WITH DIVERGENT WRENCH FAULTS

Three main types of structural traps associated with divergent wrench faults are known to be productive (see also D'Onfro and Glagola, 1983): (1) fault slices within the principal displacement zone, (2) forced folds parallel to the high side of the principal displacement zone, and (3) en echelon fault blocks flanking either or both sides of the zone. The forced folds have the greatest potential for providing both effective and abundant trap opportunities (e.g., Figs. 16, 19). Most hydrocarbon accumulations along the Cottage Grove fault system are located near anticlinal culminations along the apparent high side of the principal displacement zone, but they are productive in very low volumes in this example. Subsidiary traps commonly associated with anticlines, such as up-plunge reservoir terminations (pinch-out, truncation, etc.) and cross faults, increase the potential for hydrocarbon prospects at the forced folds (e.g., Fig. 15). External, oblique anticlines, such as the Campbell Hill anticline in Figure 16, may also provide additional fold closures, but are less common.

En echelon normal-fault blocks are productive in the Newport-Inglewood trend, California, and in several other wrench zones (Fig. 24; Harding, 1973, 1974; Barrows, 1974). The fault blocks are developed where the principal displacement zone has the characteristics of a divergent wrench fault. Effective closure is dependent on block tilt away from fault intersections and on fault seals, usually at both the oblique normal faults and the principal displacement zone (Weber et al., 1978; Smith, 1980).

A vertical linear zone of fractured, dolomitic limestone forms the reservoir for the hydrocarbon accumulation along the Scipio-Albion trend, Michigan. A sag overlies the zone and is thought to reflect the distribution of fractures (Fig. 25). Both the sag and the productive limits of the fractured reservoir have a segmented right-stepping, en echelon pattern that resembles the orientation of synthetic faults at the principal displacement zone of a left-slip system. The series of sags trends obliquely down the regional dip (inset map in Fig. 25), and there is little other discernible deformation. Divergent wrench faulting is interpreted to have formed the unique sag and its en echelon pattern while at the same time enhancing fracturing and limiting deformation to a narrow swath (Harding, 1974). Graben slices within the principal displacement zone of the Cottage Grove fault system are also intensely fractured (Nelson and Krausse, 1981).

Although potential traps occur along divergent wrench faults, it is thought that the limited extent of external structures, the paucity of folds, and the dependence on sealing faults for closure are all significant limiting factors. Trends developed where blocks move in a convergent sense (Harding, 1974) or with neither convergence nor divergence (Harding, 1973, 1976) should have larger numbers of effective structural traps.

CONCLUSIONS

Our investigations indicate that some wrench zones contain a structural assemblage that is dominated by extensional features and that distinctive elements of this assemblage are repeated in a wide variety of tectonic settings. In the past, zones such as these may not have been properly recognized as wrench faults. The structural assemblage differs in important ways from the other wrench styles and constitutes a discrete structural style: (1) faulting dominates deformation adjacent to the principal displacement zone; (2) folds are for the most part oriented parallel to the principal displacement zone; and (3) the overall zone of deformation is typically relatively narrow. The structures may be combined in several ways. Some divergent wrench systems consist solely of a principal displacement zone with normal separation; other divergent wrench zones are composed of sets of en echelon normal faults; and still other systems have both elements. One result of their structural characteristics is that the potential for effective structural traps is not as great along divergent wrench faults as it is in regions of convergent or simple strike slip. The extensional characteristics of the style also make it difficult to differentiate divergent wrench faults from normal fault blocks, especially when interpreting seismic reflection data. In these instances it is necessary to determine both profile and map characteristics.

ACKNOWLEDGMENTS

This study was originally conducted for Exxon Production Research Company and we are grateful to that company for its permission to publish. Exxon Company, U.S.A., provided the Bering Sea and Lake Basin seismic reflection profiles and the maps of the Scipio-Albion trend. Esso Exploration, Inc., supplied the Andaman Sea example. K. T. Biddle, C. A. Dengo, R. P. George, D. W. Phelps, and A. C. Tuminas, of Exxon Production Research Company, reviewed the manuscript, and R. W. Wiener provided regional data for the Illinois Basin. G. C. Bond, Kristian Meisling, K. R. Schmitt, and D. M. Worral also made helpful suggestions. Logistical support at Lamont-Doherty Geological Observatory was provided by an ARCO Foundation Fellowship to Christie-Blick. Lamont-Doherty Geological Observatory Contribution No. 3912 (Christie-Blick).

REFERENCES

ALLEN, D. R., AND HAZENBUSH, G. C., 1957, Sunset oil field: California Division of Oil and Gas, California Oil Fields—Summary of Operations, v. 43, no. 2, p. 47–50.

ALPHA, A. G., AND FANSHAWE, J. R., 1954, Tectonics of northern Bighorn basin area and adjacent south-central Montana: Billings Geological Society Guidebook, 5th Annual Field Conference, Pryor Mountains-Northern Bighorn Basin, Montana, p. 72–79.

AYDIN, A., AND NUR, A., 1982, Evolution of pull-apart basins and their scale independence: Tectonics, v. 1, p. 91–105.

AYDIN, A., AND NUR, A., 1985, The types and role of stepovers in strike-slip tectonics, *in* Biddle, K. T., and Christie-Blick, N., eds., Strike-slip Deformation, Basin Formation and Sedimentation: Society of Economic Paleontologists and Mineralogists Special Publication No. 37, p. 35–44.

BARROWS, A. G., 1974, A review of the geology and earthquake history of the Newport-Inglewood structural zone, southern California: California Division Mines and Geology Special Report 114, 115 p.

BECK, M. E., JR., 1980, Paleomagnetic record of plate-margin tectonic processes along the western edge of North America: Journal of Geophysical Research, v. 85, p. 7115–7131.

BISHOP, D. G., 1968, The geometric relationships of structural features associated with major strike-slip faults in New Zealand; New Zealand Journal of Geology and Geophysics, v. 11, p. 405–417.

BURCHFIEL, B. C., WALKER, D., DAVIS, G. A., AND WERNICKE, B., 1983, Kingston Range and related detachment faults—a major "breakaway" zone in the southern Great Basin: Geological Society of America Abstracts with Programs, v. 15, p. 536.

BUSCHBACH, T. C., AND ATHERTON, E., 1979, History of the structural uplift of the southern margin of the Illinois basin, *in* Palmer, J. E., and Dutcher, R. R., eds., Depositional and Structural History of the Pennsylvanian System of the Illinois Basin, Part 2: Illinois State Geological Survey Guidebook Series 15a, Field trip 9, Ninth International Congress, Carboniferous Stratigraphy and Geology, p. 112–115.

CALIFORNIA DIVISION OF OIL AND GAS, 1961, California oil and gas fields maps and data sheets, part 2, Los Angeles–Ventura basins and central coastal regions: California Division of Oil and Gas, p. 496–913.

CEMEN, I., WRIGHT, L. A., DRAKE, R. E., AND JOHNSON, F. C., 1985, Cenozoic sedimentation and sequence of deformational events at the southeastern end of the Furnace Creek strike-slip fault zone, Death Valley region, California, *in* Biddle, K. T., and Christie-Blick, N., eds., Strike-Slip Deformation, Basin Formation, and Sedimentation: Society of Economic Paleontologists and Mineralogists Special Publication No. 37, p. 127–141.

CHAMBERLAIN, R. T., 1919, A peculiar belt of oblique faulting: Journal of Geology, v. 27, p. 602–613.

CHEADLE, M. J., CZUCHRA, B. L., BYRNE, T., ANDO, C. J., OLIVER, J. E., BROWN, L. O., KAUFMAN, S., MALM, P. E., AND PHINNEY, R. A., in press, The deep crustal structure of the Mojave Desert, California, from COCORP seismic reflection data: Tectonics.

CHRISTIE-BLICK, N., AND BIDDLE, K. T., 1985, Deformation and basin formation along strike-slip faults, *in* Biddle, K. T., and Christie-Blick, N., eds., Strike-Slip Deformation, Basin Formation, and Sedimentation: Society of Economic Paleontologists and Mineralogists Special Publication No. 37, p. 1–34.

CLARK, S. K., AND ROYDS, J. S., 1948, Structural trends and fault systems in Eastern Interior basin: American Association of Petroleum Geologists Bulletin, v. 32, p. 1728–1749.

COCHRAN, J. R., 1983, A model for development of Red Sea: American Association of Petroleum Geologists Bulletin, v. 67, p. 41–69.

COHEE, G. V., 1962, Tectonic map of the United States; United States Geological Survey and American Association of Petroleum Geologists, 1:2,500,000.

COLEMAN, R. G., 1974, Geologic background of the Red Sea, *in* Burk, C. A., and Drake, C. L., eds., The Geology of Continental Margins: New York, Springer-Verlag, p. 743–751.

COMMITTEE FOR COORDINATION OF JOINT PROSPECTING FOR MINERAL RESOURCES IN ASIAN OFFSHORE AREAS (C.C.O.P.), 1981, Studies in east Asian tectonics and resources (SEATAR): United Nations ESCAP, C.C.O.P. Technical Publication 7a, p. 37–50.

COURTILLOT, V., TAPPONNIER, P., AND VARET, J., 1974, Surface features associated with transform faults: a comparison between observed examples and an experimental model: Tectonophysics, v. 24, p. 317–329.

CROWELL, J. C., 1974, Origin of Late Cenozoic basins in southern California, *in* Dickinson, W. R., ed., Tectonics and Sedimentation: Society of Economic Paleontologists and Mineralogists Special Publication No. 22, p. 190–204.

———, 1981, Juncture of San Andreas transform system and Gulf of California rift: Oceanologica Acta, Proceedings, 26th International Geological Congress, Geology of Continental Margins Symposium, Paris, p. 137–141.

———, 1982, The tectonics of Ridge Basin, southern California, *in* Crowell, J. C., and Link, M. H., eds., Geologic History of Ridge Basin, Southern California: Society of Economic Paleontologists and Mineralogists, Pacific Section, p. 25–42.

CROWELL, J. C., AND SYLVESTER, A. G., eds., 1979, Tectonics of the Juncture Between the San Andreas Fault System and the Salton Trough, Southeastern California—A Guidebook: Santa Barbara, California, Department of Geological Sciences, University of California, 193 p.

CURRAY, J. R., MOORE, D. G., LAWVER, L. A., EMMEL, F. J., RAITT, R. W., HENRY, M., AND KIECKHEFEIZ, R., 1979, Tectonics of the Andaman Sea and Burma, *in* Watkins, J. S., Montadert, L., and Dickenson, P. W., eds., Geological and Geophysical Investigations of Continental Margins: American Association of Petroleum Geologists Memoir 29, p. 189–198.

DAHLSTROM, C. D. A., 1970, Structural geology in the eastern margin of the Canadian Rocky Mountains: Bulletin of Canadian Petroleum Geology, v. 18, p. 332–406.

DAVIS, G. A., AND BURCHFIEL, B. C., 1973, Garlock fault: an intracontinental transform structure, southern California: Geological Society of America Bulletin, v. 84, p. 1407–1422.

DICKINSON, W. R., 1974, Plate tectonics and sedimentation, *in* Dickinson, W. R., ed., Tectonics and Sedimentation: Society of Economic Paleontologists and Mineralogists Special Publication No. 22, p. 1–27.

DOBBIN, C. E., AND ERDMAN, C. E., 1955, Structure contour map of the Montana plains: United States Geological Survey Oil and Gas Investigation Map OM 178A, Scale 1:500,000.

D'ONFRO, P., AND GLAGOLA, P., 1983, Wrench fault, southeast Asia, *in* Bally, A. W., ed., Seismic Expression of Structural Styles, v. 3: American Association of Petroleum Geologists Studies in Geology, Series 15, p. 4.2–9 to 4.2–12.

EGUCHI, T., UYEDA, S., AND MAKI, T., 1979, Seismotectonics and tectonic history of the Andaman sea: Tectonophysics, v. 57, p. 35–51.

EMMONS, R. C., 1969, Strike-slip rupture patterns in sand models: Tectonophysics, v. 7, No. 1, p. 71–87.

ERVIN, C. P., AND McGINNIS, L. D., 1975, Reelfoot rift: reactivated precursor to the Mississippi embayment: Geological Society America Bulletin, v. 86, p. 1287–1295.

EYAL, Y., AND RECHES, ZE'EV, 1983, Tectonic analysis of the Dead Sea rift region since the Late-Cretaceous based on mesostructures: Tectonics, v. 2, p. 167–185.

FITCH, T. J., 1972, Plate convergence, transcurrent faults, and internal deformation adjacent to southeast Asia and the western Pacific: Journal of Geophysical Research, v. 77, p. 4432–4460.

FOX, P. J., AND GALLO, D. G., 1984, A tectonic model for ridge-transform-ridge plate boundaries: Implications for the structure of oceanic lithosphere: Tectonophysics, v. 104, p. 205–242.

FREUND, R., 1965, A model of the structural development of Israel and adjacent areas since Upper Cretaceous times: Geological Magazine, v. 102, p. 189–205.

———, 1970, Rotation of strike slip faults in Sistan, southeast Iran: Journal of Geology, v. 78, p. 188–200.

———, 1982, The role of shear in rifting, *in* Pálmason, G., ed., Continental and Oceanic Rifts: American Geophysical Union Geodynamics Series, v. 8, p. 33–39.

FREUND, R., GARFUNKEL, Z., ZAK, I., GOLDBERG, M., WEISSBROD, T., AND DERIN, B., 1970, The shear along the Dead Sea rift: Philosophical Transactions of Royal Society of London, Series A, v. 267, p. 107–130.

GARFUNKEL, Z., 1981, Internal structure of the Dead Sea leaky transform (rift) in relation to plate kinematics: Tectonophysics, v. 80, p. 81–108.

GIEGENGACK, R., 1977, Late Cenozoic tectonics of the Tabay-Estangues graben, Venezuelan Andes, *in* Espejo, A., ed., Memorias del V Congreso Geologico Venezolano: Ministerio de Energia y Minos, Caracas, Tomo II, p. 721–737.

GUTH, P. L., 1981, Tertiary extension north of the Las Vegas Valley shear zone, Sheep and Desert Ranges, Clark County, Nevada: Geological Society of America Bulletin, Part I, v. 92, p. 763–771.

HAMILTON, W., 1979, Tectonics of the Indonesian region: United States Geological Survey Professional Paper 1078, 345 p.

HARDING, T. P., 1973, Newport-Inglewood trend, California–an example of wrenching style of deformation: American Association of Petroleum Geologists Bulletin, v. 57, p. 97–116.

———, 1974, Petroleum traps associated with wrench faults: American

Association of Petroleum Geologists Bulletin, v. 58, p. 1290–1304.

———, 1976, Tectonic significance and hydrocarbon trapping consequences of sequential folding synchronous with San Andreas faulting, San Joaquin Valley, California: American Association of Petroleum Geologists Bulletin, v. 60, p. 356–378.

———, 1983, Divergent wrench fault and negative flower structure, Andaman Sea, *in* Bally, A. W., ed., Seismic Expression of Structural Styles, v. 3: American Association of Petroleum Geologists Studies in Geology, Series 15, p. 4.2–1 to 4.2–8.

———, 1984, Graben hydrocarbon occurrences and structural style: American Association of Petroleum Geologists Bulletin, v. 68, p. 333 to 362.

———, 1985, Seismic characteristics and identification of negative flower structures, positive flower structures and positive structural inversion: American Association of Petroleum Geologists Bulletin, v. 69, p. 582–600.

HARDING, T. P., AND LOWELL, J. D., 1979, Structural styles, their plate-tectonic habitats and hydrocarbon traps in petroleum provinces: American Association of Petroleum Geologists Bulletin, v. 63, p. 1016–1058.

HARLAND, W. B., 1971, Tectonic transpression in Caledonian Spitsbergen: Geological Magazine, v. 108, p. 27–42.

HAYS, W. H., 1957, Geology of the central Mecca Hills, Riverside County, California [unpubl. Ph.D. Thesis]: New Haven, Yale University, 324 p.

HEYL, A. V., JR., AND BROCK, M. R., 1961, Structural framework of the Illinois-Kentucky mining district and its relation to mineral deposits: United States Geological Survey Professional Paper 424-D, p. D3–D6.

HOROWITZ, A., 1973, Development of the Hula basin, Israel: Israel Journal Earth Sciences, v. 22, p. 107–139.

ILLIES, J. H., 1974, Taphrogenesis and plate tectonics, *in* Illies, J. H., and Fuchs, K., eds., Approaches to Taphrogenesis: Stuttgart, E. Schweizerbart'sche Verlagsbuchhandlung, p. 433–460.

JENNINGS, C., 1967, Salton Sea sheet, Geologic map of California, Olaf P. Jenkins edition: California Division of Mines and Geology, Scale 1:250,000.

KANE, M. F., HILDENBRAND, T. G., AND HENDRICKS, J. D., 1981, Model for the tectonic evolution of the Mississippi embayment and its contemporary seismicity: Geology, v. 9, p. 563–568.

KING, P. B., 1969, Tectonic map of North America: United States Geological Survey, Scale 1:5,000,000.

KINGMA, J. T., 1958, Possible origin of piercement structures, local unconformities, and secondary basins in the Eastern Geosyncline, New Zealand: New Zealand Journal of Geology and Geophysics, v. 1, p. 269–274.

LAWVER, L. A., AND CURRAY, J. R., 1981, Evolution of the Andaman Sea (abs.): EOS, Transactions of American Geophysical Union, v. 62, No. 45, p. 1044.

LENSEN, G. J., 1959, Secondary faulting and transcurrent splay-faulting at transcurrent fault intersections: New Zealand Journal of Geology and Geophysics, v. 2, p. 729–734.

LE PICHON, X., AND FRANCHETEAU, J., 1978, A plate-tectonic analysis of the Red Sea–Gulf of Aden area: Tectonophysics, v. 46, p. 369–406.

LUYENDYK, B. P., KAMERLING, M. J., AND TERRES, R., 1980, Geometric model for Neogene crustal rotations in southern California: Geological Society of America Bulletin, Part I, v. 91, p. 211–217.

MANN, P., HEMPTON, M. R., BRADLEY, D. C., AND BURKE, K., 1983, Development of pull-apart basins: Journal of Geology, v. 91, p. 529–554.

MANSPEIZER, W., 1985, The Dead Sea Rift: Impact of climate and tectonism on Pleistocene and Holocene sedimentation, *in* Biddle, K. T., and Christie-Blick, N., eds., Strike-Slip Deformation, Basin Formation, and Sedimentation: Society of Economic Paleontologists and Mineralogists Special Publication No. 37, p. 143–158.

NELSON, W. J., AND KRAUSSE, H.-F., 1981, The Cottage Grove fault system in southern Illinois: Illinois Institute of Natural Resources, State Geological Survey Division, Circular 522, 65 p.

NICHOLSON, C., SEEBER, L., WILLIAMS, P., AND SYKES, L. R., 1985, Seismicity and fault kinematics through the eastern Transverse Ranges, California: Block rotation, strike-slip faulting and shallow-angle thrusts: Journal of Geophysical Research, in press.

POSAVEC, M., TAYLOR, D., VAN LEEVEN, TH., AND SPECTOR, A., 1973, Tectonic controls of vulcanism and complex movements along the Su-

matran fault system: Geological Society of Malaysia Bulletin, v. 6, p. 43–60.

PROFFETT, J. M., JR., 1977, Cenozoic geology of the Yerington district, Nevada, and implications for the nature and origin of Basin and Range faulting: Geological Society of America Bulletin, v. 88, p. 247–266.

RIEDEL, W., 1929, Zur mechanik geologischer Brucherscheinungen: Zentralblatt für Mineralogie und Paleontologie, v. 1929B, p. 354–368.

ROCKWELL, T., AND SYLVESTER, A. G., 1979, Neotectonics of the Salton Trough, *in* Crowell, J. C., and Sylvester, A. G., eds. Tectonics of the Juncture Between the San Andreas and the Salton Trough, Southeastern California—A Guidebook: Santa Barbara, California, Department of Geological Sciences, University of California, p. 41–52.

ROGERS, T. H., 1965, Santa Ana sheet, Geologic Map of California, Olaf P. Jenkins edition: California Division of Mines and Geology, Scale 1:250,000.

ROYDEN, L. H., 1985, The Vienna Basin: A thin-skinned pull-apart basin, *in* Biddle, K. T., and Christie-Blick, N., eds., Strike-Slip Deformation, Basin Formation, and Sedimentation: Society of Economic Paleontologists and Mineralogists Special Publication No. 37, p. 319–338.

SCHUBERT, C., 1980, Late-Cenozoic pull-apart basins, Boconó fault zone, Venezuelan Andes: Journal of Structural Geology, v. 2, p. 463–468.

SCHULMAN, N., AND BARTOV, Y., 1978, Tectonics and sedimentation along the rift valley, excursion Y2: 10th International Congress on Sedimentology, Jerusalem, International Association of Sedimentologists, Guidebook, p. 37–94.

SLOSS, L. L., 1979, Plate-tectonic implications of the Pennsylvanian System in the Illinois basin, *in* Palmer, J. E., and Dutcher, R. R., eds., Depositional and Structural History of the Pennsylvanian System of the Illinois Basin, Part 2: Illinois State Geological Survey Guidebook, Series 15a, Field trip 9, Ninth International Congress, Carboniferous Stratigraphy and Geology, p. 107–112.

SMITH, D. A., 1980, Sealing and nonsealing faults in Louisiana Gulf Coast salt basin: American Association of Petroleum Geologists Bulletin, v. 64, p. 145–172.

SMITH, J. G., 1965, Fundamental transcurrent faulting in northern Rocky Mountains: American Association of Petroleum Geologists Bulletin, v. 49, p. 1398–1409.

STEWART, J. H., 1983, Extensional tectonics in the Death Valley area, California: Transport of the Panamint Range structural block 80 km northwestward: Geology, v. 11, p. 153–157.

SYLVESTER, A. G., AND SMITH, R. R., 1976, Tectonic transpression and basement-controlled deformation in San Andreas fault zone, Salton trough, California: American Association of Petroleum Geologists Bulletin, v. 60, p. 2081–2102.

TCHALENKO, J. S., AND AMBRASEYS, N. N., 1970, Structural analysis of the Dasht-e Bayaz (Iran) earthquake fractures: Geological Society of America Bulletin, v. 81, p. 41–60.

TERRES, R. R., AND SYLVESTER, A. G., 1981, Kinematic analysis of rotated fractures and blocks in simple shear: Seismological Society of America Bulletin, v. 71, p. 1593–1605.

TRYGGVASON, E., 1982, Recent ground deformation in continental and oceanic rift zones, *in* Pálmason, G., ed., Continental and Oceanic Rifts: American Geophysical Union Geodynamics Series, v. 8, p. 17–29.

VROMAN, A. J., 1967, On the fold pattern of Israel and the Levant: Geological Survey of Israel Bulletin, v. 43, p. 23–32.

WEBER, K. J., MANDL, G., PILAAR, W. F., LEHNER, F., AND PRECIOUS, R. G., 1978, The role of faults in hydrocarbon migration and trapping in Nigerian growth fault structures: Offshore Technology Conference Paper OTC 3356, p. 2643–2652.

WERNICKE, B., SPENCER, J. E., BURCHFIEL, B. C., AND GUTH, P. L., 1982, Magnitude of crustal extension in the southern Great Basin: Geology, v. 10, p. 499–502.

WILCOX, R. E., HARDING, T. P., AND SEELY, D. R., 1973, Basic wrench tectonics: American Association of Petroleum Geologists Bulletin, v. 57, p. 74–96.

WRIGHT, L. A., AND TROXEL, B. W., 1970, Summary of regional evidence for right-lateral displacement in the western Great Basin: Discussion: Geological Society of America Bulletin, v. 81, p. 2167–2173.

ZARTMAN, R. E., BROCK, M. R., HEYL, A. V., AND THOMAS, H. H., 1967, K-Ar and Rb-Sr ages of some alkaline intrusive rocks from central and eastern United States: American Journal Science, v. 265, no. 10, p. 848–870.

The American Association of Petroleum Geologists Bulletin
V. 58, No. 7 (July 1974), P. 1290-1304, 15 Figs.

Petroleum Traps Associated with Wrench Faults[1]

T. P. HARDING[2]

Houston, Texas 77001

Abstract The interaction of three factors results in several different predictable patterns of wrench-fault-related petroleum traps: (1) the evolutionary stage or magnitude of the wrench faulting; (2) the configurations of the laterally moving plates and their orientations to regional movement vectors; and (3) the structural response of the deformed terrane. In the more common fold and fault responses small strike-slip displacements develop narrow trends of en echelon anticlinal culminations which straddle an incipient or underlying wrench fault. Traps commonly are structurally complex. Wrenches with intermediate strike-slip displacement have offset-truncated half-fold culminations and structural bowings closed upplunge by the wrench fault, and less complex intact en echelon anticlinal-culmination traps away from the fault. Folds adjacent to many large-displacement wrench faults often are adversely disrupted structurally, or deeply eroded. Potential hydrocarbon traps form downstructure where basinward-plunging en echelon folds cross sedimentary wedges associated with basin margins, and where large anticlinal culminations are preserved.

Fault responses to wrenching deformation result in various patterns of en echelon fault traps. Degrees of regional-plate convergence or divergence enhance compressive or extensional structuring, respectively, and further modify prospective structures.

INTRODUCTION

This article describes the types of petroleum traps, and thereby potential prospect types, that commonly form along wrench faults; it shows the ways in which these traps have been grouped to form trends and play fairways (see also Harding, 1973b, 1974). The present work is an extension of two earlier companion papers published in the *Bulletin*. First, the development of the wrenching structural style was treated in Wilcox *et al.* (1973). A detailed subsurface example followed in Harding (1973a).

Publications concerning wrench-fault exploration usually have emphasized regional patterns made by sets of faults and the ways to recognize or project these patterns. Most recently Moody (1973) has suggested two worldwide wrench-fault patterns, and Stone (1969) applied a wrench-tectonic scheme to the Rocky Mountains utilizing an approach presented earlier by Moody and Hill (1956). Corey (1962) has discussed the effects that wrenching would have on the exploration of a hypothetical basin.

The en echelon anticline is the basic trap

(Fig. 1). Three important factors can significantly influence the structural pattern. Shifts in location of exploration fairways and changes in trap type occur as the stage of the deformation evolves from small strike-slip displacements to intermediate-magnitude displacements, and ultimately to large-scale strike slip. Additional variations in trap geometry result from differences in structural response by the terranes involved in the deformation. Trapping that is dominated by faults or by fold closures, or that results from combined folding and faulting, has been documented. The latter is the most common response to wrench deformation.

The configuration of the laterally moving blocks and the orientations of their boundaries to regional plate-movement vectors comprise the third factor. The classic case consists of blocks moving side-by-side parallel with their mutual boundaries and in line with regional-movement vectors, but areas of wrenching with components of oblique-block convergence or divergence also are common (see Wilcox *et al.*, 1973, for elaboration).

SMALL-DISPLACEMENT PLAYS

Small strike-slip displacement ideally results in a series of slightly offset en-echelon-fold culmination closures which obliquely straddle the early stage wrench fault. Displacements are insufficient to create a major shift in facies or to disrupt general structural continuity, and production may be obtained from essentially the same horizons in correlative fold elements di-

[1] Manuscript received, July 10, 1973; accepted, November 28, 1973.

[2] Esso Production Research Co., P.O. Box 2189.

The writer is indebted to the Esso Production Research Co. and the Exxon Company, U.S.A., for permission to publish this investigation. The California Division of Oil and Gas, Department of Conservation, is acknowledged for the use of its published reports of oil and gas pools in the Los Angeles and San Joaquin Valley basins. Instructive discussions were held with associates at Esso Production Research Company, and Mason L. Hill provided helpful suggestions concerning the manuscript.

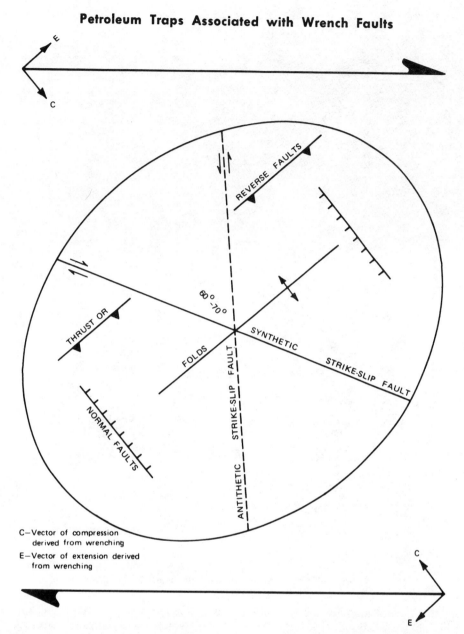

C—Vector of compression
   derived from wrenching

E—Vector of extension derived
   from wrenching

Fig. 1—Forces and composite of structures that can result from wrenching deformation combined schematically with strain ellipse. Depicts right-lateral movements; view in reverse for left-lateral.

rectly across the fault. The major exploration trend extends in a narrow band along both sides of, or above, the wrench throughout the latter's length in the sedimentary basin.

The Newport-Inglewood zone shows this basic prospect geometry and is an expression of simple, parallel, side-by-side wrenching with a fold-fault response (Fig. 2). Displacement magnitudes are indicated by consistent right-lateral offsets of fold axes in the range 600 to 2,500 ft. The presence of the wrench zone, with all its unique characteristics, at the level of the pro-

ducing horizons and within the closures themselves causes unusual structural complexities (see Harding, 1973a, for elaboration).

The Scipio-Albion trend in the Michigan basin and the Sussex–Meadow Creek trend of the Powder River basin, Wyoming, illustrate variations in the basic trap pattern caused by fault responses to early stage small-displacement strike-slip deformation. At the Scipio-Albion trend, major production comes from a narrow, laterally extensive, steeply inclined, dolomitized porosity zone (Fig. 3). The exis-

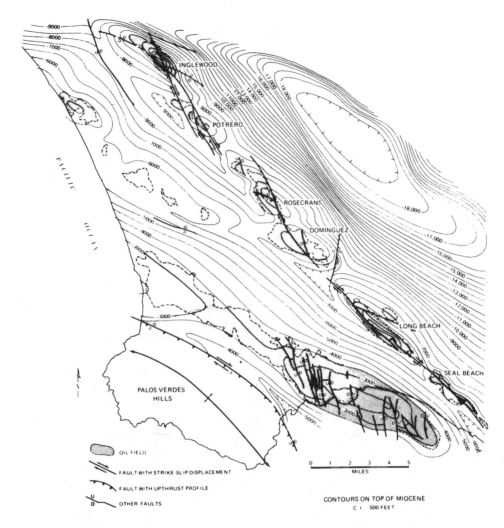

FIG. 2—Structure of northwest and central part of Newport-Inglewood trend (with oil field names) and southwestern shelf of Los Angeles basin (after Harding, 1973a).

tence of a fault is indicated by the unusually straight, ribbonlike distribution of the porous zone which controls production and by the zone's disregard for depositional facies which are crosscut transversely (J. N. Bubb, personal commun., 1965).

Strike-slip movements in the Michigan basin example are suggested by a unique sag which overlies and closely follows the productive fairway (Fig. 4a). Bubb (personal commun., 1965) has attributed the sag to contraction of rocks during dolomitization and consequent subsidence of the overlying section. The sag therefore should reflect the pattern of dolomitization and quite possibly approximate the distribution of fractures which controlled this dolomitization. The structure contours of Figure 4 reveal a consistent right-handed (defined in Hard-

ing, 1973a) en echelon pattern of the low's axial trace, which closely approximates the synthetic-fault orientation for left-lateral wrenching (view Fig. 1 in reverse). The same pattern also is apparent in the outlines of individual producing areas and their intervening dry holes. Deep-seated, slight, left-lateral strike-slip displacements, possibly along a preexisting basement fault, could account for the observed structure. In addition, an oblique divergent component would have emphasized the extensional effects of the mild deformation and would have tended to open the synthetic fractures, facilitating dolomitization. Such "stretching" or opening also would enhance development of the sag while minimizing regional, wrench-related compressional folding (anticlines are lacking and the angle or degree of "en echelonness" is low

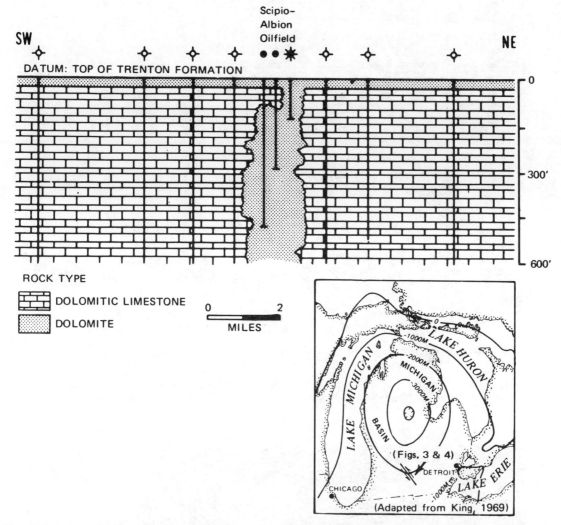

FIG. 3—Cross section across Scipio-Albion trend showing distribution of carbonate rock types in productive Middle Ordovician Trenton–Black River sequence (courtesy of J. N. Bubb) and area-index map contoured on top of Precambrian basement.

for a typical wrench-related fold set, Fig. 1). The fractures apparently die out abruptly in the overlying Utica Shale, but the sag persists throughout the remaining Paleozoic section (Ells, 1962).

The interpreted Scipio-Albion fracture pattern is analogous to that made by the northwest-trending synthetic, right-lateral, en echelon strike-slip faults of the Newport-Inglewood fault zone (Fig. 2); only the sense of en echelon overlap is the opposite. Similar opened fracture systems and associated sags have been developed with divergent wrench models in the laboratory (Fig. 5).

At the Sussex–Meadow Creek trend, oil is trapped by en echelon normal faults which ap-

pear to be secondary to deep-seated left-slip on an east-west-trending fault (Fig. 6). The en echelon normal faults have been mapped in surface exposures of Cretaceous and lower Tertiary formations (Horn, 1955), and they segment the area into separate subsurface fault-block reservoirs, some of which are shown in Figure 6. Part of the updip closure is provided by a through-going, high-angle fault which may approximate the actual wrench fault (see also Harding, 1973a, Fig. 14).

INTERMEDIATE-DISPLACEMENT PLAYS

Increases in lateral movement result in a wrench fault that is through-going and has significant strike-slip displacements at the struc-

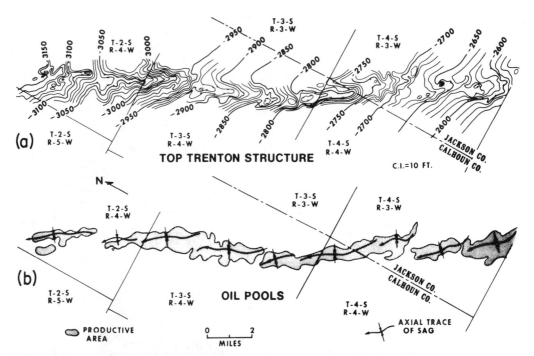

FIG. 4—(a) Structure-contour map of central and northern parts of Scipio-Albion trend contoured on top of Trenton formation which directly overlies productive interval (see Fig. 3). (b) Plot of main axial trace of sag.

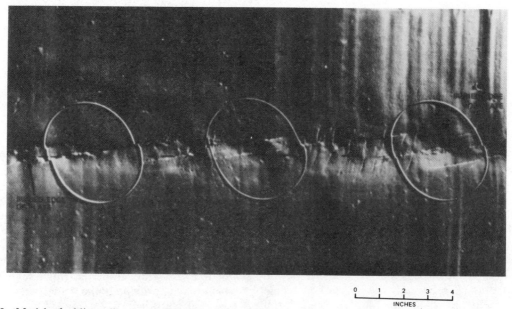

FIG. 5—Model of oblique-divergent wrench showing early stage pattern of "opened" en echelon, synthetic fractures, and related sag. Upper plate underlying 1.3-in. clay cake was moved left-laterally with 11° angle of divergence relative to position of lower, fixed plate. (Photograph courtesy of D. R. Seely.)

tural level of the petroleum traps. En echelon anticlines are offset so that only half of a fold culmination may be present at the fault, and the wrench itself becomes significant to the accumulation of hydrocarbons where it provides critical updip closure. Potential traps include complex structures directly along the wrench fault, and less complex anticlinal-culmination closures develop farther away, where en echelon fold trends project obliquely outward some distance from the principal strike-slip fault zone.

The Whittier fault and adjacent eastern shelf of the Los Angeles basin are an example of this prospect pattern (Fig. 7) and also of wrench faulting with important vertical and oblique convergent components.

The Whittier fault is widely recognized as having right-lateral displacements, possibly as much as 3 mi during its later Neogene history (Kundert, 1952; Troxel, 1954; Woodford *et al.*, 1954; Gray, 1961). The fault splays westward off the extensive northwest course of the Chino-Elsinore fault system and, in so doing, lies athwart the more typical, northwest-southeast regional displacement directions of major southern California wrench faults. This splaying has placed the northern block of the zone obliquely in the path of relative right-lateral movements along the southeastern side of the Elsinore fault (Fig. 8).

In response to the movement pattern, and apparently also as part of the continuing subsidence of the Los Angeles basin, the southwest side of the Whittier fault has obliquely underthrust the northern Puente Hills block. Throughout most of its course the Whittier fault dips 65 to 75° northeast, with the latter side relatively upthrown and the basin side downdropped. Apparent vertical separation components are greatest near the midpoint of the zone's trace, amounting to as much as 14,000 ft and diminishing to 2,000 or 3,000 ft near either end (Yerkes *et al.*, 1965). Near its southeast end the dip direction and apparent upthrown side reverse, but the fault maintains its reverse-separation character.

The more consistent and important vertical components, the reverse sense of apparent dip-slip separation, and the marked association of compressive structuring along the zone are thought to be characteristic of oblique, convergent wrenching. The characteristics contrast significantly with the structural expressions of oblique, divergent (Fig. 6) or simple, parallel,

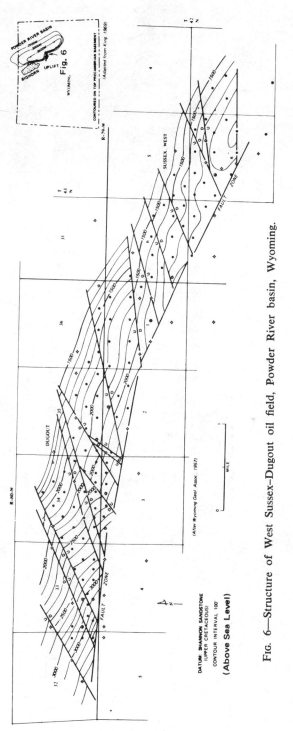

Fig. 6—Structure of West Sussex–Dugout oil field, Powder River basin, Wyoming.

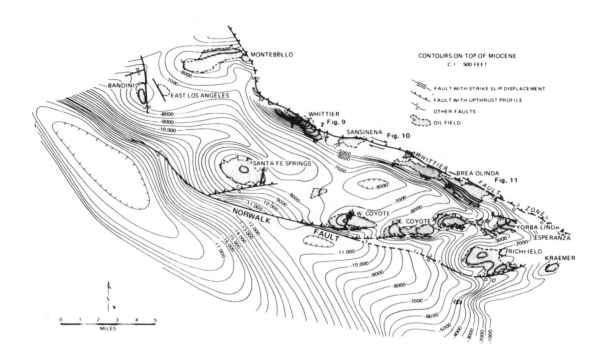

CONTOURS ON TOP OF MIOCENE
C I 500 FEET

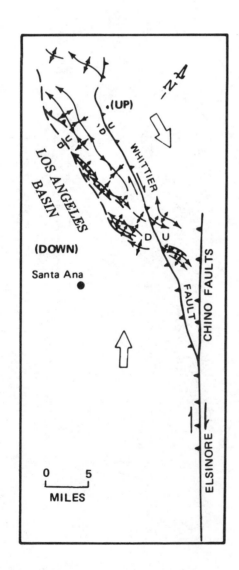

side-by-side wrench zones. Strike-slip faults of the Newport-Inglewood zone, an example of the latter type, have inconsistent dip directions, inconsistent apparent upthrown sides, both apparent reverse and normal fault-dip separations, and associated structuring that includes elements of extension as well as compression (Harding, 1973a, Figs. 5, 13, 14). The deformation at the Whittier fault has developed a strongly squeezed "pucker"; strata on either side dip steeply away at most places; and en echelon upthrust or reverse faults flank several productive anticlines on the eastern shelf (Fig. 7; Harding, 1973b, Fig. 11). At the latter faults, apparent dip-slip separations are down to the south in concert with basin subsidence, and erosion has removed many of the potentially productive reservoir zones from the high Puente Hills block.

Four types of potential prospects adjacent to the fault are illustrated by major oil fields: Montebello, part of an en echelon anticline closed upplunge by the wrench (Fig. 7); Whittier, part of an anticline aligned parallel with the wrench (Fig. 9); Sansinena, small en echelon drag folds preserved under the zone's hanging wall (Fig. 10); and Brea-Olinda, a regional

FIG. 8—Tectonic setting of Whittier fault and eastern shelf of Los Angeles basin. Large open arrows indicate deep-seated wrenching movement.

FIG. 7—Structure of eastern shelf of Los Angeles basin and adjacent Whittier fault. Figure numbers refer to illustrations of individual oil fields.

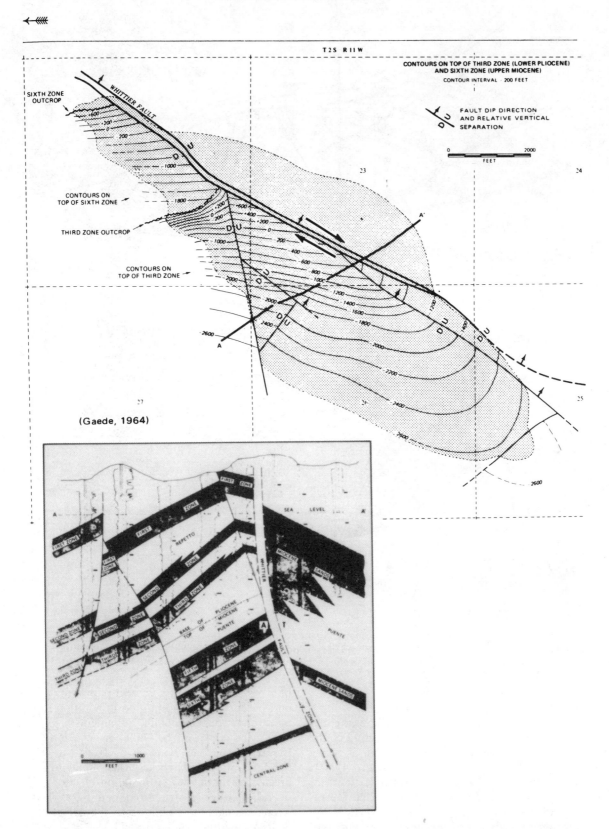

(Gaede, 1964)

FIG. 9—Structure of Whittier oil field. Northeast closure is provided by Whittier fault, upplunge closure by tar seal in outcrops of producing sands. Production on northeast side of fault is marginal and comes from tight older Miocene sands.

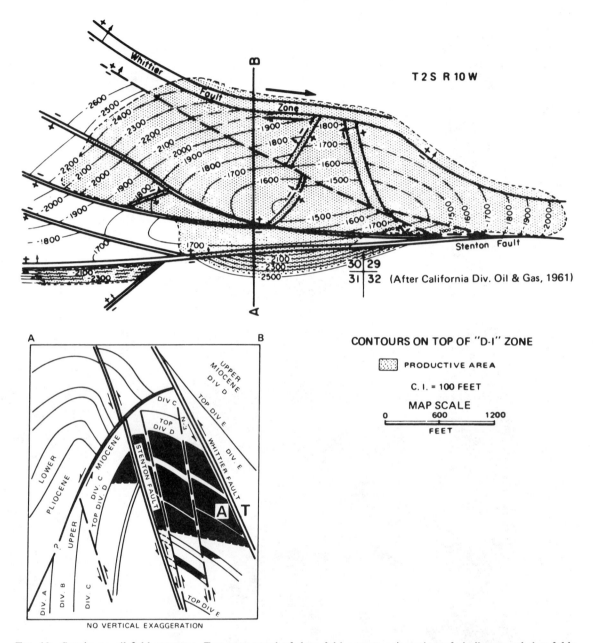

FIG. 10—Sansinena oil-field structure. East-west trend of drag fold repeats orientation of similar en echelon folds exposed along high side of Whittier fault and is supportive evidence of right-lateral wrenching deformation. Internal faulting and squeezed-fold profile document highly compressive aspect of deformation.

homocline partly closed updip by the wrench (Fig. 11).

The fault closure at the Brea-Olinda field is enhanced by updip pinchouts which may be the result of bathymetric expression on an early segment of the Whittier fault (Kundert, 1952). The elongation of the Whittier field's structure parallel with the fault is thought to be the result of the structural "pucker" or vertical up-turning of beds paralleling the zone (apparent

in cross section of Fig. 9) and represents a second potential fold orientation in addition to the more common en echelon alignments.

On the eastern shelf the en echelon fold trends plunge basinward away from the Whittier fault but locally reverse this plunge to form productive anticlinal culmination closures. The overall favorably structured belt parallels the wrench zone but is oblique to the orientations of its individual folds (Fig. 7).

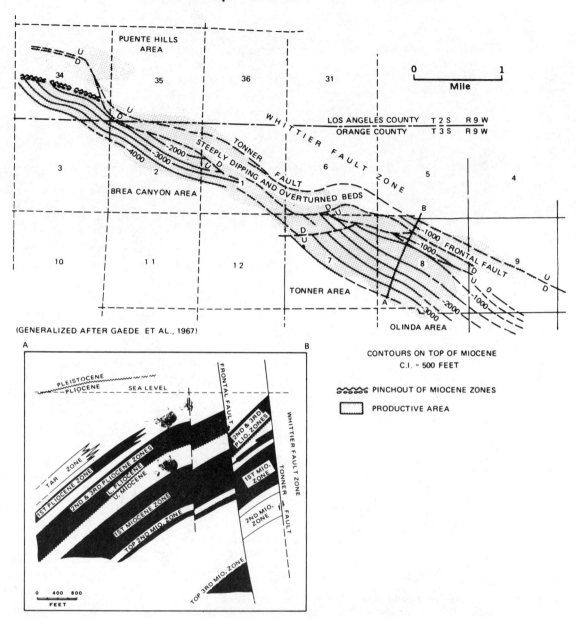

Fig. 11—Brea-Olinda oil field. Lateral closure is provided by broad, structural bowing against wrench zone, secondary cross faults, and porosity terminations. Frontal fault is anomalous and has been interpreted variously as having reverse or normal apparent dip-slip separations. The fault bounds an extensive slice whose dominant movement may have been strike slip.

LARGE-DISPLACEMENT PLAYS

Significantly different petroleum accumulations can result from greatly increased strike slip. Structural relief or slope often should rise appreciably toward the mature-stage wrench owing to the increased upbuilding and broadening of its associated anticlinal welt. Folds nearest the fault zone may be disrupted severely by the intensified structuring focused there, and the adjacent terrane may be deeply

eroded, causing the exploration fairway to shift downdip away from the fault zone.

Basin areas originally extending across the wrench fault ultimately will be truncated structurally and offset laterally considerable distances. In later stages of deposition basin margins may shift away from the fault in response to the upbuilding and exposure of its anticlinal welt. Important stratigraphic traps then will develop where younger reservoir fa-

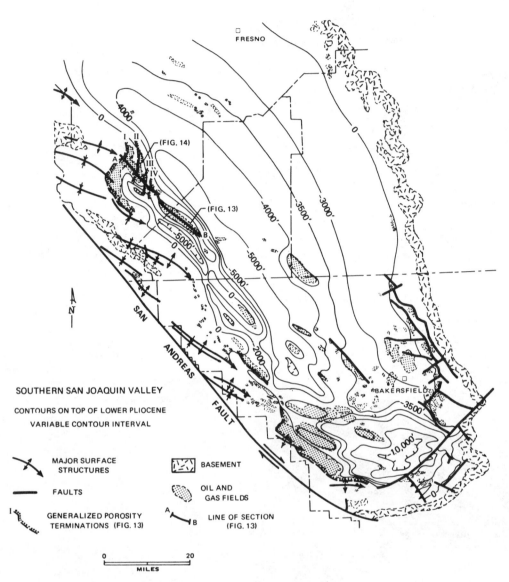

FIG. 12—Major structures and oil fields, San Joaquin Valley, California (after Hoots *et al.*, 1954). Figure numbers refer to subsequent illustrations in text, Roman numerals to porosity terminations on Coalinga Nose shown in Figure 13.

cies limits cross the basinward-plunging en echelon anticlines.

Opposite the San Joaquin Valley basin of California (Fig. 12), right-lateral strike-slip displacements as great as 175 to 190 mi since early Miocene time have been demonstrated for the San Andreas fault (Hill and Dibblee, 1953; Huffman, 1972) and considerably larger, older right-lateral movements have been suggested (Hill and Hobson, 1968; Ross *et al.*, 1973). The terrane adjacent to the San Andreas fault is tightly folded, complexly disrupted by a myriad of secondary faults, and deeply eroded. The intact and preserved part of the Tertiary oil basin lies well downstructure on the

northeast, where the en echelon fold set, plunging basinward away from the San Andreas fault, crosses multiple, prolific porosity terminations. Effective stratigraphic traps include sand shaleouts, wedgeouts, onlaps, truncations and sand channels, permeability barriers, and surface tar seals. Major production also has come from large intact anticlinal culminations, mostly farther downplunge in the fold set where reservoirs are most numerous. The Coalinga nose trend illustrates the interplay of wrench-related folding, marginward wedging, stratigraphic trapping, and downplunge structural trapping (Figs. 13, 14).

In Sumatra, relative northward encroachment

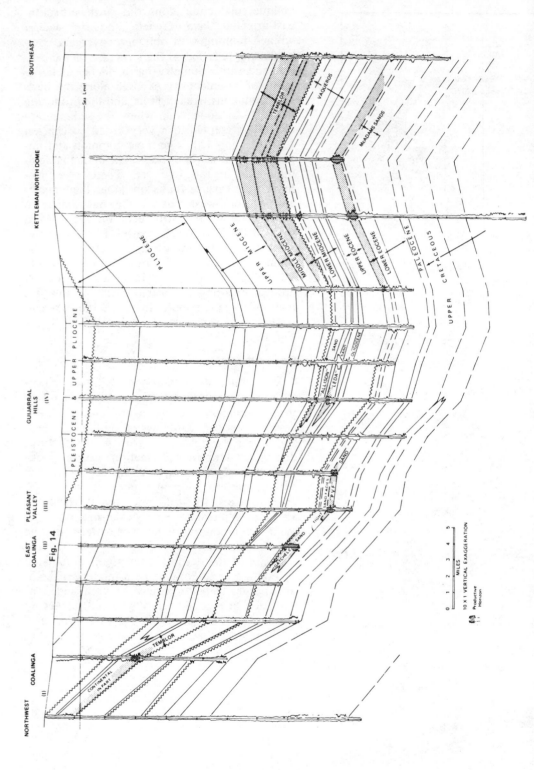

Fig. 13—Cross section upplunge of Coalinga Nose trend (generalized after Krammes, 1959); see Figure 12 for location of section and stratigraphic oil fields identified with Roman numerals. Effective stratigraphic closures included (*I*) surface tar seal, sand truncation, and permeability loss caused by transition to nonmarine facies; (*II*) sand shaleout (see also Fig. 14); (*III*) permeability barriers; and (*IV*) sand shaleout, sand wedgeout, and multiple permeability barriers in Eocene sands (latter not indicated). Productive limits are generalized; stratigraphic traps had yielded nearly 1.2 billion bbl of oil by end of 1971 (California Div. Oil and Gas, 1971).

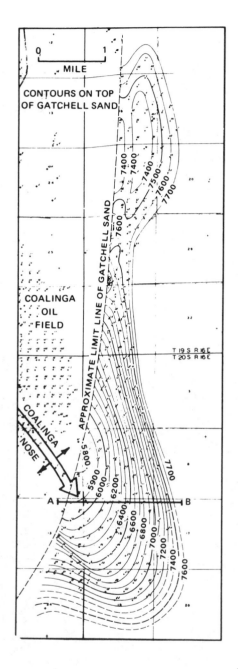

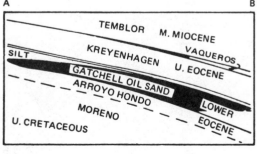

FIG. 14—Structure of major East Coalinga Extension oil field (after Ellison, 1952; cross section adapted from California Div. Oil and Gas, 1960). Relative absence of internal structural complexities at folds well away from wrench fault is shown here (compare with Figs. 9, 10).

of the Indian Ocean plate has resulted in oblique subduction along the northwest-southeast-oriented Java Trench, and right-lateral wrench faulting with oblique convergent components is reflected in the structure of the central and south Sumatra basin oil fields. Large tracts of basement are exposed along the Barisan Mountains wrench fault, and the producing fields lie far downflank, where the sedimentary cover is preserved in a very broad en echelon fold set (Fig. 15). The most common and important hydrocarbon traps are anticlinal culminations in the fold set. These often are flanked by reverse faults on either their northeast or southwest sides, somewhat similar in appearance to the structures on the eastern shelf of the Los Angeles basin (Fig. 7).

On the north flank of the Eastern Venezuela basin in Venezuela, the broad en echelon fold set along the El Pilar fault also is deeply eroded, exposing prebasinal rocks. In this late mature stage example little of the wrench-related play is preserved (Wilcox et al., 1973, Fig. 1b).

CONCLUSIONS

Wrenching deformation is a prolific generator of hydrocarbon trapping structures where the sedimentary terrane is appropriate. Early development of anticlines straddling wrench zones and the continued growth of structures during prolonged deformation assure availability of effective traps. Prospects are in systematic patterns that are repeated with a degree of regularity from one wrench region to another. With increasing strike-slip displacements evolutionary changes in pattern may be anticipated, and, depending on whether the response of the deformed terrane is dominated by faulting or folding, or by a mixture of both, the specific trap types may be predicted. Anticipation of prospects may be aided further by knowledge of components of block convergence or divergence.

318

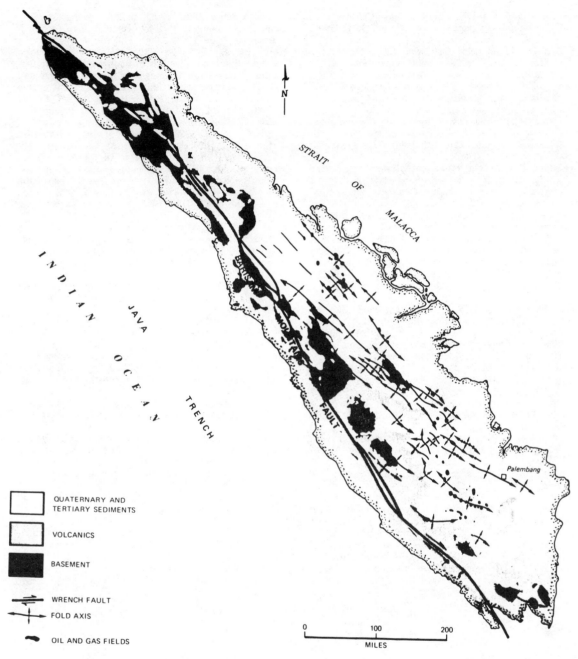

QUATERNARY AND
TERTIARY SEDIMENTS

VOLCANICS

BASEMENT

WRENCH FAULT

FOLD AXIS

OIL AND GAS FIELDS

FIG. 15—Structures and oil fields of central and south Sumatra (after Sigit, 1962; Hamilton, 1972).

REFERENCES CITED

California Division of Oil and Gas, 1960, California oil and gas fields, maps and data sheets, pt. 1, San Joaquin–Sacramento Valleys and northern coastal regions: California Div. Oil and Gas, p. 1–493.

——— 1961, California oil and gas fields, maps and data sheets, pt. 2, Los Angeles–Ventura basins and central coastal regions: California Div. Oil and Gas, p. 495–913.

——— 1971, Oil and gas statistics: California Div. Oil and Gas, California Oil Fields—Summ. Operations, v. 57, no. 2, p. 68–79.

Corey, W. H., 1962, Effects of lateral faulting on oil exploration: Am. Assoc. Petroleum Geologists Bull., v. 46, no. 12, p. 2199–2212.

Ellison, B. R., 1952, The East Coalinga Extension oil field, in Guidebook, field trip routes, oil fields, geology: Am. Assoc. Petroleum Geologists, Soc. Econ. Paleontologists and Mineralogists, Soc. Explor. Geophysicists, Joint Ann. Mtg., Los Angeles, California, p. 182–183.

Ells, G. D., 1962, Structures associated with the Albion-Scipio oil field trend: Michigan Geol. Survey Div., 86 p.

Gaede, V. F., 1964, Central area of Whittier oil

field: California Div. Oil and Gas, California Oil Fields—Summ. Operations, v. 50, no. 1, p. 59–67.

—— R. V. Rothermel, and L. H. Axtell, 1967, Brea-Olinda oil field: California Div. Oil and Gas, California Oil Fields—Summ. Operations, v. 53, no. 2, pt. 2, p. 5–24.

Gray, C. H., Jr., 1961, Geology of the Corona South quadrangle and the Santa Ana Narrows area, Riverside, Orange, and San Bernardino Counties, California: California Div. Mines Bull. 178, p. 5–58.

Hamilton, W., 1972, Preliminary tectonic map of the Indonesian region, scale 1:500,000: U.S. Geol. Survey Open File Rept.

Harding, T. P., 1973a, Newport-Inglewood trend, California—an example of wrenching style of deformation: Am. Assoc. Petroleum Geologists Bull., v. 57, no. 1, p. 97–116.

—— 1973b, Major hydrocarbon plays resulting from wrenching deformation: Soc. Brasileira Geologia, 27th Cong. Brasileira Geologia Anais (in press).

—— 1974, Acumulaciones importantes de hidrocarburos originadas por deformaciones causadas por fallas laterales: Inst. Argentino Petroleo Petrotecnia, nos. 2–3, p. 17–22. (Article continues in following issues.)

Hill, M. L., and T. W. Dibblee, Jr., 1953, San Andreas, Garlock, and Big Pine faults, California (a study of the character, history, and tectonic significance of their displacements): Geol. Soc. America Bull., v. 64, no. 4, p. 443–458.

—— and H. D. Hobson, 1968, Possible post-Cretaceous slip on the San Andreas fault zone, in W. R. Dickinson and A. Grantz, eds., Proceedings of the conference on geologic problems of San Andreas fault system: Stanford Univ. Pubs. Geol. Sci., v. 11, p. 123–129.

Hoots, H. W., T. L. Bear, and W. D. Kleinpell, 1954, Geological summary of the San Joaquin Valley, California, in R. H. Jahns, ed., Geology of southern California: California Div. Mines Bull. 170, chap. 2, pt. 8, p. 113–129.

Horn, G. H., 1955, Geologic and structure map of the Sussex and Meadow Creek oil fields and vicinity, Johnson and Natrona Counties, Wyoming: U.S. Geol. Survey Oil and Gas Inv. Map OM 164.

Huffman, O. F., 1972, Lateral displacement of upper Miocene rocks and the Neogene history of offset along the San Andreas fault in California: Geol.

Soc. America Bull., v. 83, p. 2913–2946.

King, P. B. (compiler), 1969, Tectonic map of North America, scale 1:5,000,000: U.S. Geol. Survey.

Krammes, K. F., chm., 1959, Correlation section longitudinally north-south through westside San Joaquin Valley from Coalinga to Midway Sunset and across San Andreas fault into southeast Cuyama Valley: Am. Assoc. Petroleum Geologists Pacific Coast Subcomm. Stratig. Correlations, Pacific Sec. Cross-Section Comm., sec. no. 11.

Kundert, C. J., 1952, Geology of the Whittier-La Habra area, Los Angeles County, California: California Div. Mines Spec. Rept. 18, 22 p.

Moody, J. D., 1973, Petroleum exploration aspects of wrench-fault tectonics: Am. Assoc. Petroleum Geologists Bull., v. 57, no. 3, p. 449–476.

—— and M. J. Hill, 1956, Wrench-fault tectonics: Geol. Soc. America Bull., v. 67, p. 1207–1246.

Ross, D. C., C. M. Wentworth, and E. H. McKee, 1973, Cretaceous mafic conglomerate near Gualala offset 350 miles by San Andreas fault from oceanic source near Eagle Rest Peak, California: U.S. Geol. Survey Jour. Research, v. 1, no. 1, p. 45–52.

Sigit, S., 1962, Geologic map of Indonesia, scale 1:2,000,000: U.S. Geol. Survey Misc. Geol. Inv. Map I-414.

Stone, D. S., 1969, Wrench faulting and Rocky Mountain tectonics: Mtn. Geologist, v. 6, no. 2, p. 67–79.

Troxel, B. W., 1954, Geologic guide for the Los Angeles basin, southern California, in R. H. Jahns, ed., Geology of southern California: California Div. Mines Bull. 170, Geol. Guide 3, 46 p.

Wilcox, R. E., T. P. Harding, and D. R. Seely, 1973, Basic wrench tectonics: Am. Assoc. Petroleum Geologists Bull., v. 57, no. 1, p. 74–96.

Woodford, A. O., J. E. Schoellhamer, J. G. Vedder, and R. F. Yerkes, 1954, Geology of the Los Angeles basin, in R. H. Jahns, ed., Geology of southern California: California Div. Mines Bull. 170, chap. 2, p. 65–81.

Wyoming Geological Association, 1957, Wyoming oil and gas fields symposium: Wyoming Geol. Assoc., 446 p.

Yerkes, R. F., T. H. McCulloh, J. E. Schoellhamer, and J. G. Vedder, 1965, Geology of the Los Angeles basin, California—an introduction: U.S. Geol. Survey Prof. Paper 420-A, 57 p.

The American Association of Petroleum Geologists Bulletin
V. 60, No. 3 (March 1976), P. 356-378, 12 Figs.

Tectonic Significance and Hydrocarbon Trapping Consequences of Sequential Folding Synchronous with San Andreas Faulting, San Joaquin Valley, California[1]

T. P. HARDING[2]
Houston, Texas 77001

Abstract Subsurface anticlines along the west side of the San Joaquin Valley grew progressively basinward synchronous with the history of strike-slip on the adjacent San Andreas fault. At the south end of the anticlinorium the first conclusive documentation of en-echelon folding appears in early Miocene rocks, corresponding to the start-up of significant late Tertiary displacements on the San Andreas. Absence of important late Eocene or Oligocene folding here corresponds to a previously established period of negligible strike-slip. A marked increase in the area folded occurred near the end of the Miocene, reflecting a recognized increase in rate of fault displacements. Secondary faulting of the structures apparently lags behind fold inceptions but may have the same general outward growth sequence.

At the northwest end of the anticlinorium en-echelon folding commenced at least by early Eocene or possibly Paleocene time, coeval in part with displacements on the proposed early Cenozoic "proto"-San Andreas system. Incomplete data suggest that folding proceeded both outward from the San Andreas here and longitudinally southeastward parallel with the fault, implying a southeastward propagation of the present San Andreas strand of this system during the latter's probable inception in earliest Tertiary–latest Cretaceous time.

Hydrocarbon trap closures change critically in response to the timing sequences. Subsurface folds nearest the San Andreas have grown contemporaneously with deposition of the earliest reservoirs, but some ceased growing while deformation was just commencing basinward. Such traps may be hidden below relatively undisturbed younger basin-margin sedimentary wedges. In other instances and more distant from the San Andreas, younger anticlines are expressed at the surface, and vertical closure increases with depth in response to several periods of flexing. Structures at the periphery of the anticlinorium have the latest inceptions, but closure may diminish downward at levels where the gentle late flexing was insufficient to reverse preexisting basinward dips imposed by earlier basin subsidence.

The observed relations suggest that folds and wrench faults are essentially independent, different structural responses to a common, diffuse coupling movement originating in the deeper crust. Folds are propagated generally outward within an expanding, sometimes amorphous and diffuse, deformational front. Where tightly flexed interior structures become less susceptible to additional growth, renewed flexing may be concentrated at the periphery of the advancing front. The former folds then may undergo further stress release in later stage faulting. Outermost folds, having undergone less cumulative shortening, lack this faulting and generally are larger and broader. Regional strike-slip displacements are contemporaneous with the same deep coupling drive mechanism but remain concentrated on the fold set's established, original inner line of throughgoing crustal weakness.

INTRODUCTION

The tectonic interrelations and common deformational genesis of wrench faults and en-echelon fold sets have been demonstrated by Wilcox et al (1973) and previous investigators, using laboratory models as geologic analogs. Their results were corroborated with strain ellipse and other related kinds of tectonic analyses. Other workers have reached a similar conclusion from field observations of a common association of en-echelon folds and major faults known to have had strike-slip displacement (e.g., Moody and Hill, 1956).

Large wrench faults more commonly are associated with broader sets of en-echelon folds than are similar faults that have had smaller displacements (see Harding, 1974a, for exploration ramifications). This observation suggests that folds must grow outward from the fault as displacement (i.e., deformation) accumulates during the history of the strike-slip movements. Rock mobility and degree of plate convergency obviously also can affect the widths of potentially prospective fold sets (Wilcox et al, 1973). The San Joaquin Valley basin of California (Fig. 1) was investigated to evaluate the inferred fold-timing pattern and to determine what effect it might have had on petroleum trapping (Harding, 1974b). More than 125 structures, mostly subsurface anticlines, were studied.

Early authors, using mainly surface control, have emphasized, sometimes exclusively, the mid-

[1]Manuscript received, April 9, 1975; accepted, September 11, 1975.

[2]Exxon Production Research Co., P. O. Box 2189.

The writer is indebted to the California Division of Oil and Gas, Department of Conservation, whose oil and gas pool reports provided most of the basic data investigated. A. W. Marianos and H. S. Sonneman, Exxon Company, U.S.A., and J. L. Lamb, Exxon Production Research Co., assisted in correlation of local California stratigraphic stages with the worldwide time frame and radiometric time scale. J. C. Crowell, W. R. Dickinson, and M. L. Hill reviewed the manuscript. Exxon Production Research Company and its affiliate Esso Inter-America, Inc., supported the study and gave permission for its publication.

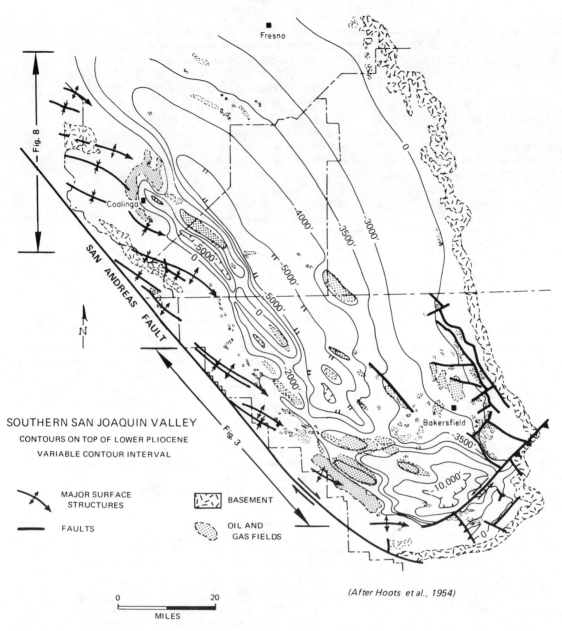

SOUTHERN SAN JOAQUIN VALLEY

CONTOURS ON TOP OF LOWER PLIOCENE

VARIABLE CONTOUR INTERVAL

⤢ MAJOR SURFACE STRUCTURES

── FAULTS

▨ BASEMENT

▨ OIL AND GAS FIELDS

0 ____ 20

MILES

(After Hoots et al., 1954)

FIG. 1—San Andreas fault, surface fold axes in adjacent Diablo and Temblor Ranges, subsurface structure (contours), and oil and gas fields of San Joaquin Valley basin. Areas treated in detail on Figures 3 and 8 bracketed by arrows.

dle Pleistocene as the period of folding in the San Joaquin Valley. This oversimplification has been encouraged perhaps by the high relief of the Temblor and Diablo Ranges adjacent to the San Andreas, by the obvious recent topographic expressions of large young structures within the valley floor, and by the statewide importance of the middle Pleistocene Pasadenan orogeny. The present work attempts to put the effects of the Pasa-

denan orogeny into proper perspective with other deformational episodes and brings a new type of data—subsurface fold timing—to bear on the general San Andreas problem.

BASIN SETTING

The region containing the San Joaquin Valley has had a long and complex geologic history. The present sedimentary-basin phase commenced at the end of the Cretaceous or the beginning of the Paleocene and continues to the present as the southern segment of the intermontane Great Valley of California. Structurally, it is an asymmetric basin whose east flank is a broad, gently dipping, relatively undeformed (except for normal faulting and sparse doming) homocline (Fig. 1). The "mobile west flank," the subject of this study, is narrower and consists of a tightly folded, complexly faulted anticlinorium that parallels the San Andreas fault. Fold axes within the anticlinorium are en echelon in a right-handed sense (Campbell, 1958), tectonically compatible with right-lateral strike-slip movements. Some of the anticlines produce oil and gas from sedimentary rocks that provide an essentially complete syntectonic time-stratigraphic record.

Significant folding periods were identified mostly from the crestward bed truncations, onlaps, and incremental thinning, or varying degrees of bed flexure discernible on detailed cross sections that are readily available for all oil and gas pools (see California Div. Oil and Gas, 1973, for most references). Correlation of the worldwide time-stratigraphic subdivisions with the provincial California stages that appear on these basic data have been updated to conform with currently accepted usage (A. W. Marianos, personal commun., 1974).

Deeply eroded outcrop areas of the anticlinorium in the Temblor and Diablo Ranges (southwest margin of Fig. 1) intervene in a narrow band between the border of the San Joaquin Valley and the San Andreas fault. Dense drilling necessary for identifying subtle episodes of fold growth is not available in this region, and lengthy erosional hiatuses further hamper results. For these reasons, and because of the need to ascertain directly the effects on hydrocarbon trapping, comprehensive dating of fold growth was limited to the basin proper.

SAN ANDREAS OFFSET HISTORY: ASPECTS CRITICAL TO FOLD-GROWTH TIMING

Displacements on the San Andreas fault, as now known, have not occurred in consistent increments, and the complexities of the offset history and its anomalies have a distinct bearing on the timing of trap development. To facilitate direct comparison, documented displacement episodes and the stratigraphic subdivisions used for dating folding have been fitted to a common or absolute time scale in Figure 2.

The most thoroughly documented displacement by the San Andreas fault is the right-lateral offset of a once-contiguous early Tertiary shoreline. This shoreline with its distinct facies is pres-

ent east of the fault at the south end of the San Joaquin Valley in the San Emigdio Mountains and is now west of the fault in the northernmost Gabilan Range, 175 to 185 mi (282 to 298 km) northwest (Hill and Dibblee, 1953; Bazeley, 1961; Addicott, 1968). Offset early Miocene volcanic rocks associated with the shoreline have been dated at 22 m.y. B.P. (Turner, 1969). An approximate 190-mi (306 km) displacement of a second set of distinct, upper Oligocene volcanic rocks dated at 23.5 m.y. B.P. by Turner et al (1970) also has been identified (Huffman, 1970).

A 145 ± 5-mi (233 ± 8 km) post-Miocene right-lateral offset of rocks between 8 and 12 m.y. old has been demonstrated by Huffman (1970, 1972). Comparison of this displacement with the previously described post-Oligocene displacements has led to the important observation (Huffman, 1970; Turner, et al, 1970; Dickinson et al, 1972) that displacements in the approximately 10 m.y. since the Miocene must have been at a significantly greater rate than in the previous 10 m.y. or so.

Pre-Miocene history is not so precise or abundantly controlled, but work in progress suggests additional important anomalies. Approximate offsets of 200 mi (322 km) since the middle Eocene (Nilsen and Dibblee, 1974) and since the late Eocene (Clarke and Nilsen, 1973) have been proposed in central California. The similarity of these offsets and the 190-mi (306 km) offset since the Oligocene has been interpreted (Huffman, 1972; Dickinson et al, 1972) as evidence of a period of relative fault quiescence during the Eocene and Oligocene (Fig. 2).

Significant older, post-Cretaceous strike-slip is indicated by unique Campanian and/or early Maestrichtian mafic conglomerates and quartz-plagioclase arkoses that are exposed west of the fault at its most northerly, landward extent. Ross et al (1973) have proposed that these rocks were derived from a gabbroic source terrace that is now across the fault and about 350 mi (563 km) southeast. Their work lends credibility to previous, more speculative, suggestions by Hill and Dibblee (1953) and by Hill and Hobson (1968) of major right-lateral displacements since the Cretaceous period.

The relations cited by the latter investigators caused Suppe in 1970 to postulate a two-stage concept of San Andreas strike-slip movements to reconstruct offset California Mesozoic basement terranes. According to Suppe (and now others), an older "proto-San Andreas" stage (second steep-curve segment, right side of Fig. 2) must have had approximately the same amount of movement as the better documented post-Oligocene stage. The early Cenozoic strike-slip, as cur-

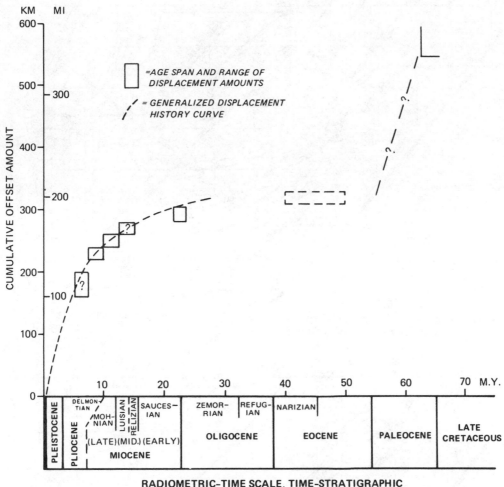

FIG. 2—Offset-time grid for right-lateral displacements on San Andreas fault in northern and central California (adapted from Dickinson et al, 1972). Radiometric dating of universal time-stratigraphic subdivisions summarized by J. L. Lamb (personal commun., 1975), and dating and equating of provincial California stages after Turner (1970) and A. W. Marianos (personal commun., 1974), respectively.

rently known, occurred sometime between the end of the Cretaceous and the end of the Oligocene (Suppe, 1970) or, according to Nilsen and Dibblee (1974), more specifically during Late Cretaceous–early Paleocene time. These displacements may have utilized other faults, bypassing the present trace of the San Andreas somewhere west of the San Joaquin Valley (Clarke and Nilsen, 1973; Howell, 1975). Some sort of bypassing is required because the maximum displacement demonstrated on the southern part of the system is approximately 160 mi (257 km; Crowell, 1962) or roughly equivalent to only the late Tertiary offsets farther north, and, in southern California, the displacements are all considered to be probably post-late Miocene, or since 8 to 14 m.y. ago (Crowell, 1973).

FOLD-GROWTH EPISODES

Timing of Folding at South End of Basin

In the southern part of the west-side foldbelt (southwest quadrant of Fig. 1), control includes horizons as old as latest Eocene and indicates that the inception of folding was progressively younger at subsurface anticlines increasingly more distant from the San Andreas fault (Fig. 3).

Significant fold development was uncommon during the end of the Eocene and the Oligocene according to available subsurface information (Fig. 3b). Evidence of Narizian to post-Narizian–

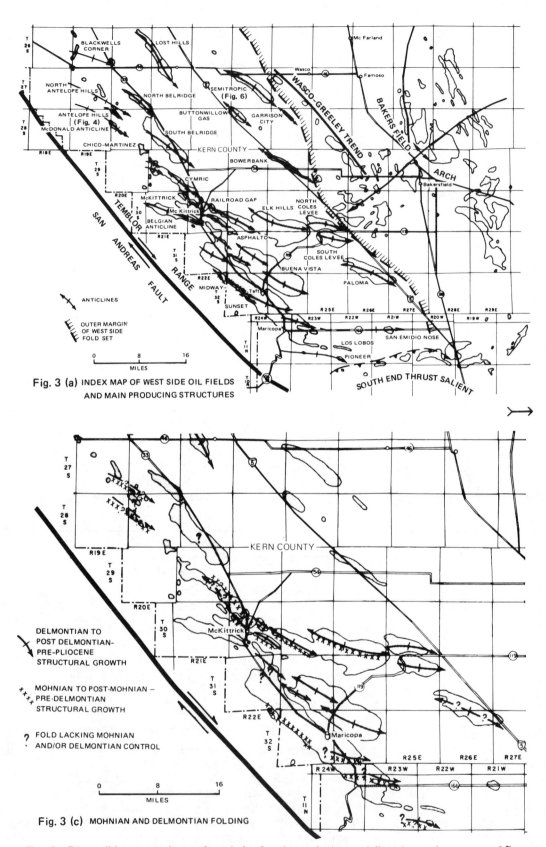

Fig. 3 (a) INDEX MAP OF WEST SIDE OIL FIELDS
AND MAIN PRODUCING STRUCTURES

Fig. 3 (c) MOHNIAN AND DELMONTIAN FOLDING

FIG. 3—Discernible structural growth periods of main producing anticlines in southwest part of San Joaquin Valley basin. Structures questionably or imprecisely dated are denoted by axes symbols combined with small question marks. In cases of ambiguous time spans all possible episodes included by time span are indicated in this manner. Approximately 110 detail subsurface cross sections were utilized in compilation.

326

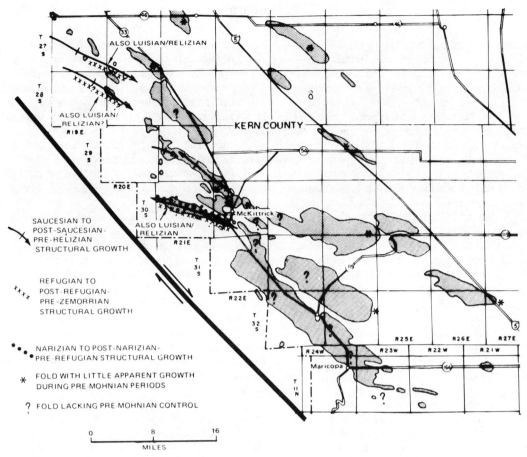

Fig. 3 (b) PRE-MOHNIAN FOLDING

SAUCESIAN TO
POST-SAUCESIAN-
PRE-RELIZIAN
STRUCTURAL GROWTH

REFUGIAN TO
POST-REFUGIAN-
PRE-ZEMORRIAN
STRUCTURAL GROWTH

NARIZIAN TO POST-NARIZIAN-
PRE-REFUGIAN STRUCTURAL GROWTH

* FOLD WITH LITTLE APPARENT GROWTH
DURING PRE-MOHNIAN PERIODS

? FOLD LACKING PRE-MOHNIAN CONTROL

0   8   16
MILES

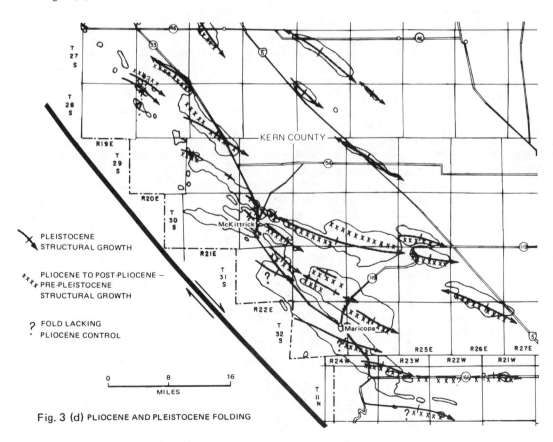

Fig. 3 (d) PLIOCENE AND PLEISTOCENE FOLDING

PLEISTOCENE
STRUCTURAL GROWTH

PLIOCENE TO POST-PLIOCENE –
PRE-PLEISTOCENE
STRUCTURAL GROWTH

? FOLD LACKING
PLIOCENE CONTROL

0   8   16
MILES

pre-Refugian folding (i.e., approximately post-late Eocene–pre-Oligocene) has been described thus far at only one south-end field, the Belgian Anticline (southernmost field with fold axes symbols on Fig. 3b; Park et al, 1957). However, severe structural complications, especially from late thrusting, make interpretation of available data here very tentative (see also Karp, 1968).

The first conclusive evidence of an en-echelon fold set appears in the Saucesian or post-Saucesian–pre-Luisian/Relizian (i.e., early Miocene to post-early Miocene, pre-middle Miocene) record. At the Antelope Hills oil field (Fig. 4), which illustrates the effects of this and several other folding episodes, the truncation of early Miocene and Oligocene beds by the base of the middle Miocene is the earliest indication of strong flexing. The truncation delineates the flanks of an older fold which lies 1,200 to 2,000 ft (366 to 610 m) south of the present anticlinal axis in the Hopkins area (right side of Fig. 4). The Button Bed sand, the basal unit of the middle Miocene, laps out crestward above the unconformity surface. The regional importance of this folding episode is underscored by pronounced truncation of beds below the middle Miocene at anticlines exposed on the west in the core of the Temblor Range (Jennings, 1958; Smith, 1964). Mild earlier folding during the Oligocene is suggested by an apparent gradual, crestward truncation of the late Eocene Kreyenhagen Shale along the unconformity at the base of the *Phacoides* sand and by an upplunge truncation of the intervening Refugian (i.e., earlier Oligocene) beds just east of the area shown in Figure 4.

A continuation of structural growth into the middle Miocene at the older anticline in the Hopkins area is demonstrated by consistent thickening of the Devilwater-Gould section down both flanks of the reconstructed fold. Similar to the pre-Button Bed truncations, thickening continues southward without disruption across a younger, parallel fold in the Williams area. The crestward thinning and/or truncation of the Antelope-McDonald shale toward both present axes below the base of the Tulare is considered the first evidence of differentiation of two anticlines. This latter event illustrates a local deviation from the general growth pattern and is thought to result from a continuation of the accumulation of the greater crustal shortenings that typify positions nearer the interior of the anticlinorium.

Delineation of the northeastward limits of regions affected by each folding period is critical to establishment of the regional timing pattern. Control for the older episodes is afforded by key scattered fields which show no discernible effects of

late Eocene to pre-late Miocene folding (located with asterisks, Fig. 3b). Similar evidence is more abundant for delineating latest Miocene and younger flexing. Several of the key structures also illustrate instances where folding once begun later was aborted. This latter phenomenon is at successively later times on anticlines progressively more distinct from the foldbelt interior and mirrors the pattern of fold inceptions.

At the Salt Creek pool of the Cymric oil field, for example (Fig. 5a), the pre-middle Miocene folding episode is well documented, but younger fold rejuvenation is notably absent. In addition, the lack of discernible crestward truncation or thinning in horizons deeper than the base of the middle Miocene unconformity is taken to indicate that pre-Miocene flexing episodes did not appreciably affect structuring this far distant from the San Andreas fault. The minor changes that are present may be attributed to regional variations in sedimentation into or out of the basin.

Six miles (10 km) downtrend at the northeast McKittrick pool (Fig. 5b), anticlinal reversal appears abruptly below an intra-Delmontian—i.e., the Pliocene and/or latest Miocene—unconformity at the base of the Olig sand. Above this there is little evidence of continued flexing. Anticlinal relief increases incrementally through the underlying late Miocene section, indicating contemporaneous structural growth, but below this there are no significant increases in fold relief. (The pre-Pliocene–Miocene unconformities are reproduced as originally shown by Weddle, 1965. Such unconformities are recognized regionally and have been projected from areas nearer the basin margin where they are more significant structurally.) A series of shallow Pleistocene surface folds only partly corresponds to the deep structuring here.

The western culmination of the large Elk Hills oil field structure completes illustration of fold timing on the trend. Discernible fold periods for which there is control include late Miocene, Pliocene-Miocene, Pliocene, and Pleistocene (Fig. 5c; Maher et al, 1975). The relative recency of much of the folding at this position on the basin flank is attested by the large size of the Pleistocene closure, which includes all the greater Elk Hills structure, and by the 1,000 ft (305 m) or more of corresponding topographic expression.

The youngest period of fold inception is illustrated by the Semitropic gas field at the far periphery of the fold set (Fig. 3a). Very late, gentle folding of the eastern basin-flank sedimentary wedge has reversed regional west dip in Pleistocene and uppermost Pliocene beds (Fig. 6) and is expressed topographically at the surface. Shallow

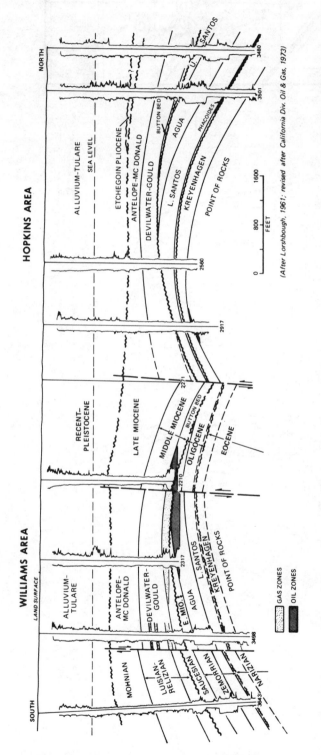

Fig. 4—Cross section across Antelope Hills oil field (see Fig. 3a for location). Multiple periods of folding, possibly as early as pre-Zemorrian to as late as Pleistocene, are indicated by successive levels of crestward truncation, downflank thickening, and changes in degree of bed flexure.

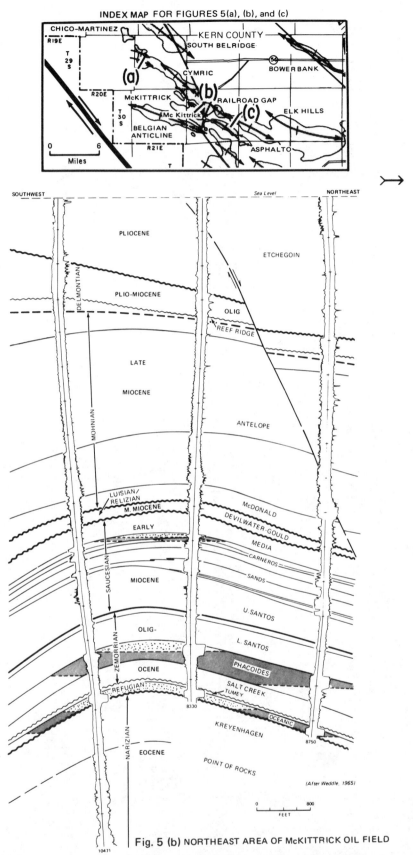

INDEX MAP FOR FIGURES 5(a), (b), and (c)

Fig. 5 (b) NORTHEAST AREA OF McKITTRICK OIL FIELD

FIG. 5—Cross sections across three producing pools in central part of west-side foldbelt opposite south end of San Joaquin Valley. Abrupt shifts in stratigraphic levels at which beds are or are not involved in folding demonstrate that these folds began and ceased growing at progressively younger times outward from San Andreas fault. Productive limits in **a** after California Division of Oil and Gas (1973). See index map for location of sections.

Folding and Trap Development, San Joaquin Valley

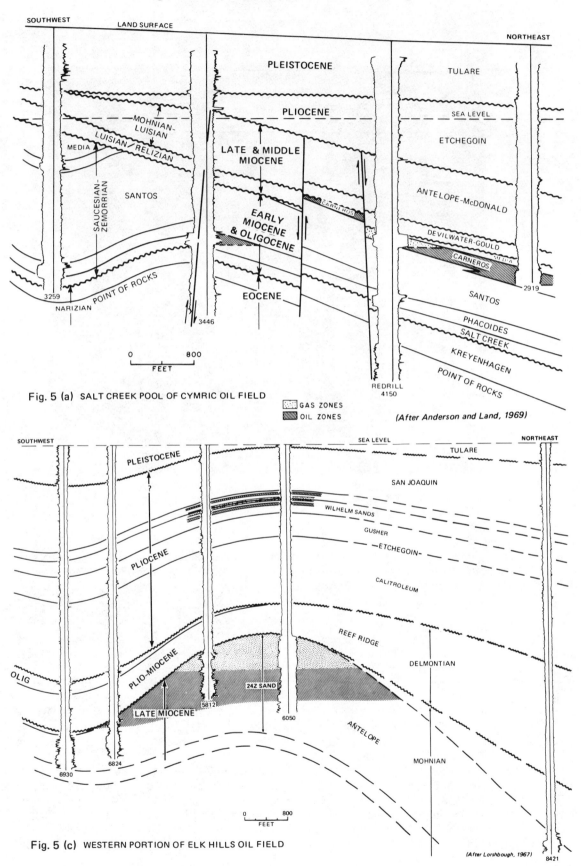

Fig. 5 (a) SALT CREEK POOL OF CYMRIC OIL FIELD

GAS ZONES
OIL ZONES

(After Anderson and Land, 1969)

Fig. 5 (c) WESTERN PORTION OF ELK HILLS OIL FIELD

(After Lorshbough, 1967)

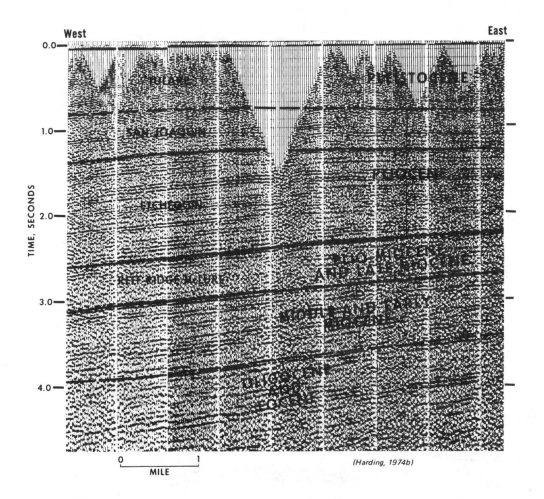

FIG. 6—Seismic profile across Semitropic gas field structure (interpretation by T. M. Hamilton). Shallowest reflectors at top of sedimentary wedge exhibit steepest east dips and largest degree of rollover. Field well control also demonstrates loss of closure with depth (Kaplow, 1938, pl. III).

reflectors show no discernible effects of deposition synchronous with or following fold growth, and, in the lower part of the sedimentary wedge, flexing was of an insufficient degree to rotate out the original west dip imposed by Pliocene and earlier basin subsidence.

At the far southeast end of the west-side producing fairway, comprehensive subsurface control does not extend below late Miocene strata, but a similar timing sequence is evident. Three examples are shown in Figure 7.

At the pool nearest the San Andreas fault (General American pool, Fig. 7a), late Miocene horizons reflect the strongest folding, but later events may be obscured partly by abrupt stratigraphic changes. Five mi (8 km) east-southeast

(Exeter and 29-D pools, Fig. 7b) and more distant from the San Andreas fault, local fold growth, for which there is control, began essentially near or at the end of the Pliocene-Miocene and continued strongly into the Pliocene or even into the Pleistocene at pools just on the east (Zulberti, 1961). Four mi (6.4 km) farther northeast at the Buena Vista oil field (Fig. 7c), significant discernible folding did not commence until probably late Pliocene.

Pleistocene folding was strongest at this outermost anticline and was aborted essentially at the two older structures. The contrast is emphasized by the 400 to 600 ft (122 to 183 m) of surface relief at Buena Vista Hills, which closely reflects the deeper structural configuration there, and by

332

the substantial dip reversal in the Tulare Formation. Continued compression at Buena Vista apparently is evidenced by an active thrust fault which has buckled production-gathering lines on the southwest side of the field (Manning, 1968).

Lack of pre-latest Miocene flexing at Buena Vista and at the Exeter and 29-D pools is inferred from detailed cross sections (Borkovich, 1958; Zulberti, 1957, 1961) showing general parallelism of late Miocene stratigraphic markers across these and adjacent structures, and from projection of the regional control that establishes the outward limits of areas affected by the older folding episodes (asterisks, Fig. 3b). Surface control in the Temblor Range west of the General American pool apparently may support these projections and also the overall timing pattern. Fletcher (1962) has interpreted structural quiescence at this position in the foldbelt from Eocene until the end of Miocene, at which time the most intense folding occurred. Minor flexing continued during the end of the Pliocene-Miocene and Pleistocene, the latter more or less amplifying preexisting structures.

Timing of Main North End Folds

At the northwest end of the producing fairway (upper left quadrant of Fig. 1), the distribution of critical control points is not nearly so comprehensive areally or stratigraphically because folds are eroded into the lower Tertiary and Mesozoic for considerably greater distances northeast of the San Andreas fault. Dense field-well drilling has been concentrated on only several trends, but these complete the Tertiary fold history.

Three main structures, the Vallecitos and Waltham Canyon synclines and the Coalinga anticline (Fig. 8), demonstrate that in at least several areas known pre-Mohnian folding was more extensive here and probably earlier in its inception than may be shown in the subsurface at the south end. In addition, strong Pleistocene refolding was also extensive and has affected the entire lengths of these flexures including uptrend positions near the San Andreas fault.

Past workers, using sediment distribution, depositional thicks, and facies patterns, have interpreted that the Vallecitos and Waltham Canyon synclines had early histories as deep, narrow, marine troughs that connected the San Joaquin Valley basin with open-ocean areas in the northwest. The Vallecitos trough on the north (Fig. 8) has been considered to have had its inception during the Paleocene and was probably a seaway through the early and late Eocene (or Oligocene, Seiden, 1964, Fig. 2) and middle Miocene (Flynn, 1963; see also Wilkinson, 1959). The San Benito-

Waltham Canyon trough is thought to have formed possibly as early as middle Miocene and persisted through the late Miocene and Pliocene (Flynn, 1963; Seiden, 1964, Fig. 6).

At the intervening Coalinga anticline, early structural growth of the upplunge Coalinga nose part of the trend (Fig. 8; left side of Fig. 9) has had an important influence on the development of stratigraphic traps (Harding, 1974a). The earliest tectonic influence of the nose for which there is abundant subsurface control is a deflection in the shaleout line of the shallow marine or shoreline deposits of the early Eocene Gatchell sand that traps the giant oil accumulation of the East Coalinga Extension field (Figs. 9, 10). Nilsen and Dibblee (1974) stated that the area on the northwest (i.e., upplunge) that now contains exposures of Franciscan and serpentine rocks undoubtedly was uplifted in post-middle Eocene time. The general contemporaneity of the anticline thus indicated and the complementary syncline or basin in the flanking Vallecitos region suggest Eocene, and perhaps Paleocene, folding of significant regional extent. The west-northwest, east-southeast orientation of the Paleocene and early Eocene isopach thick at Vallecitos shown by Hoots et al (1954) suggests further that the folds had a trend and possibly an en-echelon pattern similar to the present fold set. Thus, they were probably precursors of this fold set.

Relations at the Guijarral Hills field (Fig. 9) demonstrate that development of the Coalinga nose continued through the Oligocene. The distinctive distribution pattern and pinchout line of the field's main reservoir, the deep-water late Oligocene *Leda* sand (Cushman and Simonson, 1944), reflect strong influence (Figs. 10, 11). In addition, structure contours display greater relief on this horizon than is shown by similar maps at the base of the Miocene (Sullivan, 1962, pls. II, V). A regional east-west transverse cross section through the field by Church and Krammes (1957) indicates crestward truncation and/or thinning up both flanks of the nose in both the Tumey (lower Oligocene) and *Leda* shale sections. Their control also suggests that the zero line of the Oligocene truncation has been deflected considerably downplunge to its location in the East Coalinga Extension field (Fig. 9). Church and Krammes showed the truncation 7 mi (11 km) due west of the Pleasant Valley field, and a simple strike projection of this regional position at the valley margin would have crossed the nose considerably northwest of East Coalinga.

An episode of middle Miocene structural growth may be concluded from the lithology and distribution of the younger middle Miocene Big

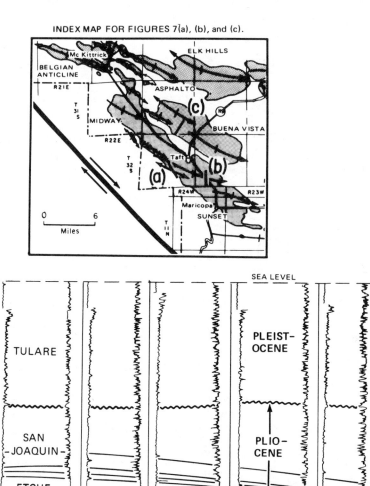

INDEX MAP FOR FIGURES 7(a), (b), and (c).

Fig. 7 (b) EXETER AND 29-D POOLS OF MIDWAY-SUNSET OIL FIELD

FIG. 7—Cross sections across three producing pools at south end of west-side foldbelt. Progressively younger beds are involved in folding northeastward from San Andreas fault. Productive limits in **b** and **c** after California Division of Oil and Gas (1973). See index map for location of sections.

334

Folding and Trap Development, San Joaquin Valley

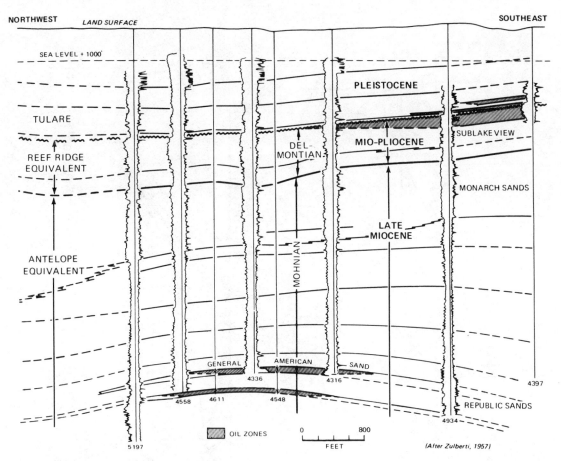

Fig. 7 (a) GENERAL AMERICAN POOL OF MIDWAY-SUNSET OIL FIELD

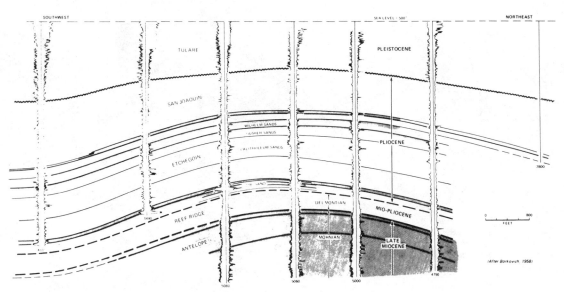

Fig. 7 (c) BUENA VISTA OIL FIELD

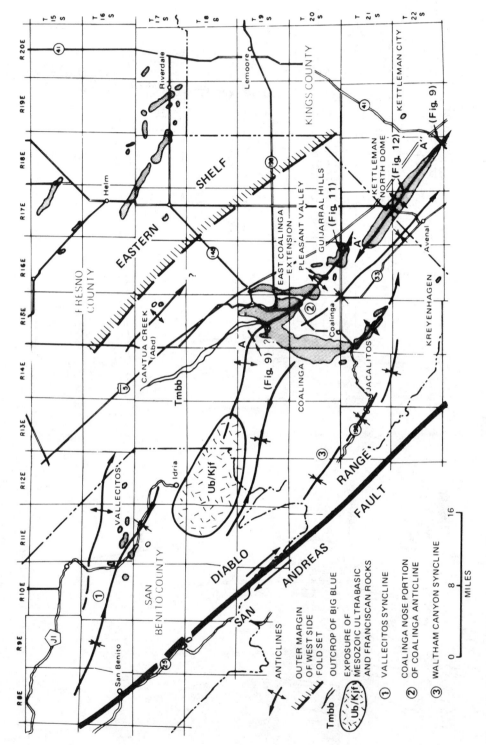

Fig. 8—Index map of subsequent text figures, major surface and subsurface structures, oil fields, and critical outcrops at north end of west-side producing trend, San Joaquin Valley basin.

Blue zone that crops out updip on the flanks of the nose (Fig. 8). According to Eckel and Myers (1946),

The Big Blue consists almost entirely of serpentine flakes and boulders and in the Big Blue Hills, about 8 miles east of the southeast end of the Idria serpentine body [indicated by random dash pattern on Fig. 8], it attains a maximum thickness of 1,000 feet. It thins northward and southward but has been traced. . . about 20 miles.

They report that the Idria serpentine diapir is the only nearby serpentine mass of sufficient size to have provided a source for the large volumes of debris. Its protrusion through the overlying sedimentary section at the crest of the anticline during the middle Miocene evidences renewed compression and folding. The upplunge gradation to a lenticular and less permeable nonmarine facies that partly controls giant middle Miocene production in the Coalinga oil field is a closely related consequence of this rejuvenation and its accompanying influx of serpentine debris.

A more detailed study of the entire Tertiary section in the Coalinga area possibly could indicate more episodes of structural growth.

The first culmination downplunge from the Coalinga nose corresponds to the Kettleman North Dome oil field (Fig. 8). At this position the Coalinga anticline trend is relatively quite distant from the San Andreas fault and constitutes the outermost element in the fold set (Fig. 1). Folding has been considered to be mostly or entirely middle Pleistocene by prior investigators (Woodring et al, 1940; Seiden, 1964; Sullivan, 1966). Relative recency of at least part of the deformation is demonstrated by the presence of more than 700 ft (213 m) of topographic expression that conforms closely in general shape to the subsurface structure, and by the 20 to 40° dips in exposed Pleistocene beds that fringe the anticline.

Similar evidences of Pleistocene growth persist northwestward to the Coalinga nose area and beyond, but Seiden (1964) reported that isopachs of the Eocene and Miocene (approximately Eocene, Oligocene, and Miocene of this report) show no evidence of older structural growth at Kettleman North Dome field. Consistent southeastward into-the-basin thickening is present without disruption across the culmination in most horizons (Fig. 12), and cross sections perpendicular to the fold axis (Sullivan, 1966) show no anomalous crestward thinnings or truncation such as those uptrend. Furthermore, structure contours at five levels demonstrate a marked, consistent loss of vertical closure in successively older horizons rather than the increases typical on culminations

that have been rejuvenated (compare progressively deeper levels of rollover on right side of Fig. 9). Vertical closure diminishes from 3,000 ft (914 m) on a lower Pliocene surface marker (Woodring et al, 1940, pl. 5) to 1,750 ft (533 m) on the lower Eocene McAdams sands (Sullivan, 1966, pl. IV). Late folding and consequent uplift of a basinward-expanding section are thought to be indicated by these relations.

TECTONIC IMPLICATIONS

Late Cenozoic Control

The periods of fold growth described in the subsurface along the west side of the San Joaquin Valley basin correspond to the known and proposed episodes of strike-slip on the San Andreas fault. It is believed that the parts of the foldbelt studied are sufficiently complete samples to make the observed relation tectonically meaningful.

Folding correlates closest with the better known history of late Cenozoic displacements and their anomalies, and this is demonstrated best in the south end of the San Joaquin Valley. The first en-echelon folding (Saucesian axes on Fig. 3b) approximates the general start-up in the early Miocene of significant, well-documented strike-slip movements on the San Andreas. By contrast, instances of late Eocene and Oligocene structural growth are notably mild and areally are quite limited here; their absence correlates with an indicated period of comparatively minor strike-slip displacements.

Folding, with shifts from one structure to another, has been essentially continuous since the early Miocene, as apparently have been the strike-slip movements of the San Andreas fault (Fig. 2). The area undergoing discernible fold growth increased noticeably near the end of the Miocene or within the Pliocene-Miocene (Mohnian and Delmontian axes, respectively, on Fig. 3c), and this increase has been amplified in subsequent periods. The accelerated folding is contemporaneous with an acceleration of fault movement commencing at about 7.4 to 12.5 m.y. B.P. (Dickinson et al, 1972; and others).

Within the pattern of outward structural growth are fold cessations upstructure that occurred while flexing continued farther down the anticlinal trend from the San Andreas fault (Fig. 5). Though not fully understood, these examples demonstrate that folding was not directly dependent on fault movement (i.e., drag) for its tectonic drive. Instead, the two features are different structural responses to a common regional coupling movement that must originate in the deeper crust and may include components of plate convergence. While major strike-slip displacements

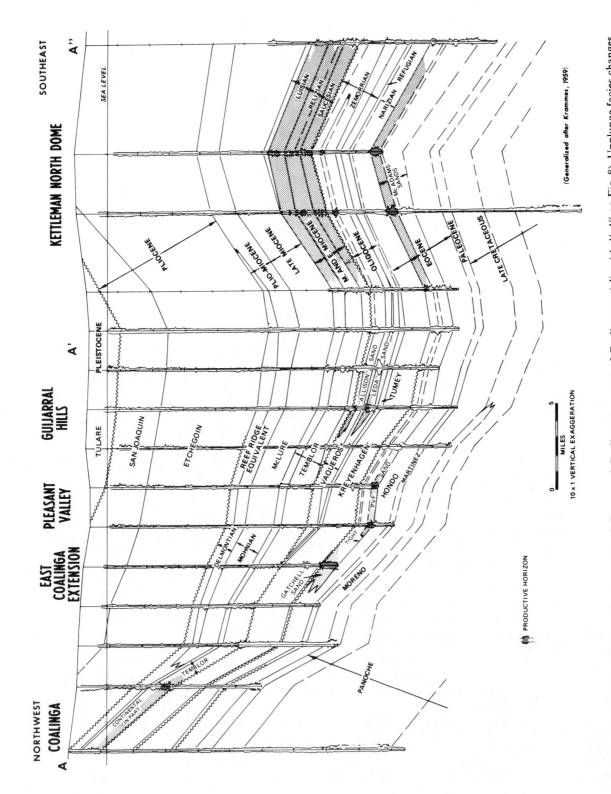

Fig. 9—Cross section along Kettleman North Dome and Coalinga nose part of Coalinga anticline (*A-A'-A"* on Fig. 8). Upplunge facies changes and truncations are indications of early growth of nose. Pre-Mohnian stratigraphic stages are projected onto figure from southeastward continuation of Krammes (1959) original cross section.

338

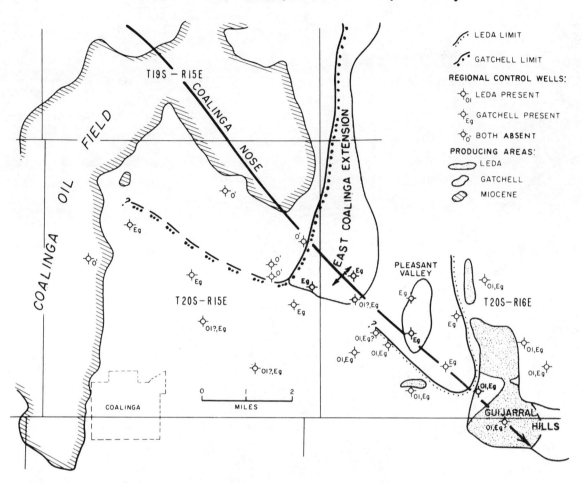

FIG. 10—Upplunge limits of Oligocene *Leda* and early Eocene Gatchell sands in vicinity of Coalinga nose. Structure is asymmetric with decidedly steeper southwest flank. Control in respective producing areas is much denser than indicated by regional well symbols (see Fig. 11).

remained concentrated on the system's inner boundary and established line of throughgoing crustal weakness, folds were propagated outward in an expanding, diffuse deformation front. In the case of the San Joaquin Valley, the resulting folding undoubtedly was enhanced by the presence of relatively mobile Franciscan basement on the northeast side of the San Andreas fault. Dickinson (1966) similarly and more elaborately has proposed that the structures formed within a passive, surficial plate in response to an underlying, broadly distributed velocity-gradient flow oriented longitudinal to the San Andreas and with gradient magnitude increasing toward the fault.

The interior structures that aborted continued growth apparently did so when they became less susceptible to renewed folding than the terrane nearer the advancing front. Producing anticlines at interior positions generally are flexed more tightly, and those nearest the San Andreas com-

monly are disrupted by relatively young thrusting that may be a late-stage reaction or strain release to the prolonged deformation. Secondary faults in the Temblor Range nearest the San Andreas are apparently older in part than those on the northeast because in some locations they appear to have been dragged and folded by later stages of the deformation (Dickinson, 1966).

Early Cenozoic Speculations

The presence of large early Tertiary folds at the northwest end of the anticlinorium may corroborate previous proposals of a pre-Miocene proto-San Andreas system. The structures demonstrate that significant deformation occurred during much of the Paleocene, Eocene, and Oligocene Epochs. Documentation of an early Cenozoic en-echelon fold pattern that can be related exclusively to strike-slip, however, is only fragmentary. Yet, when the early and late Cenozoic control is

339

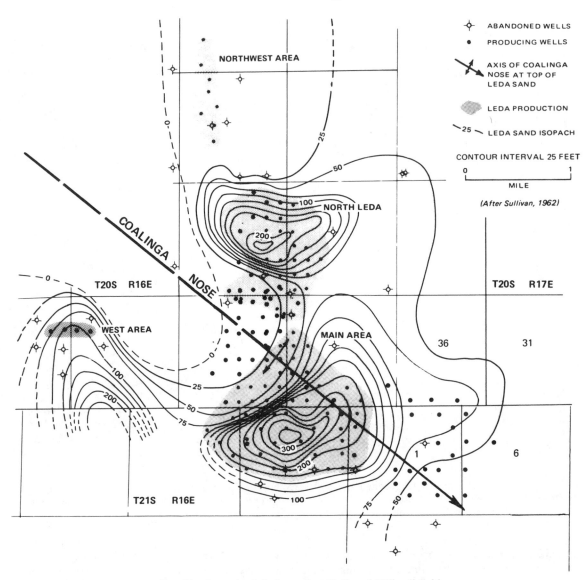

FIG. 11—Isopach of *Leda* sand at Guijarral Hills oil field.

considered together, the data support such a fold set, related temporally and spatially to the present San Andreas fault, and indicate a potentially significant longitudinal progression of fold inceptions.

At positions approximately equidistant from the San Andreas fault, the first precursors of the present fold set that can be identified are Paleocene and early Eocene in age at Vallecitos and Coalinga nose (Fig. 8), early Miocene or possibly early Oligocene farther south at Antelope Hills, early Miocene at Salt Creek, and possibly as late as late Miocene in Midway Sunset (Fig. 3). Absence of data directly adjacent to the San Andreas (where presumably the progression would be oldest) and of control below upper Eocene horizons at the last three areas is critical, but Oligocene folding, which is controlled at both ends of the fold set (Figs. 4, 5, 10), displays a documentable southeastward decrease in intensity that adds credence to the apparent progression.

One inference is that an early Cenozoic strand of the present San Andreas fault had its inception north of the San Joaquin Valley and was propagated southeastward during ensuing episodes of strike-slip and related folding. Troxel (1970) has documented such a longitudinal propagation of folds and fault rupture, but on a much smaller scale and shorter time interval, along the Death Valley strike-slip fault zone of eastern California. Southeastward propagation by the accumulation of strike-slip offsets could help to explain the

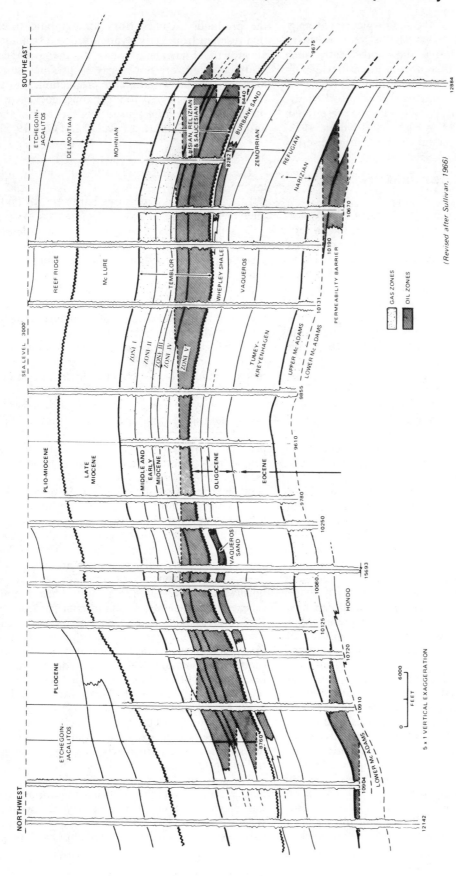

FIG. 12—Longitudinal cross section along axis of Kettleman North Dome structure. Critical northeast plunge reversal is greater on base of Reef Ridge than at top of upper McAdams sand (regional southeast plunge continues to right of figure), and consequently latter level has less vertical closure. Stratigraphic stages adapted from Krammes (1959).

large discrepancies in age and displacement magnitudes described previously between northern and southern California and, more locally, such anomalies as the differences in intensities of Oligocene folding. The options of major early Cenozoic displacements having been absorbed by splays or other faults parallel with the present San Andreas (Howell, 1975; Clarke and Nilsen, 1973) seem required to account for most of the discrepancies, however.

CONSEQUENCES TO HYDROCARBON EXPLORATION

Significant changes in hydrocarbon-trap morphology have resulted from the timing pattern. Because of cessation of folding, some productive anticlines nearest the San Andreas fault are buried beneath younger, relatively undisturbed beds. The shallow zones themselves may produce from stratigraphically controlled closures formed by updip, out-of-the-basin wedging, onlap, truncation, etc. In these cases, a vertical stacking of disparate trap types results (Fig. 7a). Elsewhere, and generally at positions farther from the fault, strong younger structuring has pronounced surface expression, and there are incremental increases in relief through descending stratigraphic levels owing to the additive effects of several periods of flexing. At the far margin of the foldbelt, very young anticlines may have their greatest amplitude near the surface but can lose structural relief with depth because of the canceling-out effect of inherited basinward dips. Successive levels of increasing closure are then in just the opposite progression from those at several interiormost traps (compare Figs. 6, 4). In other exterior settings, very late culmination closures can remain essentially consistent with depth, lacing either increases in relief from prior flexing or decreases from basinward tilting and sedimentary wedging.

Relative position within the foldbelt can have other consequences to optimum trapping conditions. Pools nearest the San Andreas typically are segmented by internal faulting, often complexly so, and correspondingly are reduced in effective size. Significant faulting is minimal or entirely absent from anticlinal culmination closures at the periphery of the fold set. The latter not only have greater intact drainage areas because of this, but, having undergone generally less cumulative crustal shortening, they are quite commonly significantly larger folds with broader dip reversals.

An additional major influence has been the early Tertiary folding at the northwest end of the anticlinorium. Middle Miocene and older reservoirs account for all of the large quantities of oil and gas that have been produced there; much of the production comes from stratigraphic traps that resulted from early, syndepositional structural growth. Equivalent zones at the younger, south-end folds are not nearly so profilic; more than 90 percent of the major production here is from late Miocene and younger zones coeval with the established folding episodes of this part of the anticlinorium (California Div. Oil and Gas, 1972).

CONCLUSIONS

The results of this study substantiate the following conditions.

1. Fold timing is a sensitive barometer of fault movements.

2. Both faults and folds had a common deformational genesis, but fold development is not directly dependent on drag at the fault for its tectonic drive.

3. The outward expansion of the deformed area causes progressive changes in zones that will have the most optimum fitting of structural-growth timing, hydrocarbon maturation, and migration.

4. Previous generalizations that folding was mostly of one age (middle Pleistocene) are incorrect and inadequate for effective hydrocarbon exploration.

Tertiary strike-slip in California is distributed across both major and minor faults. Instances of subsidiary wrench faults aligned parallel with the principal displacement zone and typically with a similar movement sense are present in wrench systems here and elsewhere. If deformational zones interfere spatially, they should disrupt the regular outward fold-growth sequence and create a local timing pattern of their own. The subsurface Wasco-Greely trend near the valley center (Fig. 3a) is considered an example. There folds with definite early Tertiary growth are in a right-handed en-echelon orientation adjacent to a northwest-southeast-trending fault. Their timing repeats the growth histories in interior parts of the San Andreas fold set and has potential regional significance as further evidence of proto-San Andreas stage wrenching. Southwest of Fresno (northwest quadrant of Fig. 1), a roughly right-handed en-echelon group of productive subsurface anticlines is also early Tertiary and possibly latest Cretaceous in age. They may have had a similar tectonic origin. Full appreciation of the significance of these examples requires additional investigation and particularly deep-seismic control along the trend connecting the two areas.

The sequential outward growth of folds within narrow bands along the faults casts doubt on certain previous tectonic proposals. These proposals suggest that the San Andreas is part of a conjugate system of northwest- and northeast- (e.g.,

Garlock fault) trending faults that have resulted from regional north-south compression in a pure shear setting (e.g., Lensen, 1959, Fig. 3). Such a system would seem to require that folds be oriented east-west across the basin, each alignment being essentially contemporaneous along its length, and that these fold trends would most likely be progressively younger northward away from the junction of the conjugate elements at the south margin of the basin. This is clearly not the case (Fig. 1). Newer plate tectonic syntheses indicate a transform plate boundary here (Atwater, 1970) whose deformation would be dominated by the diffuse crystal couple arising from northwest-southeast plate motion. This more closely satisfies the observed relations.

REFERENCES CITED

Addicott, W. O., 1968, Mid-Tertiary zoogeographic and paleogeographic discontinuities across the San Andreas fault, California: Stanford Univ. Pubs. Geol. Sci., v. 11, p. 144-165.

Anderson, D. N., and P. E. Land, 1970, Cymric oil field: California Oil Fields, v. 55, no. 1, p. 5-21.

Atwater, T., 1970, Implications of plate tectonics for the Cenozoic tectonic evolution of western North America: Geol. Soc. America Bull., v. 81, p. 3513-3536.

Bazeley, W. J. M., 1961, 175 miles of lateral movement along the San Andreas fault since lower Miocene (abs.): Pacific Petroleum Geologists Newsletter, v. 15, p. 2-3.

Borkovich, G. J., 1958 [1959], Buena Vista oil field: California Oil Fields, v. 44, no. 2, p. 5-20.

California Division of Oil and Gas, 1972, Oil, gas and geothermal production statistics, 1972: California Oil Fields, v. 58, no. 2, 171 p.

——— 1973, California oil and gas fields, Volume 1, north and east-central California: California Div. Oil and Gas, variously paged.

Campbell, J. D., 1958, En echelon folding: Econ. Geology, v. 53, p. 448-472.

Church, H. V., Jr., and K. F. Krammes, 1957, Correlation section across central San Joaquin Valley from San Andreas fault to Sierra Nevada foothills: AAPG Pacific Sec., Correlation Sec. 9.

Clarke, S. H., Jr., and T. H. Nilsen, 1973, Displacement of Eocene strata and implications for the history of offset along the San Andreas fault, central and northern California, in Conference on tectonic problems of the San Andreas fault system, Proc.: Stanford Univ. Pubs. Geol. Sci., v. 13, p. 358-367.

Crowell, J. C., 1962, Displacement along the San Andreas fault, California: Geol. Soc. America Spec. Paper 71, 61 p.

——— 1973, Problems concerning the San Andreas fault system in southern California, in Conference on tectonic problems of the San Andreas fault system, Proc.: Stanford Univ. Pubs. Geol. Sci., v. 13, p. 125-135.

Cushman, J. A., and R. R. Simonson, 1944, Foraminifera from the Tumey Formation, Fresno County, California: Jour. Paleontology, v. 18, p. 186-203.

Dickinson, W. R., 1966, Structural relationships of San Andreas fault system, Cholame Valley and Castle Mountain Range, California: Geol. Soc. America Bull., v. 77, p. 707-726.

——— D. S. Cowan, and R. A. Schwickert, 1972, Test of new global tectonics: Discussion: AAPG Bull., v. 56, p. 375-384.

Eckel, E. B., and W. B. Myers, 1946, Quicksilver deposits of the New Idria district, San Benito and Fresno Counties, California: California Jour. Mines and Geology, v. 42, no. 2, p. 81-124.

Ellison, B. R., 1952, The East Coalinga Extension oil field, in Guidebook, field trip routes, oil fields, geology: AAPG, SEPM, SEG Ann. Mtg., Los Angeles, California, p. 182-183.

Fletcher, G. L., 1962, The Recruit Pass area of the Temblor Range, San Luis Obispo and Kern Counties, California, in Geology of Carrizo Plains and San Andreas fault—guidebook: San Joaquin Geol. Soc., p. 16-20.

Flynn, D. B., 1963, The San Benito—Waltham Canyon trough—possible oil province, in Geology of Salinas Valley and the San Andreas fault—Guidebook AAPG-SEPM Pacific Secs. Spring Field Trip: AAPG Pacific Sec., p. 27-33.

Hackel, O., 1966, Summary of the geology of the Great Valley, in Geology of northern California: California Div. Mines Bull. 190, p. 217-238.

Harding, T. P., 1974a, Petroleum traps associated with wrench faults: AAPG Bull., v. 58, p. 1290-1304.

——— 1974b, Sequential growth timing of wrench structures and its influence on hydrocarbon closure developments: Soc. Brasiliera Geologia, 28th Cong. Brasileiro Geologia Anais (in press).

Hill, M. L., and T. W. Dibblee, Jr., 1953, San Andreas, Garlock, and Big Pine faults, California (a study of the character, history, and tectonic significance of their displacements): Geol. Soc. America Bull., v. 64, p. 443-458.

——— and H. D. Hobson, 1968, Possible post-Cretaceous slip on the San Andreas fault zone: Stanford Univ. Pubs. Geol. Sci., v. 11, p. 123-129.

Hoots, H. W., T. L. Bear, and W. D. Kleinpell, 1954, Geological summary of the San Joaquin Valley, California, in Geology of southern California: California Div. Mines Bull. 170, p. 113-129.

Howell, D. G., 1975, Hypothesis suggesting 700 km of right slip in California along northwest-oriented faults: Geology, v. 3, p. 81-83.

Huffman, O. F., 1970, Miocene and post-Miocene offset on the San Andreas fault in central California (abs.): Geol. Soc. America Abs. with Programs, v. 2, p. 104-105.

——— 1972, Lateral displacement of upper Miocene rocks and the Neogene history of offset along the San Andreas fault in central California: Geol. Soc. America Bull., v. 83, p. 2913-2946.

Jennings, C. W., 1958, Geologic map of California, San Luis Obispo sheet, Olaf P. Jenkins edition: California Div. Mines and Geology, scale 1:250,000.

Kaplow, E. J., 1938, Gas fields of southern San Joaquin Valley: California Oil Fields, v. 24, no. 1, p. 30-50.

Karp, S. E., ed., 1968, Geology and oil fields, west side southern San Joaquin Valley—AAPG, SEG, and SEPM, Pacific Sec. 43d Ann. Mtg. Guidebook: AAPG Pacific Sec., p. 78-81.

Krammes, K. F., chm., 1959, Correlation section longitudinally north-south through westside San Joaquin Valley from Coalinga to Midway-Sunset and across San Andreas fault into southeast Cuyama Valley, California: AAPG Pacific Sec., Correlation Sec. 11.

Lensen, G. J., 1959, Secondary faulting and transcurrent splay-faulting at transcurrent fault intersections: New Zealand Jour. Geol. and Geophys., v. 2, p. 729-734.

Lorshbough, A. L., 1961, Antelope Hills oil field: California Oil Fields, v. 47, no. 2, p. 23-35.

———— 1967, Western portion of Elk Hills oil field: California Oil Fields, v. 53, no. 1, p. 33-37.

Maher, J. C., R. D. Carter, and R. V. Lantz, 1975, Petroleum geology of Naval Petroleum Reserve No. 1, Elk Hills, Kern County, California: U.S. Geol. Survey Prof. Paper 912, 109 p.

Manning, J. C., 1968, Field trip to areas of active tectonism and shallow subsidence in the southern San Joaquin Valley, in Geology and oil fields, west side southern San Joaquin Valley—AAPG, SEG, SEPM, Pacific Secs., 43d Ann. Mtg. Guidebook: AAPG Pacific Sec., p. 133-134.

Moody, J. D., and M. J. Hill, 1956, Wrench-fault tectonics: Geol. Soc. America Bull., v. 67, p. 1207-1246.

Nilsen, T. H., and T. W. Dibblee, Jr., 1974, Stratigraphy and sedimentology of the Cantua Sandstone Member of the Lodo Formation, Vallacitos area, California, in The Paleogene of the Panoche Creek—Cantua Creek area, central California: SEPM Pacific Sec., Guidebook, p. 36-68.

Park, W. H., P. E. Land, and D. D. Bruce, 1957, Belgian Anticline oil field: California Oil Fields, v. 43, no. 1, p. 5-12.

Ross, D. C., C. M. Wentworth, and E. H. McKee, 1973, Cretaceous mafic conglomerate near Gualala offset 350 miles by San Andreas fault from oceanic crustal source near Eagle Rest Peak, California: U.S. Geol. Survey Jour. Research, v. 1, p. 45-52.

Seiden, H., 1964, Kettleman Hills area: San Joaquin Geol. Soc. Selected Papers, v. 2, p. 46-53.

Smith, A. R. (compiler), 1964, Geologic map of California, Bakersfield sheet, Olaf P. Jenkins edition: California Div. Mines and Geology, scale 1:250,000.

Sullivan, J. C., 1962, Guijarral Hills oil field: California Oil Fields, v. 48, no. 2, p. 37-51.

———— 1966, Kettleman North Dome oil field: California Oil Fields, v. 52, no. 1, p. 5-21.

Suppe, J., 1970, Offset of late Mesozoic basement terrains by the San Andreas fault system: Geol. Soc. America Bull., v. 81, p. 3253-3258.

Troxel, B. W., 1970, Anatomy of a fault zone, southern Death Valley, California (abs.): Geol. Soc. America Abs. with Programs, v. 2, p. 154.

Turner, D. L., 1969, K-Ar ages of California coast range volcanics—implications for San Andreas fault displacement (abs.): Geol. Soc. America Abs. with Programs, p. 70.

———— 1970, Potassium-argon dating of Pacific Coast foraminiferal stages: Geol. Soc. America Spec. Paper 124, p. 91-129.

———— G. H. Curtis, F. A. F. Berry, and R. Jack, 1970, Age relationships between the Pinnacles and Parkfield felsites and felsite clasts in the southern Temblor Range, California—implications for San Andreas fault displacements (abs.): Geol. Soc. America Abs. with Programs, v. 2, p. 154.

Weddle, J. R., 1965, Northeast area of McKittrick oil field: California Oil Fields, v. 51, no. 2, p. 5-20.

Wilcox, R. E., T. P. Harding, and D. R. Seely, 1973, Basic wrench tectonics: AAPG Bull., v. 57, p. 74-96.

Wilkinson, E. R., 1959, Vallecitos field: California Oil Fields, v. 45, no. 2, p. 17-33.

Woodring, W. P., R. B. Stewart, and R. W. Richards, 1940, Geology of the Kettleman Hills oil field, California: U.S. Geol. Survey Prof. Paper 195, 170 p.

Zulberti, J. L., 1957, Republic sands of Midway-Sunset field: California Oil Fields, v. 43, no. 2, p. 21-33.

———— 1960, Exeter and 29-D pools of Midway-Sunset oil field: California Oil Fields, v. 46, no. 1, p. 41-50.

———— 1961, Lakeview pool of Midway-Sunset oil field: California Oil Fields, v. 47, no. 1, p. 29-38.

HYDRODYNAMIC TRAPS

BULLETIN OF THE AMERICAN ASSOCIATION OF PETROLEUM GEOLOGISTS
VOL. 37, NO. 8 (AUGUST, 1953), PP. 1954-2026, 44 FIGS.

ENTRAPMENT OF PETROLEUM UNDER HYDRODYNAMIC CONDITIONS[1]

M. KING HUBBERT[2]

Houston, Texas

ABSTRACT

The anticlinal or so-called "gravitational" theory, despite its effectiveness as a basis for petroleum exploration, represents but a special case of oil and gas accumulation, and is valid only when the associated ground water is in hydrostatic equilibrium. Since this need not be the case, a more general formulation, valid for both hydrostatic and hydrodynamic conditions, is required.

Oil and gas possess energy with respect to their positions and environment which, when referred to unit mass, may be termed the potential at any given point of the fluid considered. When the potential of a specified fluid in a region of underground space is not constant, an unbalanced force will act upon the fluid, driving it in the direction in which its potential decreases. Hence, oil and gas in a dispersed state underground migrate from regions of higher to those of lower energy levels, and come ultimately to rest in positions which constitute traps, where their potentials assume locally minimum or least values. In nearly all cases traps for petroleum are regions of low potential which are enclosed jointly by regions of higher potential and impermeable barriers.

Oil and gas migration occurs through a normally water-saturated environment. If the water is at rest, the oil and gas equipotential surfaces will be horizontal, the impelling forces will be directed vertically upward, and the traps will be the familiar ones of the anticlinal theory. If the water is in motion in a non-vertical direction, the oil and gas equipotentials will be tilted downward in the flow direction with those for oil inclined at an angle greater than those for gas. The impelling forces for oil and for gas will not be parallel and the two fluids will migrate in divergent directions to traps which in general will not coincide and may, in fact, be separated entirely, a trap for oil being incapable of holding gas, and *vice versa*.

Under hydrodynamic conditions accumulations of oil or gas will invariably exhibit inclined oil- or gas-water interfaces with the angle of inclination given by

$$\tan \theta = \frac{dz}{dx} = \frac{\rho_w}{\rho_w - \rho_0}\frac{dh}{dx},$$

where dz/dx is the slope of the interface, ρ_w the density of the water and ρ_0 that of the oil (or gas), and dh/dx the component of slope of the potentiometric surface of the water in the horizontal direction x. Stable oil and gas accumulations may be found in anticlines but they may equally well occur in structural terraces, noses, monoclines, and other unclosed structures entirely devoid of lithologic barriers to updip migration.

Not only are these effects theoretically expectable but they have been found to occur, with tilts ranging from tens to hundreds of feet per mile, in almost every major oil-producing area. If many such accumulations are not to be overlooked, we must supplement our customary knowledge of structure and stratigraphy with the three-dimensional ground-water hydrology of every petroliferous basin.

[1] Publication No. 29, Exploration and Production Technical Division, Shell Oil Company, Houston, Texas.

Manuscript received, November 5, 1952.

This paper was presented on March 26, 1952, before the American Association of Petroleum Geologists at its annual meeting in Los Angeles, and also during October and November, 1952, as a Distinguished Lecture before the affiliated societies of the Association.

A study of this kind is necessarily a cooperative enterprise, and the writer expresses his indebtedness to his many colleagues who in one way or another have assisted in its development. For specific assistance he is indebted to Henry Rainbow for the clarification of obscure mathematical points, to Robert Nanz and W. R. Purcell for information on capillary-pressure phenomena, and to Arne Junger, Charles H. Fay, and C. C. Templeton for critical discussions of the manuscript; but he is especially indebted to his research assistant, Jerry Conner, for performing the extensive series of confirmatory experiments cited, and to J. B. Woolley for invaluable editorial assistance in the final compilation of · the paper.

[2] Chief consultant (general geology), Shell Oil Company.

ENTRAPMENT UNDER HYDRODYNAMIC CONDITIONS

INTRODUCTION

During the first 5 years following Colonel Drake's discovery of oil at Titusville, Pennsylvania, in 1859, the idea that oil and gas in a ground-water environment are impelled by the forces of buoyancy into a stable stratified arrangement in porous or fractured strata on the crests of anticlines was clearly formulated; and by 1890 the "anticlinal theory" had become firmly established as the controlling principle of oil accumulation. (For a review of early literature see J. V. Howell, 1934.) Other geometrical arrangements such as fault and stratigraphic traps have subsequently been added, but these have mostly been in accord with the basic principle that oil and gas accumulations occur in a normally water-saturated environment in the highest local position to which these fluids can migrate. As a special case the possibility of a "dry" or water-free sand has also been admitted wherein the oil would occur in the lowest positions.

Implicit in this formulation, although rarely expressed, is the assumption that the environmental ground water is at rest and the resulting fluid equilibrium is one of complete hydrostatics with the interfaces between separate fluids, except as modified by capillary forces, forming horizontal surfaces.

The only notable departure from this line of thought was that represented by a minority literature extending from 1909 into the 1930's (M. J. Munn, 1909; E. W. Shaw, 1917; R. Van A. Mills, 1920; John L. Rich, 1921, 1923, 1931, 1934; V. C. Illing, 1938, 1939) in which the "hydraulic theory" of oil and gas accumulation was developed. Beginning with the basic premise that the migration of oil and gas in a hydrostatic environment is inhibited by capillary impediments, the flow of water was invoked as an essential condition for oil and gas migration and accumulation, the migration occurring always, except for an oblique upward drift, in the direction of the flow of the water. Entrapment of oil, according to this theory, would occur in any position where resistance to further migration exceeded the propulsive force exerted by the flowing water. Accumulations might thus be formed in anticlines where the oil could be dragged up the dip on one side but not down the dip on the other. They might occur by filtering action with the water flowing from coarse sands into silts or shales; or on unclosed structures such as noses or structural terraces where, with the water flowing up the dip, the oil might be arrested by the increased drag produced by a decrease in the angle of dip.

During the 1920's the hydraulic theory, although never completely accepted, exercised considerable influence on contemporary geologic thought. This rapidly subsided, however, and by the mid-1930's geologic thinking had reverted largely to the premises of hydrostatics. Recently the hydraulic theory has shown signs of revival, and has in fact been accepted as the basis of petroleum migration and accumulation in one recent textbook of petroleum geology (Tiratsoo, 1952, pp. 45–90).

The present inquiry stems from a study of some years ago (Hubbert, 1940) in which during an investigation of the motion of underground fluids—particularly ground water—it became necessary to determine the behavior of any fluid in an environment dominated by the presence of another in some state of motion. In particular it was determined that, whereas under hydrostatic conditions two fluids will arrange themselves with a horizontal interface and the less dense fluid uppermost, under dynamical conditions with one or both fluids in motion, the steady-state interface would in general be inclined at an angle which might assume any value up to the vertical.

Since oil and gas are minority fluids in a ground-water dominated environment, with the ground-water commonly in some state of motion, it is evident that the foregoing results should apply to problems of petroleum geology. In what follows therefore an endeavor will be made to establish the general theory of migration and entrapment of oil and gas under hydrodynamic conditions in which the more familiar hydrostatic relationships will emerge as but special cases.

GENERAL PRINCIPLES

Present-day accumulations of petroleum and natural gas are found invariably in or adjacent to sedimentary rocks. According to all available evidence, these fluids have originated from the organic matter deposited in sediments— principally marine—at the time of their deposition. (See Smith, 1952.) The organic-rich sediments are usually the fine-textured rocks, shales, and limestones of great areal and volumetric extent; whereas the present-day accumulations are usually found in highly restricted volumes in the coarse-textured rocks, sandstones, and porous or fractured limestones. The term organic-rich applied to a shale is of course relative; the amount of organic matter present is small compared to the great porosity (up to 80 per cent) of such a rock at the time of its deposition. It is inferred, therefore, that oil and gas must have originated in a highly dispersed state, from which they have been impelled to their present positions of concentration and entrapment.

In general, the sedimentary rocks are porous with the pore space forming an intricately branching three-dimensional network. Furthermore, below shallow depths from the earth's surface the pore spaces of the rocks are normally filled with water, so that the origin, migration, and final accumulation of petroleum must take place in an otherwise water-saturated environment. We envisage, therefore, petroleum in its initial dispersed state as consisting of numerous discontinuous volume elements, each entirely surrounded by water and the solid framework of the rock in which it occurs.

We now make use of one of the fundamental principles of mechanics applicable to all manner of mechanical systems of whatever degree of complexity: namely, that if such a system is at rest and not already in a configuration for which its potential energy is a minimum it will move spontaneously until such a configuration is achieved. Mechanical equilibrium therefore is characterized by a configuration for which the kinetic energy is zero and the potential energy of the system for all small displacements compatible with the constraints of the system either remains constant or increases. The equilibrium positions of a marble in a bowl, or of a mass suspended by a spring, are familiar illustrative examples.

Applying this principle to petroleum in its dispersed state, we recognize that each element possesses an amount of mechanical potential energy with respect to its environment which, in general, will vary with position. The element will accordingly be acted upon by an unbalanced force tending to impel it from regions where its energy is higher to those where it is lower. It will therefore tend to migrate from higher- to lower-energy regions and will come stably to rest in any region which is surrounded entirely by higher energy levels, or jointly by higher energy levels and impermeable barriers. A petroleum trap is therefore such a low-energy region, and the search for petroleum reduces in large part to the determination in underground space of the positions of these local low-energy regions.

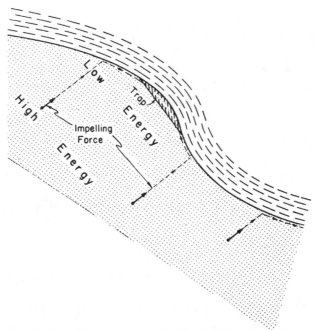

FIG. 1.—Dispersed oil in water environment migrating from region of high energy to one of low energy where impermeable barrier forms trap and accumulation is being built up.

FLOW OF GROUND WATER

Since petroleum migration occurs in a ground-water environment, the determination of the energy field for petroleum can most effectively be done in terms of that of the ambient ground water. We shall direct our attention, therefore, to the general behavior of ground water which initially we shall regard as a homogeneous fluid.

The water which saturates the pore spaces of the rocks of the earth below shallow depths possesses potential energy in the earth's gravitational field and can be in static equilibrium only when at a configuration for which this energy is a minimum. Since the pore spaces of the rocks form a three-dimensional interconnected network, those rocks are to some degree permeable to the flow of water. Ultimate equilibrium of water would therefore occur only if the upper surface of the ground water, or more strictly the water table, were a horizontal surface. Actually, the water table follows closely the earth's topographic variations, with the result that the water is not in equilibrium and so must flow continuously, descending into the ground in regions where the topography is high and emerging in areas where the topography is low. If no water were added to such a system and none withdrawn except by flow, movement would continue until the ground-water table was everywhere at the same level. Actually, water is repeatedly added to the system at the higher elevations by precipitation, keeping it perpetually out of equilibrium, and thus maintaining a general ground-water circulation.

If the ground water is of constant density and in motion in one region of underground space, it will be shown later that it must also be in motion throughout all space not isolated by impermeable barriers. If, on the contrary, the water is inhomogeneous and consists of different bodies of contrasting density, such as fresh water and salt water, it is possible for one kind of water, say fresh water, to be flowing while a contiguous body of salt water remains in hydrostatic equilibrium. Since water upon entering the ground is fresh and is not in equilibrium, it follows from the foregoing that fresh water at whatever depth, unless isolated by impermeable barriers, should be in some degree of motion. Hence, in the absence of more positive evidence, the occurrence of fresh or brackish water underground is presumptive of a dynamic state. Saline water, on the contrary, may or may not be in equilibrium.

GROUND-WATER POTENTIAL

More specific evidence of the state of motion of ground water in any region is afforded by the energy conditions which prevail. Taking a local viewpoint, an element of water at any point possesses potential energy with respect to its environment which, when referred to unit mass, we may speak of as its potential, Φ. If throughout a given region of space the potential of water at all points were constant, no work would be required to move an element of water from one point to another by a frictionless process. Hence no unbalanced force would be exerted upon the element by its environment, and there would be no tendency for the water to flow. If, on the contrary, the potential across the region is variable, work would be required to transport an element of water from a point of lower to one of higher potential. In this case an unbalanced force must therefore be exerted upon the fluid by its environment, tending to impel it in the direction of most rapid decrease of the potential.

The potential Φ of water at a given point may be thought of as the amount of work that would be required to transport a unit mass of this fluid from some arbitrarily chosen standard position and state to the position and state of the point considered. For the standard state we may take a closed chamber containing water at elevation z_0 and pressure p_0. The final state is that of the point of interest whose elevation is z and whose fluid pressure is p. The work done in transporting a unit mass of water from the initial system to the final one is composed of two parts, work against gravity to lift unit mass of water from elevation z_0 to z, and work against pressure required to pump unit mass from a chamber of pressure p_0 to one of pressure p. The sum of those two terms is the potential Φ which is given by

$$\Phi = g(z - z_0) + (p - p_0)v, \tag{1}$$

where v is the specific volume, or volume per unit mass, of the fluid, and g the acceleration of gravity. However,

$$v = \frac{\text{volume}}{\text{mass}} = \frac{1}{\rho}, \tag{2}$$

where ρ is the fluid density, which, when inserted into equation (1), gives

$$\Phi = g(z - z_0) + \frac{p - p_0}{\rho}. \tag{3}$$

Since the choice of the standard state is arbitrary, it will be convenient to choose sea-level and atmospheric pressure as our reference state and to set

$$z_0 = 0; \quad p_0 = 1 \text{ atmosphere}.$$

Then the potential at any point would be equal to the work required to transport unit mass of water from sea-level and atmospheric pressure to the elevation and pressure of the point considered. The potential then simplifies to

$$\Phi = gz + \frac{p}{\rho} \tag{4}$$

where p is now the gauge pressure (absolute pressure less the pressure of the atmosphere) at the point considered. For high-velocity flow a kinetic-energy term would also be required, but for the slow, creeping motion of underground flow this term is so small in comparison with the other two that it may be neglected.

Since the elevation z and the fluid pressure p are, in principle, determinable at any point in underground space, and since the fluid potential depends only upon these quantities and the density of the fluid, then the potential also is determinable at any point in space capable of being occupied by the given fluid.

The fluid potential may be represented in a manner more easily comprehended if we imagine an open-top manometer to be terminated at the point where the value of the potential is desired. Let z be the elevation of this point and p the pressure of the fluid. In response to the pressure p the fluid will rise statically in the tube to some height h above the standard datum, or $(h-z)$ above the point

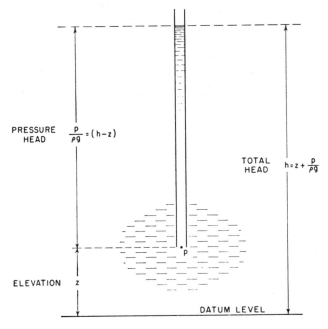

PRESSURE HEAD $\dfrac{p}{\rho g} = (h-z)$

TOTAL HEAD $h = z + \dfrac{p}{\rho g}$

ELEVATION z

DATUM LEVEL

FIG. 2.—Relation between head and pressure.

of measurement. The relation between this rise and the pressure p will be given by the hydrostatic equation

$$p = \rho g(h - z). \tag{5}$$

When this is substituted into equation (4) we obtain for the value of the fluid potential

$$\Phi = gz + \frac{p}{\rho} = gz + \frac{\rho g(h - z)}{\rho} = gh. \tag{6}$$

Thus a manometer becomes a measuring device for determining the potential at any point, and the value of the potential at the point is simply the work required to lift a unit mass the height h against the attraction of gravity.

The generality of the fluid potential, as here defined, merits attention. The potential is determinable not only in the space occupied by the given fluid, but *also at any point capable of being occupied by that fluid.* Thus fresh-water potentials have values not only in space saturated with fresh water, but also in space occupied by other fluids such as air, salt water, or oil or gas. In each case the potential is the amount of work required to transport unit mass of the fluid to that point from the standard state.

FIELD OF FORCE

The force that would be exerted upon a unit mass of water at any given point is determinable if the values of the potential throughout space are known. Since Φ depends only upon the variables z and p, in regions where p varies continuously, Φ also must vary continuously and a family of surfaces along each of which Φ is constant must exist. Let the potentials of two such surfaces be Φ and $\Phi+\Delta\Phi$, and let their normal distance of separation be Δn. Let Δs be a slant distance between the two surfaces at an angle θ to the normal, and let the vector $\mathbf{E}$, of unknown magnitude and direction, be the force per unit mass exerted upon the fluid by its environment.

The work required to transport a unit mass of water from the surface of potential Φ to that of $\Phi+\Delta\Phi$ is independent of the path and equal to $\Delta\Phi$. This is also the product of the force which must be applied by the distance moved, and the

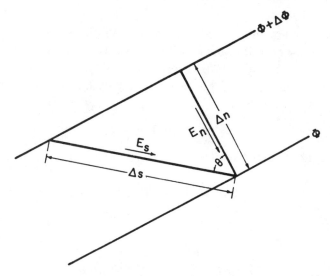

Fig. 3.—Relation between force field and potential gradient. Equipotential surfaces Φ and $\Phi+\Delta\Phi$ are perpendicular to plane of paper, $\mathbf{E}$ is force-intensity vector, and E_s its component in direction s.

applied force is the negative of the component of $\mathbf{E}$ in the given direction. Thus, along paths Δs and Δn, respectively,

$$\Delta\Phi = -E_s\Delta s = -E_n\Delta n,$$

or

$$E_n = -\frac{\Delta\Phi}{\Delta n}; \quad E_s = -\frac{\Delta\Phi}{\Delta s}.$$

However,

$$\Delta s = \Delta n/\cos\theta$$

M. KING HUBBERT

from which

$$E_s = E_n \cos \theta. \tag{7}$$

Since this is the rule of vector resolution, it follows that E_n is equal in direction and magnitude to the unknown vector $\mathbf{E}$. Hence the force per unit mass exerted upon an element of fluid by its environment must be perpendicular to the equipotential surfaces and in the direction of decreasing potential. It is thus given by

$$\mathbf{E} = -\frac{\Delta \Phi}{\Delta n} = - \operatorname{grad} \Phi = - g \operatorname{grad} h = \mathbf{g} - \frac{1}{\rho} \operatorname{grad} p, \tag{8}$$

and its component in any arbitrary direction by

$$E_s = -\frac{\partial \Phi}{\partial s} = - g \frac{\partial h}{\partial s} = g_s - \frac{1}{\rho} \frac{\partial p}{\partial s}. \tag{9}$$

Thus at every point in space about which the fluid potential is not constant a force of intensity $\mathbf{E}$ per unit of mass will act upon the fluid if placed at that point, tending to drive it in the direction of decreasing potential. Conversely, if no such force exists, then the fluid will be in a state of hydrostatic equilibrium and we shall have

$$\mathbf{E} = - \operatorname{grad} \Phi = 0, \tag{10}$$

whereby throughout such a region

$$\Phi = gz + \frac{p}{\rho} = gh = \text{const.} \tag{11}$$

Also, since

$$\operatorname{grad} \Phi = \mathbf{g} - (1/\rho) \operatorname{grad} p = 0,$$

then

$$(1/\rho) \operatorname{grad} p = \mathbf{g}, \tag{12}$$

which is the fundamental equation of hydrostatics.

The physical meaning of the force-intensity vector $\mathbf{E}$ can most simply be appreciated by means of the fourth of equations (8)

$$\mathbf{E} = \mathbf{g} - (1/\rho) \operatorname{grad} p.$$

This is a vector equation which asserts that at any given point a unit mass of the specified fluid will be acted upon by a force $\mathbf{E}$ which is the vector sum of two independent forces, gravity, and the negative gradient of the pressure divided by the density of the fluid. Since ordinarily these two primary forces are non-parallel, the resultant $\mathbf{E}$ will be parallel with neither. In the flow of underground water only rarely does $\mathbf{E}$ have a magnitude greater than a small fraction of that of the vector $\mathbf{g}$. Under these conditions the negative pressure-gradient vector can depart

354

but slightly from the vertical and the equipressure surfaces will be nearly horizontal with the pressure increasing downward. The resultant vector **E**, however, can have any direction in space—vertically downward or upward, horizontal, or any arbitrary inclination. Since the fluid will tend to flow in the direction of **E**, it is clear that the flow can be in any direction whatever with respect to the negative

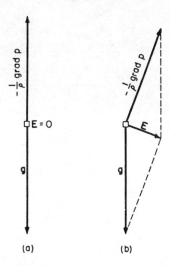

Fig. 4.—Physical interpretation of force-intensity vector **E**. (*a*) Hydrostatic case **E**=o; (*b*) hydrodynamic case.

gradient of the pressure: from lower to higher or from higher to lower pressure, parallel with the equipressure surfaces or in any other direction—a fact of great importance in view of the widely held opinion that fluids flow only from higher to lower pressures.

In general, at every point capable of being occupied by the fluid, the vector **E** will have a definite value, and the ensemble of these values throughout the field represents a *field of force*. Lines drawn through the field in such a manner as to be tangent to **E** at each point comprise the family of lines of force representing the field.

One of the most important properties of this field is that it is irrotational, that is to say, that if a unit mass of water were transported in the field around any closed curve by a frictionless process, the total work done would be zero. Thus along a path *s* the work done by such a transport in distance *ds* would be

$$dw = - E_s ds,$$

and around the closed path

$$w = - \oint_s E_s ds = + \oint_s \frac{\partial \Phi}{\partial s} ds = \mathrm{o}, \qquad (13)$$

since the value of Φ at the beginning and at the end of the path is the same.

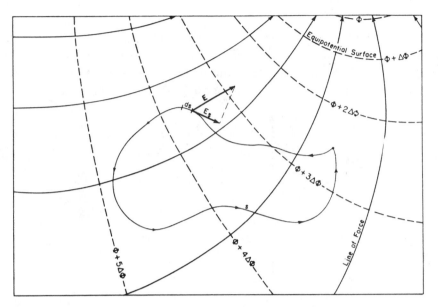

FIG. 5.—Line integral in field of force. Total work done in traversing closed path is zero.

It is this property which precludes any kind of flow that would amount to a perpetual-motion mechanism.

To sum up the energy and force relations pertaining to the flow of water through underground space, we envisage a potential field as represented by a family of equipotential surfaces. Orthogonal to these surfaces there exists a family of lines of force directed always from higher to lower values of the potential, with no force line capable of closing upon itself. Each line must therefore originate at some highest (local) value of the potential and continue "downhill" with respect to the potential until some point of lowest local value is reached; or else lines must enter any given region of space across one part of its boundary and continue until they make their exit across another part.

FIELD OF FLOW

Independently of energy or force considerations we may inquire into the kinematics of the flow of water through the water-saturated porous rocks of underground space. For this we adopt a macroscopic viewpoint whereby the geometrical quantities considered are statistical in that small distances, areas, and volumes are taken large as compared with the grain or pore size of the rocks. In a space through which water is flowing, at any given instant the flow at each point will have a definite direction. Lines drawn through the field in such a manner as to be tangent to the flow direction at each point comprise the family of flowlines of the system. The rate of flow at a point may be defined as the quantity of water per unit area crossing a surface normal to the flow direction in unit time, and may be referred to as the *specific discharge* at that point. The rate of flow could be either a mass rate or a volume rate. Since liquids are usually measured by volume rather than by mass, we shall adopt the volume rate and define the flow vector **q** by

$$q = \lim_{\Delta S \to 0} \frac{\Delta Q}{\Delta S},$$ (14)

where ΔQ is the volume of fluid crossing a macroscopic element of area ΔS normal to the flow direction in unit time, and where the limit taken is understood to mean the value approached as ΔS begins to approach the grain size of the rock, but before the ratio $\Delta Q / \Delta S$ begins to become erratic.

The flow field is therefore represented by a family of flowlines which at every point are tangent to the specific-volume-discharge vector **q**. The only kinematic requirement imposed upon this field is that it shall not violate the conservation of matter. Thus if we construct in the field any completely closed surface S, which is fixed with respect to the rock (assumed to be rigid), and if the fluid enters the space enclosed by S across one part of the surface, an equal amount must simultaneously leave across another part of the surface. This is true because the space of the flow field is assumed to be water-saturated and the water is assumed to be of constant density. Then, since the space enclosed by S is already filled with water, water can enter from one side only by expelling an equal volume on the other. The volume leaving across an element of surface area dS in unit time will be

$$dQ = q_n dS,$$ (15)

where q_n is the component of the flow vector upon dS parallel with its outward normal n. Then over the entire surface the net outward discharge is given by

$$Q = \int dQ = \iint_S q_n dS = 0.$$ (16)

Because of this fact it is impossible to have a flow field in a saturated space that converges into or diverges from any given closed region. The conservation principle does not forbid circulatory flow, however, so that so far as this requirement is concerned, there is nothing to prevent the fluid from flowing in circles or other closed paths.

RELATION BETWEEN FIELD OF FORCE AND FIELD OF FLOW

Darcy's law.—The final restriction on the flow field is imposed when we relate it to the field of force. We already have seen that at every point a force **E** will be exerted per unit mass on the water at that point. In response to this force the water will be driven through the pore spaces of the rock in the general direction of the force. The more exact relation between the flow vector **q** at a given point and the force vector **E** is expressed by Darcy's law:

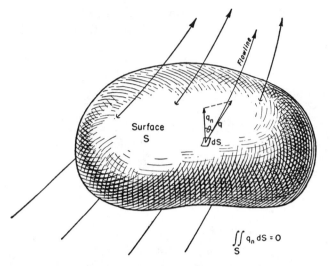

FIG. 6.—Total outward discharge over closed surface S is zero.

$$\mathbf{q} = \sigma\mathbf{E} = -\sigma \,\text{grad}\, \Phi, \qquad (17)$$

where σ is a proportionality factor depending both upon the statistical geometry of the rocks about the point and upon the properties of the fluid. This equation is analogous both physically and mathematically to Ohm's law in electricity, which for conduction through three-dimensional space may be written:

$$\mathbf{i} = \sigma_e\mathbf{E}_e = -\sigma_e \,\text{grad}\, V, \qquad (18)$$

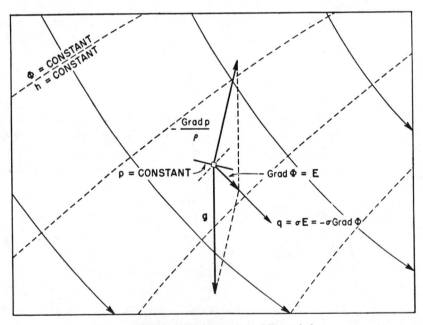

FIG. 7.—Physical interpretation of Darcy's law.

where **i** is the current crossing unit area (or charge crossing unit area in unit time), σ_e the electrical conductivity, $\mathbf{E}_e$ the force per unit of charge, and V the electrical potential, or energy per unit charge.

By analogy with electricity, the factor σ in equation (17) will be called the *fluid conductivity*. If for the same magnitude of **E** the flow rate **q** at the same point should have different magnitudes in different directions, the medium would be anisotropic with respect to conductivity and σ would be a tensor quantity. If the flow rate is not a function of direction, then the medium is isotropic and σ is a simple scalar; although, if the rocks are inhomogeneous, σ may vary widely with respect to position.

For present purposes no great error will result if it is assumed that rocks underground are inhomogeneous but isotropic and that σ accordingly is a scalar.

The value of σ for any given material can be determined by a simple flow experiment, and the variables which have been lumped together into the single factor σ can be obtained explicitly by a series of such experiments in which each is varied separately while all the others are kept constant. In this manner it has long since been shown experimentally (and can also be shown theoretically) that:

$$\left.\begin{aligned} q &\propto \rho, \\ q &\propto 1/\mu, \\ q &\propto d^2, \\ q &\propto N, \end{aligned}\right\} \tag{19}$$

where ρ is the density and μ the dynamic viscosity of the fluid, d is the mean diameter of the sand grains (or other proportional length characterizing the size-scale of the grain or pore structure), and N a dimensionless factor of proportionality whose numerical value is a function of the statistical interior shape or geometry of the rock.

When these values are substituted into equation (17) we obtain

$$\mathbf{q} = (Nd^2)(\rho/\mu)\mathbf{E} = -(Nd^2)(\rho/\mu)\,\text{grad}\,\Phi = -(Nd^2)(\rho/\mu)g\,\text{grad}\,h, \tag{20}$$

as equivalent expressions of Darcy's law. In fact, if we let

$$K = (Nd^2)(\rho/\mu)g,$$

then for the special case of vertical flow we obtain

$$q_z = -K\frac{dh}{dz}, \tag{21}$$

which is the relationship originally established experimentally by Henry Darcy (1856, pp. 590–94) nearly a century ago, and is the basis for the name "Darcy's law."

Field properties.—Regarding σ in equation (17) to be a scalar, then at every point the flow vector **q** will be parallel to the force vector **E**, and the lines of flow will coincide with the lines of force throughout the field of flow. The ultimate properties of the flow field then combine the restrictions previously enumerated for each of the two fields separately. Because of the irrotational property of the

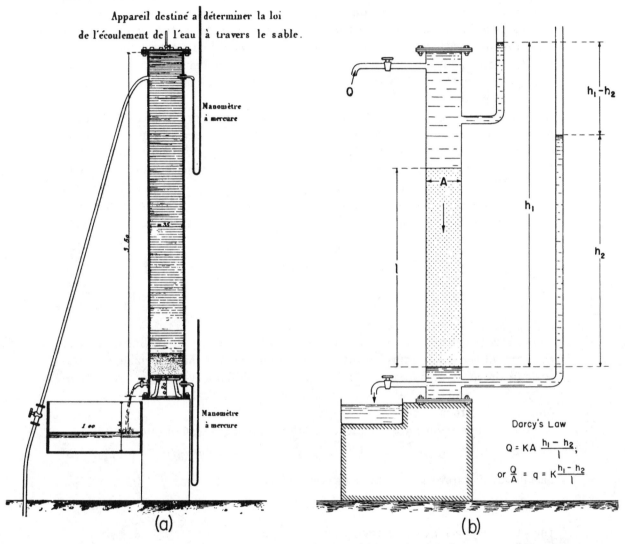

Fɪɢ. 8.—Henry Darcy's experiment on flow of water through sands. (*a*) Darcy's original apparatus with mercury manometers. (*b*) Equivalent apparatus with water manometers and Darcy's own statement of law expressed in notation of present paper.

force field the lines of flow can no longer close upon themselves; because of the conservation-of-matter requirements the flow can not terminate inside the field, or converge upon or diverge from any closed region. Hence, since the flowlines are everywhere perpendicular to the equipotential surfaces, no equipotential surface can close completely upon itself. Were this to occur such a surface would enclose a region either of higher or lower potential than that outside. The flow would then be entirely outward from or inward toward such a region, which would violate the principle of the conservation of matter.

Neither can there be a stagnant region of no flow contiguous to a region where the flow is not zero, for in the stagnant region the potential would be constant, whereas in the adjacent region it would vary and would provide a potential discontinuity or gradient across the boundary without a corresponding fluid flow; or, if along the boundary the potentials are constant, then flowlines would have to terminate upon the stagnant region, and thus violate the conservation of matter.

This is the basis for the statement made earlier that fresh water encountered at whatever depth, unless completely isolated by impermeable barriers, is presumed to be in motion. Otherwise, such a body of water would be stagnant while in communication with shallower water which has potential gradients and is therefore flowing.

It may be noted that the field restrictions imposed by conservation-of-matter requirements are only valid under conditions of complete saturation of rocks of unvarying spatial dimensions. In an unsaturated space the lines of force may converge, or they may end abruptly against an impermeable boundary. Likewise in a saturated space of variable spatial dimensions—a compacting shale, for example—potential maxima (or minima) can exist and we might then have flow completely outward or inward, merely because the reference space itself is contracting or expanding while the volume of the water remains constant.

In view of these several restrictions, with the exceptions just noted, the flow of ground water through any region reduces to a simple transit of the region. Along a single flowline the water will enter the region at a point of high potential on the boundary and then will flow with continuous loss of potential until it leaves the region by crossing the boundary at an exit point of low potential.

In the case of the earth, permeability vanishes gradually with increase of depth and, neglecting magmatic waters rising from the earth's interior in regions of active vulcanism, the only way water can enter or leave the underground reservoir is by crossing the earth's surface. At the earth's surface, or more specifically at a point at the water table, the fluid head h is equal to the elevation z, and on this surface the value of the potential at any point is equal to gz. Consequently the areas of highest potential are those where the water-table elevation is highest and the areas of lowest potential are those where the water table is lowest. The water therefore enters the earth at these areas of high elevation and high potential and flows with continuous loss of potential until it ultimately emerges at some point of lower elevation and lower potential.

EFFECT OF INHOMOGENEITIES

Although the rocks with which we are concerned may be assumed without serious error to be isotropic with respect to fluid conductivity, they are far from homogeneous. From equations (17) and (19) it will be seen that

$$\sigma = (Nd^2)(\rho/\mu), \qquad (22)$$

where Nd^2 is a geometrical property of the rocks and is known as the *permeability* and ρ and μ are the significant properties of the fluid. Of these variables the shape factor N, the fluid density ρ, and the viscosity μ have comparatively narrow ranges; the grain diameter, d, however, varies widely, ranging from the order of 10^{-4} cm. for coarse clay to 10^{-1} cm. for coarse sand (Pettijohn, 1949, p. 13). This would give a variation in the value of σ of the order of 10^6-fold.

Since sedimentary rocks occur in stratified sequences with coarse-textured, highly permeable strata, such as sands and some limestones, interbedded between much less permeable strata of shales and other fine-textured rocks, it follows that the principal freedom of motion of underground water must be through the more

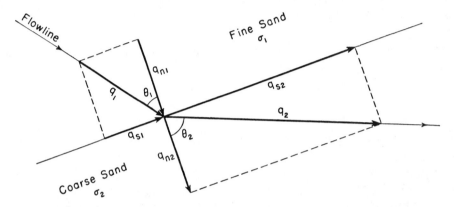

FIG. 9.—Refraction of flowline at interface of two sands of different permeabilities.

permeable rocks and nearly parallel with the stratification. That this must be so can be seen by examining the boundary conditions which must be satisfied when the flow occurs across any arbitrary surface.

Let such a surface divide the flow region into subregions 1 and 2 with conductivities σ_1 and σ_2, respectively. The conservation of matter requires that adjacent to this surface the normal component of the flow in region 1 must be equal to that in region 2, or that

$$q_{n1} = q_{n2}. \tag{23}$$

Since the potential Φ has the same value at contiguous points on the opposite sides of the surface, then along the surface in a direction s

$$-\left(\frac{\partial \Phi}{\partial s}\right)_1 = -\left(\frac{\partial \Phi}{\partial s}\right)_2. \tag{24}$$

These, however, are the components in the direction s of the force intensity-vector **E** on the respective sides. Hence,

$$E_{s1} = E_{s2}, \tag{25}$$

and by Darcy's law

$$q_{s1} = \sigma_1 E_{s1,} \atop q_{s2} = \sigma_2 E_{s2.} \Bigg\} \qquad (26)$$

If we now let θ_1 be the angle in region 1, which the flowline at the point of incidence makes with the surface normal, and θ_2 the corresponding angle in region 2, and let s be taken in the direction of the orthogonal projection of the flowlines upon the surface considered, we have

$$\tan \theta_1 = q_{s1}/q_{n1,} \atop \tan \theta_2 = q_{s2}/q_{n2.} \Bigg\} \qquad (27)$$

Then by substituting equations (23) and (26) into (27) and taking their ratio, we obtain

$$\frac{\tan \theta_2}{\tan \theta_1} = \frac{\sigma_2}{\sigma_1}, \qquad (28)$$

which is the law of refraction of the flowlines across such a surface

If, on the opposite sides of the surface, the conductivity is the same, then

$$\sigma_2 = \sigma_1; \quad \theta_2 = \theta_1,$$

and the flowlines will be unaffected; if, on the contrary, the surface is an interface between highly contrasting strata, say a sand with conductivity σ_1 and a silt of σ_2, the behavior will be quite different. The flow can be strictly perpendicular to the interface, tangential to it, or oblique. In the first case $\theta_2 = \theta_1 = 0$, and no refraction will occur. In the second case $\theta_1 = \theta_2 = 90°$, and no fluid will cross the interface. In the third case, both θ_1 and θ_2 fall between 0 and 90 degrees. Then, since σ_1/σ_2 is a large ratio, it follows from equation (28) that $\tan \theta_1/\tan \theta_2$ must be large also, which can only be satisfied when θ_1 is very nearly 90° and θ_2 very nearly zero. Hence, in this case the flowlines in crossing from the sand to the silt (or *vice versa*) will refract from nearly tangential to the interface in the sand to nearly perpendicular to it in the silt. In an alternating series of such strata, any flow oblique to the bedding would occur in sharply zigzag paths traversing the more permeable strata nearly parallel with the bedding, and the less permeable strata more nearly at right angles.

Extreme cases of this phenomenon occur when the interface is between a region of finite and another of infinite conductivity; or between a region of finite and another of zero conductivity. In the first case, which is realized by a sand bounding a water-filled open space, the potential in the open space is sensibly constant, while in the sand, if flow occurs, there must be a potential gradient. To satisfy both of these conditions simultaneously it is necessary for the interface to be an equipotential surface. The flowlines can therefore approach this interface from the sand only perpendicularly.

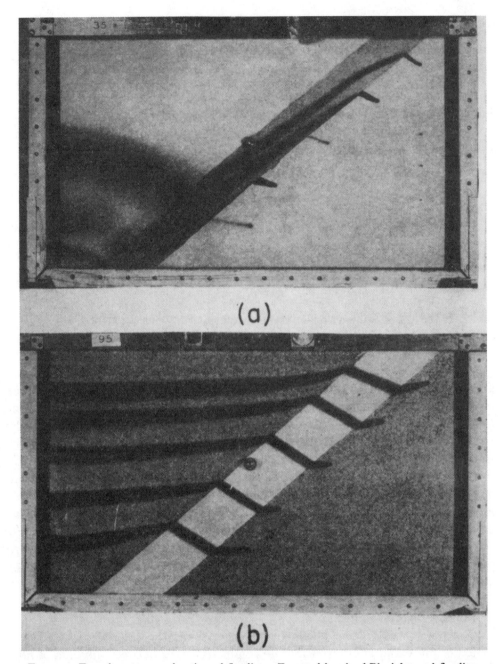

FIG. 10.—Experiments on refraction of flowlines. Front of box is of Plexiglas and flowlines are marked by injections of dye in sand. Flow is from right to left. (*a*) Thin bed of high-permeability sand between low-permeability sands. (*b*) Thin bed of low-permeability sand between high-permeability sands.

In the second case mentioned, where one space is ideally impermeable, it follows from equation (23) that the normal components of the flow on both sides of the interface must be zero. This in turn requires that in the region of finite permeability the flow must either be zero or else the flowlines must be tangential to the boundary. Upon all impermeable boundaries the equipotential surfaces must correspondingly terminate perpendicularly.

DETERMINATION OF STATE OF FLOW

In connection with the present inquiry, it is important to be able to determine whether the water is static or in motion in any given permeable stratum. This can readily be done by means of widely spaced wells which can be made to serve as manometers terminated in the stratum of interest. If the wells are allowed to fill with water from the stratum to a height of static equilibrium, then the elevation of the water level in a given well with respect to sea-level is the desired quantity h for that point in the stratum. If, as is often the case, it is not desirable to permit the well to fill with water, h can be computed from an accurately measured shut-in pressure at a point whose elevation is accurately recorded also. Thus by equation (5)

$$h = z + \frac{p}{\rho g},$$

where z is the elevation with respect to sea-level of the point at which the pressure is measured, p the gauge pressure, and ρ the density of the water in the stratum.

In a more permeable stratum, such as sand bounded above and below by relatively impermeable strata, the flow of water through the sand will be essentially two-dimensional, with the flowlines and the equipotential surfaces parallel and perpendicular, respectively, to the bedding. Then if we obtain the values of the

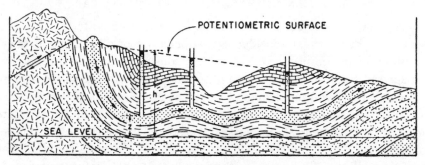

Fig. 11.—Regional flow of water through sand from higher to lower outcrop, showing continuous drop in potential.

potential, or of the head, h, at three points forming the apices of a triangle along the upper surface of the stratum, the approximately mean value of the potential gradient in that triangle can be determined. If all three values of the potential are the same, the water will be static; if any two are different and there is no sharp transition from fresh to salt water between the two wells, and no impermeable barriers, the water will be flowing.

A device that is useful in plotting the regional flow through a given stratum is the *potentiometric surface*. To every point on the upper surface of a given stratum there corresponds a head *h*. If this height from the standard datum is plotted in the vertical line through each point of measurement and a smooth surface passed through each point so plotted, the surface obtained may be known as the potentiometric surface for the stratum considered, because at every point its elevation is a measure of the potential of water in the upper surface of the stratum to which it refers, irrespective of the elevation of the stratum. Such a surface in the literature of ground-water hydrology has commonly been referred to as a *piezometric* (or pressure-measuring) *surface*, but this is a misnomer since there is an infinity of pressures, but only one potential corresponding to each point on the surface.

If the potentiometric surface for a given stratum is horizontal, the potential of the water in that stratum will be constant and the water will be at rest. If the potentiometric surface is sloping, then the water will be in motion, with the horizontal component of its flow in the approximate direction of the steepest downward slope of this surface.

These are qualitative relations only, and it must not be assumed that the horizontal component of the flow in any given direction is proportional to the downward slope of the potentiometric surface in that direction. As projected on a map, however, except for steeply dipping strata, the flowlines will everywhere be nearly perpendicular to the contours of the potentiometric surface with the flow in the "downhill" direction of this surface. Moreover, the slope of the potentiometric surface is a quantity of which we shall be able to make direct use in our subsequent discussions.

MIGRATION AND ENTRAPMENT OF OIL AND GAS

This summary treatment of the mechanics of ground-water motion provides an essential background for considering the behavior of oil and gas in a ground-water environment. Simply from volumetric considerations, oil and gas in their initial dispersed state must exist in an otherwise water-saturated environment as small discontinuous volume elements in the pore spaces of the rocks. Such an element must possess mechanical or nonthermal energy, which, when referred to unit mass, we may also speak of as its potential. As in the case of water, we may think of the potential of a given fluid as representing the work required to transport unit mass of that fluid from a standard state to the point considered. This again will comprise the work against gravity, that against pressure, and an additional energy term not heretofore encountered, the interfacial energy between the hydrocarbon, water, and the rock. The potential for a given liquid petroleum at a given point is therefore

$$\Phi_0 = gz + \frac{p}{\rho_0} + \frac{p_c}{\rho_0}, \qquad (29)$$

where the terms have the same meaning as in the case of water except that *p* is the pressure of the ambient water at the point, p_c the additional pressure inside the oil over that outside due to capillarity, and ρ_0 the density of the oil.

If the fluid of interest is gas, the potential will differ from that of a liquid in that additional work must be done to compress the gas from its volume at the pressure of the standard state to that of the final state. We will suppose this to be carried out in such a way that the variable density is a function of the pressure only. Under this condition the potential for gas becomes

$$\Phi_g = gz + \int_0^p \frac{dp}{\rho} + \int_p^{p+p_c} \frac{dp}{\rho}, \tag{30}$$

where ρ is the variable density of the gas during compression.

The impelling forces that would act upon an isolated element of oil or gas, respectively, if placed at a given point are the negative gradients of the respective potentials:

$$\mathbf{E}_0 = -\operatorname{grad} \Phi_0 = \mathbf{g} - (1/\rho_0) \operatorname{grad} p - (1/\rho_0) \operatorname{grad} p_c, \tag{31}$$

$$\mathbf{E}_g = -\operatorname{grad} \Phi_g = \mathbf{g} - (1/\rho_g) \operatorname{grad} p - (1/\rho_g) \operatorname{grad} p_c, \tag{32}$$

where ρ_g is the density of the gas at the point of reference. These are vector equations asserting that such an isolated element will be acted upon by three separate forces, one due to gravity, one to variations in the ambient pressure, and the third produced by variations in the capillary pressure from point to point.

CAPILLARY ENERGY AND FORCE

It will be convenient to dispose of the capillary term separately before devoting our attention to the other two terms. If a small volume of oil, which is still large as compared with the grain volume of the rock, is injected into a rock otherwise saturated with water, there will be a difference of pressure p_c across the interface between the oil and the water. If the rock is preferentially wet by water with respect to oil, p_c will be positive; that is, the pressure inside the oil will be greater by the amount p_c than that inside the water; if the rock is preferentially wet by oil, p_c will be negative. In all but rare cases the rocks are preferentially wet by water and p_c is positive.

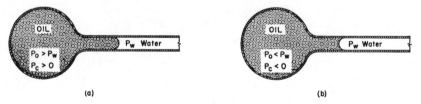

Fig. 12.—Diagrams to illustrate variation of capillary pressure with wettability. (*a*) In preferentially water-wet container, capillary pressure in oil is positive. (*b*) In preferentially oil-wet container, capillary pressure in oil is negative.

The volume element will tend to dispose itself in such a manner that its capillary energy, p_c/ρ_0, will be a minimum. Since ρ_0 is constant this requires that the configuration be such that p_c (neglecting a small change of pressure due to gravity) be constant and at its minimum value. The pressure p_c between two non-miscible fluids whose interfacial tension is γ is related to the principal radii of curvature of the interface by Plateau's equation

$$p_c = \gamma \left(\frac{1}{r_1} + \frac{1}{r_2} \right),$$ (33)

with the higher pressure on the side of the center of sharpest curvature. In a rock preferentially wet by water, in order for p_c to be a minimum, therefore, it is necessary for the sum of the reciprocals of the radii of curvature of the interface to be a minimum, which in turn requires that the oil occupy the largest accessible voids in the pore structure.

Should the wettability be reversed, that is, should the rock be preferentially wet by oil, the capillary pressure p_c would be negative, and in order for the capillary-potential term p_c/ρ_0 to approach a minimum, it would be necessary for $-p_c$ to assume a maximum value. This in turn would require a disposition having minimum radii of curvature, and the oil would distribute itself into the smallest accessible crevices and voids by precisely the same action that water is absorbed by blotting paper.

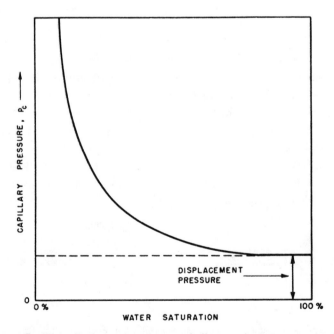

Fig. 13.—Capillary displacement pressure of oil or gas in water-saturated sand.

The pressure p_c of present interest is therefore the minimum possible value of the capillary pressure, and is the initial *displacement pressure* obtained when oil is injected into a water-saturated rock. The value of this pressure is given approximately by the equation

$$p_c = \frac{C\gamma \cos \alpha}{d}, \tag{34}$$

where C is a dimensionless factor of proportionality, α the contact angle in the water phase which the oil-water interface makes with the solid boundary, and d the mean grain diameter of the rock.

For an order of magnitude, $\gamma \cos \alpha$ may be taken to be about 25 dynes/cm., and the factor C, as determined from the work of Purcell (1949) and unpublished work of R. H. Nanz, has a value of about 16; whence by substitution in equation (34) the value of p_c is seen to be about $(400/d)$ dynes/cm.[2] This permits us to determine approximately the magnitude of the capillary displacement pressures in sediments of various degrees of coarseness, as shown in Table I. Here the size scales employed are those recommended by the National Research Council Committee on Sedimentation (see Pettijohn, 1949, p. 13).

TABLE I. CAPILLARY DISPLACEMENT PRESSURE OF WATER AGAINST OIL
IN SEDIMENTS OF VARIOUS GRAIN SIZES

Sediment	Grain Diameters d (Millimeters)	Capillary Pressure p_c (Atmospheres)
Clay	Less than 1/256[1]	Greater than 1
Silt	1/256 to 1/16	Between 1 and 1/16
Sand	1/16 to 2	Between 1/16 and 1/500
Granules	2 to 4	Between 1/500 and 1/1,000

[1] The value 1/256 mm. for clay particles is a maximum; much finer clays are known. In a clay with a particle size 10^{-4} for example, p_c would be about 40 atmospheres.

The impelling force due to capillarity behaves in a similar manner. By equations (31) and (34) this force for oil is given by

$$\mathbf{E}_{0c} = -\frac{\operatorname{grad} p_c}{\rho_0} = \frac{C\gamma \cos \alpha}{\rho_0} \frac{\operatorname{grad} d}{d^2}, \tag{35}$$

which shows that the oil will be acted upon by a net capillary force which will tend to impel it in the direction of the steepest rate of increase of the grain size of the sediment, with the magnitude of the force, for a fixed value of grad d, extremely sensitive to the grain size because of the factor d^2 in the denominator.

Or again, if we let the increase in grain diameter be proportional to the grain diameter itself by some fixed factor β, we should then have

$$\mathbf{E}_{0c} = \frac{C\gamma \cos \alpha}{\rho_0} \frac{\beta d}{d^2} = \frac{C'}{d}, \tag{36}$$

where

$$\beta = (1/d) \operatorname{grad} d$$

and

$$C' = \frac{\beta C \gamma \cos \alpha}{\rho_0}.$$

If we arbitrarily set $\beta = 1/500$ per centimeter, corresponding with a rate of change of grain size that would double the grain diameter in about 350 cm., the factor C' assumes the value of approximately 1 cm.²/sec². Then for a shale with particle diameter 10^{-5} cm., E_{0c} would have the value of 10^5 dynes per gram, or 100 times gravity; for a fine sand with $d = 10^{-2}$ cm. this would reduce to 100 dynes per gram, or 0.1 of gravity.

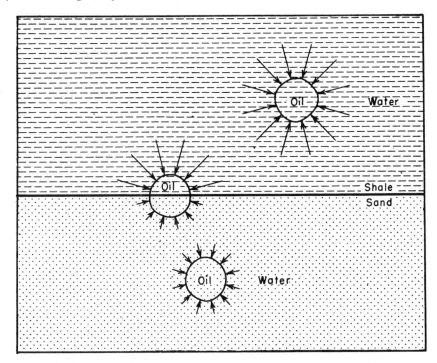

Fig. 14.—Diagram showing how capillary pressure of water against oil in preferentially water-wet environment facilitates passage of oil globules from fine- to coarse-textured rocks.

The steepest rate of sedimentary gradation occurs across the bedding rather than along it, and more often than not, as between a sandstone and a shale, definite discontinuities occur. In the case of discontinuities, equation (35) does not apply and we deal instead with a finite difference obtained by means of equation (34)

$$\Delta p_c = C \gamma \cos \alpha \left[\frac{1}{d_2} - \frac{1}{d_1} \right], \tag{37}$$

where Δp_c is the difference in the capillary displacement pressures in the two media, say shale and sandstone, and d_2 and d_1 their respective grain diameters.

Reference to Table I shows that the capillary pressure of oil in a shale is the order of tens of atmospheres, while in a sand it drops to the order of tenths. Hence a slug of oil extending across such a boundary would be expelled from the shale into the sand by an unbalanced pressure of the order of tens of atmospheres.

This formidable energy barrier, therefore, makes a shale-sand interface appear as a surface of unidirectional conductivity to oil (or gas). Across such a boundary the oil can flow in the direction from the shale to the sand without hindrance other than viscous drag; in the opposite direction it can not flow at all unless a pressure is applied to the oil in the sand greater than the opposing capillary pressure against the oil in the shale. If both rocks are completely saturated with a single-phase fluid—water, oil, or gas—this difference of capillary pressure and energy does not exist and the fluid can flow with equal facility in either direction. This phenomenon permits us to regard a sand-shale boundary as an impermeable barrier to oil trapped in the sand, but not an impermeable barrier to the passage of water in either direction.

MIGRATION IN RESERVOIR BEDS

Without further consideration of the problems of primary migration from the source rocks to the reservoir rocks, let us now direct our attention to migration and accumulation in the reservoir rocks themselves. Most commonly a reservoir rock is a coarse-textured stratum interbedded between rocks of finer texture. It is normally saturated with water which, in general, is in some state of motion with the water flowing usually nearly parallel with the bedding.

In such an environment we consider an element of petroleum in a state of migration. This element will possess a potential, Φ_0, which within the reservoir rock will be a function of position and environment, but, for reasons already stated, comparatively free from the effects of capillarity except at the boundaries. Across the boundary, due to capillarity, in the direction from the coarse- to fine-textured rocks, there will be a sudden large increase of Φ_0, permitting us to regard such a boundary as seen from the reservoir-rock side as impermeable or impenetrable to oil, though not necessarily so to water.

The element will be acted upon by an impelling force per unit of mass, $\mathbf{E}_0 = -\operatorname{grad} \Phi_0$, in whose direction it will migrate until it encounters an impermeable boundary. Along such a boundary it will then be deflected in the direction of the largest tangential component of $\mathbf{E}_0$, until a position is reached at which this component vanishes. Usually this will be a position toward which the tangential components of force from all directions along the boundary will converge. It will therefore be a position toward which oil will migrate from all directions, and from which it can not escape, and hence a petroleum trap. Since $\mathbf{E}_0$ and all of its positive components are directed only from regions of higher to those of lower potential, it follows that regions toward which the force components converge must also be local positions of minimum or least values of the potential.

At any arbitrary point in such an environment the potential of a given petroleum, as obtained from equation (29) with the capillary term omitted, is given by

$$\Phi_0 = gz + \frac{p}{\rho_0},\qquad(38)$$

and the impelling force by

$$\mathbf{E}_0 = -\text{ grad }\Phi_0 = \mathbf{g} - \frac{1}{\rho_0}\text{ grad }p.\qquad(39)$$

Formally, therefore, the potential at every point is uniquely determined by the elevation z and the environmental fluid pressure p; and the impelling force by the vectors $\mathbf{g}$ and $-$ grad p. Regrettably, however, despite the fact that the fluid pressure is one of the most easily measured quantities in a drill hole, it is a quantity whose distribution in space is not readily determinable. It is more informative, therefore, to refer Φ_0 and $\mathbf{E}_0$ to the more tractable quantities Φ_w and $\mathbf{E}_w$, the potential and impelling force of the ambient ground water.

From equations (6) and (9)

$$\Phi_w = gz + \frac{p}{\rho_w},$$

$$\mathbf{E}_w = -\text{ grad }\Phi_w = \mathbf{g} - \frac{1}{\rho_w}\text{ grad }p.$$

If we solve the first for p and substitute this into equation (38), we then obtain

$$\Phi_0 = \frac{\rho_w}{\rho_0}\Phi_w - \frac{\rho_w - \rho_0}{\rho_0}gz,\qquad(40)$$

which expresses the potential of oil at every point in terms of the potential Φ_w of water at that point, and of the elevation z.

If we solve the second equation for $-$grad p, we obtain

$$-\text{ grad }p = \rho_w(\mathbf{E}_w - \mathbf{g}),$$

which when substituted into equation (39) gives

$$\mathbf{E}_0 = \mathbf{g} + \frac{\rho_w}{\rho_0}(\mathbf{E}_w - \mathbf{g}),\qquad(41)$$

whereby the impelling force for oil is expressed in terms of that of water, $\mathbf{E}_w$, and of gravity, $\mathbf{g}$.

Thus by means of equation (40) we can map the family of oil equipotential surfaces, $\Phi_0 =$ constant, whose negative gradient will then give the force vector $\mathbf{E}_0$ and the corresponding lines of force; or, alternatively, we can obtain, by means of equation (41), $\mathbf{E}_0$ directly to which the corresponding equipotential surfaces will be perpendicular.

For purposes of calculation the oil potential, being a scalar quantity, is the easier one to work with, but for the present it may help us to visualize the essential relationships if we examine the impelling force $\mathbf{E}_0$. This, as will be seen from equation (41), is a linear vector function of the two primary vectors, $\mathbf{g}$ and $\mathbf{E}_w$. The first is the acceleration of gravity which, by definition, is directed vertically downward; the second, $\mathbf{E}_w$, is the total impelling force per unit of mass acting upon the water, and has the direction of the flow. In terms of the forces exerted upon unit mass of oil the first term, $\mathbf{g}$, of equation (41) is the force per unit mass exerted by gravity; the second term, $+(\rho_w/\rho_0)(\mathbf{E}_w-\mathbf{g})$, is the force per unit mass exerted by the now suppressed gradient of the pressure.

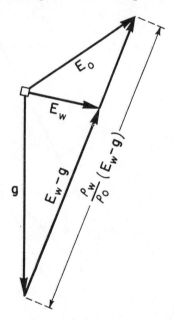

FIG. 15.—Vector diagram of forces acting on element of oil in hydrodynamic environment.

If we plot these two terms graphically (Fig. 15), plotting first from a common origin the vectors $\mathbf{g}$ and $\mathbf{E}_w$, the difference $(\mathbf{E}_w-\mathbf{g})$ will then be a vector whose origin will be at the terminus of $\mathbf{g}$ and whose terminus at the terminus of $\mathbf{E}_w$. The vector $(\rho_w/\rho_0)(\mathbf{E}_w-\mathbf{g})$ will be a vector colinear with $(\mathbf{E}_w-\mathbf{g})$ but with magnitude amplified by the factor ρ_w/ρ_0. Finally, the vector

$$\mathbf{E}_0 = \mathbf{g} + \frac{\rho_w}{\rho_0}(\mathbf{E}_w - \mathbf{g})$$

will be a vector whose origin is at the origin of $\mathbf{g}$ and $\mathbf{E}_w$ and whose terminus coincides with that of $\rho_w/\rho_0(\mathbf{E}_w-\mathbf{g})$.

Since $\mathbf{g}$ is fixed, it is clear that the vector $\mathbf{E}_0$ is a function of the variables ρ_0 and

$\mathbf{E}_w$. In other words, at a given point in space the force per unit of mass that would act upon an element of oil if placed at that point depends in direction and in magnitude upon both the density of the oil and the impelling force $\mathbf{E}_w$ acting upon the water. For a fixed value of ρ_0 it will be seen that if $\mathbf{E}_w$ is either zero or vertical, $\mathbf{E}_0$ will be vertical and, depending upon the magnitude and direction of $\mathbf{E}_w$, may also be zero. In all other cases, where $\mathbf{E}_w$ is neither zero nor vertical, $\mathbf{E}_0$ will be tilted away from the vertical in the direction of the horizontal component of $\mathbf{E}_w$.

If we now let $\mathbf{E}_w$ have some fixed value in a non-vertical direction, we shall be able to see what effect a change in the density of the oil will produce. As the value of ρ_0 is made to vary, the terminus of the vector $\mathbf{E}_0$ will move along a line colinear with the vector $(\mathbf{E}_w - \mathbf{g})$. If the oil density is equal to that of water, then

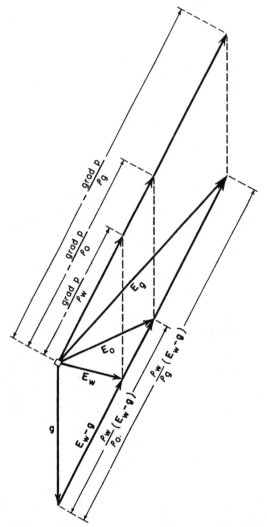

FIG. 16.—Impelling forces on water, oil, and gas in hydrodynamic environment.

$\rho_w/\rho_0 = 1$, $\mathbf{E}_0$ will coincide with $\mathbf{E}_w$, and the oil will tend to migrate in the direction of the flow of the water. If the oil is less dense than water, the vector $\mathbf{E}_0$ will terminate on the line of the vector $(\mathbf{E}_w - \mathbf{g})$ extended, and will therefore be tilted away from the vertical by an angle less than that of the vector $\mathbf{E}_w$. If the density of the oil (or gas) is made to approach zero, $\mathbf{E}_0$ (or $\mathbf{E}_g$) will become unlimitedly large and will approach parallelism with the vector $(\mathbf{E}_w - \mathbf{g})$.

Going in the opposite direction, as the density is increased beyond that of water, $\mathbf{E}_0$ will still terminate on the same line but will be tilted downward with respect to $\mathbf{E}_w$.

If in the same space (Fig. 16) we have two different hydrocarbons, say oil and gas, each with a distinctly different density, but with the gas less dense than oil

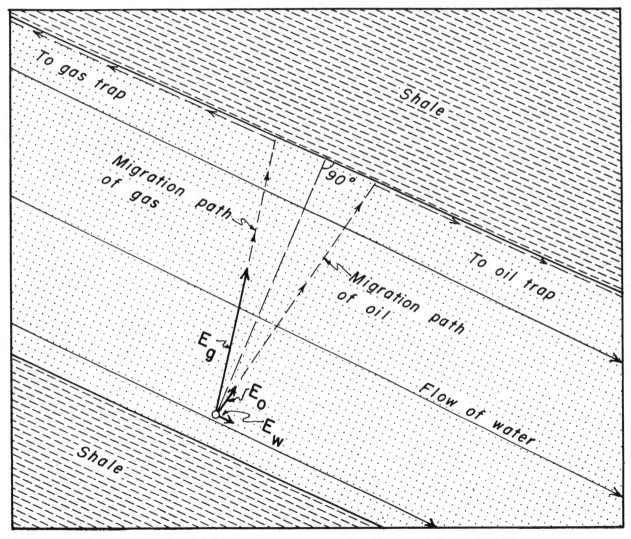

FIG. 17.—Divergent migration of oil and gas in hydrodynamic environment.

and the oil less dense than water, then at the same point in space we should have the three vectors $\mathbf{E}_g$, $\mathbf{E}_0$, and $\mathbf{E}_w$, all in the same vertical plane and tilted away from the upward vertical in the same horizontal direction, but with the angle of tilt greatest for $\mathbf{E}_w$, intermediate for $\mathbf{E}_0$ and least for $\mathbf{E}_g$. It is thus seen that in a dynamic ground-water environment with the water flowing in a non-vertical direction, oil and gas at the same point will be acted upon by forces differing both in magnitude and direction, and accordingly will migrate in divergent directions.

This is illustrated in a simple manner in Figure 17 for the case of strata with a homoclinal dip of angle δ, with the water flowing directly down the dip. The normal to the bedding will be tilted away from the vertical by the angle δ. Both the vectors $\mathbf{E}_g$ and $\mathbf{E}_0$ will also be tilted away from the vertical in the downdip direction by amounts depending upon the magnitude of $\mathbf{E}_w$, $\mathbf{E}_0$ by an angle greater than $\mathbf{E}_g$. For a certain range of values of $\mathbf{E}_w$ we can have the tilt of $\mathbf{E}_0$ greater than δ and that of $\mathbf{E}_g$ less than δ. Under such a condition, while the gas would migrate to the upper impermeable boundary of the stratum and then be deflected *up the dip*, the oil would be deflected *down the dip*, and we should have a case where along the same impermeable surface oil and gas would simultaneously migrate in diametrically opposite directions. Then by a slight decrease or increase of $\mathbf{E}_w$ both fluids could be made to migrate in the same direction either updip or downdip.

The angle of tilt of the oil vector $\mathbf{E}_0$ away from the vertical, or of the corresponding oil equipotential surface from the horizontal, can be obtained geometrically from the analysis of the vector equation (41), or analytically from the scalar equation (40). In Figure 18, which represents the vector diagram corresponding to equation (41), let the z-axis be vertical and positive upward, and the x-axis

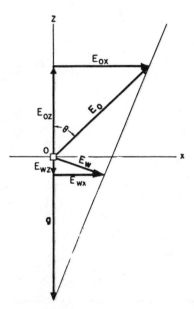

Fig. 18.—Angle of tilt of oil vector from vertical, obtained by vector analysis.

positive in the direction of the horizontal component of $\mathbf{E}_w$, and let θ (positive upward) be the angle of tilt. Then, by resolving the various vectors into their x- and z-components, we obtain

$$\tan \theta = -\frac{E_{0x}}{E_{0z}} = -\frac{\left[\mathbf{g} + \frac{\rho_w}{\rho_0}(\mathbf{E}_w - \mathbf{g})\right]_x}{\left[\mathbf{g} + \frac{\rho_w}{\rho_0}(\mathbf{E}_w - \mathbf{g})\right]_z} = \frac{-E_{wx}}{\frac{\rho_w - \rho_0}{\rho_w}g + E_{wz}} \cdot \quad (42)$$

The slope of the equipotential surface, $\Phi_0 = \text{const}$, is equal to $\tan \theta$, which is given by

$$\frac{dz}{dx} = +\tan \theta = \frac{-E_{wx}}{\frac{\rho_w - \rho_0}{\rho_w}g + E_{wz}} \cdot \quad (43)$$

Then, since

$$E_{wx} = -g\frac{\partial h_w}{\partial x},$$

$$E_{wz} = -g\frac{\partial h_w}{\partial z},$$

$$\frac{dz}{dx} = \frac{\frac{\partial h_w}{\partial x}}{\frac{\rho_w - \rho_0}{\rho_w} - \frac{\partial h_w}{\partial z}}, \quad (44)$$

which is the general equation of the slope of the oil equipotential surfaces in terms of the horizontal and vertical components of grad h_w.

Or if we start with equation (40),

$$\Phi_0 = \frac{\rho_w}{\rho_0}\Phi_w - \frac{\rho_w - \rho_0}{\rho_0}gz,$$

we can substitute

$$\Phi_0 = gh_0; \quad \Phi_w = gh_w;$$

and then by canceling the g's obtain

$$h_0 = \frac{\rho_w}{\rho_0}h_w - \frac{\rho_w - \rho_0}{\rho_0}z, \quad (45)$$

which has the advantage that every term is expressed in units of length. Here z

is the elevation of the point considered, h_w is the height above datum to which water would stand statically in a manometer terminated at the point considered, and h_0 the corresponding height that oil of density ρ_0 would stand.

Then between any two nearby points the increment dh_0 of h_0 will be

$$dh_0 = \frac{\rho_w}{\rho_0}\, dh_w - \frac{\rho_w - \rho_0}{\rho_0}\, dz, \tag{46}$$

where dh_w and dz are the corresponding increments of h_w and z. Along an equipotential surface $h_0 = \text{const}$, $dh_0 = 0$, and equation (46) reduces to

$$\frac{\rho_w}{\rho_0}\, dh_w - \frac{\rho_w - \rho_0}{\rho_w}\, dz = 0. \tag{47}$$

However, in general, h_w is a function of both x and z, and

$$dh_w = \frac{\partial h_w}{\partial x}\, dx + \frac{\partial h_w}{\partial z}\, dz. \tag{48}$$

Substituting this into equation (47), then gives

$$\frac{\rho_w}{\rho_0}\frac{\partial h_w}{\partial x}\, dx + \frac{\rho_w}{\rho_0}\frac{\partial h_w}{\partial z}\, dz - \frac{\rho_w - \rho_0}{\rho_w}\, dz = 0,$$

from which

$$\left(\frac{\rho_w - \rho_0}{\rho_w} - \frac{\partial h_w}{\partial z} \right) dz = \frac{\partial h_w}{\partial x}\, dx,$$

or

$$\frac{dz}{dx} = \frac{\dfrac{\partial h_w}{\partial x}}{\dfrac{\rho_w - \rho_0}{\rho_w} - \dfrac{\partial h_w}{\partial z}}, \tag{49}$$

which is the same result as that of equation (44) obtained from analysis of the forces involved.

In the case of water flowing directly down the dip, the tilt, as we have just seen, has a critical value when it is equal to the negative of the dip, for with tilts less than or greater than this amount the migration of oil will be updip or downdip, respectively. This critical value occurs when the oil equipotential surfaces become tangent to the flowlines of the water and when $\mathbf{E}_0$ and $\mathbf{E}_w$ are mutually perpendicular. This is also the condition which prevails when water is flowing underneath an accumulation of petroleum, for the oil in that case will be at constant potential and the oil-water interface will be an oil equipotential surface which also serves as an impermeable boundary to the water flow field.

If we represent this critical angle of tilt by θ_c, its value can be found from equation (49) by noting that when $\theta = \theta_c$,

$$\frac{\partial h_w}{\partial x} = \text{grad } h_w \cos \theta_c; \qquad \frac{\partial h_w}{\partial z} = \text{grad } h_w \sin \theta_c.$$

Making these substitutions in equation (49) then gives

$$\frac{dz}{dx} = \tan \theta_c = \frac{\sin \theta_c}{\cos \theta_c} = \frac{\text{grad } h_w \cos \theta_c}{\dfrac{\rho_w - \rho_0}{\rho_w} - \text{grad } h_w \sin \theta_c}.$$

Then by transposing and combining terms we obtain

$$\sin \theta_c = \frac{\rho_w}{\rho_w - \rho_0} \text{grad } h_w, \qquad (50)$$

which expresses the critical angle of tilt, or the tilt of the oil-water interface, in terms of the densities of the two fluids and of grad h_w.

If dz is the increase in elevation and dh_w the increase of head in a distance ds along a water flowline, and dx the horizontal projection of ds, then

$$\sin \theta_c = \frac{dz}{ds}; \qquad \text{grad } h_w = \frac{dh_w}{ds}.$$

With these substitutions equation (50) will become

$$\frac{dz}{ds} = \frac{\rho_w}{\rho_w - \rho_0} \frac{dh_w}{ds}, \qquad (51)$$

which, when multiplied by ds/dx, becomes

$$\tan \theta_c = \frac{dz}{dx} = \frac{\rho_w}{\rho_w - \rho_0} \frac{dh_w}{dx}. \qquad (52)$$

Here dh_w/dx is the increase in head of the flowing water in a distance ds along the flowline, or along the oil-water interface, whose horizontal projection is dx. Since dh_w is the increase in elevation of the potentiometric surface for the water in the horizontal distance dx, it follows that dh_w/dx is the slope of this surface. Hence the tilt of the oil-water interface is equal to the slope of the water potentiometric surface multiplied by the sensitive amplification factor $\rho_w/(\rho_w - \rho_0)$.

While the foregoing force relationships and equations of tilt are essential in understanding the mechanics of petroleum migration, in the actual delineation of traps the scalar potentials are more convenient to work with. The potential of oil in terms of that of water, and of the elevation, as given by equation (40), is:

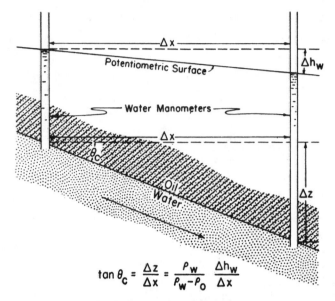

$$\tan \theta_c = \frac{\Delta z}{\Delta x} = \frac{\rho_w}{\rho_w - \rho_0} \frac{\Delta h_w}{\Delta x}$$

FIG. 19.—Relation between tilt of oil-water interface in hydrodynamic trap and slope of potentiometric surface.

$$\Phi_0 = \frac{\rho_w}{\rho_0} \Phi_w - \frac{\rho_w - \rho_0}{\rho_0} gz.$$

Then if we convert each fluid potential to an expression containing the fluid head, and divide by g, we obtain the relationship

$$h_0 = \frac{\rho_w}{\rho_0} h_w - \frac{\rho_w - \rho_0}{\rho_0} z$$

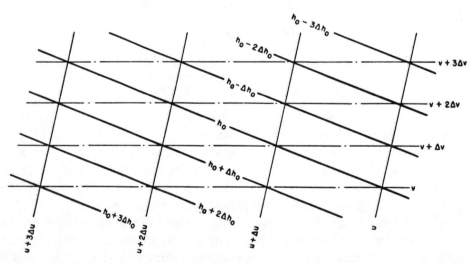

FIG. 20.—Graphical relation of surfaces $h_0 =$ const to those of $u =$ const and $v =$ const, for equal intervals $\Delta h_0 = \Delta u = \Delta v$.

of equation (45), each term of which has the dimensions of length. If h_w and z are known at all points in space, we can construct the families of surfaces $h_w = $ const and $z = $ const, and from these we can determine the desired surfaces $h_0 = $ const.

The surfaces $z = $ const are simply a family of horizontal surfaces. The family $h_w = $ const coincides with the water equipotential surfaces and is normal to the flowlines. The intersection of a surface $h_w = $ const with $z = $ const will be along a horizontal curve for which both h_w and z will be constant, and hence, by equation (45), h_0 will be constant also. It follows therefore that these three families of surfaces can only intersect along a single family of horizontal curves. They are therefore all tangent at any given point to a common horizontal axis, or perpendicular to a common vertical plane. In geological terms the surfaces $h_w = $ const and $h_0 = $ const may differ widely in dip but they always have the same strike.

Equation (45) is amenable to simple graphical treatment if we set

$$\left.\begin{array}{c} \dfrac{\rho_w}{\rho_0}\, h_w = u, \\[2ex] \dfrac{\rho_w - \rho_0}{\rho_0}\, z = v. \end{array}\right\} \tag{53}$$

Then

$$h_0 = u - v, \tag{54}$$

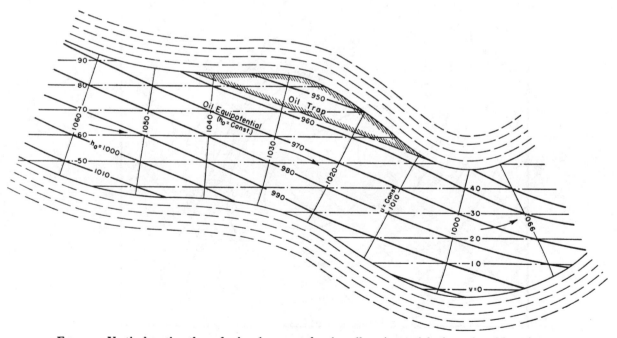

FIG. 21.—Vertical section through plunging nose, showing oil equipotentials, h_0, and position of trap, as determined graphically from *u* and *v* surfaces.

of which each term is still a length. If h_w and z are known, then by equations (53) u and v are known also. If we now plot the two families of surfaces $u =$ const and $v =$ const for constant intervals $\Delta u = \Delta v$, these intersecting surfaces will form a family of $\Delta u \, \Delta v$-solenoids with horizontal axes and parallelogram cross sections. Consider one of these solenoids bounded by the surfaces u and $u + \Delta u$, and v and $v + \Delta v$. The value of h_0 at the intersections of the surfaces u and v and of $u + \Delta u$ and $v + \Delta v$ will be $u - v$. Therefore a surface passing through this diagonal of the $\Delta u \, \Delta v$-solenoid will also be a surface $h_0 =$ const, and the family of such surfaces will be the family $h_0 =$ const of constant interval

$$\Delta h_0 = \Delta u = \Delta v.$$

On any surface passed through this three-dimensional field the traces of the surfaces along which u, v, and h_0, respectively, are constant will be corresponding families of curves whose mutual relations are likewise expressed by equation (54). Of particular significance, we may regard the upper boundary of a reservoir stratum as being such a surface on which can be mapped the curves $u =$ const and $v =$ const. From these the curves $h_0 =$ const can be drawn directly. If in three-dimensional space the values of Φ_0 and h_0 decrease in the upward direction, which

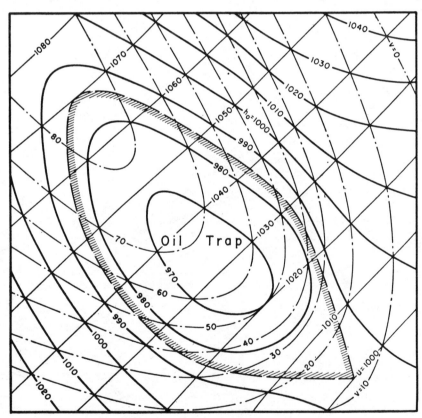

FIG. 22.—Map of traces of intersections of constant-value surfaces, h_0, u, and v, with upper surface of sand showing potential minimum, which is trap for oil.

is true with but rare exceptions, the oil will migrate to the upper boundary and will then be deflected along this boundary in the direction of decreasing potential. Since the curves $h_0 =$ const are proportional by the constant factor $1/g$ to Φ_0, it follows that along this surface the migration will be normal to the curves $h_0 =$ const and in their direction of decrease. If these curves close completely upon themselves encompassing an area of low potential, or low values of h_0, this area must represent a trap, for once inside it the oil can not escape without flowing from a region of lower to one of higher potential.

For primary data, the curves $z =$ const along the upper surface of a reservoir stratum are simply the familiar structure contours on that surface. The curves $h_w =$ const are the contours of the water potentiometric surface. From these two maps, with the aid of equations (53) we can construct the curves $u =$ const and $v =$ const. The first will be everywhere parallel with the potentiometric contours, and the second with the structure contours, but with different spacings and contour intervals.[3]

In the case of gas the strict theory is somewhat more complex, for the potential of gas involves the work against compression, as well as that of injection against the ambient pressure at the point of interest. However, since the absolute

---

[3] It has subsequently been pointed out by the writer's colleagues, A. S. Dickinson and J. J. McMahon, that by a slight modification of the foregoing procedure a still further simplification can be achieved. Beginning with equation (45),

$$h_0 = \frac{\rho_w}{\rho_0} h_w - \frac{\rho_w - \rho_0}{\rho_0} z,$$

if we divide each term by $(\rho_w - \rho_0)/\rho_0$, we obtain

$$\frac{\rho_0}{\rho_w - \rho_0} h_0 = \frac{\rho_w}{\rho_w - \rho_0} h_w - z, \tag{53$'$}$$

each term of which still has the dimensions of length.

Now if we let

and

$$\left. \begin{array}{l} u = \dfrac{\rho_0}{\rho_w - \rho_0} h_0, \\[3mm] v = \dfrac{\rho_w}{\rho_w - \rho_0} h_w, \end{array} \right\} \tag{54$'$}$$

then by substitution in equation (53$'$) we get

$$u = v + z. \tag{55$'$}$$

Here the surfaces $u =$ const coincide with oil equipotential surfaces, those for which $v =$ const coincide with the water equipotential surfaces, and the surfaces $z =$ const are a family of horizontal surfaces of elevation z.

The traces of the intersections of each of these surfaces with the upper boundary of a reservoir stratum are given by a corresponding family of curves of which the $v =$ const curves are water equipotentials, and the $z =$ const curves are the structure contours. The curves $u =$ const are then the oil equipotentials and are obtained graphically from the v and z curves in the same manner as indicated above. Areas of closure of the u-curves containing minimum values of u represent oil traps.

This procedure has an advantage over that described earlier in that it permits the direct use of the structure contour map, and further, that the u-values have a geometric as well as an energetic significance. The curves $v =$ const may be regarded as the contour lines on a hypothetical v-surface. Then the curves $u =$ const would be the isopach contours between the v-surface and the z-surface which is the top of the stratum. By this interpretation a trap would correspond to an isopach minimum between these two surfaces. Furthermore the volume of the trap can be determined from the u-contours in the same manner as the volume of a hydrostatic trap is obtained from the structure contours.

value with respect to our standard datum is in this case of little interest, slight error will be committed if we treat gas as if it were a liquid of the same density which the gas would have under the same conditions of temperature and pressure. With this simplification we may employ for gas equations (40), (45), and (53), which have been used for oil, with the understanding that ρ_0 is now the density of the gas, and h_0 the height above datum that a fluid of constant density ρ_0 would be supported in a manometer terminated at the point of interest.

Because the factors ρ_w/ρ_0 and $(\rho_w - \rho_0)/\rho_0$ differ considerably for fluids of different density, particularly oil and gas, it will be found that for a stratum of fixed structural configuration and ground-water motion the pattern of curves $h_0 = \text{const}$ will, in general, differ markedly for these separate fluids. In the special case, where the water is at rest, h_w will have a constant value and the family of surfaces $h_0 = \text{const}$ for both oil and gas will be horizontal. When mapped on the surface of the stratum the contours $h_0 = \text{const}$ will therefore coincide with the lines $z = \text{const}$, and the traps with the so-called "structural highs" of the anticlinal or gravitational theory. If the water is in motion, however, the family of curves $h_w = \text{const}$ will not coincide with the structure contours; neither will the traps for oil and for gas coincide with each other, although they may have various degrees of overlap; they may even be separated entirely, a trap for oil being incapable of holding gas and a trap for gas unable to retain oil. Furthermore, for oil and gas of given densities the positions of the traps can be changed drastically by varying the direction and magnitude of grad h_w, that is, the direction and rate of flow of the water. In fact it is only in the special case of the water at rest that the traps for oil and gas coincide. In this instance the impelling forces for both fluids are directed vertically upward, the equipotential surfaces for both are horizontal, and the traps are the familiar ones of the anticlinal or gravitational theory.

MAGNITUDE OF EFFECTS

At this point it is of interest to consider the magnitude of the effects which may be expected. For an order of magnitude we may assume that the extreme values of h_w for a given formation of regional extent are approximately equal to the highest and lowest elevations of its outcrop, or of the topography overlying it. Between these points the average value of the magnitude of grad h_w will be $(h_{max} - h_{min})/(\text{distance})$, which is of the order of magnitude of the regional topographic slopes. In regions like the Gulf Coast this may be as little as 2 or 3 feet per mile; in regions like the Great Plains it may be 10–20 feet per mile; whereas in intermontane basins it may be as great as 100 feet per mile or more. For 10 feet per mile, grad h_w will have a magnitude of $1/500$ or 2×10^{-3}, while for 100 feet per mile it will be ten times as great, or 2×10^{-2}.

Now, since the vector $\mathbf{E}_w$ is given by

$$\mathbf{E}_w = - g \text{ grad } h_w$$

the magnitude of $\mathbf{E}_w$ for a gradient of 100 feet per mile would be about 0.02 g, or about 20 dynes per gram, since g has the magnitude 980 dynes per gram.

Also, since

$$-\frac{\text{grad } p}{\rho_w} = (\mathbf{E}_w - \mathbf{g})$$

it follows that $(\text{grad } p)/\rho_w$ can differ in magnitude very slightly from g, and in direction, for a value of $\partial h_w/\partial x$ of 100 feet per mile, it can be deflected from the vertical only about 1 degree.

The tilt of an oil- or gas-water interface as given by equation (52) is

$$\frac{dz}{dx} = \frac{\rho_w}{\rho_w - \rho_0}\frac{dh}{dx},$$

where ρ_0 is the density of the oil or gas considered. The factor $\rho_w/(\rho_w - \rho_0)$ is a sensitive amplification factor whose value increases from 1 to infinity as ρ_0 is increased from 0 to ρ_w. For gases at different pressures its value may range from about 1 to 2, and for oils from about 2 to infinity, though most commonly its value lies in the range from about 7 to 15.

If we take a representative value of this tilt factor for oil to be 10, corresponding to an underground fluid density of 0.9 gm./cm.^{3}, then a slope of the potentiometric surface of 10 feet per mile would produce an oil-water tilt of 100 feet per mile; a slope of 100 feet per mile would produce a tilt of 1,000 feet per mile. A structure with a leeward closing dip less than the tilt of the oil-water interface can not hold oil.

FRESH-WATER—SALT-WATER RELATIONSHIPS

Until now we have treated water as a homogeneous fluid. Actually, however, underground waters contain highly varied amounts of dissolved minerals, and range in this respect from potable waters of less than 1,000 parts per million of dissolved solids to highly concentrated brines of a few hundred thousand parts per million. The principal significance which this has mechanically is that it causes the fluid density to vary and in this manner affects the energies and the forces involved. If the fluid is of constant density, this has no effect on the form of the potential and force equations so long as the true value of the water density is used, but how are we to deal with a case where in a single stratum the waters vary from fresh to highly saline with a corresponding change in density?

Just as in the case of water and oil, it is possible to have the flowing fresh water in contact with static salt water, and the equations of tilt of the interface are the same. If at rest, the fresh water will overlie the salt water with the interface horizontal, but if the fresh water is set in motion, this interface will tilt upward in the direction of the flow at an angle whose tangent is given by

$$\frac{dz}{dx} = \frac{\rho_w}{\rho_w - \rho_s}\frac{dh_w}{dx}, \tag{55}$$

where ρ_w is the density of the fresh water and ρ_s that of the salt water. If the salt water has the salinity of normal sea water, then ρ_s will have the value 1.025 gm./cm.³ and the tilt factor, $\rho_w/(\rho_w - \rho_s)$, will have the value -40.

In an underground basin salt water can be trapped against flowing fresh water in structures which are of the same kind as those for oil, except upside down. Thus under static conditions salt water will be trapped in the lowest parts of closed basins, but under dynamical conditions it can only be trapped with a tilted fresh-water—salt-water interface against dips steeper than the angle of tilt. Because of

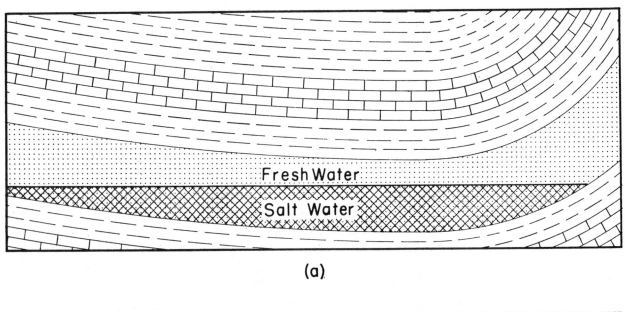

(a)

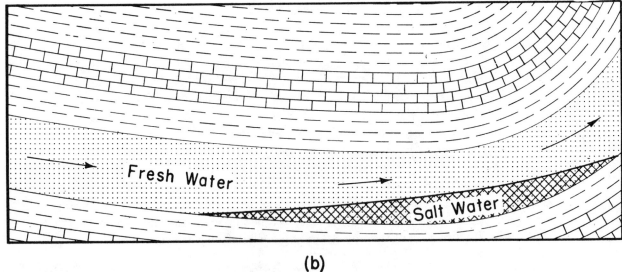

(b)

FIG. 23.—Position of salt water in syncline (*a*) under hydrostatic conditions, (*b*) under hydrodynamic conditions.

the small density contrast the tilt amplification factor is large. Thus to trap normal sea water against flowing fresh water with a 10-foot per mile gradient, a dip of 400 feet per mile would be required, and for larger gradients the dips required would be proportionately steeper.

If our object is to determine whether the water is at rest or in motion, then it appears that a (constant) local value of its density is the proper one to use. The potential of the water at any given point is then

$$\Phi = gz + \frac{p}{\rho},$$

where ρ is the value of the water density at the point, and the water will be in equilibrium if for all possible small displacements of an element away from the position it now occupies its potential remains constant or increases. It will not be in equilibrium if in some direction its potential decreases, or if grad Φ is not equal to zero.

To illustrate this point consider a regional sand extending under a sedimentary basin in which the salinity of the water increases markedly with depth. If this water is in stable mechanical equilibrium and hence at rest, it will arrange itself with a density stratification wherein the surfaces of constant density will be horizontal with the density increasing downward. On the flank of this basin suppose that we have accurate pressure and density measurements in a row of wells along a line extending down the dip. Let a well in the middle of the row be at the point of investigation and the density of the water at this well be taken as a reference density. Then at each of the wells of the series the potential of the water having the reference density can be determined using this density and the locally measured pressure.

If we plot the potential obtained in this manner at each separate well against its distance along the profile, this will be found to plot as a curve concave upward. If the water is at rest, as postulated, this curve will have a minimum point at the reference well, or will be tangent at this point to a horizontal line. If, on the contrary, the water has a component of motion along the line of wells, the curve so plotted will be tangent to a sloping line at the reference well and the flow will be in the direction of its downward slope.

If the problem of interest is the determination of the oil potentials in a water of variable density in terms of the potential of water, the procedure outlined originally is still valid *provided the water potential be that of a reference water of constant density*, independently of the manner in which the density of the actual water may vary. In terms of the potential of such a water, that of oil at any given point, as expressed by equation (40), is still given by

$$\Phi_0 = \frac{\rho_w}{\rho_0} \Phi_w - \frac{\rho_w - \rho_0}{\rho_0} gz,$$

where ρ_w is an arbitrarily chosen standard density.

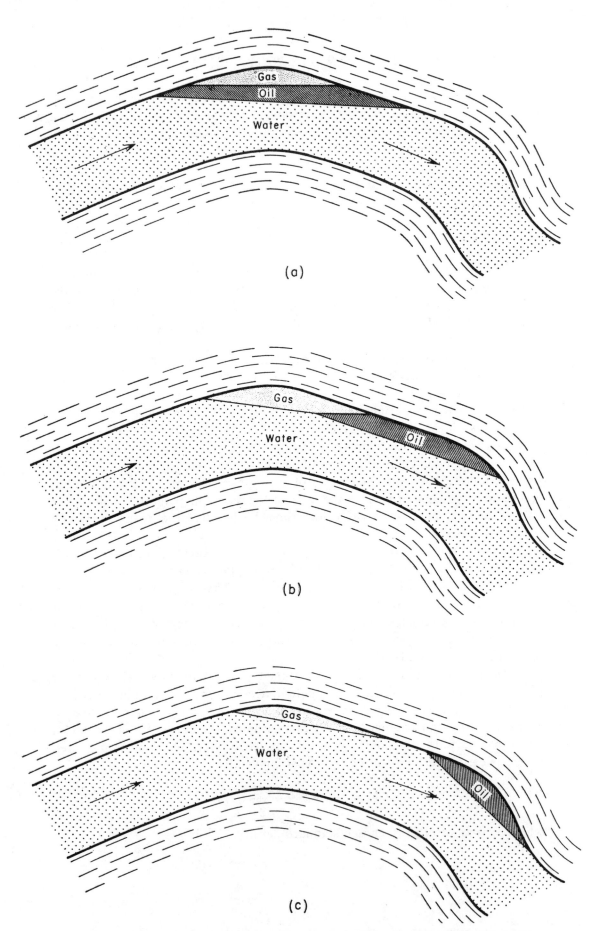

FIG. 24.—Types of hydrodynamic oil and gas accumulation in gently folded thick sand. (*a*) Gas entirely underlain by oil; (*b*) gas partly underlain by oil; (*c*) gas and oil traps separated.

That the value of Φ_0, given by this equation, is the correct one and is independent of the choice of ρ_w can be seen by direct substitution of the primary value for Φ_w:

$$\Phi_w = gz + \frac{p}{\rho_w} \cdot$$

This gives

$$\Phi_0 = \frac{\rho_w}{\rho_0} \left(gz + \frac{p}{\rho_w} \right) - \frac{\rho_w - \rho_0}{\rho_0} gz,$$

from which the water density ρ_w cancels out, leaving

$$\Phi_0 = gz + \frac{p}{\rho_0} \cdot$$

This is the primary definition of Φ_0 and its value depends only upon the density of the oil, the elevation of the point, and the pressure.

Thus, in computing the oil equipotentials, we are at liberty to use a reference water of ideally constant density (incompressible and free from thermal expansion). For a local problem there are certain advantages in choosing for this standard reference density a value of ρ_w which is about the mean of the density of the water underground in that locality, but for large regions probably the most convenient choice for all purposes would be a density which approximates that of fresh water at surface temperatures, namely, $\rho_w = 1.000$ gm./cm.^{3}.

If the water underground differs in density from the standard water, and is at rest, the equipotential surfaces for the standard water will be horizontal, and the oil equipotential surfaces, by equation (40), will be horizontal also. If the water is flowing, the surfaces $\Phi_w = $ const will be tilted, but not quite perpendicular to the flowlines, and the oil equipotentials will be tilted with respect to these in the manner indicated heretofore.

TYPES OF HYDRODYNAMIC TRAPS

The types of accumulation of oil and gas to be expected under hydrodynamic conditions are influenced by all of the structural and stratigraphic complexities which are already familiar in the case of hydrostatic traps, but with the additional complication that these also influence the flow pattern of the water which in turn determines the angle of tilt and hence the positions of the hydrodynamic traps. As illustrative examples of the types of hydrodynamic traps that may be expected, only a few of the simplest cases will be cited.

The most obvious is an anticlinal or domal structure. If water is flowing through a regional sand in such a structure, the oil and gas equipotential surfaces will each be tilted downward in the direction of the flow, those for oil by an angle greater than those for gas. If the dip on the downstream side of the structure is steeper than the tilt of the oil equipotential surface, then the structure will serve as a trap for both oil and gas. Gas or oil separately may then be trapped in the structure with a tilted water interface, high on the upstream and low on the downstream side. If both fluids are trapped together, the oil may rest upon the water with a tilted interface, while a gas cap may rest entirely upon the oil with a static, horizontal gas-oil contact [Fig. 24 (a)]; or, the gas cap may rest partly upon the oil on the downstream side and partly upon water on the upstream side, as shown in Figure 24 (b). In this case the gas-water contact will be tilted at a small angle, the oil-water contact at a greater angle, while the gas-oil contact will be horizontal. It is also possible that the gas may be entirely underlain by water with a gently tilted interface, with the oil in a completely separate trap in a down-structure nose, as shown in Figure 24 (c).

These three configurations, (a), (b), and (c), could represent the conditions under which an oil of given density would be trapped by (a) weak, (b) moderate, or (c) strong flow of ground water. Alternatively, with the same flow of ground water in each case, they could also represent the conditions under which (a) light, (b) medium, or (c) heavy oil would trap.

Again the tilt of the oil equipotential surfaces may be greater, and that of the gas surfaces less, than the closing dip on the downstream side of the structure. In this case the structure will hold gas with the appropriate tilt but will not hold oil. At still higher values of grad h_w the gas equipotential surfaces also will become tilted more than the angle of the closing dip. When that stage is reached the structure will hold neither oil nor gas.

Another type of anticlinal accumulation may occur in which the amplitude of the fold is much larger than the thickness of the sand. Under hydrostatic conditions the oil may completely fill the sand to a considerable distance down each limb of the fold, but the oil-water contact on each limb will occur at the same level. If a gas cap is present, the gas-oil contact will also be horizontal.

If the water is flowing, it can not pass directly under the accumulation since this fills the sand from top to bottom. It must therefore flow around the accumulation like a river around an island. In this case the oil-water interface will be a ring-shaped surface, but along this surface the appropriate tilt must occur. As seen in Figure 25 (a), a vertical cross section parallel with the direction of regional flow, the oil column will stand higher on the upstream limb than on that downstream by an amount

$$\Delta z = \frac{\rho_w}{\rho_w - \rho_0} \Delta h_w,$$

where Δh_w is the difference in the head of water at the two contacts.

If a gas cap exists, it may rest completely upon the oil with a horizontal gas-oil contact, as shown in Figure 25 (a), or it may rest upon the water in the upstream limb and upon the oil in the downstream limb of the structure, as shown

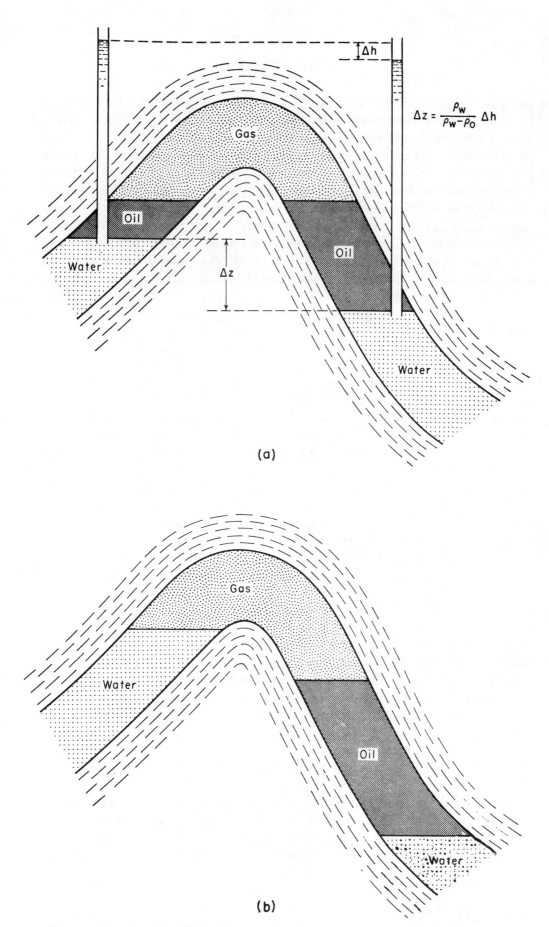

$$\Delta z = \frac{\rho_W}{\rho_W - \rho_O} \Delta h$$

(a)

(b)

Fig. 25.—Entrapment in thin sand in large-amplitude fold. (a) Oil extending farther down structure on downstream flank. (b) Oil confined to downstream flank, upstream flank containing gas and water only. (This is shown as result of more energetic ground-water flow, but it could also be result of heavier oil.)

391

in Figure 25 (*b*), the oil being entirely absent in the upstream limb. Or, as before, the structure may not be able to hold oil at all, and under more extreme conditions it may not be able to hold gas either.

A second type of structure in which oil and gas may be trapped is a structural terrace or anticlinal nose on which there is a steepening of the dip in the downdip axial direction. With the water flowing down the dip parallel to the axis of the structure, the equipotential surfaces for both oil and gas will be tilted in this direction also. If the angle of tilt for either of these fluids is intermediate between the lesser and the greater dips of the structure along its axis, that fluid will be trapped in the region where the change in dip occurs. As in the anticlinal trap, either gas or oil separately, or a combination of both may be entrapped in such a structure.

Another possible type of hydrodynamic trap is in a homoclinally dipping sand in which the permeability increases markedly down the dip. With water flowing down the dip, grad h_w will be greater in the sands of low permeability than in those in which the permeability is high. This will cause the oil (or gas) equipotential surfaces to have a steeper tilt in the low-permeability region than in the region where it is high. It will thus create a family of curved equipotential surfaces with upward concavity. Under suitable conditions of permeability and flow rate, entirely without regard for the capillary effects produced by the change of permeability, these can terminate in completely closed curves against the upper surface of the stratum and so form a trap, as shown in Figure 26.

Because of these effects areas of regional homoclinal dip become extraordinarily sensitive to the hydrological conditions. With the water flowing down the dip, traps on noses, terraces, and homoclines are possible; with the water static, some kind of conventional closure is required; with the water flowing up the dip, not only are the unclosed structures unfavorable, but even closed structures may not have steep enough reversed dips to trap oil under the conditions prevailing.

VARIATIONS OF POSITIONS OF TRAPS WITH CHANGES OF FLOW STATE

Possibly a better idea may be gained as to the influence of the state of flow on the positions of traps by a computed example. We assume a structure-contour map to be given on a sand of regional extent in a petroliferous area, and we then compute the positions of the oil and gas traps for different hydrological conditions. The results of such a computation are shown in Figures 27 to 30, for the hydrostatic case and for three separate hydrodynamic states. For the latter a constant value of 10 feet per mile, which is a relatively small magnitude, has been assumed for −grad h, and the cases differ only in the direction of the assumed flow. A density of 0.87 gm./cm.³ has been assumed for oil and of 0.4 for gas, giving for the two tilts 83 and 17 feet per mile, respectively.

Although the data used are well within the range of probability, it will be seen that the effect upon the positions of the traps in the different cases is striking. Major emphasis has been placed on the oil traps, and where combination traps are possible an arbitrary amount of gas has been shown simply for generality.

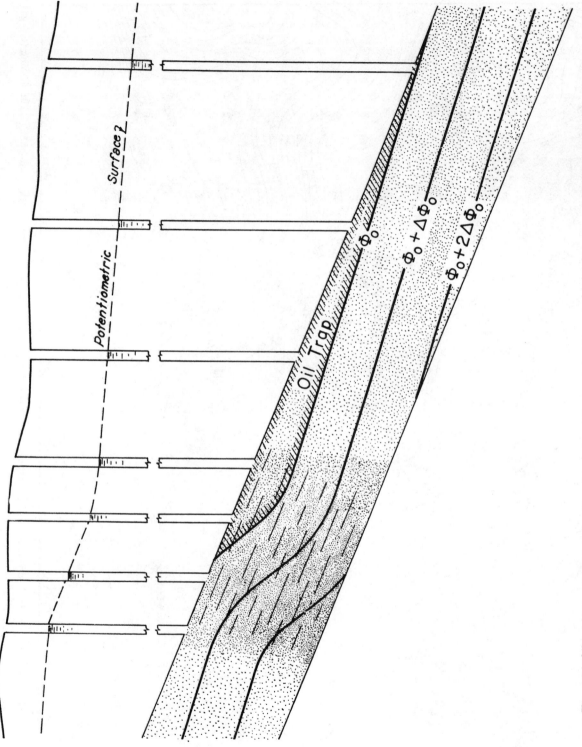

FIG. 26.—Oil entrapment downstream from low-permeability region in homocline. Increase of potential gradient through region of low permeability causes oil equipotentials to slope upward and close against top of sand, thus giving rise to trap. Tilted oil-water interface found on drilling differentiates this from hydrostatic-stratigraphic trap. Hugoton gas field appears to be of this type.

FIG. 27.—Traps in given region under hydrostatic conditions. Scale, 1 inch = 3 miles. Contour interval, 20 feet. Dark shading, oil. Light shading (G), gas.

Fig. 28.—Oil and gas traps in same region as that shown in Figure 27, with −grad *h* equal to 10 feet per mile southeast. Scale: 1 inch=3 miles. Contour interval, 20 feet. Dark shading, oil. Light shading (G), gas.

FIG. 29.—Same as Figure 28 except with flow southwest.

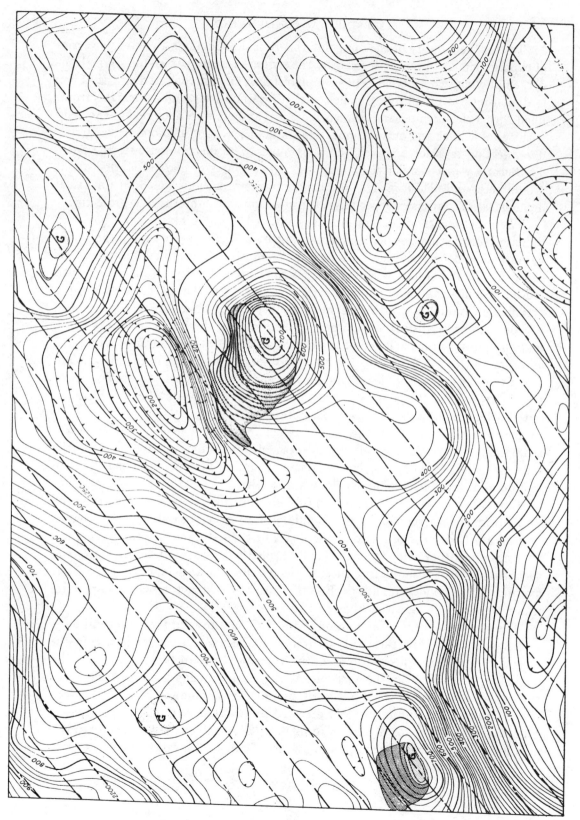

FIG. 30.—Same as Figure 28 except with flow northwest.

EXPERIMENTAL CONFIRMATION

To test the validity of the foregoing theoretical deductions an extensive series of experiments has been performed by the writer's research assistant, Jerry Conner, employing two different types of apparatus. The first consisted of a sand box of dimensions 32 by 18 by $2\frac{1}{2}$ inches, with a plywood back and a Plexiglas front, filled with a clean white sand. Screened open chambers were provided at each end with suitable inlet and outlet connections, and observation-manometer taps; and dye-injection taps for mapping the streamlines were distributed over the back. The second was a simpler cell consisting of two plates of glass separated about 1 millimeter and held in a rigid watertight framework. This was built for visual demonstration purposes, especially for projection in a lantern-slide projector.

The experiments of present interest consisted in charging the sand box with fluids of contrasting density and observing their steady-state equilibrium configurations under various states of flow. In one series of experiments the box was charged with a bottom stratum of a sugar solution of density 1.1 gm./cm.^{3}, dyed red for visibility, and overlain by colorless tap water. Under static conditions the two fluids assumed a stratified arrangement with a horizontal interface. Then without otherwise disturbing the heavy bottom fluid the upper fluid was set in motion, flowing from each end of the box to an outlet in the center of its top. In accordance with prediction the interface became tilted upward in the direction of the flow of the upper, less dense fluid, forming a cusp. At higher rates of flow these tilts increased and the cusp became sharper until finally its apex reached the outlet after which both fluids were in motion.

Short of the time when the cusp reached the outlet, however, except for a temporary transient stage between adjustments for different rates of flow, the configuration observed was one of dynamic equilibrium between one fluid in motion and a contiguous fluid entirely static (Fig. 1).

In another series of experiments a rubber stratum was embedded in the sand simulating the impermeable upper boundary of a reservoir sand. This was shaped into an anticline near one side of the box and then steepened into a structural nose near the other side. The box was first filled with water under pressure, and then colored alcohol simulating oil was injected into the anticline. After being left overnight the alcohol arranged itself at the crest of the structure with a horizontal water interface, as shown in Figure 32 (*a*).

The water was next set in motion in the direction toward the structural nose. For slow rates of flow the "oil" accumulation became tilted downward in the direction of the flow just so long as the tilt continued to be less than the dip angle the oil remained in the anticline. When, by increasing the rate of flow, the tilt was made to exceed the dip, the oil was migrated completely out of the anticline and down the dip until it was arrested by the steeper dip of the nose, as in Figure 32 (*c*). Here, for a fixed rate of flow, it remained in completely stable equilibrium. At a still higher rate of flow the tilt was made to exceed the steeper

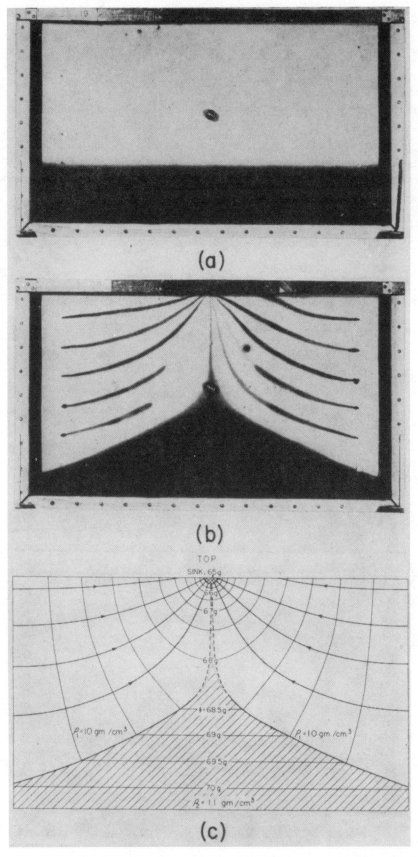

FIG. 31.—Experimental relations between flowing fresh water and static sugar solution (colored).
(a) Fresh water at rest; (b) fresh water flowing; (c) measured fresh-water equipotentials for state (b).

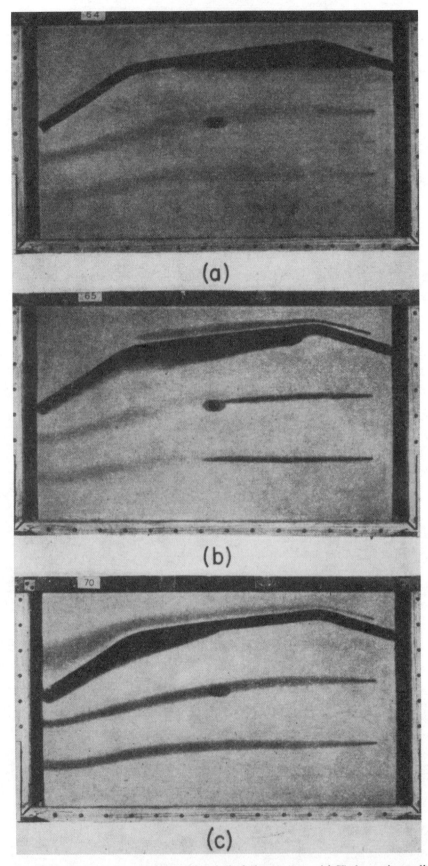

FIG. 32.—Experimental oil traps, using colored alcohol and water. (a) Hydrostatic conditions. (b) Flushing of alcohol from anticline by water flowing left. (c) Alcohol trapped on nose.

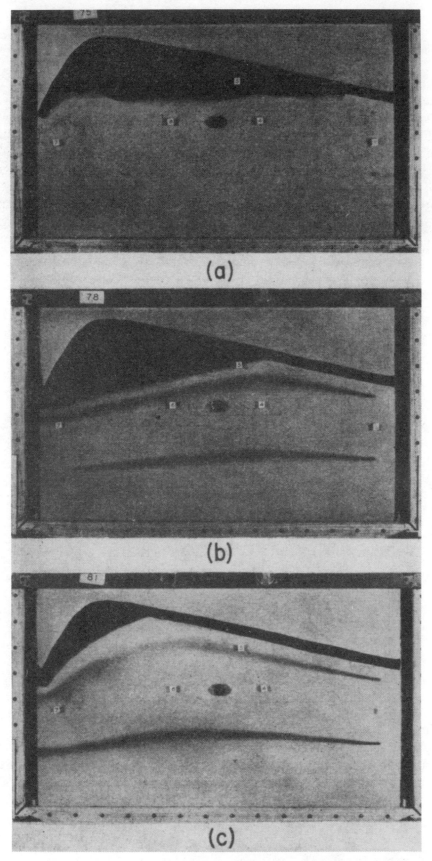

FIG. 33.—Flow-box experiment in symmetrical anticline. (*a*) Hydrostatic conditions (ragged appearance of interface caused by color fading). (*b*) Tilted interface with water flowing left. (*c*) Steeper tilt caused by higher rate of flow.

dip of the terrace and the oil was migrated farther downdip out of that structure also.

Hydrodynamic tilts in an asymmetrical anticline are shown in Figure 33.

In addition to visual observation and photographic recording of these experiments, manometer readings were also taken of the flowing water. From these it was possible to determine the magnitude of grad h_w at points along the two-fluid interface whose angles of tilt were known from direct observation. Employing the fluid densities and the observed values of grad h_w, it was possible to compute by means of equation (50) the theoretical angle of tilt. This computed value was then plotted graphically against the observed value for fifteen separate observations. The results, which are presented in Figure 34, show complete agreement within the limits of experimental error, which confirms the validity of the predictions.

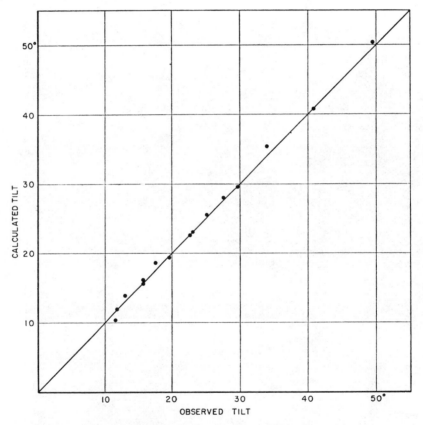

FIG. 34.—Experimental confirmation of tilt equation.

Qualitative experiments of the same kind have been performed with the simpler plate-glass apparatus. The basic theory for this equipment is the same as for the sand, but the rate of flow in the former is very much greater than in the latter. Furthermore, capillary effects can be kept so small that a gas cap can also be demonstrated. With oil alone (simulated here also by colored alcohol) all of the phenomena observed in the sand box can be duplicated. When a gas cap of air is added, under static conditions the conventional horizontally stratified anticlinal accumulation of water, oil, and gas is observed. Then with the water set in motion the oil-water interface becomes tilted, but the gas-oil interface remains horizontal. At a higher rate of flow the gas cap rests partly upon the oil and partly upon the water, with the oil extended down the leeward side of the structure. The gas-oil contact remains horizontal, but the gas-water contact is slightly tilted. At a still higher rate of flow a complete separation of oil and gas occurs, the oil becoming entrapped in the nose downdip while the gas remains in a tilted configuration in the anticline. These experiments are illustrated in Figure 35 (*a*), (*b*), (*c*), (*d*).

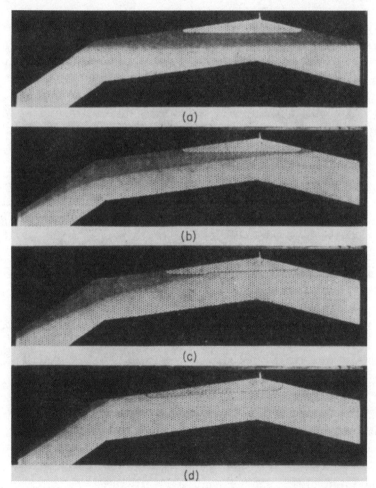

Fig. 35.—Experiments with plate-glass apparatus. (*a*) Hydrostatic conditions. (*b*) Dynamic conditions; oil-water interface tilted, gas-oil interface horizontal. (*c*) Higher rate of flow, oil highly tilted with gas cap resting partly on oil and partly on water; gas-oil interface horizontal, gas-water interface tilted. (*d*) Complete separation of oil and gas traps under still higher rate of flow, with oil trapped in down-structure nose and gas in anticline but tilted.

In this case it is also easy to demonstrate fluid migration under hydrodynamic conditions. If gas bubbles are injected into the system downdip from the nose accumulation of oil, these will migrate to the upper impermeable boundary of the sand and then proceed updip completely through the oil accumulation to the trap for gas near the crest of the anticline, directly opposite the direction of flow of the water.

GEOLOGICAL EVIDENCE OF HYDRODYNAMIC EFFECTS

In our discussion of the mechanics of ground-water motion, it was shown that at the ground-water table the ground-water potential has the value $\Phi_w = gz$, where z is the elevation of the water table. Furthermore, in all but extremely desert areas the water-table surface follows rather closely the topographic surface. Ground water flows invariably from regions of higher to lower potentials and follows principally the more permeable strata. If permeable strata in a sedimentary basin crop out around the margin with high elevations on one side and much lower elevations on the other, then a difference of potential across the basin is provided and water circulation through these strata should result.

Even if the strata do not crop out but extend from regions of higher to lower topography, a circulation may still result with the water entering and leaving through the somewhat permeable overburden.

The geological configuration of most sedimentary basins is favorable for such a regional ground-water circulation, at least through some of the more permeable strata of regional continuity. Yet despite the extensive literature that now exists on the ground-water hydrology of various areas, confirmatory data on such large-scale regional circulation are comparatively scarce, since the available ground-water studies have largely been limited to potable waters of shallow depths, and often of quite restricted areas. A notable exception, however, occurs in the case of the classical studies of N. H. Darton (1905, 1909, 1918a, 1918b) of the ground-water flow in the Dakota sandstone group of the Great Plains area of the United States.

In his pioneer work on the geology of this large area, Darton noted that this group of sandstones crop out as steeply dipping hogbacks along the Rocky Mountain front at elevations of about 6,000 feet above sea-level. They crop out again in a ring around the Black Hills at elevations ranging from about 3,000 to 4,000 feet. From these western high-level outcrops the group descends steeply underneath the Denver and Williston basins, reappearing on the east in an outcrop or shallow subcrop belt extending from southwestern Kansas through eastern Nebraska and South and North Dakota, at elevations of about 1,000 to 1,500 feet above sea-level. Darton reasoned correctly that the water should be flowing through this group of strata in a generally easterly direction and that if this were true the static heights to which water should rise in cased wells terminated in these sands should progressively decline in an easterly direction.

With this theoretical inference as a guide he began recording water levels in wells and constructing a potentiometric map for this group over the area east of the Rocky Mountains and the Black Hills from southern Colorado and Kansas to the northern boundary of South Dakota. The composite result of Darton's compilations over a period of 20 years or more, with a slight amount of additional data obtained from the Denver office of the Ground-Water Branch of the U. S. Geological Survey, is presented in Figure 36.

From this map it will be noted that the potentiometric surface has a persistent easterly slope over the entire area east of the Rocky Mountains and the Black Hills, indicating that the water is entering the Dakota sands at their higher western outcrops and discharging at the surface from their lower outcrops on the east. It will also be noted, although the map is very sketchy in these areas, that both north and south of the Black Hills the water is flowing eastward around this uplift from the Powder River Basin.

From this map the approximate magnitude of the slope of the surface can also be determined. Across the basin at about the latitude of Cheyenne a total drop of 4,800 feet occurs in a distance of about 450 miles, giving an average slope across the basin of about 10 feet per mile. In some localities the slope is somewhat greater than this and in others less.

Not many data on the oil fields of the Denver Basin have as yet been compiled, but in the gap between the Laramie Range and the Black Hills are the two older fields, Lance Creek and East Lance Creek. In his report on Lance Creek in Volume II of the American Association of Petroleum Geologists symposium on *Structure of Typical American Oil Fields*, Wilson B. Emery (1929, pp. 609-10) pointed out that in the Dakota sandstone in this field there was a gas cap on top of the structure, but oil had been found only on the south and east flanks and that both the gas cap and the oil were tilted toward the northeast, with the oil originally extending about 100 feet below the lowest closed contour.

Subsequent data on this field confirm Emery's observation, except that the tilt appears to be southeasterly rather than northeasterly. Both this field and East Lance Creek also produce from the Sundance of Jurassic age. In this formation the tilt for Lance Creek is estimated to be about 100 feet per mile and East Lance Creek about 60 feet per mile, both in a southeasterly direction.

Near the southern edge of Darton's map in southwestern Kansas and the panhandles of Oklahoma and Texas lies the large Hugoton gas field. Although this occurs in the limestones and dolomites of the Lower Permian rather than in the Cretaceous, the regional relationships are similar so that an easterly flow of the water in the Permian is to be expected unless precluded by impermeability.

According to data presented by Garlough and Taylor (1941), the Hugoton reservoir has no structural closure but occurs in rocks dipping homoclinally eastward at from 15 to 25 feet per mile. The carbonates are reported to undergo a facies change to the west, grading into red shales and silts about 20 miles west of the field. On this basis the field has been classed as a stratigraphic trap. However, according to the same authors, the gas-water contact has an elevation of 500 feet

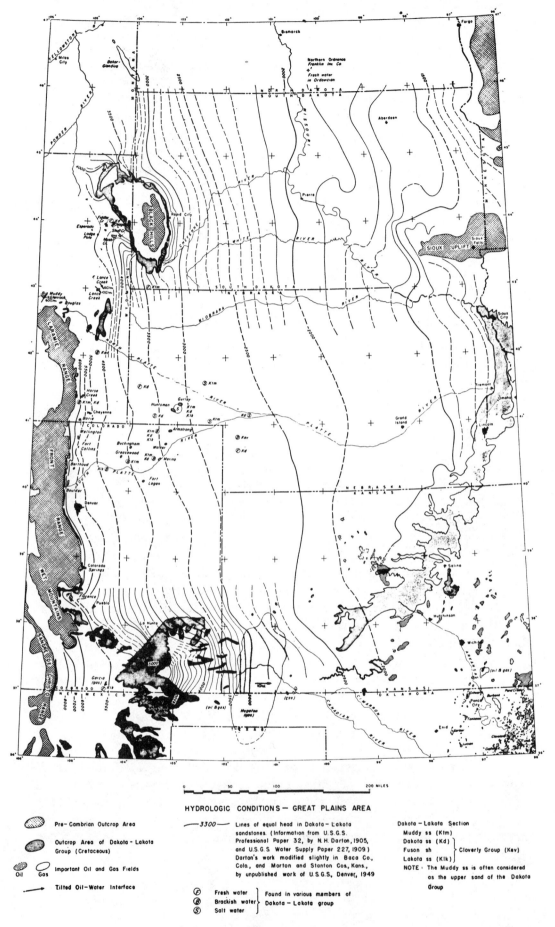

FIG. 36.—Map of Great Plains, showing potentiometric contours of water in Dakota sandstone
(after N. H. Darton, 1905, 1909, 1918a, 1918b).

above sea-level in the westernmost wells, about 250 feet in a well near the center, and 120 feet in a well near the eastern edge. Moreover, the initial pressure throughout the field had the very nearly constant value of 435 lbs./in.²

These data indicate the existence of a dynamic situation with the water flowing eastward under a decreasing potential gradient from west to east. In the western part of the field the tilt of the gas-water interface averages about 17 feet per mile, while in the eastern half it is about 7. At the pressure of 435 lbs./in.² the gas density is so low that the tilt amplification factor, $\rho_w/(\rho_w - \rho_g)$, is essentially unity, so that the tilts are also equal to the respective slopes of the potentiometric surface.

It appears therefore that Hugoton is at least in part a hydrodynamic trap, with the gas equipotential surfaces concave upward and terminating against the impervious or impenetrable strata above. A trap of this type has been shown diagrammatically in Figure 26.

The literature of petroleum geology contains numerous instances of oil or gas fields with tilted water interfaces in localities where the complementary ground-water hydrology is not known except by inference. Many more occurrences are known of oil and gas fields on structures with no known closures, either structural or stratigraphic. Fields of the latter type were first described by Edward Orton (1888, pp. 94–95) who pointed out that in southeastern Ohio oil and gas fields did not always occur in anticlines but were often found associated with "arrested anticlinals" or structural terraces, where the regional southeastward dip flattened for a mile or two and then resumed its normal rate. As one such example the Macksburg field was cited. Here the strata enter the field from the northwest, dipping southeast at a rate of 20–30 feet per mile; they then become almost horizontal for about 3 miles beyond which they resume their normal dip. In this field production of oil and gas from five different sands in a series 1,500 feet thick was obtained, the gas from the updip and the oil from the downdip parts of the field. Orton considered whether this localization of the oil and gas might be due to the "grain and composition" of the sands; but for five separate sands deposited at different times, each to have the right grain and composition to make them repositories for oil and gas within the same geographical limits, he regarded as incredible, and concluded that the entrapment must be due to the structure.

In a comprehensive report on the Cushing field of northeast Oklahoma, Carl H. Beal (1917) described as a puzzling phenomenon the tilted oil-water interfaces and the unusual arrangement of the water, oil, and gas in each of the three productive sands. One of Beal's cross sections is reproduced in Figure 37. In each of the two upper sands, Layton and Wheeler, a typically dynamic situation is shown with a high-level gas-water contact on the east flank and the oil exclusively down the west flank of the structure, with the author's notation "water (surface inclined)." In this section the conditions in the third, or Bartlesville, sand show normal hydrostatic relationships. In the field maps, however, dynamic effects are evident in all three sands with oil-water tilts of the order of 100 feet per mile.

Beal's best explanation for these odd effects was that the oil had migrated into the trap from the west, producing the asymmetrical arrangement. By comparison of Beal's cross section with Figure 25, however, it will be seen that the observed effect could readily be produced by water flowing westward through the sands from the Ozark uplift.

One of the best examples described in the literature of an oil accumulation in a completely unclosed structure is that of Wheat field in the Delaware Basin of West Texas, as described by John Emery Adams (1936). According to Adams (Figs. 38 and 39), this accumulation occurs in a sand of the Delaware Mountain series which in this region dips homoclinally east at an average rate of about 100 feet per mile. The field is located on a structural terrace on which the dip decreases to about 50 feet per mile for a distance of about 3 miles and then steep-

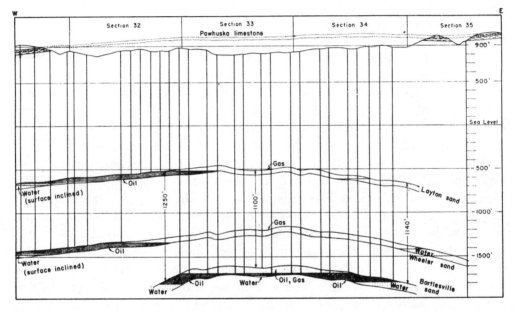

FIG. 37.—West-east cross section of Cushing field, Oklahoma, showing asymmetrical arrangement of gas, oil, and water, and westward tilted water surface (after Carl H. Beal, 1917).

ens again to the east. There is no evidence of faulting, and dry holes on the west, northwest, and southwest indicated good permeability. These facts preclude the interpretation of the field being a fault or stratigraphic trap, and in the light of present information it appears to be a hydrodynamic trap produced by water flowing eastward from the Delaware Mountains.

Another excellent example of a steeply tilted field is afforded by the Frannie field (Fig. 40) in the Big Horn Basin of Wyoming. This accumulation occurs in the Tensleep sandstone on the southwest slope of the Big Horn Mountains, and is tilted southwestward away from the mountains at about 600 feet per mile. The water in the Tensleep is of low salinity with only about 3,400 parts per million of total dissolved solids. Independent hydrologic studies in this region show the

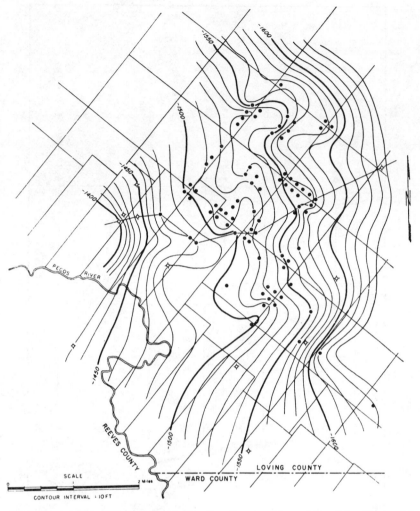

Fig. 38.—Wheat field in Delaware Basin, Loving County, Texas (after John Emery Adams, 1936).

waters in the Tensleep to be flowing in the direction in which the field is tilted and the observed tilt to be in good agreement with these data.

Another example occurs in the East Coalinga Extension oil field in the San Joaquin Valley of California, a map of which, compiled by L. S. Chambers (1943), is shown in Figure 41. The productive sand in this case is the Gatchell sand of Eocene age which wedges out to the westward along a north-south line. The field extends along this line for about 8 miles. It will be noted that the elevation of the oil-water contact decreases from south to north from about −7,000 feet to −7,700 feet in a distance of 8 miles, giving an average northward tilt of about 90 feet per mile. It is inferred therefore that the waters in this sand, which incidentally are nearly fresh having less than 1,000 parts per million of dissolved solids, are in a state of motion in a northerly direction.

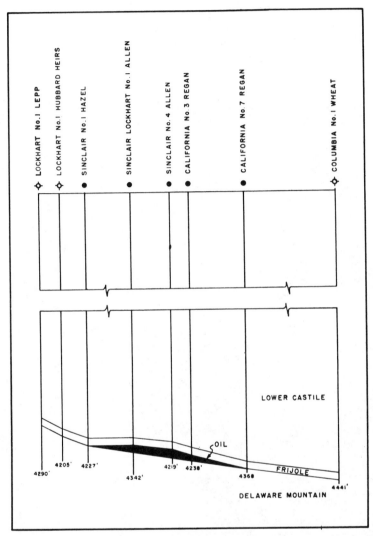

FIG. 39.—Cross section of Wheat field, Delaware Basin, Loving County, Texas (after John Emery Adams, 1936).

Another example is the Norman Wells field of northwest Canada, as described by J. S. Stewart (1948). According to Stewart's map of the field (Fig. 42), the structure is a southwestward plunging nose with only about 100 feet of closure. Toward the northwest the productive area barely extends beyond the −850-foot contour, which is the lowest closed contour on the structure. Toward the southwest, however, the productive area extends down the dip to the −1,600-foot contour, a distance of 750 feet below the lowest closed contour.

Stewart's explanation of this is indicated in his south-north cross section of the field shown in Figure 43. The reservoir rock is shown as a reef limestone overlain and underlain by shales. At the foot of the accumulation a horizontal

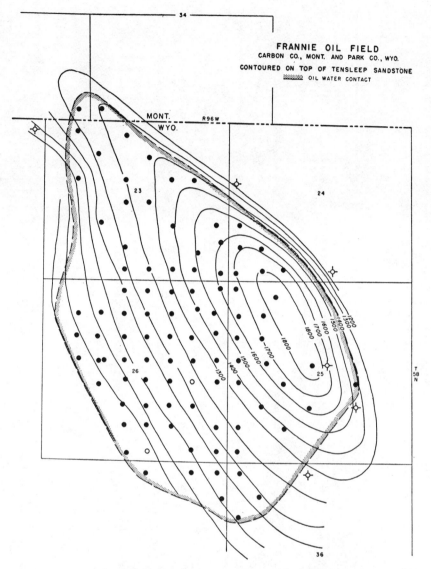

FIG. 40.—Frannie oil field, Big Horn Basin, Wyoming, showing tilt southwest.

"water level" is indicated, but elsewhere an inclined boundary is shown between an "oil-saturated zone" and an "unsaturated zone.", Since the "unsaturated zone" undoubtedly contains water in the pore spaces not occupied by oil, the only meaning that can be ascribed to this inclined boundary is that it represents an inclined oil-water interface. Figure 43 shows that the field tilts toward the southwest by an amount of about 750 feet in a distance of $2\frac{1}{2}$ miles, or about 300 feet per mile.

It appears therefore that this is a dynamic tilt with the water flowing from the northeast. This inference is given additional support by the regional topog-

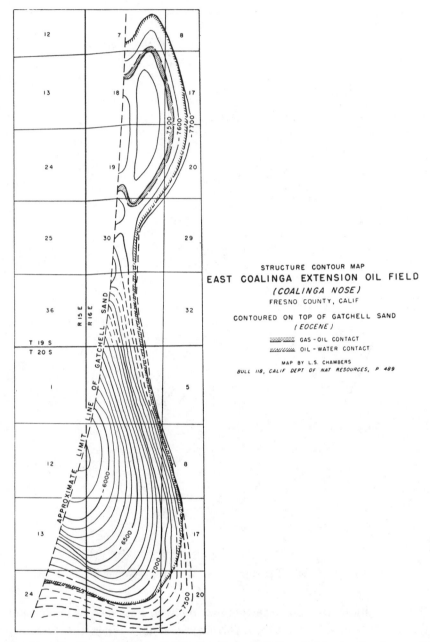

FIG. 41.—East Coalinga Extension oil field, San Joaquin Valley, California, showing northward tilt (after L. S. Chambers, 1943).

raphy, since in an oblique aerial photograph of the field, also shown by Stewart, the Franklin Mountain range can be seen in the background a comparatively short distance northeast.

Data pertaining to the oil fields of Burma have been presented by H. R. Tainsh (1950). These fields occur in a long narrow structural valley about 150

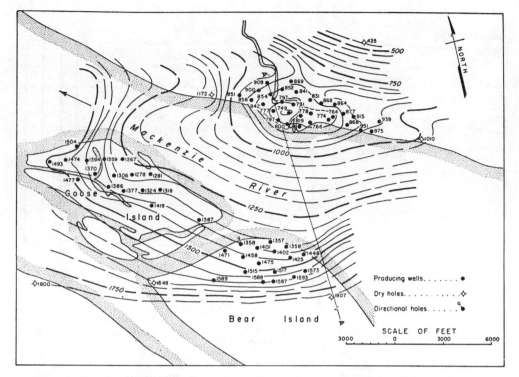

FIG. 42.—Norman Wells oil field, Northwest Territories, Canada (after J. S. Stewart, 1948).

miles wide and 700 miles long, extending north and south and open at the south. The structure and topography are favorable for ground-water circulation from the sides toward the axis of the valley and thence in a generally southward direction. The oil-field waters are dilute as compared with sea water, the concentration of total dissolved solids in the waters of the Chauk field being about 7,000 parts per million, and of the Lower sand in the Yenangyaung field about 8,500. The concentration is reported to increase with depth.

In a structure map of the Lanywa and Chauk fields (actually two parts of a single field), Tainsh showed the position of the original water margin of the field on the 3,000-foot sand. The structure is a long, narrow anticline striking about N. 20° W., bounded on the east by an axial fault, and plunging south. The water contact in the main part of the structure is inclined southward from about −2,350 feet at the north end to −2,900 feet at the south, a drop of 550 feet in about 5 miles. An even steeper tilt is shown in the south nose of the structure in a block cut off by transverse faults.

Concerning the Yenangyaung field, which is shown as an anticline cut by several transverse faults, Tainsh (p. 848) remarked:

The shallower pools are found near the crest and are up to 2,600 feet in width. The oil pools of the Okhmintaungs are crestal in some fault blocks, but *on the east flank in other blocks, with crestal gas caps*. The pools of the Padaung stage are found *on the east flank with large original gas caps* . . . ; a few still deeper pools have been found rather closer to the crest. [Italics by M.K.H.]

413

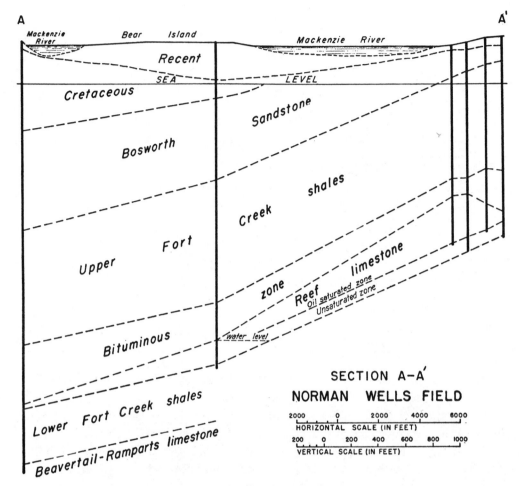

FIG. 43.—South-north cross section of Norman Wells field (after J. S. Stewart, 1948).

These phenomena are just what should occur under hydrodynamic conditions, which, in general, may vary considerably from sand to sand.

An example of an oil field with a tilt which has been induced artificially by man-made disturbances prior to its discovery is afforded by the Cairo field in Union County, Arkansas, as described by Lawrence A. Goebel (1950). This field on a small domal structure about 2 miles northeast of the much larger Schuler field was discovered in 1948, eleven years after the discovery of Schuler. The Cairo field produces from the Smackover limestone, and Schuler from the Smackover and the overlying Cotton Valley sands.

The discovery well near the crest of the structure found a good oil column with the water surface at 7,540 feet below sea level. An offset well, $\frac{1}{4}$ mile north, encountered the reservoir rock 31 feet above the water level in the first well. The rocks were oil wet but produced salt water and only a scum of oil, and the well was abandoned as a dry hole. As more wells were drilled, those on the north, east, and southeast which encountered the reservoir rock above the −7,540-foot level found it to be oil wet, but failed to produce; those in a 90° sector on the southwest were productive. Furthermore, the water contact in these wells was found at progressively lower depths toward the southwest.

After all the evidence was weighed the explanation advanced by Goebel was that prior to the disturbance of the fluids in the Smackover by the production of Schuler, the oil in the Cairo structure had been at the top of the structure with a horizontal oil-water contact at about −7,540 feet. During the period since the discovery of Schuler, large volumes of fluids had been withdrawn, causing a radially inward migration toward Schuler of the ground water in the surrounding area. Accompanying this water migration, the oil in Cairo was envisaged to have been migrated, as shown in Figure 44.

This explanation appears to be substantially correct and it gives us a case history of a tilted oil field which has assumed its tilted position in response to a dynamic environment of only 11 years' duration. This observation is of particular importance in view of the explanation that has often been advanced to account for tilted oil fields, namely, that the fields were tilted by some postulated tectonic disturbance and have not subsequently had time to regain their equilibrium.

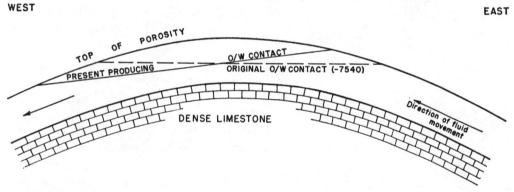

FIG. 44.—Migration of Cairo oil field, Arkansas, by fluid withdrawal from Schuler oil field (after Lawrence A. Goebel, 1950).

Both on the basis of the behavior of Cairo, and from calculations which can be made from the fluid properties and the permeability of the reservoir rocks, the time required for most oil fields to approach a new equilibrium after a disturbance is of the order of decades, whereas geological disturbances, either tectonic or hydrologic, have rates measurable in thousands to millions of years. The only way, therefore, that a tilted oil-water interface can be sustained indefinitely except by a dynamic ground-water environment is for the oil to be restrained by some kind of an impervious seal, such as an asphalt stratum. Thus far no such tilted fields have been demonstrated to exist, yet the theoretical possibility of such an occurrence must be admitted.

CONCLUSION

At the outset of the present paper we proposed to investigate the forces which cause petroleum to migrate and the characteristics of the positions in which it will become entrapped under the general conditions when the environmental ground water is in some state of motion. This we have done and our formulations reduce, as they should, in the special case for which the water is at rest, to the results with which we are already familiar: vertical and parallel impelling forces, and traps for both oil and gas in the spaces between downwardly concave impermeable barriers and horizontal surfaces.

If the water is in motion, however, in a non-vertical direction, which often is the case, our formulation leads to consequences which are by no means familiar. Oil and gas equipotentials are no longer horizontal but are inclined, with the angle of inclination of the equipotentials for oil greater than that of the equipotentials for gas. The paths of migration for oil and gas in the same space are no longer vertical, nor are they parallel, the paths for oil being deflected away from the vertical by an angle greater than that of the paths for gas. Likewise the traps for oil and gas no longer coincide and may in fact be separated entirely. In the latter event a trap for oil will not hold gas nor a trap for gas, oil; the fluids will migrate to their respective traps instead.

The oil- and gas-water interfaces will not be horizontal but inclined at an angle given by

$$\tan \theta = \frac{dz}{dx} = \frac{\rho_w}{\rho_w - \rho_0} \frac{dh}{dx},$$

where ρ_0 is the density of the oil or gas, respectively.

Under such circumstances oil or gas entrapments will not occur in the conventional positions. They may occur in anticlines in asymmetrical positions with the water high on one side and low on the other, or in completely unclosed structures such as noses or terraces, with the water flowing down the dip. Again, if the closing dip of an anticline in the downstream direction is less than the angle of tilt, this structure will not hold the specified fluid under the conditions prevailing.

These theoretical deductions have been confirmed experimentally, and the predicted phenomena have also been sought in the field. Not only have they been found, but the frequency of their occurrence has exceeded expectations, and in every major oil-producing area so far examined, hydrodynamic conditions in at least some reservoir formations, with oil-field tilts ranging from tens to hundreds of feet per mile, have been observed. We are thus led to the suspicion that many off-structure accumulations of oil and gas, which, on the basis of hydrostatic premises, have been classified as fault or stratigraphic traps, may in fact be hydrodynamic traps instead.

From such considerations it becomes evident that in the prospecting for petroleum in any area, as complete a knowledge as possible, in three-dimensional space, of the ground-water hydrology is of importance comparable with a knowledge of the stratigraphy and the structure. If conditions can be demonstrated to be very nearly hydrostatic, then our customary procedures are appropriate; if hydrodynamic conditions prevail, it is important that these be determined in detail, stratum by stratum, over the given basin in order that the positions of the traps may better be determined.

For this purpose regional geology and topography constitute the initial and most readily available information. Next comes the information obtainable from wells of which the most informative are widely spaced wildcats. Water samples for analysis and density determination should be taken in such wells in every regional sand or permeable formation. In addition, in the same formations, accurate shut-in pressures, together with the precise elevation of the point of measurement, should also be taken. This information is essential for the computation of the potential Φ_w, or the head h_w, by means of the equations

$$\left. \begin{aligned} \Phi_w &= gz + \frac{p}{\rho_w}, \\[2mm] h_w &= z + \frac{p}{\rho_w g}, \end{aligned} \right\}$$

where p is the undisturbed pressure in the formation and z the elevation of the point of measurement.

The systematic assembling of data of this kind is appropriately a cooperative enterprise for the whole petroleum industry, and such data should be taken and exchanged between various groups in the same manner that well-log information is now exchanged. It will be found that our present procedures in taking pressure measurements in wildcat wells are inadequate, both as to frequency and accuracy. Since pressure measurements are most often made incidental to drill-stem tests, there is need for an improvement in the pressure measurements and procedures in making such tests. This includes both an improvement in the precision of pressure measurements, and also a change of the routine so that shut-in pressures may be taken prior to the drastic disturbance produced by the withdrawal of fluids, rather than afterward.

In the light of the evidence before us it appears essential that in addition to our customary procedures in petroleum geology, involving principally stratigraphy and structure, we must now add regional ground-water hydrology if many otherwise obscure accumulations of petroleum are not to be overlooked.

REFERENCES CITED

ADAMS, JOHN EMERY, 1936, "Oil Pool of Open Reservoir Type," *Bull. Amer. Assoc. Petrol. Geol.*, Vol. 20, pp. 780–96.

BEAL, CARL H., 1917, "Geologic Structure in the Cushing Oil and Gas Field, Oklahoma, and Its Relation to the Oil, Gas, and Water," *U. S. Geol. Survey Bull. 658.* 64 pp.

CHAMBERS, L. S., 1943, "Coalinga East Extension Area of the Coalinga Oil Fields," *Geologic Forma-*

tions and Economic Development of the Oil and Gas Fields of California, California Dept. Nat. Resources Bull. 118, Pt. 3, pp. 486–90.

DARCY, HENRY, 1856, *Les fontaines publiques de la ville de Dijon*, pp. 590–94; also Fig. 3 in *Atlas*. Victor Dalmont, Paris.

DARTON, N. H., 1905, "Preliminary Report on the Geology and Underground Water Resources of the Central Great Plains," *U. S. Geol. Survey Prof. Paper 32.* 433 pp.

———, 1909, "Geology and Underground Waters of South Dakota," *ibid., Water-Supply Paper 227.* 156 pp.

———, 1918a, "Artesian Waters in the Vicinity of the Black Hills, South Dakota," *ibid., Water-Supply Paper 428.* 64 pp.

———, 1918b, "The Structure of Parts of the Central Great Plains," *ibid., Bull. 691*, pp. 1–26.

EMERY, WILSON B., 1929, "Lance Creek Oil and Gas Field, Niobrara County, Wyoming," *Structure of Typical American Oil Fields*, Vol. II, Amer. Assoc. Petrol. Geol., pp. 604–13.

GARLOUGH, JOHN L., and TAYLOR, GARVIN L., 1941, "Hugoton Gas Field, Grant, Haskell, Morton, Stevens and Seward Counties, Kansas, and Texas County, Oklahoma," *Stratigraphic Type Oil Fields*, Amer. Assoc. Petrol. Geol., pp. 78–104.

GOEBEL, LAWRENCE A., 1950, "Cairo Field, Union County, Arkansas," *Bull. Amer. Assoc. Petrol. Geol.*, Vol. 34, No. 10, pp. 1954–80.

HOWELL, J. V., 1934, "Historical Development of the Structural Theory of Accumulation of Oil and Gas," *Problems of Petroleum Geology*, Amer. Assoc. Petrol. Geol., pp. 1–23.

HUBBERT, M. KING, 1940, "The Theory of Ground-Water Motion," *Jour. Geol.*, Vol. 48, No. 8, pp. 785–944.

ILLING, V. C., 1938a, "The Migration of Oil," *The Science of Petroleum*, Vol. I, Oxford Univ. Press, pp. 209–17.

———, 1938b, "An Introduction to the Principles of the Accumulation of Petroleum," *The Science of Petroleum, ibid.*, pp. 218–20.

———, 1939, "Some Factors in Oil Accumulation," *Jour. Inst. Petroleum*, Vol. 25, pp. 201–25.

MILLS, R. VAN A., 1920, "Experimental Studies of Subsurface Relationships in Oil and Gas Fields *Econ. Geol.*, Vol. 15, No. 5, pp. 398–421.

MUNN, MALCOLM J., 1909a, "Studies in the Application of the Anticlinal Theory of Oil and Gas Accumulation [Sedwickley Quadrangle, Pa.]," *Econ. Geol.*, Vol. 4, pp. 141–57.

———, 1909b, "The Anticlinal and Hydraulic Theories of Oil and Gas Accumulation," *ibid.*, pp. 509–29.

ORTON, EDWARD, 1888, "The Origin and Accumulation of Petroleum and Natural Gas," *Report of the Geological Survey of Ohio*, Vol. 6, Chap. ii, pp. 60–100.

PETTIJOHN, FRANCIS J., 1949, *Sedimentary Rocks*, p. 13. Harper and Bros., New York.

PURCELL, W. R., 1949, "Capillary Pressures—Their Measurement Using Mercury and the Calculation of Permeability Therefrom," *Trans. Amer. Inst. Min. and Metal. Eng.*, Vol. 186, pp. 39–48.

RICH, JOHN L., 1921, "Moving Underground Water as a Primary Cause of the Migration and Accumulation of Oil and Gas," *Econ. Geol.*, Vol. 16, No. 6, pp. 347–71.

———, 1923, "Further Notes on the Hydraulic Theory of Oil Migration and Accumulation," *Bull. Amer. Assoc. Petrol. Geol.*, Vol. 7, pp. 213–25, and *Nat. Petrol. News*, Vol. 15, pp. 75–76.

———, 1931, "Function of Carrier Beds in Long-Distance Migration of Oil," *Bull. Amer. Assoc. Petrol. Geol.*, Vol. 15, pp. 911–24.

———, 1934, "Problems of the Origin, Migration, and Accumulation of Oil," *Problems of Petroleum Geology*, Amer. Assoc. Petrol. Geol., pp. 337–45.

SHAW, E. W., 1917, "The Absence of Water in Certain Sandstones of the Appalachian Oil Fields (discussion)," *Econ. Geol.*, Vol. 12, pp. 610–28.

SMITH, PAUL V., JR., 1952, "Preliminary Note on Origin of Petroleum," *Bull. Amer. Assoc. Petrol. Geol.*, Vol. 36, No. 2, pp. 411–13.

STEWART, J. S., 1948, "Norman Wells Oil Field, Northwest Territories, Canada," *Structure of Typical American Oil Fields*, Vol. III, Amer. Assoc. Petrol. Geol., pp. 86–109.

TAINSH, H. R., 1950, "Tertiary Geology and Principal Oil Fields of Burma," *Bull. Amer. Assoc. Petrol. Geol.*, Vol. 34, No. 5, pp. 823–55.

TIRATSOO, E. N., 1952, *Principles of Petroleum Geology.* 449 pp. McGraw-Hill Book Company, New York.

Reprinted by permission of McGraw-Hill Book Company, from *Petroleum Exploration Handbook: A Practical Manual Summarizing the Application of Earth Sciences to Petroleum Exploration,* edited by Graham B. Moody, chapter 6, pp. 6-1 through 6-68, first edition, 1982 reissue.

ENTRAPMENT OF PETROLEUM

By DAVID G. WILLIS

Manager, Research Computer Systems, Lockheed Missiles and Space Division. Research Associate in Geophysics, Stanford University

Atherton, Calif.

In the hundred years which have elapsed since the drilling of the Drake well in Pennsylvania, there have been discovered throughout the world many thousands of commercial accumulations of oil and natural gas. They have been found at depths ranging from the surface of the ground to more than 15,000 feet and in rocks ranging in age from Precambrian to Pleistocene. In the case of oil they have varied in size from a few hundred to more than 20 billion barrels of liquid petroleum. Each of these accumulations is the result of a particular combination of geological and physical circumstances by which the oil or gas has been held in a stable configuration or trap underground.

The principal problem of petroleum exploration is to determine the location of undiscovered traps for oil and gas. Hence, among the most valuable tools which can be possessed by the petroleum exploration geologist are a basic understanding of the physical principles governing the entrapment of oil and gas and a wide familiarity with the various geological factors which have contributed to the formation of already discovered traps.

In this chapter are presented a review of our modern understanding of the physics of petroleum and natural gas entrapment and its historical development, and a detailed examination of the various geological factors which may contribute to the formation of traps together with examples demonstrating the wide variety of possible combinations of these factors which may occur.

The idea that oil and gas underground, being lighter than the surrounding water, should be moved by buoyant forces to the structurally highest position in a permeable rock has dominated the prevalent scientific thinking ever since underground petroleum accumulations were discovered. Its first expression was in the "anticlinal theory," which became clearly formulated (Howell, 1934) after the Drake discovery in 1859 and which found wide acceptance as an exploration tool prior to 1900. Since anticlines, with their concave downward geometry, are perhaps the most obvious of the possible geologic configurations which might form a trap, it is not surprising that the attention of petroleum explorers was first directed to them. Later, with the discovery of other kinds of traps in which geologic factors such as faults, permeability pinch-outs, and unconformities were responsible for producing an adequate trapping geometry, the anticlinal theory was extended to include a wider variety of possible geologic circumstances. In its expanded form this has been termed the "structural theory" or "trap theory" of petroleum accumulation and can be summarized by the statement that a trap will be formed wherever rocks permeable to oil and gas underlie rocks impermeable to oil and gas with their common boundary forming a surface which is concave downward.

Although the anticlinal and structural theories were widely accepted and applied ever since their formulation, an alternative minority viewpoint was propounded and discussed in the literature for almost 30 years commencing in about 1910 (Munn, 1909; Mills, 1920; Rich, 1921, 1934). This was the "hydraulic theory" of petroleum accumulation. Its basic tenets were that the flushing action of moving ground water was a major factor causing the migration of petroleum and that accumulation would occur wherever natural forces, either buoyancy or a decrease in permeability of the rocks, would impede the force of the water. Thus accumulations might be found on structural terraces or the sides of anticlines or associated with changes in permeability.

Recently, a comprehensive physical theory of the entrapment and accumulation of petroleum and natural gas has been developed by M. King Hubbert (1940, 1953). He has shown that the basic principle of the structural theory, whereby oil and gas migrate to the highest position of the reservoir rock, is a special case and is valid only under hydrostatic conditions, in which the environmental ground water is not in motion. In a hydrodynamic environment accumulations of oil or gas will occur in which the oil-water or gas-water contact is tilted in the direction of water motion. In some cases this tilt may be so great that a configuration which would trap oil or gas under hydrostatic conditions will now be incapable of holding those fluids. A review of already discovered petroleum and natural gas accumulations has revealed many examples in which these effects are strongly apparent. The following discussion of the physical principles of petroleum and natural gas accumulation is based primarily upon the work of Hubbert.

THE PHYSICAL PRINCIPLES GOVERNING ENTRAPMENT OF OIL AND GAS

Commercial accumulations of petroleum and gas invariably occur in or adjacent to sedimentary rocks, mostly of marine origin. These rocks are to a greater or lesser extent porous and permeable, and they are almost entirely saturated with water from very near the surface of the ground down to depths at which their porosity disappears. There exists strong evidence, based on the work of P. V. Smith, Jr. (1954), indicating that petroleum or natural gas has originated as a myriad of small particles or globules which were widely disseminated throughout certain of these water-saturated sedimentary rocks soon after their deposition. Migration of these initial small particles into large accumulations of oil and gas took place through the water-saturated environment. Although we do not know the exact process by which this migration took place, it is at least possible to describe the forces which were acting on the small globules and, what is of particular interest, the positions and configurations in which they approach equilibrium and become stable.

It is a fundamental physical principle implicit in the second law of thermodynamics that any mechanical system, no matter how complicated, will always evolve through a succession of configurations such that the sum of its kinetic and potential energies will become smaller and smaller by gradual conversion to heat. Furthermore, it will approach and eventually come to rest in an equilibrium configuration such that the kinetic energy is zero and the potential energy is at a minimum with respect to all small changes compatible with the constraints of the system. There exist many familiar examples of this process such as the slowing and stopping of a swinging pendulum, water flowing downhill into a lake, or a rolling ball coming to rest. Since underground fluids, water, oil, and natural gas in their environment constitute a mechanical system, they behave in this way. Although the motions of these fluids are so slow that kinetic energies are for all practical purposes zero, each element of fluid will possess an amount of mechanical potential energy which is principally a function of its position. It will in general be acted on by an unbalanced force, tending to move it from regions in which its potential energy is high to those in which its potential energy is low. It will cease moving and will become stable whenever it reaches a position in which its potential energy is at a minimum and it is surrounded by regions of higher potential energy. In particular, the original small particles of petroleum or natural gas will come to rest and accumulate in their respective positions of equilibrium where they are surrounded by regions in which their potential energies are higher. We can, therefore, define a trap for oil or gas as an underground locality in which the potential energy of those fluids is at a minimum with respect to their immediate surroundings. Such a configuration is shown in schematic form in Figure 6-1.

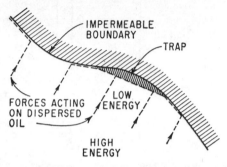

FIGURE 6-1. Schematic diagram of a trap showing the manner in which dispersed oil migrates from a high-energy region to a low-energy region.

To determine the location and configuration of traps it is first necessary to know the potential energies of the fluids of interest—water, oil, and gas in their underground environment. Inasmuch as the accumulation and migration of oil and natural gas take place in a principally water-saturated environment, the potential energy of water should be considered first.

Potential Energy of Water

Each small element of water underground will possess mechanical potential energy by virtue of both its elevation and the pressure existing within it. When it is referred to an element of unit mass, this potential energy is called the potential of the water. The value of the potential can be determined by calculating the work which would be required to bring a unit mass of water from some standard elevation and pressure to its present elevation and pressure. If the standard is taken as sea level and atmospheric pressure, it is first necessary to lift the unit mass to its present elevation z above sea level against a force equal to the acceleration of gravity g. It is then necessary to inject the unit mass of water, which has a volume equal to the reciprocal of its density ρ, against the local water pressure p. Therefore, the total work required for these two operations, and hence the potential ϕ of the water, is given by the expression

$$\phi = gz + \frac{p}{\rho}$$

A different interpretation of the potential can be obtained by consideration of the height to which water will rise in an open-topped manometer whose base is terminated at the point at which the potential is to be measured (Figure 6-2). When this height is referred to a standard datum of elevation which is again taken as sea level, it is called the "head" of the water and can be denoted by h. The value of the head will be given by the sum of the elevation of the point of interest z and the height of the static fluid $p/\rho g$ which can be supported by the local pressure p, giving

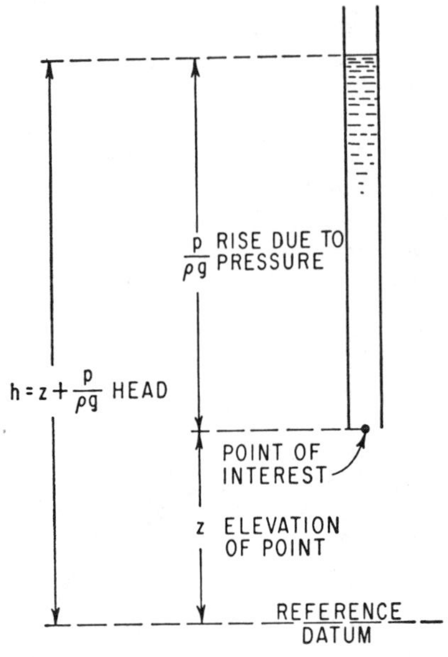

$$h = z + \frac{p}{\rho g}$$

If this expression is compared with the previous one, it is evident that the potential can be obtained simply by multiplying the head at any point by the acceleration of gravity, or

$$\phi = gh$$

Therefore, a manometer can be employed to measure the potential of water at any point. This is of particular value if we wish to determine the distribution of potential within a permeable subsurface stratum. Water wells terminated in the stratum can serve as manometers, and the static height h measured above sea level or other standard datum to which water will rise in the wells provides a direct measurement of the potential. Under hydrostatic conditions, when the water is not in motion, the head, and hence the potential, will have the same value wherever it is measured.

FIGURE 6-2. Relationship between head and pressure in an open-topped manometer.

Forces Acting on Water

The potential of water will be a scalar variable with a definite value at each point in an undergound space. Furthermore, since it is a function only of elevation and pressure and since the pressure will in general vary continuously, the potential itself will be either a constant or a continuous variable. If it is a continuous variable, surfaces will exist underground on each of which the potential assumes a constant value and which can be termed equipotential surfaces. The potential throughout a region can be described by a family of such equipotential surfaces. If the potential of water is known throughout some region of space, the forces which will act on a unit mass of water throughout that region can be determined. Each element of unit mass will be acted on by a force whose direction is that of the greatest rate of decrease of potential and whose magnitude is equal to the rate of decrease in that direction. Hence the force per unit mass **E** acting on the water will be a vector which is normal to the equipotential surfaces and which can be represented at any point by the negative of the gradient of the potential, or

$$\mathbf{E} = -\operatorname{grad} \phi$$

The physical situation shown in Figure 6-3 depicts the force field acting on the water by lines of force normal to the equipotential surfaces. An alternative physical conception is obtained if we take the gradient of the expression for the potential for water, or

$$\mathbf{E} = -\operatorname{grad} \phi = -\operatorname{grad}\left(gz + \frac{p}{\rho}\right) = \mathbf{g} - \frac{1}{\rho}\operatorname{grad} p$$

This is simply a vector equation which asserts that the force acting on a unit mass of water will be equal to the vector sum of the force per unit mass of gravity acting vertically downward and the negative gradient of the fluid pressure multiplied by the reciprocal of fluid density. There are thus two force components whose vector sum constitutes the net force acting on the fluid. The two components, as shown for two different cases in Figure 6-4, provide a simple physical conception of the situation. The first component $\mathbf{g}$ represents the force of gravity acting vertically downward on a unit mass of the water. The second component $(1/\rho)\operatorname{grad} p$ represents the force produced by the negative gradient of the pressure acting on the volume $1/\rho$ occupied by the unit mass of water. In the hydrostatic case, illustrated in Figure 6-4a, the

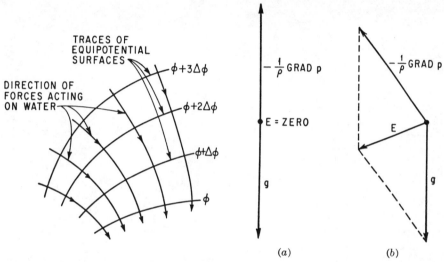

FIGURE 6-3. Section through an underground region showing traces of equipotential surfaces with lines of force normal to them and in the direction of decreasing potential.

FIGURE 6-4. Forces acting on a unit mass of water in (a) hydrostatic and (b) hydrodynamic environments.

potential is everywhere constant and these two forces are exactly equal and opposite. The net force acting on the water is therefore zero. In a hydrodynamic situation, on the other hand, the pressure component will not exactly balance the gravitational component and a net unbalanced force will act on the unit mass of water as illustrated in Figure 6-4b.

The relationship between the actual motion of the water and the field of force acting on the water is provided by Darcy's law. This is not of direct concern here, and it is sufficient to note that the direction of motion of the water will virtually always coincide with the direction of the force acting on each element of water.

These principles provide a way of determining the state of motion of the water within a permeable subsurface stratum. In such a situation, if the strata immediately above and immediately below the stratum of interest are relatively impermeable to water, the fluid flow will be essentially two-dimensional and will be restricted to the stratum of interest. At each point in the stratum there will exist a value of the potential for the water and a corresponding value of h, or "head," defined by the elevation above sea level to which water would rise in a static well terminated within the stratum. The values of h measured with respect to a standard datum will define a surface which is called the potentiometric, or "potential-measuring," surface. The water will be flowing in a "downhill" direction on this surface from regions of higher

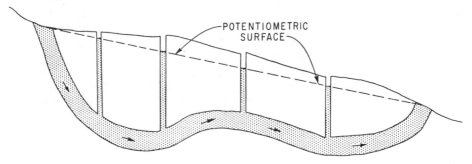

FIGURE 6-5. Schematic geologic section showing water entering the outcrop of a permeable stratum at a high elevation, flowing through it, and leaving it at a lower elevation, with the "head" decreasing in the direction of flow.

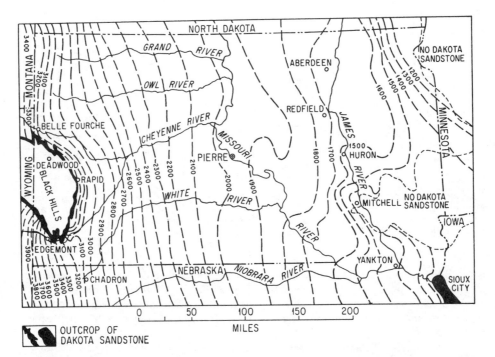

FIGURE 6-6. Regional map of South Dakota showing contours on the potentiometric surface for the Cretaceous Dakota sandstone. Water is entering the outcrop of the Dakota around the Black Hills and flowing in a "downhill" direction on the potentiometric surface. (*Redrawn after N. H. Darton, U.S. Geol. Survey Water Supply Paper 227, p. 61.*)

potential or h to regions of lower potential or h. The situation is depicted in Figure 6-5, which shows water entering the outcrop at a higher elevation, flowing through a permeable stratum, and leaving the surface at a lower elevation. The potentiometric surface is defined by wells terminated within the stratum. If sufficient measurements exist from wells within the same stratum over a large region, a contour map

424

can be constructed which accurately depicts the regional motion of the ground water within that stratum. If, as is frequently the case, it is not convenient to allow the water to enter a well and rise to a static height which could be measured, the potential, or head, can still be determined by a measurement of the underground fluid pressure within the stratum of interest together with the elevation at which the pressure was measured.

Unfortunately few data of this kind exist. One of the few extensive compilations of potentiometric measurements is that performed by N. H. Darton (1905, 1909, 1918, 1918) in his classic investigations of ground water in the Cretaceous Dakota sandstone of the Great Plains region of the United States.

A map of a portion of this region is shown in Figure 6-6 on which are plotted equipotential lines as determined by Darton. It is clear from this map that water is entering the outcrop of the Dakota sandstone in the vicinity of the Black Hills and flowing eastward across the entire state of South Dakota.

Water of Variable Salinity

Implicit in the foregoing discussion is the assumption that the density of the water is constant throughout the region under consideration. In many situations, however, the ground water will contain varying amounts of dissolved solids which will, in turn, cause its density to vary from point to point. In such a case it is not possible to use a scalar potential to describe the forces acting on the water. The vector expression

$$\mathbf{E} = g - \frac{1}{\rho} \operatorname{grad} p$$

will always correctly describe the force acting on a unit mass of fluid under any conditions. Hence, if the pressure and density throughout a region are known, the force field can always be determined by its use.

However, a scalar potential of ϕ related to the force $\mathbf{E}$ by the expression

$$\mathbf{E} = -\operatorname{grad} \phi$$

will exist only if the density of the fluid is constant or, as a special case, if the density varies only as a function of pressure.

In a region of variable salinity and hence variable density the best procedure is to approximate the force field acting on the water by determining the potential for a reference water of constant density by the expression

$$\phi = gz + \frac{p}{\rho}$$

This potential will define the force which would act at any point on water having the reference density and will coincide with the actual force acting on the water only where the actual density is the same as the reference density. If the actual density is nearly the same as the reference density, the force field, as defined by the potential, will be approximately correct. If the actual density varies too widely throughout a region, it is possible to separate it into smaller regions and define a different potential for a different reference water within each of them and thereby obtain a closer approximation to the actual force field.

Potential Energies of Oil and Gas

The potentials of oil and gas can be determined in a fashion very similar to that of water. However, since these fluids in their initial dispersed state exist in a principally water-saturated environment, their energies will be modified by the effects of surface tension, which must therefore be taken into consideration for both oil and gas. In the case of gas, the additional effect due to the compressibility of that fluid must also be considered.

In the consideration of the effects of surface tension one of the most important factors is the wettability of the rocks containing the fluids of interest. In rocks which are hydrophilic or preferentially water wet, the water will tend to remain in contact with the rock grains, leaving the oil or gas as small globules or particles within the pore spaces. In rocks which are hydrophobic and preferentially oil wet, the oil or gas will tend to disperse itself along the surfaces of the rock, leaving the water as globules within the pore spaces. In almost every situation of interest the rocks are preferentially water wet and the oil exists as globules within the pore spaces.

In general, the fluid pressure within small particles or globules of oil or gas will be somewhat higher than that of the surrounding water because of the surface tension existing between the globules and the water. This difference is called the capillary pressure P_c and can be calculated from Plateau's equation

$$P_c = \gamma \left(\frac{1}{r_1} + \frac{1}{r_2} \right)$$

in which γ is the interfacial tension between the fluids and r_1 and r_2 are the principal radii of curvature of the surface.

For spherical globules this reduces to

$$P_c = \frac{2\lambda}{r}$$

in which r is the radius of the globule.

It is evident that small globules with small radii of curvature will have a high capillary pressure and hence a high potential energy per unit mass whereas larger globules with larger radii of curvature will have a lower capillary pressure and lower potential energy per unit mass. In fine-grained rocks with small pores the globules will, of necessity, assume geometric shapes having small radii of curvature and will therefore have a high potential energy per unit mass. In coarse-grained large-pore rocks, on the other hand, the globules will assume geometric shapes with larger radii of curvature and hence lower potential energy. Therefore, in a water-saturated environment the fine-grained rocks are regions of high potential energy for oil and gas globules whereas the coarse-grained rocks are regions of lower potential energy for those fluids. At boundaries between fine- and coarse-grained rocks, for example shale-sand interfaces, a sharp gradient will exist in the potential energy of those fluids and a consequent strong force tending to impel them from the fine-grained rock into the coarse-grained rock. Conversely, for oil or gas globules existing within the coarse-grained rock the interface will present an energy barrier which renders it effectively impermeable to those fluids. Aside from this effect and as long as there are no sharp changes in the grain size of the rocks through which movement of oil and gas takes place, the effects of capillary pressures can be ignored without serious error in determining potentials for oil and gas.

Therefore, in the case of oil we can determine its potential energy per unit mass by computing the amount of work required to bring it from standard elevation and pressure to the elevation z and pressure p at the point of interest, or

$$\phi_o = gz + \frac{p}{\rho_o}$$

in which ϕ_o denotes the potential of the oil and ρ_o its density.

Since gas is a compressible fluid, its potential differs from that of a liquid in that additional work must be done in compressing it from its initial standard pressure to the local pressure at the point of interest. If this process is carried out in such a way that the varying density ρ of the gas is a function of pressure alone, the expression for the potential of gas becomes

$$\phi_g = gz + \int_0^p \frac{1}{\rho}\, dp$$

By a procedure entirely analogous to that used for water, the forces acting per unit mass on oil or gas are determined by the negative gradients of their respective potentials at any particular point, giving the vector expressions

$$\mathbf{E}_o = -\operatorname{grad} \phi_o = \mathbf{g} - \frac{1}{\rho_o} \operatorname{grad} p$$

$$\mathbf{E}_g = -\operatorname{grad} \phi_g = \mathbf{g} - \frac{1}{\rho_g} \operatorname{grad} p$$

Again, in each case the force is composed of two components: one, the force of gravity acting vertically downward on the unit mass of fluid, and the other, the negative gradient of the fluid pressure acting on the volume occupied by the unit mass of

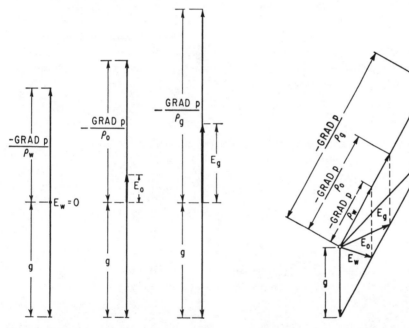

FIGURE 6-7. Forces acting on gas, oil, and water at the same point in a hydrostatic environment.

FIGURE 6-8. Forces acting on gas, oil, and water at the same point in a hydrodynamic environment.

fluid. It is of interest to examine the special case of hydrostatics, in which the environmental ground water is not in motion (Figure 6-7). In this case, there will be no net force acting on a unit mass of water, the vertically downward force of gravity being exactly balanced by an upward force in the direction of decreasing pressure acting on the volume occupied by a unit mass of water. However, for oil and gas, which both have less density than water and which therefore occupy a larger volume per unit mass, the upward force due to the negative pressure gradient will be greater and these fluids will tend to rise vertically, the gas much more strongly than the oil. This is the familiar situation of hydrostatics and represents a valid interpretation of the mechanics of oil and gas accumulation whenever the environmental ground water is not in motion.

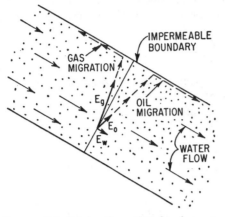

FIGURE 6-9. Schematic section showing conditions under which oil and gas could migrate in divergent directions in an underground permeable stratum.

The physical relationship between the three different forces which would act on the three fluids water, oil, and gas at the same point in a hydrodynamic situation is shown in Figure 6-8. It is clear that the first component **g** will be the same in each case while the second component, that due to the pressure gradient, will have the same direction in each case but will have a magnitude which is dependent upon the density of the particular fluid. Therefore, the resultant force vector will, in general, be different in both magnitude and direction for each of the three fluids, and they will tend to be driven in three separate directions.

It is easy to imagine a situation in which the oil and gas will move through a stratum in opposite directions and therefore end up in separate traps. This is illustrated in Figure 6-9 in which water is flowing down the dip in such a way that globules of oil will be moved down the dip whereas the gas will migrate up the dip.

Delineation of Traps

A trap will exist in any region in which the potential of oil or gas is at a minimum with respect to its surroundings. In order to delineate the positions and configurations of these regions, it is convenient to express the potential of oil or gas in terms of that of the environmental ground water. The potentials of oil and water are given by

$$\phi_w = gz + \frac{p}{\rho_w}$$

$$\phi_o = gz + \frac{p}{\rho_o}$$

in which the subscripts denote the particular fluid.

If the first of these equations is solved for p and substituted into the second,

$$\phi_o = \frac{\rho_w}{\rho_o}\phi_w - \frac{\rho_w - \rho_o}{\rho_o}gz$$

which gives the potential of oil in terms of that of water and the elevation.

A further simplification can be obtained by substituting the expressions

$$\phi_w = gh_w$$
$$\phi_o = gh_o$$

and dividing by g to give

$$h_o = \frac{\rho_w}{\rho_o}h_w - \frac{\rho_w - \rho_o}{\rho_o}z$$

which is the expression for the hydraulic head of oil in terms of the hydraulic head of water and the elevation at any point. In this expression, every term has the units of length; z is the elevation of the point, h_w is the height above datum to which water would stand in a manometer terminated at the point, and h_o the height that oil of density ρ_o would stand.   We see that the value of h_o is equal to the sum of two terms. The first is h_w multiplied by a constant coefficient depending upon the densities of the fluids, and the second the elevation multiplied by another constant depending on the densities of the fluids.   This relationship permits the determination of the equipotential surfaces for oil provided those for water are known.   In order to do this it is convenient to divide the above equation by the factor

$$\frac{\rho_w - \rho_o}{\rho_o}$$

which gives the expression

$$\frac{\rho_o}{\rho_w - \rho_o} h_o = \frac{\rho_w}{\rho_w - \rho_o} h_w - z$$

If two quantities u and v are defined by

$$u = \frac{\rho_o}{\rho_w - \rho_o} h_o$$

$$v = \frac{\rho_w}{\rho_w - \rho_o} h_w$$

then by substitution,

$$u = v - z$$

The surfaces represented by constant values of u will coincide with the oil-equipotential surfaces, whereas those corresponding to constant v will coincide with water-equipotential surfaces.   The surfaces along which z is constant are simply horizontal

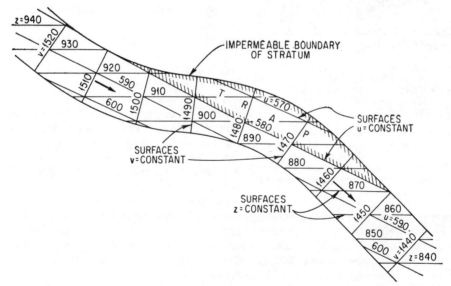

FIGURE 6-10. Section through an underground permeable stratum showing the relationship between surfaces of constant value u, v, and z.   A trap is formed by the closed region of minimum value u.

surfaces with elevation z.   The graphical expression of this relationship as it might exist in an underground permeable stratum is shown in Figure 6-10.   It is clear that the oil-equipotential surfaces will be inclined and tilted at some angle to the horizontal in the direction in which the water is flowing.   Since the oil will be impelled by a force perpendicular to the oil-equipotential surfaces and in the direction of decreasing potential, a trap will be formed wherever impermeable strata form a surface which is concave in the direction of increasing oil potential.

A straightforward method for determining the positions and configurations of traps underground is provided by the families of curves defined by the intersection of the surfaces of constant value u, v, and z with the upper surface of the permeable stratum. It is only necessary to have a structure-contour map of the region on the top of the stratum of interest and to know the configuration of the potentiometric surface referred to the upper surface of the stratum. The family of surfaces of constant z will form traces on the upper surface of the stratum which are simply the contour lines of the structure map. The lines of constant value v will coincide with the water-equipotential lines, and the values v on these lines can be obtained simply by multiplying the head by a suitable density factor. The difference between the values of v and z can be determined at the intersection of these two families of lines, and lines of constant value u along the upper surface of the reservoir can be drawn directly. These lines of constant value u will coincide with the oil-equipotential lines, and oil will be deflected along the upper boundary of the reservoir in the direction of decreasing u. Wherever the lines form a closed region of minimum u or oil-potential value, this area will represent a trap, since the oil cannot escape without moving from lower to higher potential. This procedure is illustrated in Figure 6-11.

This example has been worked out in terms of oil equipotentials for the delineation of oil traps. The problem of delineating gas traps is theoretically more complicated, by virtue of the fact that gas is not a constant-density fluid. The potential for gas strictly should depend upon the variations in density with local pressure.

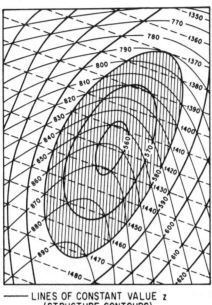

LINES OF CONSTANT VALUE z
  (STRUCTURE CONTOURS)
--- LINES OF CONSTANT VALUE v
  (WATER EQUIPOTENTIALS)
LINES OF CONSTANT VALUE u
  (OIL EQUIPOTENTIALS)
TRAP

FIGURE 6-11. Map showing relationship between traces of surfaces of constant value u, v, and z intersecting the upper surface of a permeable stratum. A trap is formed by closed regions of minimum value u.

However, for practical purposes only slight error will result from considering gas as a constant-density fluid, treating it in exactly the same way as oil, and substituting the gas density ρ_g for the oil density ρ_o in all the expressions used. In this case the value h_o would represent the height to which a fluid with the density of gas would rise in a manometer terminated at the point of interest.

Configurations of Traps

By the foregoing procedure it is possible to delineate the traps for oil or gas in any region from structure-contour maps and a knowledge of the hydrodynamic situation within the strata. The oil- or gas-equipotential surfaces will be inclined or tilted in the direction of water motion, and a trap will be formed by any impermeable surface which is concave in the direction of increasing potential for oil or gas. It is of particular interest to determine the magnitude of the tilts of the oil- or gas-equipotential surfaces in terms of the slope of the potentiometric surface for water. There are several ways to obtain an expression for these quantities, but perhaps the most straightforward is that illustrated in Figure 6-12. A stable oil-water interface is shown in which the oil is at rest and the water is in motion below it. Since the oil is at rest, this surface itself will be an oil-equipotential surface. Two water manometers are separated by a horizontal distance Δx and are terminated in the water immediately below the interface, where the fluid pressures are p_1 and p_2, respectively. The difference in static level of the water in the manometers is given by Δh_w, and the difference of elevation of their bases is denoted by Δz. The difference in pressure Δp between p_1 and p_2 can be determined from the heights of static water supported by the manometers, or

$$p_2 - p_1 = \Delta p = \rho_w g (\Delta z - \Delta h_w)$$

Alternatively, except for minor differences caused by capillary effects, the fluid pressures within the oil will be very nearly equal to those in the adjacent water and Δp can be computed from the difference in elevation within the static oil. Hence,

$$\Delta p = \rho_o g \, \Delta z$$

Equating these two expressions, dividing by $\rho_w g \, \Delta x$, and combining terms give

$$\frac{\Delta z}{\Delta x} = \frac{\rho_w}{\rho_w - \rho_o} \frac{\Delta h_w}{\Delta x}$$

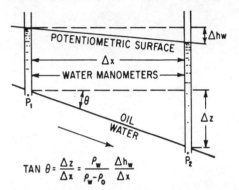

$$\text{TAN } \theta = \frac{\Delta z}{\Delta x} = \frac{\rho_w}{\rho_w - \rho_o} \frac{\Delta h_w}{\Delta x}$$

FIGURE 6-12. Diagram showing manner by which the equation describing the tilt of the oil-water interface in a hydrodynamic situation can be derived. Two water manometers are terminated in the water just below the interface.

which is the equation for the tilt of the oil-water interface in terms of the hydrodynamic gradient or slope of the potentiometric surface. The tilt of a gas-water interface will be given by exactly the same expression provided the density of the gas ρ_g is used in place of that of oil.*

In either case the interface and hence the equipotential surface for the particular fluid are equal to the slope of the water-equipotential surface multiplied by a dimensionless coefficient. This coefficient $\rho_w/(\rho_w - \rho_o)$ is therefore called the tilt-amplification factor. It may range from a value of very nearly 1 for some of the lighter gases to infinity (corresponding to complete instability for fluids the same density as water). In general the value will be between 1 and 2 for gases and between 7 and 15 for oils of various densities. For an average figure oil commonly has a density not far from 0.9 gram per cubic centimeter, giving a tilt-amplification factor of 10.

The magnitude of the hydrodynamic gradient or slope of the potentiometric surface is largely dependent upon the regional topography and the locations of outcrops of the permeable formations through which water may enter or leave. Thus in general, regions of low relief such as the Gulf Coast may have hydrodynamic gradients from zero to 2 or 3 feet per mile. In regions of great relief such as the sedimentary basins of the Rocky Mountains the gradients may be as high as 100 feet per mile. All gradations between these situations are possible.

* The foregoing method of deriving the Hubbert tilt equation is somewhat simpler than that given in either of the Hubbert (1940, 1953) papers. However, it was first derived jointly by Hubbert and his assistant Jerry Conner and is frequently used by Hubbert himself in private discussions and lectures.

Applying the tilt-amplification factors to these gradients, it is evident that oil-equipotential surfaces may be tilted as much as 800 to 1,000 feet per mile in extreme situations whereas tilts of gas-equipotential surfaces may be as much as 100 to 200 feet per mile in the same situation.

As a general consideration it is evident that if water is flowing through some stratum in a particular direction, the oil-equipotential surfaces will be tilted in that direction a certain amount and gas-equipotential surfaces will be tilted a lesser amount in the same direction. Only under hydrostatic conditions will the equipotential surfaces for gas and oil, and hence the traps for these fluids, exactly coincide.

Some of the configurations which may occur can be illustrated by considering the various positions of oil and gas traps associated with the same underground geometry

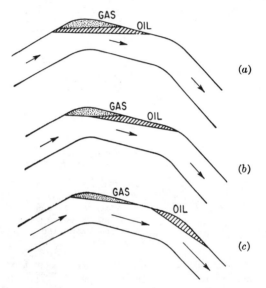

FIGURE 6-13. Schematic section showing several configurations of oil and gas traps which may occur under different hydrodynamic conditions with the same underground geometry.

but under different conditions of fluid flow. Such a situation is depicted in Figure 6-13. Under conditions of a relatively gentle hydrodynamic gradient in a permeable stratum the oil-water interface will be tilted slightly in the direction of flow. Since the oil and gas are both static, the gas-oil contact will be horizontal (Figure 6-13a). If the hydrodynamic gradient is somewhat stronger, the oil may be tilted sufficiently to cause a gas-water contact to occur, which will be tilted in the same direction as, but a lesser amount than, the oil-water contact (Figure 6-13b). Under a very strong hydrodynamic gradient the oil and gas traps may be entirely separated, with the gas on the crest of the structure and the oil in a structural terrace on its side (Figure 6-13c). With even stronger gradients the same structure may become incapable of holding either gas or oil.

A different possibility for a hydrodynamic trap is illustrated in Figure 6-14. In this case water is flowing downdip through a homoclinally dipping stratum containing a low-permeability region. In this local region the hydrodynamic gradient will become greater, causing the oil-equipotential surfaces to tilt more steeply. Where these surfaces resume their normal tilt, a region of upward concavity will occur. It is therefore possible for an equipotential surface to form a closed region against the flat upper surface of the stratum, forming a trap as shown.

Summary

It is apparent from the foregoing discussion that the locations and configurations of traps are determined jointly by the underground geometry of the rocks and the state of motion of the environmental ground water. Under hydrostatic conditions a trap will be formed wherever rocks permeable to oil or gas are overlaid by an impermeable surface which is concave downward. Under hydrodynamic conditions the impermeable surface must be concave in the direction of increasing potential for the particular fluid. As an additional possibility arising from permeability changes within

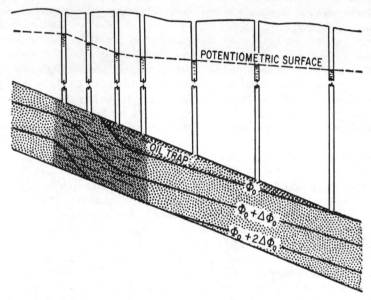

FIGURE 6-14. Schematic section showing the manner in which a trap may be formed by water flowing downdip in a homoclinal stratum through a region of low permeability. (*Redrawn from Hubbert, Bull. AAPG, vol. 37, p. 2001.*)

the rocks, the equipotential surfaces themselves may form a concave geometry terminating against the upper surface of a homoclinally dipping permeable stratum.

THE CLASSIFICATION OF TRAPS

In the previous section were described the physical principles governing the accumulation of oil and gas. It was shown that a trap is simply an underground region in which the potential energy of the oil or of the gas is at a minimum with respect to its surroundings. The geometry of the rocks together with the state of motion of the environmental groundwater jointly determine the location and configuration of these underground regions.

There are perhaps a dozen different geological factors which can contribute to making geometry favorable for a trap, and throughout the world traps have been found in which these factors have been combined in hundreds of different ways. With any group of phenomena as complex as this, schemes of classification are very useful in aiding our understanding. Many classifications of oil and gas traps based on many different kinds of criteria, ranging from age, size, and depth of the accumulation to extremely detailed and exhaustive classifications based on geologic conditions, have been described in the literature (Clapp, 1929; Wilhelm, 1945; Knebel and Rodriguez-Eraso, 1956).

To be of value to the petroleum exploration geologist, a scheme of classification should be based on those criteria which are useful in the exploration for new traps. By far the most widely accepted scheme of classification is of this kind and is based on the distinction between structural and stratigraphic factors, structural factors being those which have arisen by virtue of the deformation of the rocks and stratigraphic factors being those which have arisen by virtue of the conditions under which the rocks themselves were formed. Strictly speaking, almost every trap arises from a combination of both structural and stratigraphic factors. For example, in any sedimentary section, the stratified layers form many surfaces which are capable of acting as impermeable barriers to the movement of oil. One of these stratigraphic surfaces is an important element in almost every trap. Therefore, these traps are in part stratigraphic. Similarly, even a simple tilting of the strata into a homocline constitutes a structural deformation. Since there exist very few traps not containing at least this much of a structural element, almost every trap is also in part structural.

Although almost every trap is strictly both structural and stratigraphic, it has become customary to classify traps as structural if the major elements in their geometry have arisen through folding, faulting, or doming of uniformly stratified rocks and to classify traps as stratigraphic if the major elements in their geometry arise from lateral variations in the strata because of differences in sedimentation, the existence of unconformities, or the existence of anomalous organic deposits. Many traps exist in which structural and stratigraphic factors appear to be of equal importance, and these are commonly classed as combination traps. This three-way classification of structural, stratigraphic, and combination traps is widely used throughout the petroleum industry and is quite useful in obtaining a general understanding of the basic reasons for the existence of various accumulations. Yet in spite of its widespread use, the scheme is rather general and imprecise. In many cases the classification of a trap is to a large degree arbitrary and is a matter of personal opinion.

In the present discussion, therefore, there is no attempt to classify traps as such. Instead, the various possible geologic factors which can contribute to the formation of a trap are discussed in turn, and examples are presented of traps in which each of these factors is significant. At the same time the various examples will serve to show the multitude of ways in which the various factors may combine to form different traps. In this discussion the following breakdown of the various factors which may contribute to the formation of traps is used:

Structural factors:
    Folds
    Faults
    Intrusive structures—domes
Stratigraphic factors:
    Lateral variations in sedimentation
    Unconformities
    Organic deposits
Miscellaneous factors:
    Fractured reservoirs
    Secondary chemical alteration
    Asphaltic seals
Hydrodynamic effects

Structural Factors

Traps which arise as the result of structural factors are characteristically much more easily discovered than those which arise from stratigraphic factors. Folds, faults,

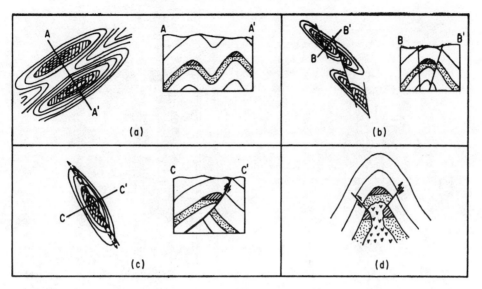

FIGURE 6-15. Schematic diagrams of traps which may be associated with anticlinal structures: (*a*) parallel folding, (*b*) echelon folding and characteristic transcurrent faulting, (*c*) more highly deformed fold with development of thrust faulting, (*d*) diapir fold.

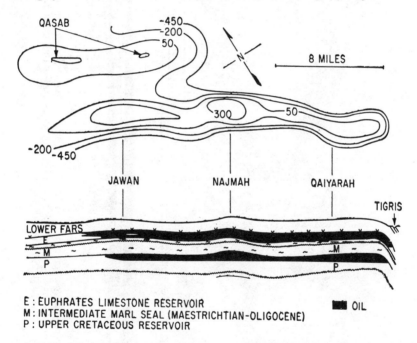

FIGURE 6-16. Qaiyarah oil fields of northern Iraq showing traps associated with parallel anticlinal folds. (*Redrawn from Dunnington, in "Habitat of Oil," p. 1238, AAPG, 1958.*)

and domes are often revealed in the surface geology. They represent strong anomalies in the distribution of subsurface rocks and are therefore easily revealed by geophysical methods. Consequently, when a new basin or province is explored, the structural traps are usually found earliest.

A single structural factor is also likely to produce more than one trap. It may cut across or affect the entire sedimentary section and thus form a trap in any potential

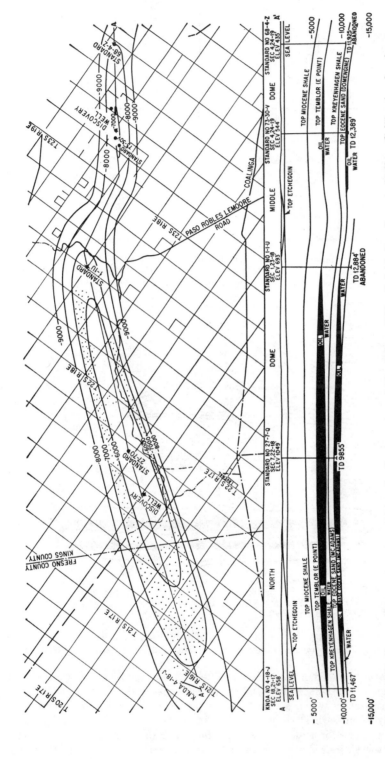

FIGURE 6-17. Structure contour map and longitudinal section of Kettleman Hills oil field, San Joaquin Valley, California, showing traps associated with unfaulted echelon folds. Structure contours are on the top of the Miocene Temblor shale. (*Redrawn from Porter*, "1962 *Guidebook*," *pp. 170–171, Pacific Section AAPG.*)

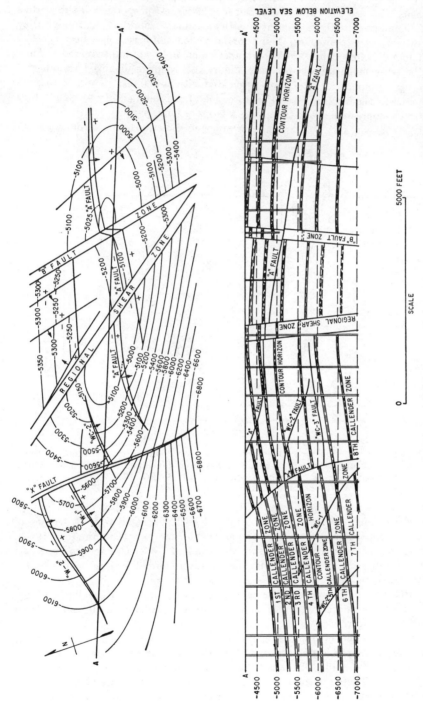

FIGURE 6-18. Structure map and section of the Dominguez oil field. This anticlinal structure lies in echelon relation to several others along the Newport-Inglewood trend in the Los Angeles Basin of California. Note the characteristic transcurrent faulting (regional shear zones). *(Redrawn from Graves, Map 32, Calif. Div. Mines Bull. 170, 1954.)*

reservoirs which may exist within the section. Examples are shown which illustrate dozens of separate traps associated with a single structural feature (Figures 6-18, 6-28, 6-30, and 6-32).

Folds. Of all the various kinds of traps which have been discovered by far the most common, and also the ones accounting for the largest proportion of the world's reserves of oil and gas, are those associated with anticlinal structures. Although they are comparatively easy to discover, their abundance does not arise solely because they have been found before other kinds of traps which are yet undiscovered. They appear to be truly much more common. Figure 6-15 depicts idealized sketches of some of the most common forms of traps associated with anticlines. These structures commonly arise as the result of lateral compression of the sedimentary section. Frequently, if the compression is not too great, a series of long, parallel, gentle folds will be formed. Wherever such folds plunge in opposite directions away from an axial

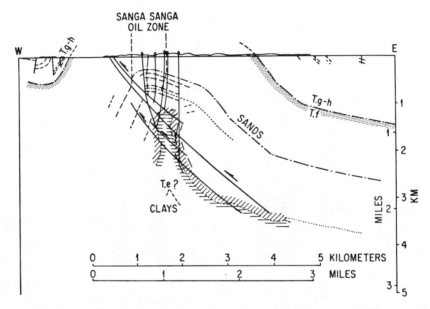

FIGURE 6-19. Structure section through the Sanga Sanga oil field on the Balikpapan trend in eastern Borneo. A number of minor thrust faults have developed near the crest of the fold. (*Redrawn from Weeda, in "Habitat of Oil," p. 1345, AAPG, 1958.*)

high, a trap may be formed as shown in Figure 6-15a. This kind of trap is typified by many of the huge oil fields of the Middle East. The Qaiyarah structures of northern Iraq depicted in Figure 6-16 are excellent examples (Dunnington, 1958). In these fields oil has been accumulated in Miocene and Upper Cretaceous limestone reservoirs. Within each of the reservoirs the oil-water contact is tilted to the southeast, giving indication of a hydrodynamic situation.

Often, anticlinal folds, instead of being adjacent, are formed in echelon arrangement. Such structures are frequently, but not always, cut by transcurrent faulting across their axes (Figure 6-15b).

An interesting example of traps associated with echelon folds in which transcurrent faulting has not developed is afforded by the Kettleman Hills oil field which is located in the San Joaquin Valley of California. This is depicted in Figure 6-17 (Porter, 1952; Woodring, 1940). There are three separate structures, only two of which are shown in the figure. Production is from multiple sands interstratified with shales of Miocene and Eocene age. The North Dome field is by far the largest and is expected to produce more than 500 million barrels of oil. The oil-water contacts in these reservoirs are also tilted, in this case toward the northwest, again giving evidence of hydrodynamic effects.

Another example of a series of echelon folds is afforded by the Newport-Inglewood trend of oil fields in the Los Angeles Basin of California. The Dominguez oil field, which is located on this trend, is depicted in Figure 6-18 (Graves, 1954). Here the

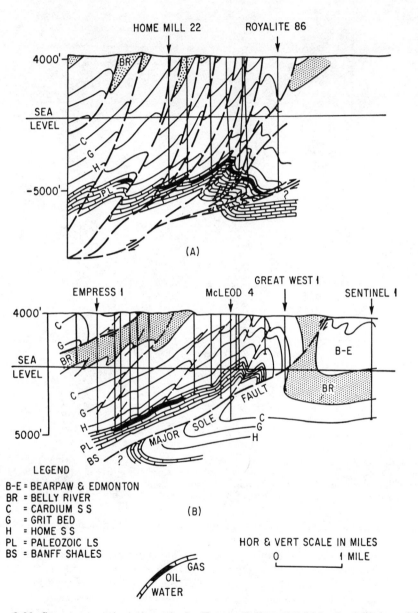

FIGURE 6-20. Structure sections through the Turner Valley oil field in the folded foothills belt of the Rocky Mountains of Western Canada: (A) section across the north end of the field, (B) section across the center of the field. (*Redrawn from Link, Bull. AAPG, vol. 33, pp. 1479, 1480.*)

characteristic transcurrent faulting across the axis is well developed together with minor thrust faulting parallel to the axis. In this case the faulting has, to some degree, influenced the accumulation of oil, and some of the fault blocks are much more productive than others. This field illustrates the way in which multiple traps may be formed by folding which extends through a large thickness of strata. In the

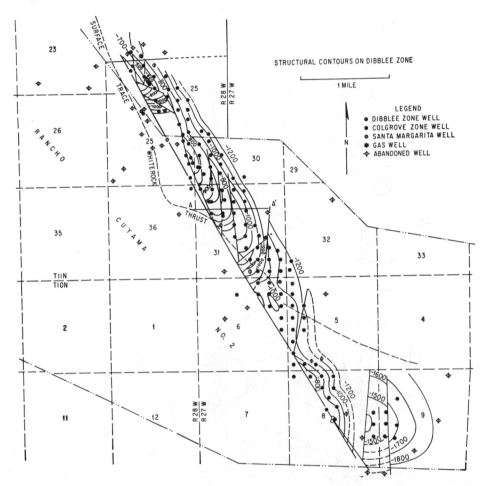

FIGURE 6-21. Structure contour map of the Russell Ranch oil field, Cuyama Valley, California. Contours are on the top of the Dibblee sand, which is the major reservoir in the field. This is an example of an anticlinal fold which has been strongly modified by faulting. (*Redrawn from Schwade, Carlson, and O'Flynn, in " Habitat of Oil," p. 94, AAPG, 1958.*)

Dominguez structure there are eight different pay zones. Production is from Miocene and Pliocene sands interspersed with shales.

In regions of more severe deformation, anticlinal structures may be overturned or cut by thrust faulting as shown in Figure 6-15c. An example of traps associated with an anticlinal structure in which some thrust faulting has taken place is afforded by the Sanga Sanga oil field located on the Balikpapan oil trend in east Borneo [Figure 6-19 (Weeda, 1958)]. Production here is from Tertiary sandstones.

The results of a much more intensive deformation are shown in Figure 6-20, which depicts structural sections through the Turner Valley oil field in the folded foothills of

the Canadian Rocky Mountains (Link, 1949; Gallup, 1951). The accumulation is in
a Paleozoic limestone reservoir which has been cut by many related thrust faults.

Another example of an anticlinal structure which has been strongly modified by
several kinds of faulting is that of Russell Ranch oil field (Schwade et al., 1958) in
the Cuyama Valley of California, which is depicted in Figures 6-21 and 6-22. The
producing zones, which are sandstones of Lower Miocene age, have been cut by

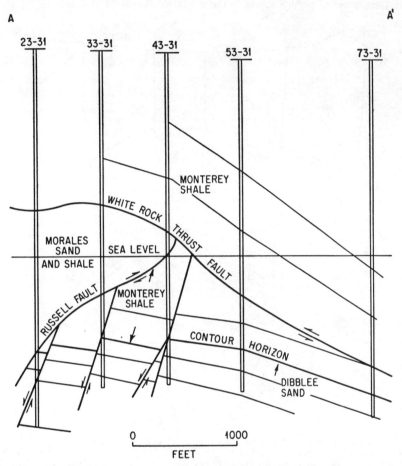

FIGURE 6-22. Structure section *A-A'* through the Russell Ranch oil field, Cuyama Valley,
California.   Note how the underlying structure has been covered and hidden by the White
Rock thrust fault.   (*Redrawn from Schwade, Carlson, and O'Flynn, in "Habitat of Oil,"
p. 95, AAPG, 1958.*)

multiple normal faults, and the entire structure has been covered by the White Rock
thrust fault which probably occurred during a later period of deformation.

When folds are formed by compression of competent strata underlain by soft incompetent rocks such as salt or weak shales, diapiric anticlines may be formed by the soft
rocks being squeezed up and intruded into the overlying strata along the axis of the
structure.   This situation is depicted in idealized form in Figure 6-15d.   Diapiric
structures are prevalent in the Aquitain Basin of southern France, the Carpathian
regions of Eastern Europe, and the Caucasus region of Russia.   In Figure 6-23 is
depicted a section through the St. Marcet gas field in the southern Aquitain Basin

(Schneegans, 1948). Here plastic Triassic anhydrite and gypsum have been squeezed up into the overlying competent Cretaceous limestones and sands. One of the interesting characteristics of this kind of structure is the thrust faulting which develops on the edges of the diapiric mass. Figure 6-24 illustrates a section through the Moreni diapiric anticline (Walters, 1946; Small, 1959) in Rumania. Here plastic salt has been squeezed up and intruded along the axis of the long anticlinal structure. Accumulations occur in porous strata butting against the diapiric mass.

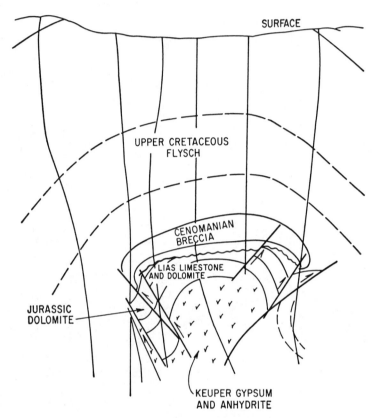

FIGURE 6-23. Structure section through the St. Marcet gas field in the southern Aquitaine Basin of France. Gas is produced from Upper Cretaceous Cenomanian breccia and from Jurassic dolomite. Oil has also been found in the structure underlying the gas. Note the thrust faults which have developed on both flanks of the diapir. (*Redrawn from Schneegans, Bull. AAPG, vol. 32, p. 209.*)

An extremely interesting example of a very complex anticlinal structure is afforded by the Penal oil field located in Trinidad which is illustrated by the section in Figure 6-25 (Bitterli, 1958). Here the structure has been overturned and two thrust faults have developed. Production is from the Herrera sandstone which, because of the overturn, is penetrated at three different depths in some of the wells. In addition to the overturning and faulting, a small diapiric intrusion occurs in the center of the structure.

Faults. Fault surfaces provide many different possibilities for the development of the proper geometry for a trap. Often there will be a zone of gouge, or broken rock, associated with the fault surface which is capable of acting as an impermeable barrier to oil. Frequently an impermeable stratum is displaced into a position where it can serve as a barrier to oil moving in a permeable stratum. However, faults do not always seem to be effective seals, and there is evidence in many cases that fault surfaces have served as conduits for the movement of oil up through the strata. Figure 6-26 depicts schematic diagrams of a number of possible traps associated with faults.

One of the most common types of fault trap is that formed by normal faulting. This may occur either by two or more normal faults intersecting homoclinally dipping strata or by a single normal fault intersecting a structural nose or terrace. These

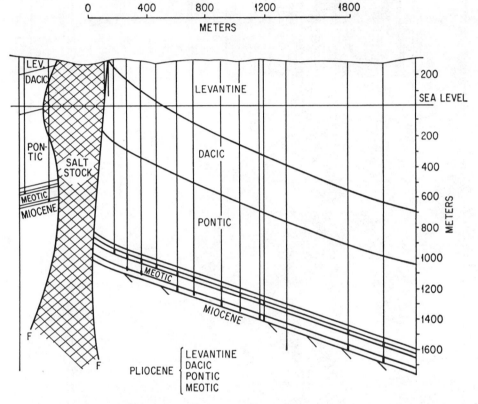

FIGURE 6-24. Section through the Moreni diapiric anticline in the Ploesti region of Rumania. Traps are formed by permeable strata butting against the diapiric mass. (*Redrawn from Walters, Bull. AAPG, vol. 30, p. 333.*)

situations are indicated in Figure 6-26a and b. In either case two possibilities exist for the relationship between the dip of the fault surface and the dip of the strata. If the strata dip in a direction opposite to that of the fault plane, the trap will be on the upthrown side of the fault, whereas if the dips are in the same direction, the trap will be on the downthrown side of the fault. For some reason which is not fully understood accumulations are much more frequently found on the upthrown side of normal faults. An excellent example of this kind of trap is afforded by the West Guara field (Hedberg et al., 1947) in the Greater Oficina area of eastern Venezuela (Figures 6-27 and 6-28.) This prolific field contains 47 separate productive traps in the same structural segment, illustrating again the probability of finding multiple traps associated with structural features. Production is from the Oficina sands of Oligocene and

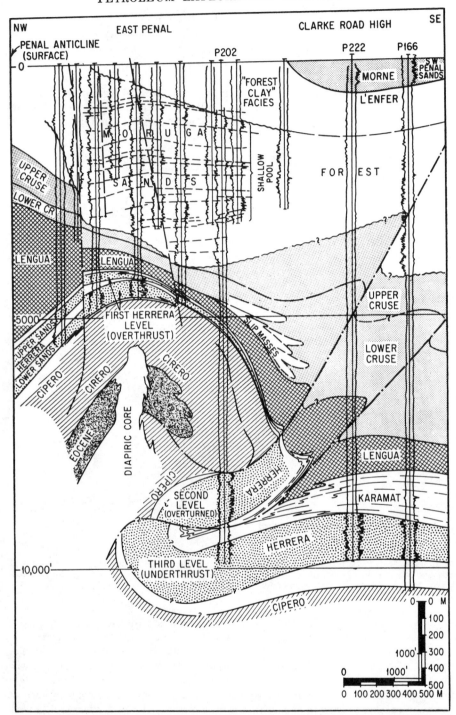

FIGURE 6-25. Section through the Penal oil field, Trinidad, B.W.I. Production is obtained from the shallow Moruga sands and the deeper Herrera sand, which is penetrated at three levels in some of the wells. Thrust faulting, overturning, and diapiric intrusion are all evident in this complex structure. (*Redrawn from Bitterli, Bull. AAPG, vol. 42, p. 145.*)

Miocene age. Samples taken from some of these sands show permeabilities as high as 10,000 millidarcys.

An example of multiple traps formed on the downthrown side of a series of normal faults is afforded by the Velasquez field (Morales, 1958) in the middle Magdalena Valley of Colombia which is depicted in Figures 6-29 and 6-30. Oil here is produced from early Tertiary sandstones.

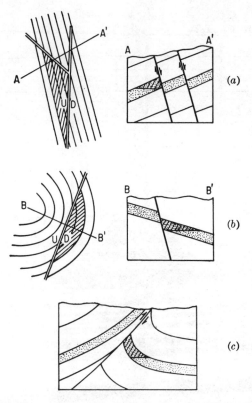

FIGURE 6-26. Schematic diagrams of typical traps associated with fault planes: (a) intersecting normal faults in homoclinal strata, traps on upthrown side; (b) normal fault intersecting a broad structural nose or terrace, trap on the downthrown side of the fault; (c) trap underlying a thrust fault formed by permeable strata terminating updip against fault plane.

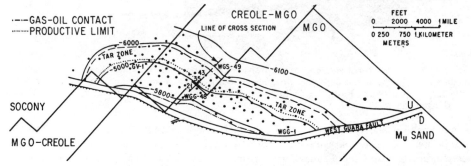

FIGURE 6-27. Structural contour map of the West Guara oil field, Greater Oficina area, Venezuela. Multiple traps are formed by a broad structural nose intersected by the normal West Guara Fault. (Redrawn from Hedberg, Sass, and Funkhouser, Bull. AAPG, vol. 31, p. 2159.)

445

Another example of multiple traps on the downthrown side of a normal fault is provided by the series of accumulations which occur for several miles along the Steinberg fault in the Inner-Alpine Vienna Basin of Austria and Czechoslovakia (Janoschek, 1958). Typical of these accumulations is the Gaiselberg field in which oil and gas are produced in Middle and Upper Miocene sandstones which have been faulted down

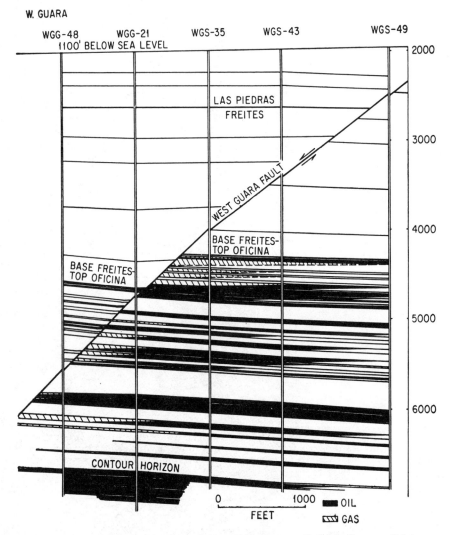

FIGURE 6-28. Structural section through the West Guara oil field, Greater Oficina area, Venezuela. Forty-seven separate productive traps have been formed on the upthrown side of the West Guara Fault. (*Redrawn from Hedberg, Sass, and Funkhouser, Bull. AAPG, vol. 31, p. 2160.*)

against the Eocene flysch basement. This field is illustrated in Figures 6-31 and 6-32. Again, multiple reservoir sands have formed several traps in a single structure. The map indicates a number of subsidiary normal faults which form a series of independent reservoir segments in each of the sands.

Traps are very frequently formed on the downthrown side of reverse or thrust faults as illustrated in Figure 6-26c. However, these traps are characteristically difficult to

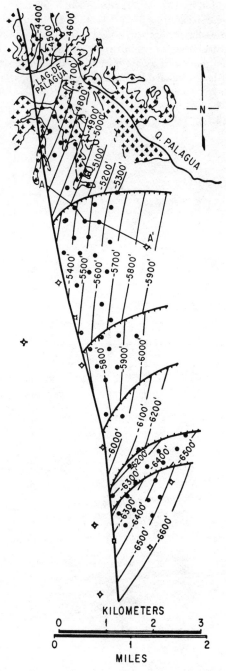

FIGURE 6-29. Structural contour map of the Velasquez oil field in the Middle Magdalena Valley of Colombia. Traps are formed by intersecting faults in approximately homoclinal strata. (*Redrawn from Morales and the Colombian Petroleum Industry, in "Habitat of Oil,"* p. 692, AAPG, 1958.)

discover, since they are hidden by the structure and the deformation on the upthrown side of the fault. An excellent example of this kind of accumulation is afforded by a series of traps which have been formed underneath the Oakridge thrust fault in the eastern portion of the Ventura Basin of California. In this region Miocene, Oligocene, and Eocene rocks have been displaced over Pliocene sands and shales and oil has been trapped in the upturned edges of these porous Pliocene sands where they terminate against the fault surface (Ware and Stewart, 1958). Several large anticlinal struc-

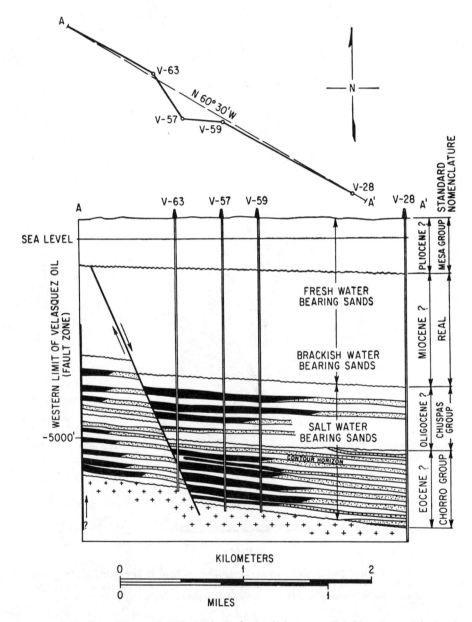

FIGURE 6-30. Structure section *A-A'* through the Velasquez oil field. Accumulation has occurred in traps on the downthrown side of the fault. (*Redrawn from Morales and the Columbian Petroleum Industry in " Habitat of Oil," p. 693, AAPG, 1958*).

tures have been formed in the overthrust block, and large accumulations are found within them. A north-south section through the South Mountain oil field and the Bridge Pool underlying the Oakridge thrust is shown in Figure 6-33.

An interesting prospecting technique has been developed in this region which greatly aids in the discovery of these traps. Wells are directionally drilled underneath

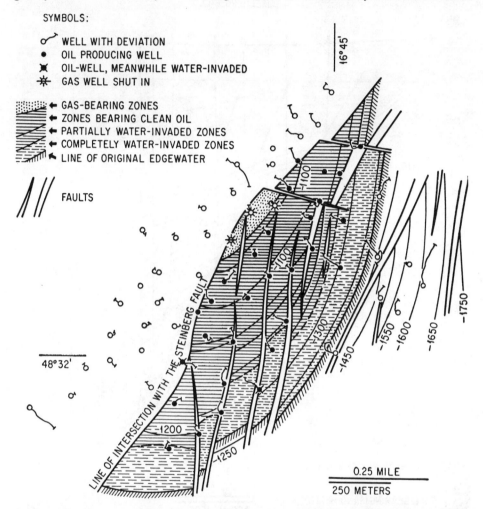

SYMBOLS:

    WELL WITH DEVIATION
    OIL PRODUCING WELL
    OIL-WELL, MEANWHILE WATER-INVADED
    GAS WELL SHUT IN

    GAS-BEARING ZONES
    ZONES BEARING CLEAN OIL
    PARTIALLY WATER-INVADED ZONES
    COMPLETELY WATER-INVADED ZONES
    LINE OF ORIGINAL EDGEWATER

    FAULTS

0.25 MILE
250 METERS

FIGURE 6-31. Structural contour map of the Gaiselberg oil field in the Inner Alpine Vienna Basin together with production data. Traps are formed in a broad structural nose on the downthrown side of the Steinberg Fault. (*Redrawn from Janoschek, in "Habitat of Oil," p. 1148, AAPG, 1958.*)

and parallel to the fault surface, so that a single test has a good probability of intersecting any accumulation that may be present. This technique has been successful in the discovery of several pools underneath the Oakridge thrust during the last few years.

An example of a series of traps in which both faulting and folding have played a significant part is shown by the Potrero oil field, which is another of the major structures along the Newport-Inglewood trend in the Los Angeles Basin of California (Willis and Ballantyne, 1943). This is illustrated in Figures 6-34 and 6-35. In this case, a series of multiple, independent traps is formed by several transcurrent faults cutting across an anticlinal structure. In particular, note the traps which are formed on the northwest plunging nose of the anticline. These traps have been formed in Miocene and Pliocene sandstones.

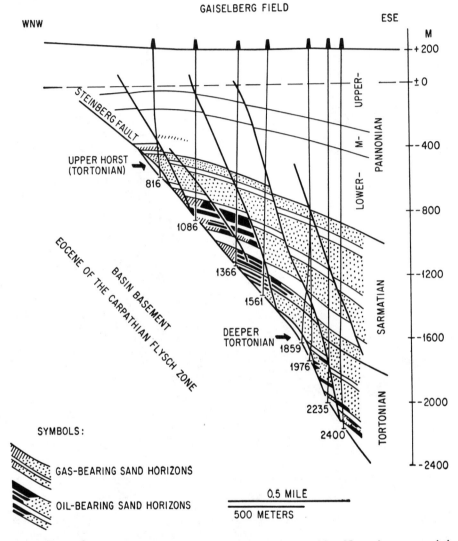

FIGURE 6-32. Structural section through the Gaiselberg oil field. Note the traces of the numerous normal faults subsidiary to the main Steinberg Fault. (*Redrawn from Janoschek, in "Habitat of Oil," p. 1147, AAPG, 1958.*)

Intrusive Structures. Deep-seated intrusion of the sedimentary section by plastic or fluid rocks provides the possibility for many different kinds of trapping configurations. Intrusions are often formed as the result of igneous activity in which the intruded material is a molten or semimolten rock. Although there undoubtedly exist many traps associated with igneous intrusions, they all appear to be unproductive, probably because of the high temperatures to which the rocks have been subjected.

A much more favorable situation arises in the case of intrusive salt structures or salt domes, which have originated at much lower temperatures and which have formed productive traps in many regions throughout the world.

The mechanism by which salt-dome structures are formed is fairly well understood for several reasons. A tremendous amount of subsurface information has been available for study and analysis, principally from well records in the Gulf Coast region of the United States, where these structures have been intensively exploited. In addition, the classic model experiments of Nettleton (1936, 1943) and of Parker and

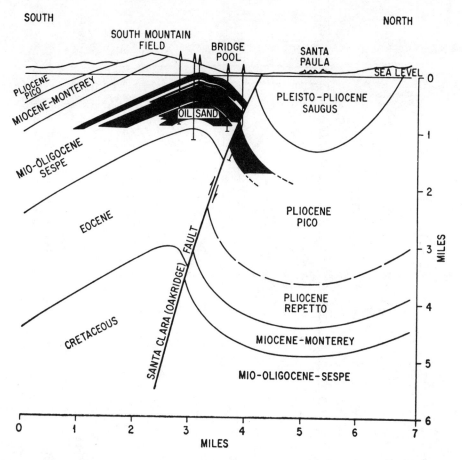

FIGURE 6-33. Schematic north-south structural section through the South Mountain oil field and the Bridge Pool in the eastern Ventura Basin, California. The trap in the Bridge pool is formed by Pliocene sands terminating updip against the Oakridge Thrust Fault. (*Redrawn from Ware and Stewart, "Guidebook," p. 180, AAPG Pacific Section, 1958.*)

McDowell (1951, 1955) have provided a convincing demonstration of the mechanism of salt-dome formation on a laboratory scale.

It is believed that a thick layer of sedimentary salt is overlain and buried by other sediments. The salt, being less dense than the overlying sediments, is in a mechanically unstable position with respect to them. Any small irregularities in the upper surface of the salt will have a tendency to allow the salt to penetrate and intrude upward into the overlying sediments. The salt, by virtue of its relative buoyancy, will then move upward as a plug or a dome, sometimes many thousands of feet, penetrating and piercing the sediments as it moves.

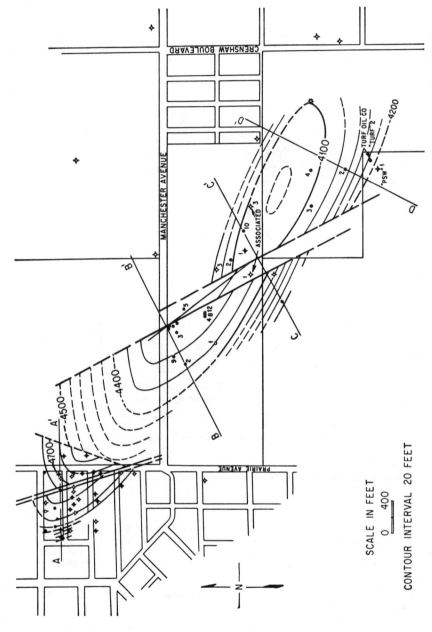

SCALE IN FEET

0   400

CONTOUR INTERVAL 20 FEET

Figure 6-34. Structure contour map of the Potrero oil field in the Los Angeles Basin of California. This field is another of the echelon folds along the Newport-Inglewood trend in the Los Angeles Basin (Fig. 6-18). *(Redrawn from Willis and Ballantyne, Calif. Div. Mines Bull. 118, p. 312.)*

Structures of this kind are closely related to diapiric structures but with the distinction that in this case the energy is supplied primarily by the relative buoyancy of the salt whereas in diapiric structures the energy is supplied primarily by lateral compression of the sediments. All gradations between the two kinds of structures can be found. In many cases the rate at which the salt rises appears to be delicately balanced against the rate at which the overlying sediments are deposited, and as a result salt-dome structures are found in all stages of development, ranging from those which are too deep to be penetrated currently by the drill to structures in which the salt lies immediately below the surface of the ground.

Intrusive salt structures have been found in many regions throughout the world, and they have been responsible for the formation of many oil traps in the Gulf Coast region of the United States, the Isthmus of Tehuantepec in Mexico, the Hannover Basin of northern Germany, and the Emba region bordering the Caspian Sea in Russia. Intrusive salt structures have also been found in the Paradox Salt Basin of Utah and more recently in the Canadian Arctic islands (Fortier et al., 1954), but as yet no oil has been found associated with these. In these various regions the composition of the salt varies widely, with halite being the principal constituent associated with varying quantities of various chloride salts, anhydrites, carbonates, and clastic material. In the Gulf Coast region where salt intrusions, or domes, have been studied more intensively than anywhere else in the world, the salt is almost pure halite together with 5 to 10 per cent anhydrite and very minor quantities of other minerals (Hanna, 1934).

One of the extremely interesting characteristics of salt-dome structures which should be noted is the wide variation which can occur in the shape of the salt mass. One of the very common occurrences is the overhanging dome, which is particularly important from the exploration standpoint, as traps may exist below the overhang. Such a situation requires that wells must be drilled for varying distances through the salt before the traps can be reached.

Another very interesting characteristic of salt domes is the frequent occurrence of a cap rock. This is a layer of insoluble material covering the entire top surface of the salt mass. If present, it may be as much as 2,000 feet thick. The major constituent of the cap rock is usually anhydrite, but gypsum, limestone, dolomite, and free sulfur are frequently present. It was originally thought that cap-rock material represented a stratum which overlay the original salt and had been retained as it moved upward. However, the evidence available at present indicates that the cap rock is probably an insoluble residue which has accumulated as the salt mass was continuously dissolved and washed away by circulating ground water. This activity probably occurred more slowly than the concurrent upward movement of the salt plug, thereby precluding collapse and caving of the overlying strata.

Salt-dome structures are relatively easy to locate underground, especially if they are shallow. They represent such a strong anomaly in the physical properties of the rocks that geophysical methods are usually effective. In particular, gravitational measurements which indicate the distribution of deep-seated masses will often clearly reveal the less dense salt. However, in spite of the ease with which salt domes can be located, the multitude of various traps which can be associated with them is not nearly so easily discovered.

In the Gulf Coast region of Texas and Louisiana during the last 15 or 20 years large reserves of petroleum have been discovered in a great many relatively small fault and stratigraphic traps around the flanks of previously known piercement salt-dome structures. These discoveries have required extremely careful and detailed study of large numbers of well logs obtained from previous exploratory and development drilling.

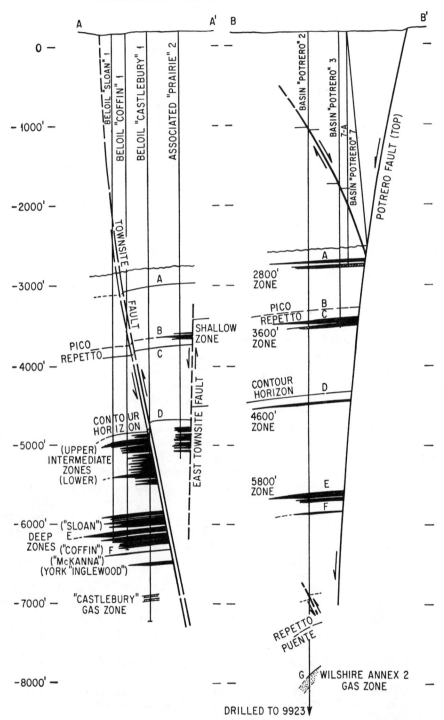

FIGURE 6-35. Structure sections through the Potrero oil field. Multiple traps are formed
118, p. 313.)

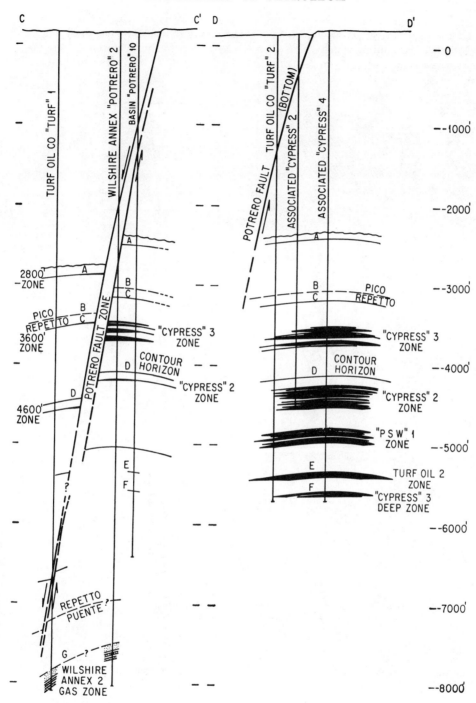

against the fault planes. (*Redrawn from Willis and Ballantyne, Calif. Div. Mines Bull.*

Idealized diagrams of traps which are commonly associated with intrusive salt-dome structures are shown in Figure 6-36. These may range from structural domes with little or no faulting which occur when the sediments likely to contain oil are far above the salt intrusion (Figure 6-36a) through more highly faulted structural domes

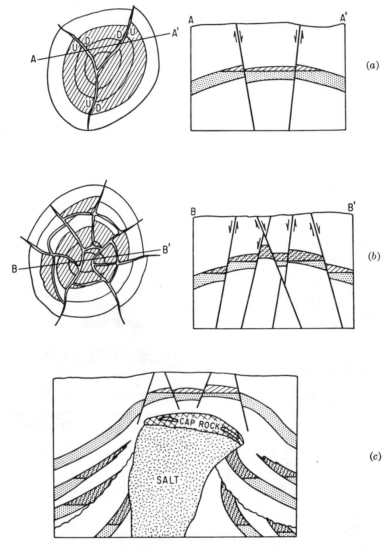

FIGURE 6-36. Schematic diagrams of traps associated with salt domes: (a) structure contours over a deep-seated dome showing typical normal fault and graben development, (b) structure contours over a shallower dome showing characteristic radial and tangential faulting, (c) section through a piercement dome showing development of traps on the flanks.

showing the characteristic radial and peripheral faulting which occurs nearer the salt intrusion (Figure 6-36b) to piercement structures in which the intrusion actually penetrates the sediments likely to contain oil and a great variety of traps are formed on the flanks of the dome (Figure 6-36c). The cap rock of the intrusion itself sometimes has developed excellent permeability and porosity, thereby forming a trap.

The examples which follow have been selected so as to show many of these various ways in which traps may be associated with salt-dome structures. In Figure 6-37 is depicted a structure contour map of the Brookhaven oil field located in the central portion of the interior salt basin of Mississippi (Womack, 1950). The contours are on

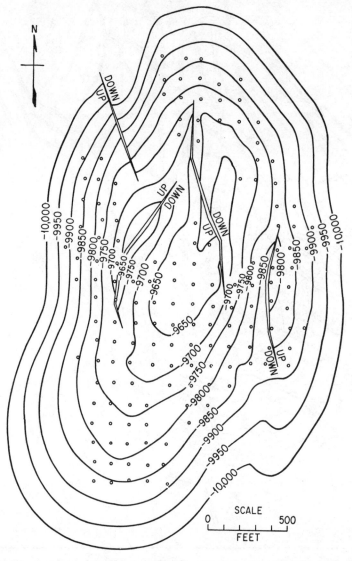

FIGURE 6-37. Structure contours on the basal Tuscaloosa sand in the Brookhaven oil field in Mississippi. This field overlies a deep-seated salt dome and shows the typical normal faulting. (*Redrawn from Womack, Bull. AAPG, vol. 34, p. 1522.*)

the basal Tuscaloosa sandstone of Upper Cretaceous age, which is the reservoir rock of the field. Although the underlying salt has never been penetrated by the drill, its existence is inferred both from the structure and from a gravity minimum over the field shown by geophysical measurements. The three normal faults on the top of the structure form a central graben area which is typical of these deep-seated salt-dome structures.

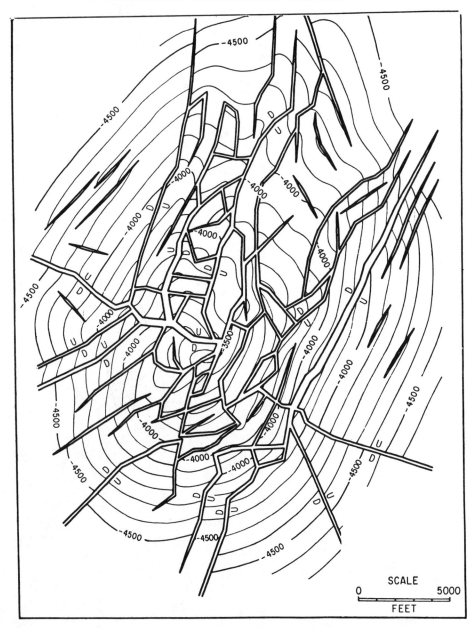

FIGURE 6-38. Structure contour map of the Hawkins oil field, Wood County, Texas. Contours are on the Woodbine sand, which is the producing zone of the field. The structure is formed over a deep-seated salt intrusion, and the characteristic radial and tangential faulting is apparent. (*Redrawn from Wendlandt, Shelby, and Bell, Bull. AAPG, vol. 30, and Parker and McDowell, Bull. AAPG, vol. 35, p. 2080.*)

A more intricate system of normal faulting is shown by the Hawkins oil field which is located in northeastern Texas (Wendlandt et al., 1946). Production is from the Woodbine sand of Cretaceous age. A structure map on the top of the Woodbine is depicted in Figure 6-38. The typical radial and tangential normal faults are evident.

Another example of the highly intricate system of normal faulting which may develop near the salt mass is afforded by the Reitbrook dome in the Hannover Basin of northern Germany (Reeves, 1946). A section through this structure is depicted in

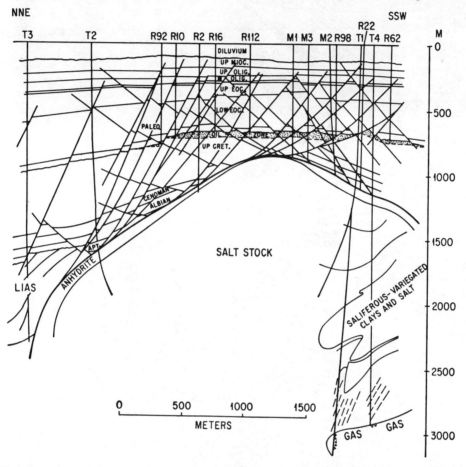

FIGURE 6-39. Structure section through the Reitbrook salt dome in the Hannover Basin of northern Germany, showing characteristic intricate normal faulting. (*Redrawn from Reeves, Bull. AAPG, vol. 30, p. 1564, and Behrmann, in British Intelligence Objectives Subcommittee, Oil Fields Investigation, Part III, Fig. 95.*)

Figure 6-39. Oil is produced principally from Marly limestones of Upper Cretaceous age in which the porosity is primarily due to fracturing. Some gas and oil are produced from the overlying Oligocene and Eocene sands. Note that the Cretaceous lies unconformably on truncated Lias beds, indicating that the salt structure had probably started to form prior to a period of pre-Cretaceous erosion.

When the salt actually pierces the sediments likely to contain oil, a great many possibilities exist for traps. Figure 6-40 depicts a structure map on the salt-dome material together with a number of structural cross sections in the vicinity of the South Liberty dome in the Gulf Coast region of Texas (Halbouty and Hardin, 1951).

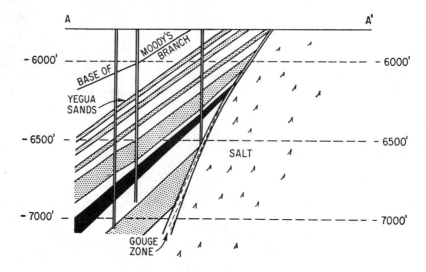

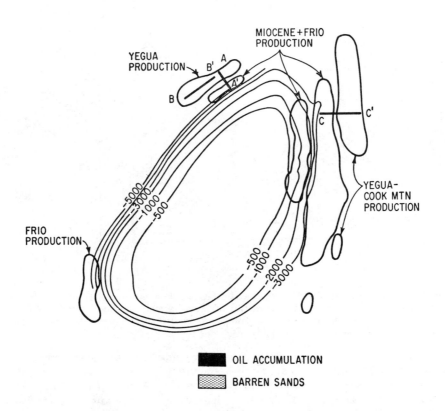

FIGURE 6-40. The South Liberty salt dome and associated oil accumulations, Liberty
three sections illustrate some of the various traps which may be developed on the flanks
1939–1977.)

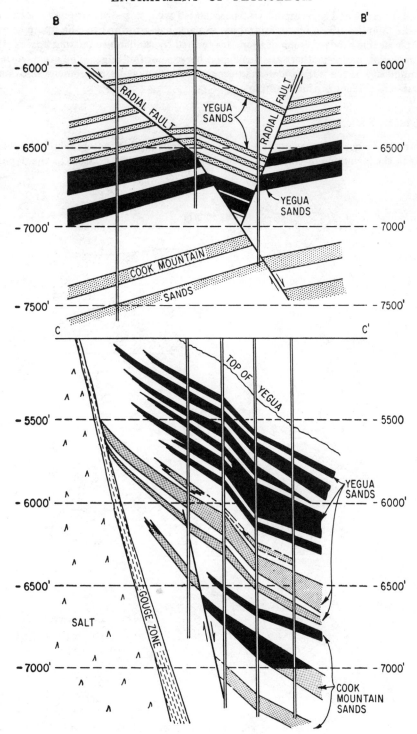

County, Texas. The structure contours are drawn on the top of the dome material. The
of a piercement dome. (*Redrawn from Halbouty and Hardin, Bull. AAPG, vol. 35, pp.*)

Over 50 different and separate oil traps associated with this structure have been discovered. Many of these are stratigraphic in part, being formed by the updip pinchout of permeable sands. Some of them are formed by sandstones butting against the dome material, and yet others are developed because of faulting. The great variety and complexity of the traps which may occur on the flanks of a piercement salt-dome structure are evident in this example.

Stratigraphic Factors

There are, in general, three different kinds of stratigraphic factors which can contribute to the formation of traps. The first of these is lateral variation in the deposi-

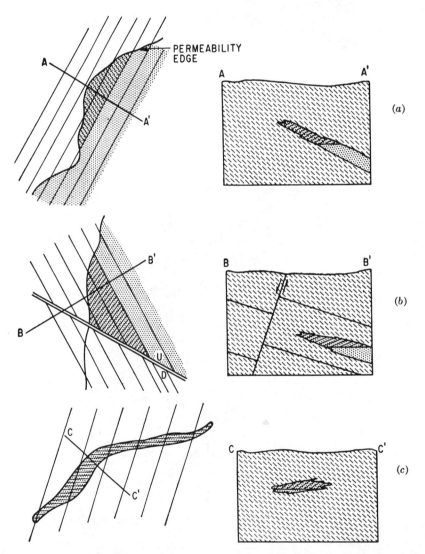

FIGURE 6-41. Schematic diagrams of traps associated with lateral sedimentary variations: (a) edge of a permeable stratum terminating updip in a homocline along a curving trend, (b) updip edge of a permeable stratum intersecting a fault, (c) shoestring sand entirely surrounded by impermeable rocks.

tion of sedimentary rocks. For example, a porous sandstone stratum which is deposited between layers of impermeable shale and which laterally pinches out or grades into impermeable shale will form a lateral boundary for the migration of oil or gas and may therefore contribute to the formation of a trap. A second class of stratigraphic factors are those which are associated with unconformities. In the

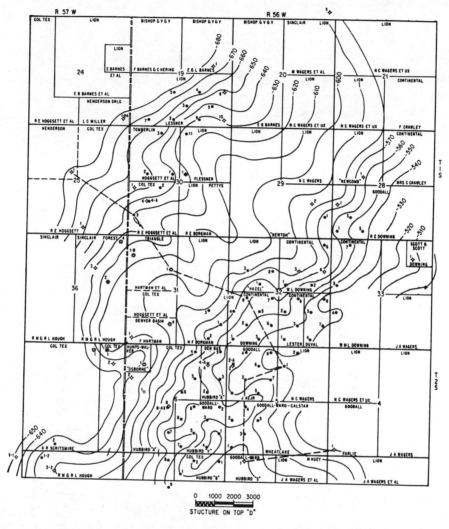

FIGURE 6-42. Structure contour map of the Little Beaver oil field in the Denver Basin, Colorado. Contours are on the top of the *D* sand body which terminates updip along a curving edge. (*Redrawn from Fentress, Bull. AAPG, vol. 39, p. 167.*)

third category are the organic sedimentary structures, biostromes and bioherms, or organic reefs. All these stratigraphic features are characteristically much more difficult to detect than structural features, principally because they constitute minor anomalies in the physical properties of the rocks in which they occur. They are seldom revealed in the surface geology and are frequently not reflected in geophysical measurements.

The most effective way of discovering traps associated with stratigraphic features is by the detailed analysis and comparison of the records obtained from the drilling of many wells in a region. Each stratum can be studied individually. Its lateral variations can be mapped, and unconformities can be detected. Since these kinds of data are available only when the exploration of a region is well underway, it is not surprising

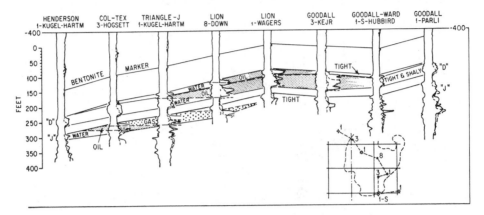

FIGURE 6-43. Structure section across the Little Beaver oil field showing accumulations in the *J* and *D* sands. (*Redrawn from Fentress, Bull. AAPG, vol. 39, p. 162.*)

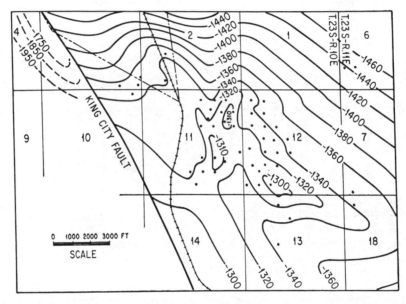

FIGURE 6-44. Structure contour map of the Lombardi sandstone in the San Ardo field in the Salinas Valley of California. The accumulation is controlled by the structural high and by the pinch-out of the Lombardi and Orradre sands along the King City Fault. (*Redrawn from Baldwin, Bull. AAPG, vol. 34, p. 1982.*)

that the structural traps in a new region are characteristically discovered before the stratigraphic traps.

Lateral Variations in Sedimentation. Several of the various ways in which lateral variations in sedimentation may contribute to the formation of traps are shown in idealized form in Figure 6-41. Perhaps the most common situation of this kind is the lateral "pinch-out" of a sand body. It is necessary that both the underlying and overlying strata be impermeable. If a homoclinally dipping sand body terminates updip along a curved edge, as shown in Figure 6-41a, a trap will be formed. If the updip edge is linear, it is necessary for some other factor such as a fault (Figure 6-41b) to be present before a trap exists.

A trap may also be formed by a lenticular sandstone being entirely surrounded by impermeable strata. The most common example of this situation is in "shoestring" sands in which the sand body is much longer than it is wide (Figure 6-41c). Deposits of this kind probably originate as stream channels or as offshore sand bars similar to those which occur today along the Atlantic and Gulf Coasts of the United States.

Although these kinds of traps are most commonly associated with sandstones and shales, they are not limited to these rocks. They might occur in any clastic sediments

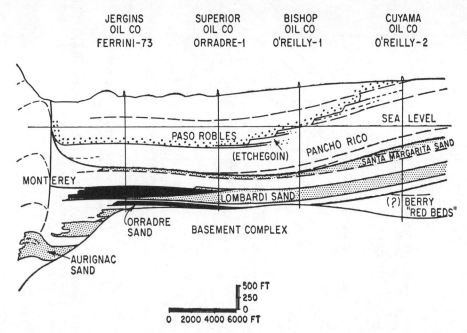

FIGURE 6-45. Southwest-northeast section through the San Ardo oil field. The Lombardi and Orradre sands grade laterally into Monterey shale. (*Redrawn from Baldwin, Bull. AAPG, vol. 34, p. 1986.*)

in which variations in permeability are possible, including organic clastic deposits, shell beds, breccias, arkoses, and many other kinds of rocks.

Some excellent examples of traps associated with sedimentary variations in sandstones and shales are depicted in some of the sections around the South Liberty salt-dome structure as shown in Figure 6-40. These sedimentary variations were undoubtedly caused by the presence of some topographic feature related to the dome itself at the time of deposition of the sediments.

Another good example of traps associated with lateral sedimentary variations is afforded by the Little Beaver oil field in the Denver Basin of Colorado which is depicted in Figures 6-42 and 6-43. Production in this field is obtained from two separate traps associated with updip pinch-outs in sandstones of the Dakota formation of Cretaceous age (Fentress, 1955).

A similar situation occurs in the San Ardo oil field located in the Salinas Valley of central California, which is depicted in Figures 6-44 and 6-45. Here, the Lombardi and Orradre sandstones of Miocene age grade laterally into siliceous Monterey shale

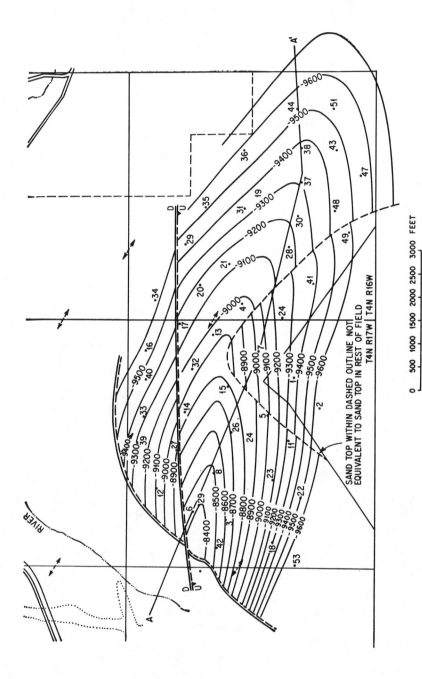

FIGURE 6-46. Contour map of the Castaic Junction oil field in the eastern Ventura Basin, California. Traps are formed in Miocene sands which pinch out updip to the northwest on a plunging anticlinal nose. (*Redrawn from Dudley, "Guidebook," p. 174, AAPG Pacific Section, 1958.*)

to the west.   The gradation in lithology is apparently related to ancient topography associated with the King City fault, which is just to the west of the productive limits of the reservoirs.   It is interesting to note that the oil-water contact in the Lombardi sand is tilted to the northwest, indicating a hydrodynamic situation (Baldwin, 1950).

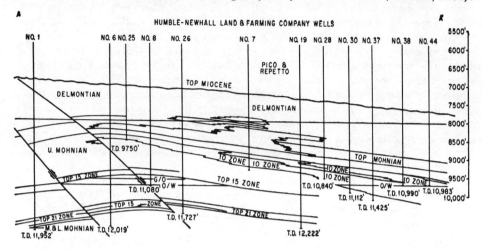

FIGURE 6-47. Structure section *A-A'* through the Castaic Junction oil field.   (*Redrawn from Dudley, "Guidebook," p. 175, AAPG Pacific Section, 1958.*)

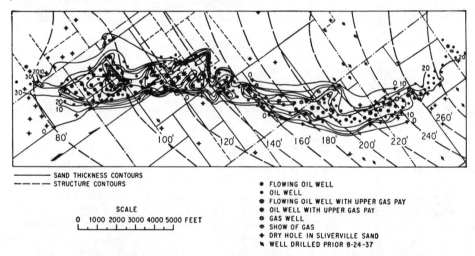

FIGURE 6-48. The Music Mountain oil field, McKean County, Pennsylvania.   The solid lines are isopach contours showing the thickness of the producing Sliverville sand; the dashed lines are structure contours on the top of the underlying third Bradford sand.   The producing sand body is apparently an ancient offshore sand bar.   (*Redrawn from Fettke, in "Stratigraphic Type Oil Fields," pp. 493, 498, AAPG, 1941.*)

Another example of traps associated with lateral variation in sand deposition is found in the Castaic Junction oil field located at the eastern end of the Ventura Basin in California (Dudley, 1958) (Figures 6-46 and 6-47).   Traps are formed by lensing sands of Miocene age which pinch out to the northwest across a southeastward plunging anticlinal nose.

An example of a trap formed by a shoestring sand entirely surrounded by shales is afforded by the Music Mountain oil field in northwestern Pennsylvania (Fettke, 1941), which is depicted in Figure 6-48. Isopach contours of the thickness of the producing Sliverville sandstone of Upper Devonian age are shown. Dashed structure contours are shown on the top of the Bradford third sand, which lies 240 feet below the Sliverville. It is apparent that the accumulation bears very little relation to the structure and is controlled primarily by the presence of the sand body.

Unconformities. In general, there are two ways in which unconformities may contribute to the formation of traps. These are depicted in idealized form in Figure 6-49. Perhaps the most common occurrence is the truncation of a permeable stratum by the unconformity with subsequent deposition of an impermeable stratum on the truncated edge. Of course, for a trap to be formed it is necessary that the permeable stratum be both overlain and underlain by impermeable strata (Figure 6-49a). The second way in which unconformities may contribute to the formation of traps arises when permeable strata are laid down over and lapping against a tilted unconformity surface as shown in Figure 6-49b. Again, an impermeable stratum is required to be deposited over the

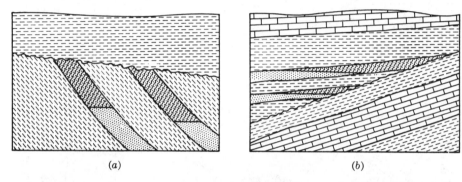

(a)   (b)

FIGURE 6-49. Schematic diagrams of traps associated with unconformities: (a) truncation of permeable strata and subsequent deposition of impermeable rocks on truncated edge, (b) deposition of "buttress sands" lapping on to the unconformity surface.

permeable one. Sandstones which occur in this configuration are frequently termed "buttress sands."

Perhaps the classic example of a trap formed by unconformity surfaces is that of the east Texas oil field (Minor and Hanna, 1941). This is illustrated in Figures 6-50 and 6-51. It is a single trap in the prolific Woodbine sandstone of northeastern Texas. Both the upper and lower surfaces of the trap are represented by unconformities. The regional Sabine uplift which culminates some 80 miles to the southeast has created a broad structural nose which together with the two unconformity surfaces forms a very large trap. This field is the largest yet discovered in the Western Hemisphere and has an expected ultimate recovery of about 6 billion barrels of oil.

It is interesting to note that traps associated both with lateral variations in sedimentation and with unconformity surfaces have a tendency to occur along the same trend lines. It has been suggested that regions such as these are likely to occur along the edge of a basin shelf in the transition zone between thinly deposited strata on the shelf itself and the more thickly deposited strata toward the middle of the basin. This transition zone is therefore frequently termed a "hinge line" and is regarded as a favorable location for the occurrence of traps.

An example of a probable hinge-line region is afforded by the huge Pembina accumulation in western Canada (Patterson and Arneson, 1957) which is depicted in Figures 6-52 and 6-53. Multiple traps are present, and factors which have contributed

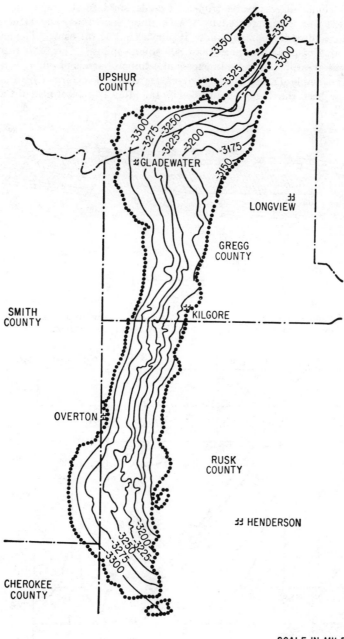

FIGURE 6-50. Structure contours of the producing Woodbine sand in the east Texas oil field. The trap is formed in the Woodbine by two bounding unconformity surfaces which have been gently arched by the Sabine uplift to the east. (*Redrawn from Minor and Hanna, in "Stratigraphic Type Oil Fields," p. 617, AAPG, 1941.*)

toward them include buttress sands, truncated sands, and lateral changes in lithology. The largest reservoir is represented by the Cardium sandstone, and although it is locally irregular and variable in thickness, it pinches out to the east. The entrapped oil covers an area of probably more than 800 square miles. The total recoverable reserves have been estimated to be in excess of 1 billion barrels of oil.

An excellent example of a field in which several traps have been formed by unconformity surfaces as well as other factors is afforded by the Midway-Sunset field of California,

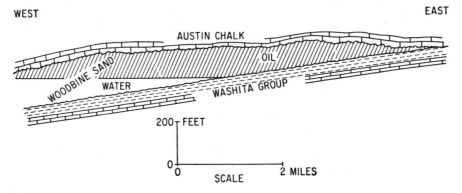

FIGURE 6-51. Schematic structure section across the east Texas oil field showing intersection of unconformity surfaces above and below the Woodbine reservoir.

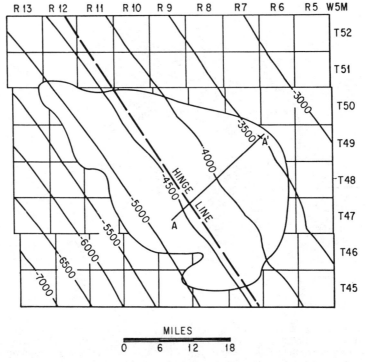

FIGURE 6-52. Map of the Pembina oil field, Alberta, Canada. The outline indicates the probable productive limits of the field. The contours are on the Mississippian Banff formation, and a "hinge-line" region is suggested by the steepening of the regional dip. (*Redrawn from Patterson and Arneson, Bull. AAPG, vol. 41, p. 947.*)

of which a generalized structure section is shown in Figure 6-54. In this field, anticlinal structure, truncations, buttress sands, and pinch-outs have all contributed to the formation of multiple traps in Pliocene and Miocene sandstones (Hoots et al., 1954).

Organic Deposits. Organic sedimentary structures have been responsible for forming the traps for many significant accumulations of petroleum. They are frequently

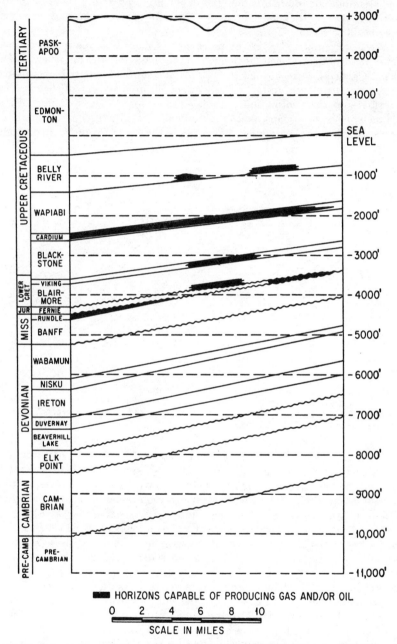

FIGURE 6-53. Schematic section across the Pembina oil field showing stratigraphic relations without structure. Horizons capable of producing oil or gas are indicated in black. (*Redrawn from Patterson and Arneson, Bull. AAPG, vol. 41, p. 940.*)

divided into two classes called biostromes and bioherms (Cumings and Shrock, 1928). Biostromes are simply bedded layers containing a high percentage of shells and other similar organic remains. If they are permeable, biostromes may form traps in the same way as other permeable sedimentary strata. On the other hand, bioherms, or reefs as they are commonly called, are moundlike structures which are the result of growth and accumulation of organic material in one localized area. These structures may form traps wherever they have developed sufficient permeability and are covered by an impermeable stratum.

Organic reefs are found today in the process of formation at many places in the world, and study of them has shed much light on the characteristics of ancient or fossil reefs which, in general, were formed in a similar fashion. Modern bioherms have popularly been termed "coral reefs." However, corals are only one of the organisms which contribute to their formation. Among others are algae, sponges, crinoids, mollusks, brachiopods, bryozoans, and foraminifera. In spite of the wide variety of different organisms and the consequent differences in the structure and form of modern

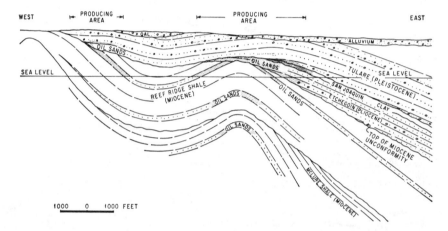

FIGURE 6-54. Schematic section through the Midway-Sunset oil field, San Joaquin Valley, California. Traps are formed by truncation, buttress sands, pinch-outs, and anticlinal structure in Pleistocene, Pliocene, and Miocene sands. (*Redrawn from Hoots, Bear, and Kleinpell, Calif. Div. Mines Bull. 170, chap. 9, p. 31.*)

reefs, there seem to be a number of common and essential factors for their formation. Reasonably warm water on the order of 65 to 70°F is required. The organisms also need some light, and as a result, reefs grow only in relatively shallow water, certainly not more than 50 fathoms. Because of this depth requirement reefs are frequently, although not always, associated with shore lines. Their size and shape may vary greatly. At one extreme are structures many miles in length and width such as the Great Barrier Reef along the northern coast of Australia. This extends for a length of over 1,000 miles and averages nearly 100 miles in width. At the other extreme are structures such as the typical South Pacific coral atoll with its sandy island and surrounding circular reef. Reefs may also occur in isolated moundlike masses.

Ancient reefs have been found throughout the world in sediments of every geologic age from the Cambrian onward. Although many of these were formed by different organisms from those found in modern reefs, their characteristics seem to be quite similar. Although they are not always productive, reef structures have accounted for some of the most prolific oil fields which have yet been discovered. This may perhaps be due to the nature of the permeability and porosity which are commonly found in

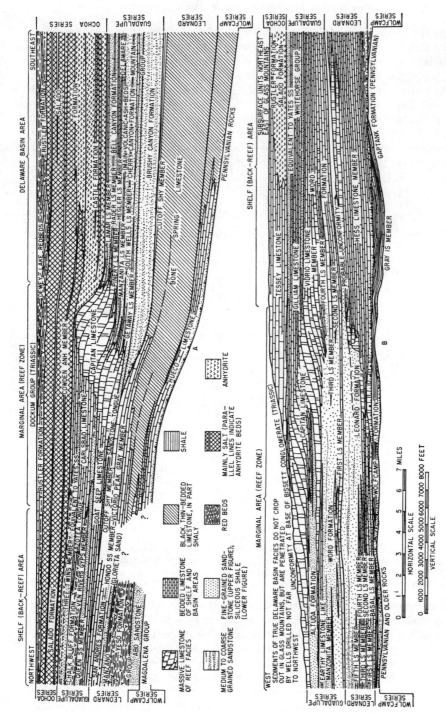

FIGURE 6-55. Stratigraphic diagrams of the Capitan Reef and adjacent formations in west Texas. The upper section is from the Guadalupe Mountains and is drawn normal to the strike of the reef trend. The lower section is from the Glass Mountains and is drawn obliquely to the strike of the reef trend. (*Redrawn from King, Bull. AAPG, vol. 26, p. 542.*)

473

them. In the development of the reef itself, many of the organisms involved grow in a branching structure which produces large numbers of cavities. Often much of the reef material is the result of deposition and accumulation of coarse permeable clastic material which has been broken off by wave action. Some of the organisms which live in the reef are of the kind which bore through the solid material, forming natural cavities with extremely high permeability. All these factors can be enhanced by secondary effects after the reef has been buried. For example, solution of the reef material may occur through the agency of circulating ground water, or as is frequently the case the reef may become dolomitized. Not all the factors involved contribute to the formation of high permeability and porosity, and there are a number of processes

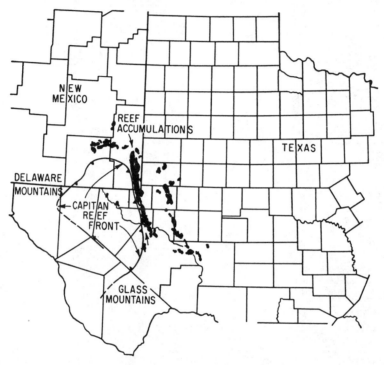

FIGURE 6-56. Regional map of west Texas and southeastern New Mexico showing oil and gas fields producing from Permian upper Guadalupe strata and the location of the Capitan Reef front. The line of fields adjacent to the reef front produces from the Capitan limestone. (*Redrawn from Galley, in "Habitat of Oil," p. 438, AAPG, 1958.*)

such as recrystallization of the reef material and the growth of other kinds of organisms which may tend to fill up cavities and voids which are present. In general, however, reefs are usually highly permeable and porous, giving rise to prolific accumulations.

On the other hand reef traps characteristically are very difficult to discover. They usually do not represent large or significant changes in the physical properties of the rocks in which they occur and are therefore not easily detected by geophysical or structural methods. Many of the known reef traps have probably been discovered by accident in the search for other kinds of traps. A possible tool which can be used in the exploration for reefs is based on the probability of their occurrence in association with shore lines. Analysis of already discovered reef traps has revealed that sediments of the same age on the basinward, or fore-reef, side of a bioherm are characteristically a deeper water marine deposit whereas those on the landward, or back-reef, side are frequently evaporites, red beds, or shallow-water or continental deposits. Many of these features can be correlated with those found in modern reef structures. The location of an undiscovered reef might well be revealed by very careful studies of the regional facies relationships of various strata (Cloud, 1952).

One of the most spectacular examples of a reef structure which has formed several prolific oil fields is afforded by the Capitan Reef in the Permian Basin of west Texas and southeastern New Mexico (Lloyd, 1929; King, 1938, 1942, 1948; Galley, 1958). Two sections through the reef drawn from its surface outcrops are depicted in Figure 6-55, and a map showing its regional trend is presented in Figure 6-56. As shown on the map, it forms a circular arc commencing in the Glass Mountains of west Texas, extending along the Pecos River and into southeastern New Mexico and thence back into west Texas, where it culminates at present in the massive exposed topographic feature known as El Capitan. The map shows oil accumulations in upper Guadalupe strata which include the Capitan Reef formation. The first line of oil fields immediately adjacent to the reef front occur within the reef itself. The Capitan Reef is particularly interesting, since it has been studied extensively and its outcrop constitutes one of the best known examples of an exposed fossil reef structure.

Another example of a series of prolific reef traps is the Devonian reef reservoirs of Alberta Province in western Canada (Downing and Cooke, 1955). Figure 6-57 shows structure contours on the surface of a portion of the Leduc trend, and Figure 6-58 a section indicating the relation of the reef to surrounding sediments. Some of the individual fields along this trend are expected to produce over 500 million barrels of oil.

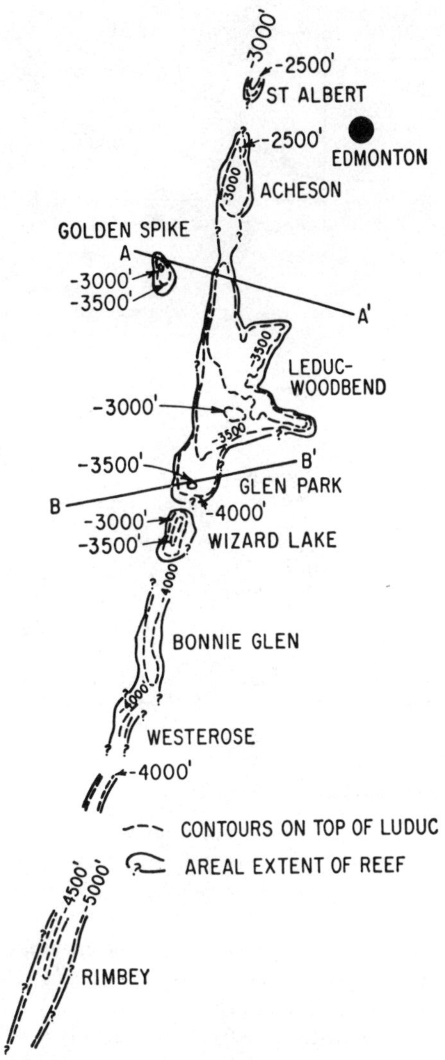

Another spectacular example of oil accumulation in a series of reef traps is afforded by the Golden Lane trend (Guzman and Mina U., 1956; Rockwell and Rojas, 1953) on the eastern coast of Mexico which is depicted in Figure 6-59. Here production is from the El Abra reef limestone of Cretaceous age and is expected to reach well over a billion barrels of oil. A similar reeflike structure accounts for the Poza Rica field shown several miles to the west of the Golden Lane trend. This field itself is expected to produce more than a billion barrels of oil.

FIGURE 6-57. Map of the Rimbey-Leduc-Acheson Devonian reef trend in Alberta, Canada. The areal extent of the reefs together with structural contours on their surface are shown. Only the structurally highest portions of the reefs are productive. (*Redrawn from Downing and Cooke, Bull. AAPG, vol. 39, p. 198.*)

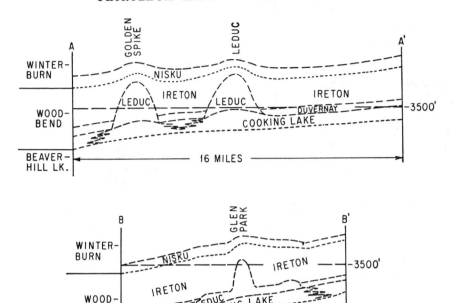

FIGURE 6-58. Schematic cross sections of Devonian reefs in the Leduc area. (*Redrawn from Downing and Cooke, Bull. AAPG, vol. 39, p. 200.*)

Miscellaneous Factors

There are a number of factors capable of producing traps which are of sufficiently unusual character that they can best be discussed separately from the structural and stratigraphic factors which have already been described. Specifically, these are the fractured reservoirs, traps formed by secondary chemical alteration of limestones to dolomites producing changes in permeability, and traps formed by the sealing of permeable strata by asphalt near the surface of the ground.

Fractured Reservoirs. A great many oil fields produce from fractured reservoirs. In many of these, the fracturing is only a secondary factor in the formation of the trap, whereas in a few it is a primary and essential feature.

An interesting example of a trap formed principally by a fractured reservoir is the huge Spraberry accumulation in west Texas (Figure 6-60). This field covers an area of nearly 1,000 square miles, and as shown in the figure, it bears no direct relationship to the structure. The oil is produced from the Spraberry sandstone of Lower Permian age. The average permeability of the unfractured reservoir rock is less than 1 millidarcy, and production is obtained principally by means of an extensive system of vertical fractures occurring within the formation. Most of the wells within the field have required artificial stimulation by means of hydraulic fracturing before significant production could be obtained.

Another interesting example of a fractured reservoir is afforded by the old Florence oil field which produces from fractures within the Upper Cretaceous Pierre shale at the southern end of the Denver Basin in Colorado (de Ford, 1929; McCoy et al., 1951).

A section through this old Florence oil field is depicted in Figure 6-61. Production here is entirely from fractures within the shale, and no water accompanies the oil. Apparently, solid unfractured shale surrounds the entire reservoir and forms the trap.

Secondary Chemical Alteration. Another factor sometimes contributing to the formation of traps is the increase in porosity and permeability which often occurs when

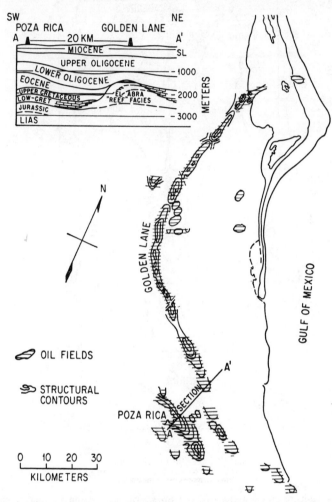

FIGURE 6-59. The Golden Lane trend on the eastern coast of Mexico. Schematic contours are shown on the top of the Cretaceous El Abra reef limestone. (*Redrawn from Guzman and Mina U., Bull. AAPG, vol. 40, p. 1492, and Rockwell and Rojas, ibid., vol. 37, p. 2555.*)

limestones undergo secondary alteration to dolomite (Landes, 1946). This process appears to occur by the replacement of some of the calcium within the limestone by magnesium carried by moving ground water. The exact mechanism and the manner in which permeability and porosity are thereby created are not well understood.

Excellent examples of traps occurring within patches of permeable dolomitized limestone are afforded by the individual pools of the Lima-Indiana oil field of Ohio and Indiana (Figure 6-64). The oil occurs in the Ordovician Trenton limestone and is almost entirely controlled by local dolomitization (Orton, 1888, 1889; Carman and

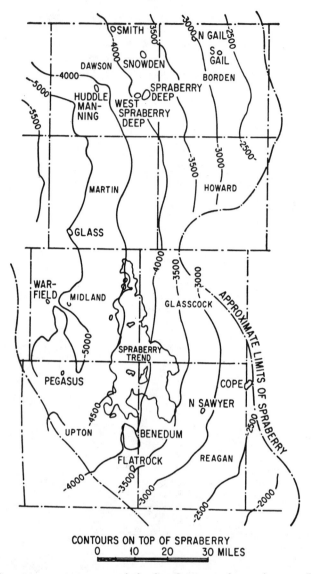

FIGURE 6-60. Structure contour map of the Spraberry sandstone in west Texas showing areas of oil accumulation. (*Redrawn from Hubbert and Willis, Proc. 4th World Petrol. Congr., Sect. I/A/1, p. 69.*)

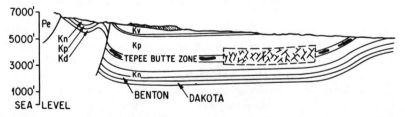

FIGURE 6-61. Section through the Florence oil field, Colorado. The trap is formed by fractures within the Cretaceous Pierre shale. No water is produced with the oil. (*Redrawn from McCoy et al., Bull. AAPG, vol. 35, p. 1008.*)

478

Stout, 1934). The accumulations are also interesting in that they appear to be influenced by hydrodynamic effects over a large region, as described below.

Asphaltic Seals. One unusual factor which can contribute to the formation of traps is the plugging of the pores of an otherwise permeable rock by asphaltic material. This situation is likely to occur where a sandstone or other impermeable rock outcrops at the surface and oil which has moved up the dip comes in contact with the atmosphere and is oxidized and thickened to form an asphaltic material. Some of the accumulations associated with the Midway-Sunset oil field in the San Joaquin Valley of California (Figure 6-54) have been formed in this way.

Hydrodynamic Effects

In the first section of this chapter the physical principles governing the accumulation of oil were reviewed and it was shown that under hydrostatic conditions in which the environmental ground water is not in motion, traps will be formed by any impermeable structural surface which is concave downward overlying a reservoir rock.

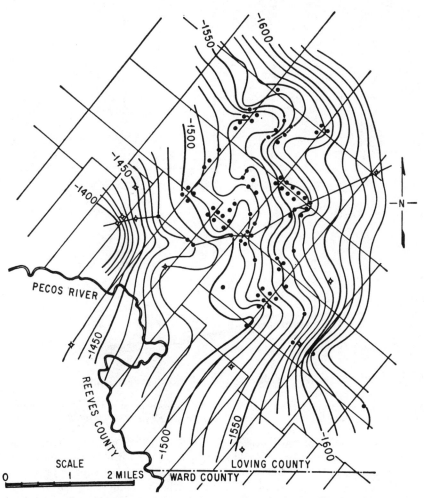

FIGURE 6-62. Structure contour map of the Wheat field in the Delaware Basin of west Texas. The trap is probably formed by a hydrodynamic gradient across the open structural terrace. (*Redrawn from Adams, Bull. AAPG, vol. 20, pp. 780–796.*)

However, under conditions of hydrodynamics in which the environmental ground water is in motion, traps will be formed by structural geometry which is concave in the direction of increasing potential for oil or gas. In general, the effects of hydrodynamics may range from a simple tilting of the oil-water or gas-water interface in the direction of water flow to a profound modification of the position of the trap to the extent that normally closed structures will not hold oil and that normally unclosed structures such as anticlinal noses will form excellent traps.

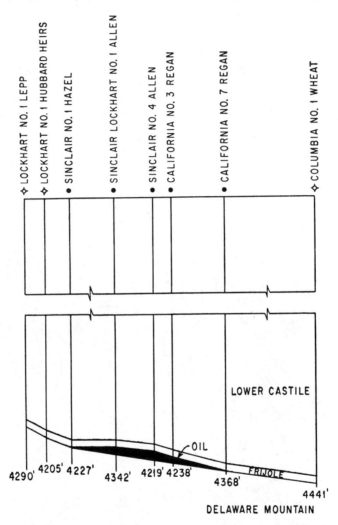

FIGURE 6-63. Section through the Wheat field. Wells to the west of the accumulation show good permeability and porosity. (*Redrawn from Adams, Bull. AAPG, vol. 20, pp. 780–796.*)

Some of the examples of traps already presented show the effects of hydrodynamic conditions. In the Kettleman Hills oil field (Figure 6-17), the oil-water contact shows a tilt of 50 feet per mile in a northwesterly direction, indicating that water is flowing through the reservoir sands in that direction. The water in these sands is of low salinity, providing further evidence of a hydrodynamic situation. The San Ardo oil field in the Salinas Valley of California (Figures 6-44, 6-45) also shows the effects of hydrodynamics. The oil-water contact in the Lombardi sand is tilted approximately

40 feet per mile to the northwest. A map showing contours on the oil-water contact has been presented by Baldwin (1950).

In these examples hydrodynamic effects are represented by relatively gentle tilts of the oil-water contact. However, in each case the difference in the area covering the

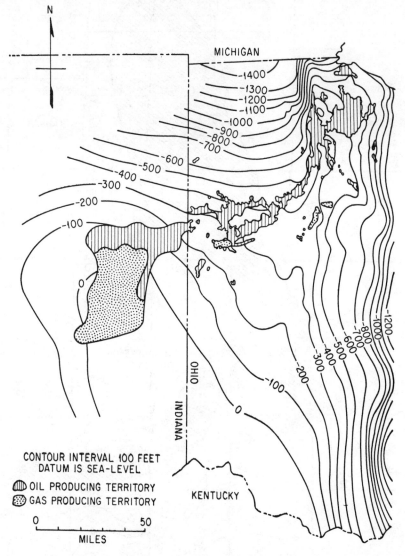

FIGURE 6-64. Map of the Lima-Indiana oil field of Indiana and Ohio showing structure contours on the surface of the Trenton limestone. Production is locally controlled by dolomitization and permeability of the Trenton but regionally appears to be affected by a hydrodynamic gradient. (*Redrawn from Carman and Stout, in "Problems of Petroleum Geology," p. 522, AAPG, 1934.*)

existing trap and that which would cover a hydrostatic trap containing the same amount of oil represents many thousands of acres and is therefore extremely significant in the exploitation of the accumulations.

There exist a number of more spectacular examples of hydrodynamic traps. One of these is the Wheat field in the Delaware Basin of west Texas (Adams, 1936). This is

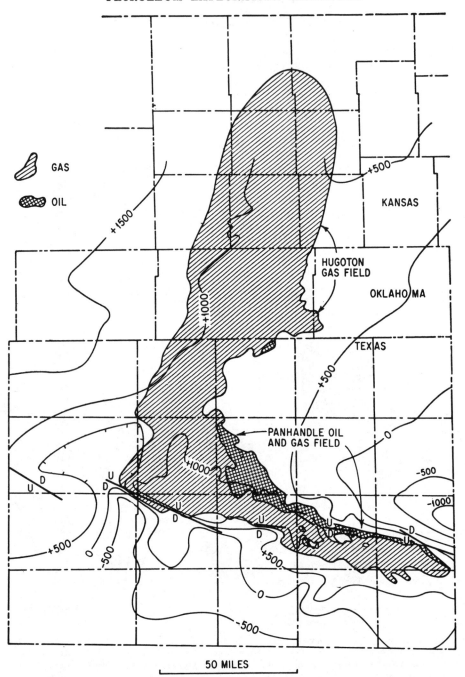

FIGURE 6-65. Map of the Hugoton gas field and the Panhandle oil and gas fields. Structure contours are shown on the top of the "Big Lime." The traps are apparently strongly modified by hydrodynamic effects with both the oil-water and gas-water contacts tilted eastward. The gas-oil contact is horizontal, and the oil is in the lowest structural position of both reservoirs. (*Compiled from Lilly, Bull. AAPG, vol. 42, p. 1235; Hemsell, ibid., vol. 23, p. 1055; and Cotner and Crum, in "Geology of Natural Gas," p. 387, AAPG, 1935.*)

illustrated in Figures 6-62 and 6-63 showing a structure map and a section through the accumulation. Oil is trapped in a sandstone stratum of the Delaware Mountain series of Permian age. The regional dip is to the east, and the field is located on a structural terrace having no structural closure to the west. There is no indication of faulting, and wells, as shown in the diagram, on the west and southwest of the field show excellent permeability within the reservoir rock. These facts have been pointed out by Adams in his discussion of the field, and it appears highly probable that the accumulation is a hydrodynamic one. This interpretation is consistent with the regional topography, since the Delaware Mountains to the west would be a natural source of water flowing eastward into the Delaware Basin.

Another very interesting example is afforded by the old Lima-Indiana oil field which extends for a distance of about 150 miles from central Ohio into eastern Indiana (Figure 6-64). It actually consists of a great many separated areas of production. The oil has accumulated in the 600-foot-thick Trenton limestone of Ordovician age. The field was discovered in 1885 and was described in detail by Edward Orton (1888, 1889). The oil-water contact occurs at about 650 feet below sea level in central Ohio and gradually rises to the southwest until it is about 50 feet above sea level in Indiana. This evidence, together with the presence of gas in the structurally highest portion of the field in eastern Indiana, indicates the strong possibility of hydrodynamic tilt. Another major factor in determining the location of the accumulation is the dolomitization of the Trenton limestone. The actual producing areas appear to be closely correlated with this chemical alteration. However, they appear to be interconnected as evidenced by the related oil-water contacts. Therefore the principal effect of the dolomitization is apparently the creation of sufficient local permeability to permit commercial production. The over-all tilt of the oil-water contact is on the order of 2 or 3 feet per mile. This effect could be produced by a hydrodynamic gradient of a few tenths of a foot per mile, a situation which could easily be present in this region.

Another interesting example of a probable hydrodynamic trap of very large dimensions is afforded by the Hugoton gas field and the adjacent Panhandle oil and gas field which occur in western Kansas and in the Panhandles of Oklahoma and Texas (Rogatz, 1939; Hemsell, 1939; Garlough and Taylor, 1941; Cotner and Crum, 1935; Lilly, 1958). A map of this region is depicted in Figure 6-65 showing the oil- and gas-producing areas and regional structural contours on the top of the Panhandle Big Lime formation of Permian age. Production is obtained from Permian and Pennsylvanian rocks. The initial gas pressure within both the Hugoton and the Panhandle fields was very nearly constant throughout, ranging from approximately 430 to 440 psi. No structural closure exists to the west of the Hugoton field, and this trap has been classed as a stratigraphic one on the basis that the limestone and dolomite reservoir rocks grade into red shales and silts some 20 miles west of the field. However, according to Garlough and Taylor (1941), the gas-water contact varies in elevation from about 500 feet above sea level on the west edge of the field to about 120 feet above sea level on the eastern edge of the field with an average tilt of 12 feet per mile. Since the gas, initially, was at nearly constant pressure, the inclined gas-water contact was an equipressure surface. Thus the equipressure surfaces in the water must have been tilted downward toward the east. This could happen only if water were flowing down the dip from the Rocky Mountain region to the west. This is strong evidence of a hydrodynamic condition which could arise from water flowing down the dip from the Rocky Mountain region to the west. A possible mechanism for the accumulation may be a lower permeability in the reservoir rocks just to the west of the field, creating a situation similar to that depicted in Figure 6-14. The associated Panhandle oil and gas field shows a tilt of the oil-water contact to the east on the order of 10 feet per mile. Considering the two fields together, the oil is found in the structurally lowest portion of both with an essentially horizontal contact between the gas and the oil, as should be expected if the two fields jointly form a single trap. It is interesting to note that in all probability any oil which ever entered the trap migrated in it to the lowest portion, ending up in Panhandle oil accumulation.

SUMMARY

At the beginning of this chapter, it was pointed out that among the most valuable tools which can be possessed by the petroleum exploration geologist are an understanding of the physical principles covering the accumulation of petroleum and natural gas together with a wide and thorough acquaintance with the geological conditions under which oil and gas have accumulated in already discovered traps.

It was shown that traps for oil and gas consist of underground regions in which the potential energies of these fluids are at a minimum with respect to their surroundings. Under hydrostatic conditions in which the environmental ground water is not in motion, traps will be formed in rocks permeable to oil and gas wherever they are bounded by impermeable surfaces which are concave downward. Under hydrodynamic conditions in which the environmental ground water is in motion, traps will be formed whenever the impermeable surface is concave in the direction of increasing potential for the particular fluid. The respective traps for oil and gas will not coincide except under hydrostatic conditions.

There are about a dozen different geologic factors which may contribute to the proper geometry for an oil or gas trap. There are literally hundreds of different ways in which combinations of these factors may occur, and examples have been presented of some of the most common of these. Undoubtedly traps are yet to be found which can combine these factors and perhaps new ones in new and unexpected ways.

REFERENCES

Adams, John Emery: Oil Pool of Open Reservoir Type, *Bull. AAPG*, vol. 20, pp. 780–796, 1936.
Baldwin, Thomas A.: San Ardo—A Stratigraphic Analysis of a California Oil Field, *Bull. AAPG*, vol. 34, pp. 1981–1989, 1950.
Bitterli, Peter: Herrera Subsurface Structure of Penal Field, Trinidad, B.W.I., *Bull. AAPG*, vol. 42, pp. 145–158, 1958.
Carman, J. Ernest, and Wilbur Stout: Relationship of Accumulation of Oil to Structure and Porosity in the Lima-Indiana Field, in "Problems of Petroleum Geology," pp. 521–529, AAPG, Tulsa, Okla., 1934.
Clapp, Frederick G.: Role of Geologic Structure in the Accumulation of Petroleum, in "Structure of Typical American Oil Fields," vol. 2, pp. 667–716, AAPG, Tulsa, Okla., 1929.
Cloud, Preston E.: Facies Relationships of Organic Reefs, *Bull. AAPG*, vol. 36, pp. 2125–2149, 1952.
Cotner, Victor, and H. E. Crum: Geology and Occurrence of Natural Gas in Amarillo District, Texas, in "Geology of Natural Gas," pp. 385–415, AAPG, Tulsa, Okla., 1935.
Cumings, E. R.: Reefs or Bioherms? *Bull. Geol. Soc. Am.*, vol. 43, pp. 345–347, 1932.
Cumings, E. R., and Robert R. Shrock: Niagaran Coral Reefs of Indiana and Adjacent States and Their Stratigraphic Relations, *Bull. Geol. Soc. Am.*, vol. 39, p. 599, 1928.
Darton, N. H.: Preliminary Report of the Geology and Underground Water Resources of the Central Great Plains, *U.S. Geol. Survey Profess. Paper 32*, 433 pp., 1905.
Darton, N. H.: Geology and Underground Waters of South Dakota, *U.S. Geol. Survey Water Supply Paper 227*, 156 pp., 1909.
Darton, N. H.: Artesian Waters in the Vicinity of the Black Hills, South Dakota, *U.S. Geol. Survey Water Supply Paper 428*, 64 pp., 1918.
Darton, N. H.: The Structure of Parts of the Central Great Plains, *U.S. Geol. Survey Bull. 691*, pp. 1–26, 1918.
de Ford, Ronald K.: Surface Structure, Florence Oil Field, Fremont County, Colorado, in "Structure of Typical American Oil Fields," vol. 2, pp. 75–92, AAPG, Tulsa, Okla., 1929.

Downing, John A., and D. Y. Cooke: Distribution of Reefs of Woodbend Group in Alberta, Canada, *Bull. AAPG*, vol. 39, pp. 189–206, 1955.

Dudley, Jr., P. H.: The Castaic Junction Oil Field, in "A Guide to the Geology and Oil Fields of the Los Angeles and Ventura Regions," pp. 173–176, Pacific Section, AAPG, Los Angeles, 1958.

Dunnington, H. V.: Generation, Migration, Accumulation, and Dissipation of Oil in Northern Iraq, in "Habitat of Oil," pp. 1194–1251, AAPG, Tulsa, Okla., 1958.

Fentress, George H.: Little Beaver Field, Colorado, A Stratigraphic, Structural, and Sedimentation Problem, *Bull. AAPG*, vol. 39, pp. 155–188, 1955.

Fettke, Charles R.: Music Mountain Oil Pool, McKean County, Pennsylvania, in "Stratigraphic Type Oil Fields," pp. 492–506, AAPG, Tulsa, Okla., 1941.

Fortier, Y. O., A. H. McNair, and R. Thorsteinsson: Geology and Petroleum Possibilities in Canadian Arctic Islands, *Bull. AAPG*, vol. 38, pp. 2075–2109, 1954.

Galley, John E.: Oil and Geology in the Permian Basin of Texas and New Mexico, in "Habitat of Oil," pp. 395–446, AAPG, Tulsa, Okla., 1958.

Gallup, W. B.: Geology of Turner Valley Oil and Gas Field, Alberta, Canada, *Bull. AAPG*, vol. 35, no. 4, pp. 797–821, 1951.

Garlough, John L., and Garvin L. Taylor: Hugoton Gas Field, Grant, Haskell, Morton, Stevens, and Seward Counties, Kansas, and Texas County, Oklahoma, in "Stratigraphic Type Oil Fields," pp. 78–104, AAPG, Tulsa, Okla., 1941.

Graves, Doyle, T.: Geology of the Dominguez Oil Field, Los Angeles County, Map Sheet No. 33, in "Geology of Southern California," *Calif. Dept. Nat. Resources, Div. Mines Bull.* 170, San Francisco, 1954.

Guzman, Edwardo J., and Frederico Mina U.: Petroleum Developments in Mexico in 1955, *Bull. AAPG*, vol. 40, pp. 1485–1497, 1956.

Halbouty, Michel T., and George C. Hardin, Jr.: Types of Hydrocarbon Accumulation and Geology of South Liberty Salt Dome, Liberty County, Texas, *Bull. AAPG*, vol. 35, pp. 1939–1977, 1951.

Hanna, Marcus A.: Geology of the Gulf Coast Salt Domes, in "Problems of Petroleum Geology," pp. 629–693, AAPG, Tulsa, Okla., 1934.

Hedberg, H. D., L. C. Sass, and H. J. Funkhouser: Oil Fields of Greater Oficina Area, Central Anzoategui, Venezuela, *Bull. AAPG*, vol. 31, pp. 2089–2169, 1947.

Hemsell, Clenon C.: Geology of Hugoton Gas Field of Southwestern Kansas, *Bull AAPG*, vol. 23, pp. 1054–1067, 1939.

Hoots, Harold W., Ted L. Bear, and William D. Kleinpell: Stratigraphic Traps for Oil and Gas in the San Joaquin Valley, in "Geology of Southern California," chap. 9, Oil and Gas, pp. 29–32, *Calif. Dept. Nat. Resources, Div. Mines Bull.* 170, 1954.

Howell, J. V.: Historical Development of the Structural Theory of Accumulation of Oil and Gas, in "Problems of Petroleum Geology," pp. 1–23, AAPG, Tulsa, Okla., 1934.

Hubbert, M. King: The Theory of Ground-water Motion, *J. Geol.*, vol. 48, no. 8, pp. 785–944, 1940.

Hubbert, M. King: Entrapment of Petroleum under Hydrodynamic Conditions, *Bull. AAPG*, vol. 37, no. 8, pp. 1954–2026, 1953.

Hubbert, M. King, and David G. Willis: Important Fractured Reservoirs in the United States, *Proc. 4th World Petrol. Congr.*, Section I/A/1, Rome, 1955.

Janoschek, Robert: The Inner-Alpine Vienna Basin, in "Habitat of Oil," pp. 1134–1152, AAPG, Tulsa, Okla., 1958.

King, Philip B.: Geology of the Marathon Region, Texas, *U.S. Geol. Survey Profess. Paper* 187, 1938.

King, Philip B.: Permian of West Texas and Southeast New Mexico, *Bull. AAPG*, vol. 26, pp. 535–763, 1942.

King, Philip B.: Geology of the Southern Guadalupe Mountains, Texas, *U.S. Geol. Survey Profess. Paper* 215, 1948.

Knebel, G. M., and Guillermo Rodriguez-Eraso: Habitat of Some Oil, *Bull. AAPG*, vol. 40, pp. 547–561, 1956.

Landes, Kenneth K.: Porosity through Dolomitization, *Bull. AAPG*, vol. 30, pp. 305–318, 1946.

Lilly, Russell N.: Developments in Texas and Oklahoma Panhandle in 1957, *Bull. AAPG*, vol. 42, pp. 1234–1247, 1958.

Link, Theodore A.: Interpretations of Foothills Structures, Alberta, Canada, *Bull. AAPG*, vol. 33, no. 9, pp. 1475–1501, 1949.

Lloyd, E. Russell: Capitan Limestone and Associated Formations of New Mexico and Texas, *Bull. AAPG*, vol. 13, pp. 645–658, 1929.

McCoy, III, Alex W., Robert L. Sielaff, George R. Downs, N. Wood Bass, and John H. Maxson: Types of Oil and Gas Traps in the Rocky Mountain Region, *Bull. AAPG*, vol. 35, p. 1000, 1951.

Mills, R. Van A.: Experimental Studies of Subsurface Relationships in Oil and Gas Fields, *Econ. Geol.*, vol. 15, no. 5, pp. 398–421, 1920.

Minor, H. E., and Marcus A. Hanna: East Texas Oil Field, Rusk, Cherokee, Smith, Gregg, and Upshur Counties, Texas, in "Stratigraphic Type Oil Fields," pp. 600–640, AAPG, Tulsa, Okla., 1941.

Morales, Luis G., and the Colombian Petroleum Industry: General Geology and Oil Occurrences of Middle Magdalena Valley, Colombia, in "Habitat of Oil," pp. 641–695, AAPG, Tulsa, Okla., 1958.

Munn, Malcom J.: The Anticlinal and Hydraulic Theories of Oil and Gas Accumulation, *Econ. Geol.*, vol. 4, pp. 141–157, 1909.

Nettleton, L. L.: Fluid Mechanics of Salt Domes, in "Gulf Coast Oil Fields," pp. 79–108, AAPG, Tulsa, Okla., 1936.

Nettleton, L. L.: Recent Experimental and Geophysical Evidence of Mechanics of Salt-dome Formation, *Bull. AAPG*, vol. 27, pp. 51–63, 1943.

Orton, Edward: The Origin and Accumulation of Petroleum and Natural Gas, *Rept. Geol. Survey Ohio*, vol. 6, pp. 1–310, 1888.

Orton, Edward: The Trenton Limestone as a Source of Petroleum and Inflammable Gas in Ohio and Indiana, *U.S. Geol. Survey 8th Ann. Rept.*, part II, pp. 475–662, 1889.

Parker, Travis J., and A. N. McDowell: Scale Models as Guide to Interpretation of Salt-dome Faulting, *Bull. AAPG*, vol. 35, pp. 2076–2094, 1951.

Parker, Travis J., and A. N. McDowell: Model Studies of Salt Dome Tectonics, *Bull. AAPG*, vol. 39, p. 2384, 1955.

Patterson, A. M., and A. A. Arneson: Geology of Pembina Field, Alberta, *Bull. AAPG*, vol. 41, pp. 937–949, 1957.

Porter, Jr., Livingston: Kettleman Hills Oil Field, in "Guidebook," pp. 170–172, Pacific Section, AAPG, Los Angeles, 1952.

Reeves, Frank: Status of German Oil Fields, *Bull. AAPG*, vol. 30, pp. 1546–1584, 1946.

Rich, John L.: Moving Underground Water as a Primary Cause of the Migration and Accumulation of Oil and Gas, *Econ. Geol.*, vol. 16, no. 6, pp. 347–371, 1921.

Rich, John L.: Problems of the Origin, Migration and Accumulation of Oil, in "Problems of Petroleum Geology," pp. 337–345, AAPG, 1934.

Rockwell, D. W., and Antonio Garcia Rojas: Coordination of Seismic and Geologic Data in Poza Rica—Golden Lane Area, Mexico, *Bull. AAPG*, vol. 37, pp. 2551–2565, 1953.

Rogatz, Henry: Geology of the Texas Panhandle Oil and Gas Field, *Bull. AAPG*, vol. 23, pp. 983–1053, 1939.

Schneegans, Daniel: Gas-bearing Structures of Southern France, *Bull. AAPG*, vol. 32, no. 2, pp. 198–214, 1948.

Schwade, Irving T., Stanley A. Carlson, and James B. O'Flynn: Geologic Environment of Cuyama Valley Oil Fields, California, in "Habitat of Oil," pp. 78–98, AAPG, Tulsa, Okla., 1958.

Small, Walter M.: Thrust Faults and Ruptured Folds in Romanian Oil Fields, *Bull. AAPG*, vol. 43, pp. 455–471, 1959.

Smith, Jr., P. V.: Studies on Origin of Petroleum: Occurrence of Hydrocarbons in Recent Sediments, *Bull. AAPG*, vol. 38, no. 3, pp. 377–404, 1954.

Walters, Ray P.: Oil Fields of the Carpathian Region, *Bull. AAPG*, vol. 30, pp. 319–336, 1946.

Ware, Jr., G. C., and R. D. Stewart: Bridge Area, South Mountain Oil Field, in "A Guide to the Geology and Oil Fields of the Los Angeles and Ventura Regions," pp. 180–181, Pacific Section, AAPG, Los Angeles, 1958.

Weeda, Jan: Oil Basin of East Borneo, in "Habitat of Oil," pp. 1337–1346, AAPG, Tulsa, Okla., 1958.

Wendlandt, E. A., T. H. Shelby, Jr., and John S. Bell: Hawkins Field, Wood County, Texas, *Bull. AAPG*, vol. 30, pp. 1830–1856, 1946.

Wilhelm, O.: Classification of Petroleum Reservoirs, *Bull. AAPG*, vol. 29, pp. 1537–1579, 1945.

Willis, Robin, and Richard S. Ballantyne: Potrero Oil Fields, in "Geologic Formations and Economic Development of the Oil and Gas Fields of California," *Calif. Dept. Nat. Resources, Div. Mines Bull.* 118, pp. 310–317, 1943.

Womack, Jr., Robert: Brookhaven Oil Field, Lincoln County, Mississippi, *Bull. AAPG*, vol. 34, pp. 1517–1529, 1950.

Woodring, W. P., Ralph Stewart, and R. W. Richards: Geology of the Kettleman Hills Oil Fields, California, *U.S. Geol. Survey Profess. Paper* 195, 170 pp., 1940.

Copyright 1986© Society of Petroleum Engineers of AIME.
First published in the SPEFE, June 1986.

Hydrodynamic Conditions of Hydrocarbon Accumulation Exemplified by the Carboniferous Formation in the Lublin Synclinorium, Poland

Ludwik Zawisza, U. of Mining and Metallurgy, Poland

Summary. Groundwater movement in a selected aquifer of a sedimentary basin in the Lublin synclinorium is considered on the basis of rock and fluid data, and conditions of migration and accumulation of hydrocarbons are presented. The problem of groundwater movement is solved analytically, taking into consideration variable density of water and variable permeability of rocks.

A modified theory is advanced to define more precisely the position of potential oil and gas traps. The theory introduced here is more universal than Hubbert's theory because it accepts both the variability of the oil, gas, and water density (static effect) and the groundwater motion (dynamic effect). The suggested method of determining the positions of potential hydrodynamic petroleum traps is comparatively simple to apply and gives good results in regions with high hydraulic gradients and high variability of salinity and hydrocarbon densities.

Introduction

This paper (1) states and illustrates equations for determining rates and directions of flow in groundwater of variable density, (2) introduces a modified theory of mapping hydrodynamic petroleum entrapments, and (3) defines the conditions of migration and accumulation of oil and gas in a selected aquifer of the Carboniferous formation in the Lublin synclinorium located in eastern Poland (Fig. 1). For hydrodynamic research, the Carboniferous Aquifer G was selected because the oil and gas fields discovered up to now in this aquifer have tilted oil/water and gas/water contacts and are connected with the structural nose (Fig. 1).

Groundwater Movement in Carboniferous Aquifer G in Lublin Synclinorium

Derivation of Equations for Flow of Groundwater of Variable Density. The Carboniferous Aquifer G is nearly horizontal and has a negligible slope (Fig. 1). The aquifer is a sandstone complex of nearly constant thickness and permeability. The permeability changes continuously and very slowly from 5 to 70 md (Fig. 2). The maximum permeability is in the center of the basin and is reduced toward the margins. Fig. 2 shows the local minimum value of permeability connected with hydrocarbon accumulations.

The salinity of groundwater in Carboniferous Aquifer G (Fig. 3) varies greatly, from 0.8 to 150 g/L. We observed two areas of high salinity, northwest of Lublin and southeast of Zamość. Also, there is the local maximum value of salinity in the region of hydrocarbon accumula-

tion southeast of Lublin. The reduction in salinity toward the margins of the Carboniferous basin is caused by infiltration of fresh water along the Carboniferous outcrops (Figs. 3 and 4). Thus, the infiltration water defines the intake areas.

The relatively well-defined geological and hydrogeological characteristics of the analyzed area offer a comparatively clear-cut basis for the mathematical reasoning.

To predict the migration and accumulation of hydrocarbons, it is important to evaluate quantitatively the magnitude and direction of formation-fluid flows in aquifer systems and geologic basins.

Groundwater motion is controlled by the properties of reservoir rocks and the hydrodynamic gradient of the region. Included are such features of basins as hypsometry of intake and discharge areas.

The basic equations of groundwater flow are Darcy's law and the continuity equation:

$$\vec{v}_w = -\sigma_w \nabla H_w \quad \dots \dots \dots \dots \dots \dots \dots \dots (1)$$

and

$$\nabla \cdot \rho_w \vec{v}_w = \frac{\partial(\phi \rho_w)}{\partial t}, \quad \dots \dots \dots \dots \dots \dots (2)$$

where

$$H_w = z + \frac{p}{\rho_w g}. \quad \dots \dots \dots \dots \dots \dots \dots \dots (3)$$

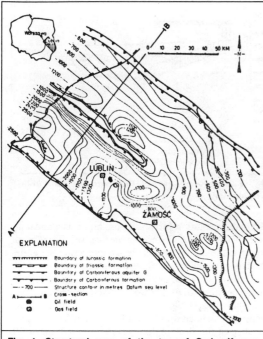

Fig. 1—Structural map of the top of Carboniferous Aquifer G.

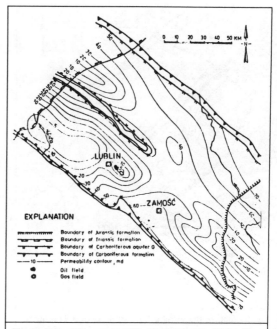

Fig. 2—Map of the permeability distribution for Carboniferous Aquifer G.

In these equations, $\vec{v}_w$ is the vector velocity of water, σ_w is the hydraulic conductivity, ∇ is the gradient operator, H_w is the hydraulic head, ρ_w is the density of water, ϕ is the porosity, t is time, g is the gravitational acceleration, p is the pressure, and z is the elevation of the point p above the standard datum (mean sea level).

With reference to regional groundwater motion, we assume that (1) the flow is steady state, (2) the fluid is noncompressible, and (3) compaction phenomena are negligible. Under these circumstances, the time derivative disappears. Using the simplified continuity equation $\nabla \cdot \vec{v}_w = 0$, we obtain

$$\nabla \cdot \sigma_w \nabla H_w = 0, \qquad \ldots \ldots \ldots \ldots \ldots \ldots \ldots \ldots \ldots (4)$$

which is Laplace's equation. Eq. 4 is the general filtration equation for steady flow of a homogeneous, noncompressible liquid in a porous medium.

When the salinity variations of groundwater in an aquifer system are considerable (i.e., greater than 10 g/L), velocity components are defined by the following equations,[1] provided that the x and y directions are taken to be horizontal and z is vertical.

$$v_x = -K_{11} \frac{g}{\mu_i} \left(\rho_{fw} \frac{\partial H_{ifw}}{\partial x} \right), \quad \ldots \ldots \ldots \ldots \ldots (5a)$$

$$v_y = -K_{22} \frac{g}{\mu_i} \left(\rho_{fw} \frac{\partial H_{ifw}}{\partial y} \right), \quad \ldots \ldots \ldots \ldots \ldots (5b)$$

and

$$v_z = -K_{33} \frac{g}{\mu_i} \left(\rho_{fw} \frac{\partial H_{inw}}{\partial z} \right), \quad \ldots \ldots \ldots \ldots \ldots (5c)$$

where v_x, v_y, and v_z are velocity components along the x, y, and z coordinates, respectively; K_{11}, K_{22}, and K_{33} are the principal directional permeabilities in the x, y, and z directions; i is any point in groundwater of variable density; μ_i is the dynamic viscosity at i; ρ_{fw} is the density of fresh water; H_{ifw} is the freshwater head at i; and H_{inw} is the environmental-water head at i. The point-water head at a point in groundwater of variable density is defined as the water level, referred to a given datum (mean sea level), in a well filled sufficiently with water of the type at the point to balance the existing pressure at the point. A freshwater head at any point i in groundwater of variable density is defined as the water level in a well filled with fresh water from i to a level high enough to balance the existing pressure at i. Freshwater heads define hydraulic gradients along a horizontal. Environmental water between a given point in a groundwater system and the top of the zone of saturation is defined as the water of constant or variable density occurring in the environment along a vertical between that point and the top of the zone of saturation. Environmental-water head at a given point in groundwater of variable density is defined as a freshwater head reduced by an amount that corresponds to the difference of salt mass in fresh water and in the environmental water between that point and the top of the zone of saturation. Environmental-water head at a point in groundwater of variable density may also be defined as a point-water head increased by an amount that corresponds to the difference of salt mass in salt water and in the environmental water between that point and the top of the zone of saturation. Environmental-water heads define hydraulic gradients along a vertical. Vertical and horizontal components of velocity in an anisotropic system with groundwater of variable density may be computed from hydraulic gradients defined by environmental-

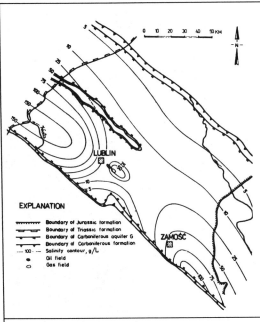

Fig. 3—Map of the salinity for Carboniferous Aquifer G.

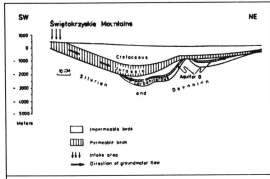

Fig. 4—Hydrogeologic section across the Lublin synclinorium.

water and freshwater heads, respectively, and from appropriate components of the permeability tensor.

If the medium is assumed to be isotropic and the permeability is replaced by the hydraulic conductivity, then Eqs. 5a and 5b may be rewritten in the given two-dimensional (2D) region—i.e., for horizontal velocity components—as

$$v_x = -\sigma_{fw}\frac{\partial H_{fw}}{\partial x}, \qquad \qquad (6a)$$

$$v_y = -\sigma_{fw}\frac{\partial H_{fw}}{\partial y}, \qquad \qquad (6b)$$

or

$$\vec{v} = -\sigma_{fw}\nabla H_{fw}(x,y), \qquad \qquad (7)$$

where σ_{fw} is the hydraulic conductivity for fresh water and H_{fw} is the freshwater head (a scalar force potential). Thus, the horizontal components of velocity for a given system with groundwater of variable density may be computed from Eqs. 6a and 6b, and Laplace's equation for the 2D region may be expressed as

$$\nabla \cdot \sigma_{fw}\nabla H_{fw}(x,y)=0. \qquad \qquad (8)$$

Boundary Conditions. The structural, permeability, and salinity maps (Figs. 1 through 3) were constructed on the basis of the rock and fluid data taken from about 80 wells. To construct the potentiometric map for the whole Aquifer G, however, it was possible to complete only 17 formation pressure measurements (Fig. 5). Because it was not sufficient to use the standard graphical-mapping method, the problem of groundwater movement was solved analytically.

To make a solution for Eq. 8 possible, it is mandatory that the value of H_{fw}, its normal derivative, $\partial H_{fw}/\partial n$, or a linear combination of both be known along the entire boundary containing the flow region.

On the basis of such input data as formation pressures, salinity variations of groundwater, and hydrochemical and permeability maps (Figs. 2 through 4), the following conditions were defined for Carboniferous Aquifer G: (1) the geometry of the flow region; (2) the distribution of the value H_{fw} along the entire boundary of the flow region; and (3) the permeability distribution. Furthermore, it was assumed that flow in the Carboniferous Aquifer G is confined, nearly horizontal, and of continuous permeability.

The geologic boundaries were approximated by a rectangular coordinate system. The origin of coordinates was located in Hole P-1 and the x-axis was led by Hole T-1 (Fig. 5). In this way, the analyzed region is limited by the rectangle having the dimensions $a \times b$ (200×100 km [124×62 miles]).

Next, for such a defined region, the boundary conditions were stated and described (Dirichlet problem), thereby illustrating the distribution of H_{fw} along the entire boundary of the flow region. Because of the bad quality of input data (not many measured pressures for defining boundary conditions), the salinity map (Fig. 3) and hydrochemical map were also used.

A fragment of the potentiometric map was contoured on the basis of 17 measured formation pressures (Fig. 5). Next, taking into consideration the salinity map (Fig. 3) and a hydrochemical map, we drew a salinity contour equal to 5 g/L and determined the areas of infiltration water by the coefficient rNa/rCl to be greater than 0.87 (Fig. 5). On the basis of these maps and the analogies from other Carboniferous aquifers in which numerous formation-pressure measurements have been made but in which, unfortunately, the oil and gas fields have not been discovered, the intake areas and boundary conditions were determined for the regions where pressure measurements have not been made (Fig. 5).

All these data were projected on the boundary of Carboniferous Aquifer G and approximated by the functions $f_1(x)$, $f_2(x)$, $f_3(y)$, and $f_4(y)$ (Figs. 5 and 6).

Solution of the Laplace Equation. The Laplace equation for the given 2D region (the potential along the z-axis is constant) in the rectangular coordinate system is

$$\nabla^2 H_{fw}(x,y)=0. \qquad \qquad (9)$$

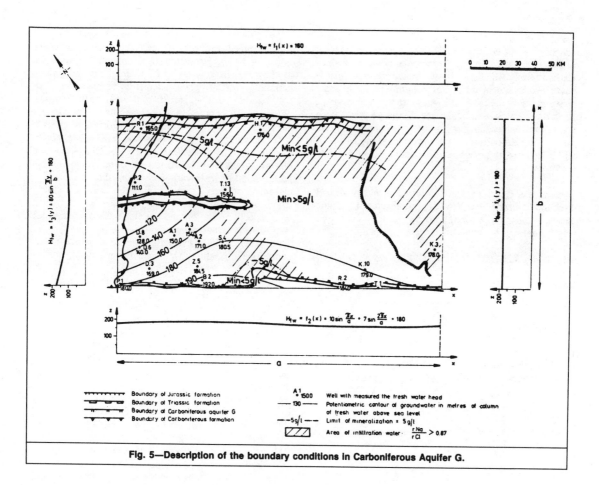

Fig. 5—Description of the boundary conditions in Carboniferous Aquifer G.

The general solution of Eq. 9 is found by a separation of variables[2]:

$$H_{fw} = \sum_{0}^{\infty} A_n e^{\pm q_n x} \cos q_n y + \sum_{1}^{\infty} B_n e^{\pm q_n x} \sin q_n y$$

. . . . (10a)

or

$$H_{fw} = \sum_{n=0}^{\infty} (A_n e^{\pm q_n y} \cos q_n x + B_n e^{\pm q_n y} \sin q_n x),$$

. . . . (10b)

where A_n, B_n, and q_n are arbitrary constants to be found from the boundary conditions.

A scheme of boundary conditions for Carboniferous Aquifer G with distribution of potential along the boundaries is shown in Figs. 5 and 6. Boundary conditions are listed below.

$$y=b, \quad H_{fw}=f_1(x)=180,$$

$$y=0, \quad H_{fw}=f_2(x)=10 \sin \frac{\pi x}{a} + 7 \sin \frac{2\pi x}{a} + 180,$$

$$x=0, \quad H_{fw}=f_3(y)=80 \sin \frac{\pi y}{b} + 180,$$

and

$$x=a, \quad H_{fw}=f_4(y)=180.$$

The solution of Eq. 9 is discussed in Ref. 3. The final solution of Eq. 9 is

$$H_{fw}=10 \frac{\sinh \frac{\pi}{a}(b-y)}{\sinh \frac{\pi b}{a}} \sin \frac{\pi x}{a}$$

$$+7 \frac{\sinh \frac{2\pi}{a}(b-y)}{\sinh \frac{2\pi b}{a}} \sin \frac{2\pi x}{a}$$

$$-80 \frac{\sinh \frac{\pi}{b}(a-x)}{\sinh \frac{\pi a}{b}} \sin \frac{\pi y}{b} + 180. \quad \ldots \ldots (11)$$

Eq. 11 is the only solution that satisfies both the Laplace equation and the boundary conditions. The graphical form of Eq. 11 is shown in Fig. 7; the potentiometric map of Carboniferous Aquifer G shows freshwater-head distribution expressed in meters of columns of fresh water above sea level. Fig. 7 also presents the directions of groundwater flow and intake and discharge areas. Two main groundwater movements are in two opposite directions from Carboniferous outcrops toward the center of the basin. Then, these two streams come together and flow northwest.

Dynamics of Groundwater Motion. The magnitude and direction of steady-state flow of groundwater of variable density at the given point of porous space for a 2D region may be defined by a modified Darcy's equation:

$$\vec{v} = -\sigma_{fw} \nabla H_{fw}(x,y), \quad \dots\dots\dots\dots\dots\dots (7)$$

where

$$\nabla H_{fw}(x,y) = \frac{\partial H_{fw}}{\partial x} \vec{a}_x + \frac{\partial H_{fw}}{\partial y} \vec{a}_y. \quad \dots\dots\dots (12)$$

For the Carboniferous Aquifer G, $\nabla H_{fw}(x,y)$ is obtained by the differentiation of Eq. 11:

$$
\begin{aligned}
\nabla H_{fw}(x,y) = \vec{a}_x & \left[10 \frac{\pi}{a} \frac{\sinh \frac{\pi}{a}(b-y)}{\sinh \frac{\pi b}{a}} \cos \frac{\pi x}{a} \right. \\
& + 14 \frac{\pi}{a} \frac{\sinh \frac{2\pi}{a}(b-y)}{\sinh \frac{2\pi b}{a}} \cos \frac{2\pi x}{a} \\
& \left. + 80 \frac{\pi}{b} \frac{\cosh \frac{\pi}{b}(a-x)}{\sinh \frac{\pi a}{b}} \sin \frac{\pi y}{b} \right] \\
- \vec{a}_y & \left[10 \frac{\pi}{a} \frac{\cosh \frac{\pi}{a}(b-y)}{\sinh \frac{\pi b}{a}} \sin \frac{\pi x}{a} \right. \\
& + 14 \frac{\pi}{a} \frac{\cosh \frac{2\pi}{a}(b-y)}{\sinh \frac{2\pi b}{a}} \sin \frac{2\pi x}{a} \\
& \left. + 80 \frac{\pi}{b} \frac{\sinh \frac{\pi}{b}(a-x)}{\sinh \frac{\pi a}{b}} \cos \frac{\pi y}{b} \right]. \quad \dots\dots\dots (13)
\end{aligned}
$$

The directions of groundwater flow (streamlines) may be determined from Fig. 7. The velocity of groundwater flow (Fig. 8) may be obtained by multiplying the hydraulic gradients for fresh water by the hydraulic conductivities pertaining to fresh water taken from relevant maps according to Eq. 7. On the basis of the map of flow velocity (Fig. 8), it is possible to evaluate the magnitude and directions of groundwater flow. The velocities of groundwater flow in the Carboniferous Aquifer G are relatively low, 0.10 to 5.5 cm/a [0.04 to 2.17 in./yr]. Assuming that the great flow velocities are unfavorable for the preservation of hydrocarbons because they cause a displacement or even a destruction of the existing accumulations, areas with relatively low velocities may be treated as prospective. The oil and gas fields discovered so far confirm that relationship.

From the results of hydrodynamic research on the Carboniferous formation, we make the following confirmations.

1. Permeability of Carboniferous Aquifer G is relatively low, 5 to 70 md (Fig. 2).
2. Groundwaters have variable salinity, 0.8 to 150 g/L (Fig. 3).
3. Formation pressures are normally hydrostatic.
4. Directions of flow are centripetal (Fig. 7).
5. Velocities of groundwater flow along the horizontal plane are relatively low, 0.10 to 5.5 cm/a [0.04 to 2.17 in./yr] (Fig. 8).
6. Accumulations of oil and gas are accompanied by a local minimum of permeability, groundwater of local high salinity, and relatively low velocities of groundwater flow (Figs. 2, 3, and 8).

Coustau et al.[4] classified sedimentary basins according to hydrodynamic conditions and related this classification to their petroleum potential from organic geochemical considerations. Three main types of basins are distinguished (Fig. 9).

1. *Juvenile basins*—not necessarily young, with compaction-induced centrifugal, lateral water movement. Some examples are Nigeria, Gulf of Mexico, Douala basin, North Sea, and northeast Sahara. Petroleum interest in the basins is very strong.

2. *Intermediate basins*—with centripetal water movement, artesian properties, and freshwater invasions. Some examples include the Persian Gulf, east Sahara, Paris basin, central Tunisia, and Sahara. Petroleum interest in such basins varies from very strong to moderate and is connected with the areas of low velocities of groundwater flow and of local high salinity.

3. *Senile basins*—with hydrostatic conditions and generally invaded by meteoric waters. Northwest Aquitaine basins and parts of the North Spanish basin are examples. There is little or no petroleum interest.

Coustau et al.[4] found that groundwater flow is initially of great importance to secondary migration and formation of pools. However, if it is too strong or lasts too long, groundwater retards the formation of pools or even destroys the existing accumulations by dismigration.

On the basis of the presented hydrodynamic characteristics and the classification of basins, we found that the Carboniferous basin is hydrodynamically active, of an intermediate type, and may be classified as prospective. Oil and gas fields that have been discovered in the Lublin synclinorium are connected with the areas of low velocities

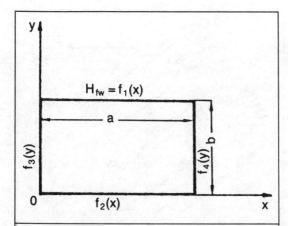

Fig. 6—Schematic of the boundary conditions in Carboniferous Aquifer G with the distribution of the potential along the boundary.

of groundwater flow (Fig. 3) and high salinity (Fig. 8). Other sedimentary basins in Poland that may also be classified as intermediate type confirm this relationship.

Methodology of Mapping Hydrodynamic Petroleum Traps in Sedimentary Basins with Groundwater of Variable Density

According to the original version of the anticlinal theory, petroleum accumulations exist under hydrostatic conditions. Under hydrostatic conditions, a buoyancy force exerts major control over the distribution of water, oil, and gas in a reservoir, causing the fluids to be segregated according to their density and the oil/water contact, and causing the gas/oil or gas/water contact to be horizontal. Under hydrodynamic conditions, however, the oil/water and gas/water contacts are not horizontal because both moving water and the buoyancy force have an effect. According to Hubbert,[5] the tilting of oil/water and gas/water contacts by moving water in some geologic structures governs whether the structures are capable of serving as traps.

Methodology of Positioning of Hydrodynamic Petroleum Traps. The potentials for water, oil, and gas at a given point of a hydrodynamic field are defined as[5]

$$\Phi_w = gz + \frac{p}{\rho_w}, \quad \ldots\ldots\ldots\ldots\ldots\ldots\ldots (14)$$

$$\Phi_o = gz + \frac{p}{\rho_o}, \quad \ldots\ldots\ldots\ldots\ldots\ldots\ldots (15)$$

and

$$\Phi_g = gz + \int_0^p \frac{dp}{\rho_g}, \quad \ldots\ldots\ldots\ldots\ldots\ldots (16)$$

where ρ_w is the density of the water, ρ_o is the density of the oil, and ρ_g is the variable density of the gas during compression. To determine the petroleum entrapments

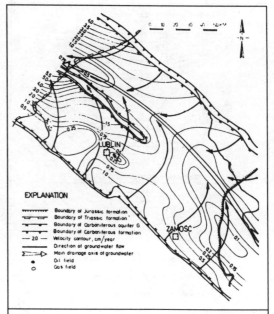

Fig. 8—Map of the flow velocity for Carboniferous Aquifer G.

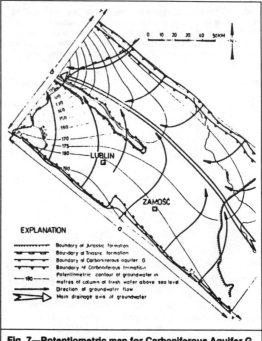

Fig. 7—Potentiometric map for Carboniferous Aquifer G.

under hydrodynamic conditions, Hubbert[5] proposed the following equations that are derived from Eqs. 14 through 16:

$$U_o = V_o + z \quad \ldots\ldots\ldots\ldots\ldots\ldots\ldots\ldots (17)$$

and

$$U_g = V_g + z, \quad \ldots\ldots\ldots\ldots\ldots\ldots\ldots\ldots (18)$$

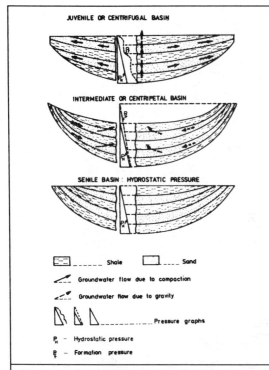

Fig. 9—Three main types of sedimentary basins classified according to hydrodynamic conditions: juvenile basins, (before invasion by meteoric waters), intermediate basins (during invasion by meteoric waters), and senile basins (after invasion by meteoric waters) (after Coustau et al.[4]).

where

$$U_o = \frac{\rho_o}{\rho_w - \rho_o} h_o, \quad\quad\quad\quad\text{(19a)}$$

$$V_o = \frac{\rho_w}{\rho_w - \rho_o} h_w, \quad\quad\quad\quad\text{(19b)}$$

$$U_g = \frac{\rho_g}{\rho_w - \rho_g} h_g, \quad\quad\quad\quad\text{(20a)}$$

and

$$V_g = \frac{\rho_w}{\rho_w - \rho_g} h_w, \quad\quad\quad\quad\text{(20b)}$$

where z is the elevation of the point considered; h_w is the height above datum to which water would stand statically in a manometer terminated at the point considered; h_o is the corresponding height to which oil of density ρ_o would rise; and h_g is the corresponding height to which gas of density ρ_g would rise. Here, the U surfaces coincide with the oil or gas equipotential surfaces, the V surfaces coincide with the water equipotential surfaces, and the z surfaces are a family of horizontal surfaces of elevation z. Areas of closure of the U curves containing minimum values of U represent petroleum traps.[5] Eqs. 17 and 18 may be used only when factors $\rho_o/(\rho_w-\rho_o)$, $\rho_w/(\rho_w-\rho_o)$, $\rho_g/(\rho_w-\rho_g)$, and $\rho_w/(\rho_w-\rho_g)$ are constant—that is, when the densities of the water, oil, and gas are constant. Thus Hubbert's method of mapping petroleum entrapments is correct only when the densities of the water, oil, and gas are constant.

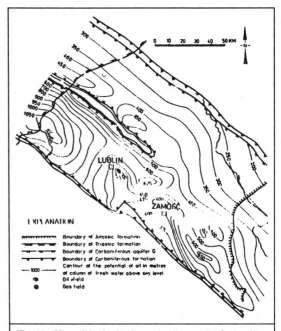

Fig. 10—Map of the hydrodynamic oil traps for Carboniferous Aquifer G.

Modified Theory. A modified theory is introduced that allows the fluid's density to vary and the formation fluids to move. Calculating p from Eq. 14 and substituting it into Eqs. 15 and 16, we find that

$$\Phi_o = \frac{\rho_w}{\rho_o}\Phi_w - \frac{\rho_w - \rho_o}{\rho_o} gz \quad\quad\text{(21)}$$

and

$$\Phi_g = \frac{\rho_w}{\rho_g}\Phi_w - \frac{\rho_w - \rho_g}{\rho_g} gz. \quad\quad\text{(22)}$$

Taking into consideration that

$$\Phi_w = g \cdot h_w, \quad\quad\quad\quad\text{(23)}$$

$$\Phi_o = g \cdot h_o, \quad\quad\quad\quad\text{(24)}$$

and

$$\Phi_g = g \cdot h_g, \quad\quad\quad\quad\text{(25)}$$

Eqs. 21 and 22 may be rewritten as

$$gh_o = \frac{\rho_w}{\rho_o} gh_w - \frac{\rho_w - \rho_o}{\rho_o} gz \quad\quad\text{(26)}$$

and

$$gh_g = \frac{\rho_w}{\rho_g} gh_w - \frac{\rho_w - \rho_g}{\rho_g} gz. \quad\quad\text{(27)}$$

Multiplying both sides of Eqs. 26 and 27 by ρ_o and ρ_g, respectively, gives

$$\rho_o gh_o = \rho_w gh_w - (\rho_w - \rho_o)gz \quad\quad\text{(28)}$$

and

$$\rho_g g h_g = \rho_w g h_w - (\rho_w - \rho_g)gz, \qquad \ldots\ldots\ldots (29)$$

which may be rewritten as

$$\Psi_o = \Psi_w - z(\gamma_w - \gamma_o) \qquad \ldots\ldots\ldots\ldots\ldots (30)$$

and

$$\Psi_g = \Psi_w - z(\gamma_w - \gamma_g), \qquad \ldots\ldots\ldots\ldots\ldots (31)$$

where Ψ_w, Ψ_o, and Ψ_g are the scalar force potentials for water, oil, and gas expressed in pressure units; γ_w is the specific weight of the water in the reservoir conditions; γ_o is the specific weight of the oil in the reservoir conditions; and γ_g is the specific weight of the gas in the reservoir conditions. Replacing Ψ_w, Ψ_o, and Ψ_g by an equivalent column of fresh water—that is, dividing Eqs. 30 and 31 by γ_{fw}—gives

$$H_o = H_{fw} - z\left(\frac{\gamma_w - \gamma_o}{\gamma_{fw}}\right) \qquad \ldots\ldots\ldots\ldots (32)$$

and

$$H_g = H_{fw} - z\left(\frac{\gamma_w - \gamma_g}{\gamma_{fw}}\right), \qquad \ldots\ldots\ldots\ldots (33)$$

where H_{fw}, H_o, and H_g are the potentials of water, oil, and gas, respectively, expressed in meters of column of fresh water, and γ_{fw} is the specific weight of the fresh water. The term $-z[(\gamma_w - \gamma_o)/\gamma_{fw}]$ expresses the additional energy of unit weight of oil at z with respect to the groundwater potential, and $-z[(\gamma_w - \gamma_g)/\gamma_{fw}]$ is the additional energy of unit weight of gas at z with respect to the groundwater potential.

Eqs. 32 and 33 apply also to reservoir fluids of variable density because the potentials for water, oil, and gas are expressed in lengths of columns of fresh water—i.e., as a liquid with constant density that is the condition for the existence of a scalar force potential.

According to the method presented, to determine potential sites for hydrodynamic entrapment of hydrocarbon, it is necessary to construct the potentiometric maps for groundwater in terms of fresh water (Fig. 7) and the maps of differential energy of oil or gas with respect to groundwater and then to add them up by superimposition.

The problem of groundwater movement and distribution of potentiometric surface in Carboniferous Aquifer G was solved mathematically (Fig. 7). The maps of differential energy for oil and gas (methane) with respect to water were made on the basis of the following equations:

$$E_{ow} = H_o - H_{fw} = -z\frac{\gamma_w - \gamma_o}{\gamma_{fw}} \qquad \ldots\ldots\ldots (34)$$

and

$$E_{gw} = H_g - H_{fw} = -z\frac{\gamma_w - \gamma_g}{\gamma_{fw}}. \qquad \ldots\ldots\ldots (35)$$

In practice, they were made by superpositional subtraction of the values shown on two maps—the map of groundwater and oil densities or, more conveniently, the map of groundwater and methane densities in reservoir conditions—and then multiplication of the obtained difference by the values shown on the structural contour map

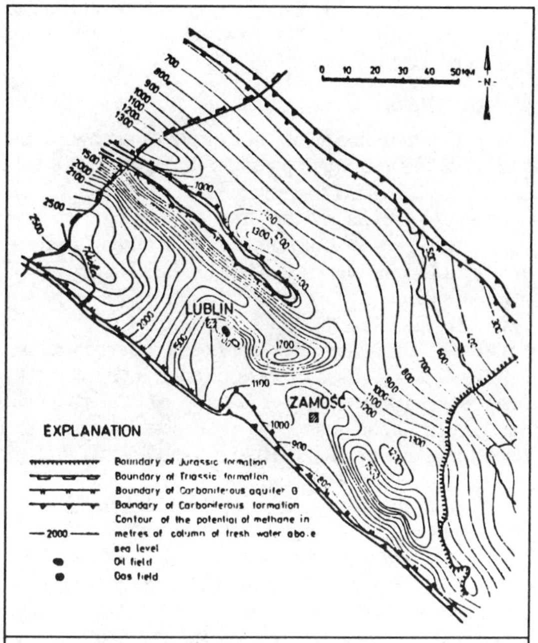

Fig. 11—Map of the hydrodynamic gas traps for Carboniferous Aquifer G.

of the given aquifer top (Fig. 1). The amplification factors in Eqs. 34 and 35 are the functions of the density contrast between water and oil or water and gas, respectively, and were calculated with reference to methane and average Carboniferous oil. The densities of oil and methane were reduced to reservoir conditions with pressure and temperature taken into consideration. The densities of oil and gas were programmed on the basis of structural maps; that is, the densities of oil or methane suitable for a given isohypse were calculated and then the maps of oil and gas density were contoured on the basis of the structural map of a given aquifer top. For the given contour line, a suitable density calculated from its depth was attributed. The map of fluid density (for oil or methane) imitates the structural plan in density units. These data may be interpreted by both manual and computer methods.

Results of Hydrodynamic Research on Carboniferous Formation in Lublin Synclinorium. The maps of hydrodynamic traps for oil and gas (Figs. 10 and 11) define the conditions of equilibrium in reservoir rocks caused by structural entrapments and the dynamic effects of reservoir fluids. These maps present the prospective areas for accumulation of hydrocarbons that are determined by local minimum values of potentials. The map of hydrodynamic oil traps (Fig. 10) presents three prospective areas: southeast of Lublin where the oil field was discovered and north and northeast of Zamość. The map of hydrodynamic gas traps (Fig. 11) is not similar to the map of hydrodynamic oil traps (Fig. 10) because the potentials of gas are greater than the potentials of oil, according to Eqs. 32 and 33. The map of hydrodynamic gas traps (Fig. 11) also presents two prospective areas: near Lublin (gas field) and southeast of Zamość. A comparison of the structural map of the top of the Carboniferous Aquifer G (Fig. 1) with the maps of hydrodynamic oil and gas traps (Figs. 10 and 11) shows that the oil and gas fields discovered thus far have tilted oil/water and gas/water contacts.

Conclusions

1. With a modified theory of hydrodynamic hydrocarbon accumulation, it was possible, through an introduction of the potentials for water, oil, and gas expressed in meters of column of fresh water (as a liquid with constant density) into hydrodynamic considerations, to compare hydraulic heads in regions with variable density of reservoir fluids.

2. The directions of groundwater flow in the Carboniferous formation are centripetal. Carboniferous Aquifer G is reinforced in water through Jurassic formations along a Carboniferous outcrop on a Carboniferous/Jurassic disconformity. The velocities of groundwater flow are relatively low, 0.10 to 5.5 cm/a [0.04 to 2.16 in./yr]. (See Figs. 4, 7, and 8.)

3. The Carboniferous basin of the Lublin synclinorium may be classified as a hydrodynamically active basin of an intermediate type. Assuming that great flow velocities are unfavorable for the preservation of hydrocarbons, areas with relatively low velocities (and high salinity) may be treated as prospective. The oil and gas fields discovered up to now confirm this relationship (Figs. 3 and 8).

4. Our method of positioning of hydrodynamic petroleum traps that accepts both the variability of reservoir fluid density (static effect) and moving water (dynamic effect) permits a more precise definition of the position of oil and gas hydrodynamic traps.

5. On the basis of our results, the suggested method of determining the positions of potential hydrodynamic petroleum traps is comparatively simple to apply and gives good results in regions with high hydraulic gradients and high variability of salinity and hydrocarbon densities (Figs. 10 and 11).

Nomenclature

$A_n, B_n,$
$\quad q_n$ = arbitrary constants
E_{gw} = additional energy of unit weight of gas (methane) at the point of measurement, z, with respect to the groundwater potential, m [ft]
E_{ow} = additional energy of unit weight of oil at z with respect to the groundwater potential, m [ft]
g = acceleration of gravity, m/s^2 [ft/sec^2]
h_g = height above datum to which gas of density ρ_g would rise, m [ft]
h_o = height above datum to which oil of density ρ_o would rise, m [ft]
h_w = height above datum to which water of density ρ_w would rise, m [ft]
H_{fw} = freshwater head (scalar force potential), m [ft]
H_g = potential of gas (methane), m [ft]
H_{ifw} = freshwater head at i, m [ft]
H_{inw} = environmental-water head at i, m [ft]
H_o = potential of oil, m [ft]
H_w = potential of water, m [ft]

i = any point in groundwater of variable density
$K_{11}, K_{22},$
$\quad K_{33}$ = principal directional permeabilities in the x, y, and z directions, md
p = pressure, kPa [psi]
t = time, seconds
U_g = gas equipotential surface, m [ft]
U_o = oil equipotential surface, m [ft]
$\vec{v}_w$ = vector velocity of water, m/s [ft/sec]
$v_x, v_y,$
$\quad v_z$ = velocity components along the x, y, and z coordinates, respectively, m/s [ft/sec]
V_g, V_o = water equipotential surfaces, m [ft]
x, y, z = rectangular coordinates
γ_{fw} = specific weight of fresh water, N/m^3 [lbf/ft^3]
γ_g = specific weight of gas in reservoir conditions, N/m^3 [lbf/ft^3]
γ_o = specific weight of the oil in reservoir conditions, N/m^3 [lbf/ft^3]
γ_w = specific weight of water in reservoir conditions, N/m^3 [lbf/ft^3]
μ = dynamic viscosity, Pa·s [cp]
ρ_{fw} = freshwater density, kg/m^3 [lbm/ft^3]
ρ_g = gas density, kg/m^3 [lbm/ft^3]
ρ_o = oil density, kg/m^3 [lbm/ft^3]
ρ_w = water density, kg/m^3 [lbm/ft^3]
σ_{fw} = hydraulic conductivity for fresh water, m/s [ft/sec]
σ_w = hydraulic conductivity, m/s [ft/sec]
ϕ = porosity, fraction
Φ_g = potential for gas, m^2/s^2 [ft^2/sec^2]
Φ_o = potential for oil, m^2/s^2 [ft^2/sec^2]
Φ_w = potential for water, m^2/s^2 [ft^2/sec^2]
Ψ_g = scalar force potential for gas, kPa [psi]
Ψ_o = scalar force potential for oil, kPa [psi]
Ψ_w = scalar force potential for water, kPa [psi]

References

1. Lusczynski, N.J.: "Head and Flow of Ground Water of Variable Density," *J. Geophys. Res.* (Dec. 1961) 4247–56.
2. Moon, P. and Spencer, D.E.: *Field Theory for Engineers*, D. Van Nostrand Co. Inc., Princeton, NJ (1961) 92–93.
3. Zawisza, L.: "Hydrodynamic Conditions of Hydrocarbon Accumulation in Carboniferous and Devonian Formations in the Lublin Synclinorium," PhD dissertation, U. of Mining and Metallurgy, Kraków, Poland (1980).
4. Coustau, H. *et al.*: "Classification hydrodynamique des basins sédimentaires utilisation combinée avec d'autres méthodes pour rationaliser l'exploration dans des basins non-productifs," *Proc.*, Ninth World Pet. Cong., Tokyo (1975) 105–18.
5. Hubbert, M.K.: "Entrapment of Petroleum Under Hydrodynamic Conditions," *Bull.*, AAPG (1953) 1954–2026.

SI Metric Conversion Factors

ft	× 3.048*	E−01	= m
in.	× 2.54*	E+00	= cm
mile	× 1.609 344*	E+00	= km

*Conversion factor is exact.   **SPEFE**

Original manuscript received in the Society of Petroleum Engineers office June 1, 1984. Paper accepted for publication Aug. 7, 1985. Revised manuscript received Dec. 2, 1985.

The American Association of Petroleum Geologists Bulletin
V. 63, No. 2 (February 1979), P. 152-181, 21 Figs., 3 Tables

Deep Basin Gas Trap, Western Canada[1]

JOHN A. MASTERS[2]

Abstract Gas accumulations are distributed in a fashion similar to most other natural resources. The high-grade deposits are comparatively small. In general, as the grade decreases the size increases.

Three of the largest sandstone gas fields in western North America are in low porosity–low permeability Cretaceous sandstone, in downdip structural locations, with porous water-filled reservoir rock updip. Examination of the details of these fields sets the stage for recognizing an enormous tight-sand gas trap in western Canada.

The Mesozoic rock section, only 1,000 ft (300 m) thick on the shelf in eastern Alberta, thickens westward to over 15,000 ft (4,570 m) in the Deep Basin in front of the Foothills overthrusts. Most of the developed sandstone gas fields are in updip porosity traps, or minor structural traps, on the shelf. The porous, generally water-saturated sands of the shelf become less porous and permeable westward and downdip, passing from the water-bearing area with local gas traps through a transition zone to a gas-bearing area. This change is demonstrated by electrical resistivity logs and confirmed by drill-stem tests.

Recent exploratory drilling in the Deep Basin has resulted in numerous discoveries in the area. Several hundred log analyses provide reliable data for measuring potential gas resources in the range of 400 Tcf. Recoverable gas at $2.00/Mcf net after royalty may reach 150 Tcf.

The quantities of gas apparently present would be a major addition to the North American energy supply.

INTRODUCTION

Geologic evidence suggests the probable existence of an enormous gas accumulation trapped in the deepest part of the Alberta syncline and its extension into British Columbia. This area, termed the "Deep Basin," covers 26,000 sq mi (67,600 sq km). The gas interval includes almost the entire Mesozoic clastic section which reaches a maximum thickness of 15,000 ft (4,570 m). Reserve potential is so large as to alter significantly the energy supply estimates for North America. This paper attempts to document the unusual geologic conditions that relate to this and similarly trapped giant gas accumulations.

RESOURCE TRIANGLE

The philosophical basis for the effort and money which have been invested in the development of the Deep Basin concept is described by the Resource Triangle in Figure 1 (Gray, 1977). It is reasonable to suggest that most natural resources are distributed as in a triangle. The high-grade deposits occupy the peak, the smallest part of the triangle. In general, as the grade decreases, the size increases. Geologists are familiar

with the concept of large, low-grade ore deposits versus small, high-grade deposits. However, we are not accustomed to thinking about oil and gas in these terms. Yet, in Canada, reserves of "high-grade" oil (i.e., oil of good gravity in high-deliverability reservoirs) amount to only 16 billion bbl. In contrast, the "low-grade" asphalt sands of Alberta contain 1,000 billion bbl.

Our fundamental thesis is that gas reserves have a similar triangular distribution. As reservoir quality deteriorates, larger and larger amounts of gas are trapped in the lower porosity sands. The good-quality reservoir fields are measured in bil-

[1]Manuscript received, April 1, 1978; accepted, September 7, 1978. Condensed versions of this paper appear in the *Oil and Gas Journal*, September 18, 1978, p. 226-241, and in *World Oil*, November 1978, p. 105-120.

[2]President, Canadian Hunter Exploration Ltd., Calgary, Alberta, Canada.

This study could not have been made without the able assistance of the technical staff at Canadian Hunter. I am particularly indebted to my partner, James K. Gray, for innumerable discussions about every facet of the study; also to my geologist colleagues who contributed many important observations and compiled much of the data, specifically David Smith, Duncan McCowan, Paul Jackson, Earl Hawkes, Lorne Larson, Keith Cole, and Bob Gies. Our Director of Research, Richard Wyman, and his assistant, Tom Davis, participated significantly in the interpretation of hydrodynamics and other phenomena at the gas-water interface. The excellent log analysis work done by Ted Connolly, Ron Hietala, and their two assistants provided the fundamental data for the study.

Special recognition is due our principal consultants, Sneider and Meckel Associates of Houston. This group, containing geologists, an engineer, and a log analyst, guided us through the complexities and confusion of a problem that was virtually an unknown exploration frontier in Canada. Without the unique insight and ability of Robert Sneider, Larry Meckel, John Farina, and Lloyd Fons, our success would have been greatly reduced.

Finally, I express my thanks and admiration to Alf Powis, President of Noranda Mines, the parent company of Canadian Hunter. Noranda agreed with the concept of low-grade gas reserves and began the Hunter operations in 1973 when much of the industry was leaving western Canada. They have supported us in a high risk pioneering venture.

Article Identification Number
0149-1423/79/B002-0001$03.00/0

Deep Basin Gas Trap, Western Canada

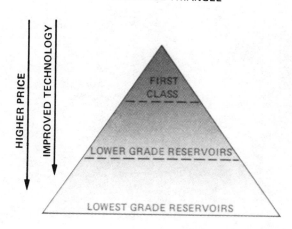

FIG. 1—Natural resources have triangular distribution. Mineral deposits range from small high-grade to large low-grade deposits. Oil and gas fields are similar. As price and technology increase it is possible to produce lower quality reservoirs. After Gray (1977).

lions of cubic feet. The low-quality fields, like the Milk River of southeastern Alberta, which is the first of its kind in Alberta, are measured in trillions of cubic feet. The Milk River gas field, now the largest in Canada with approximately 9 Tcf of recoverable gas, was actually discovered by a Canadian Pacific Railroad well drilled for water in 1883, but was not economically producible until 1973. The largest gas field in western North America, the San Juan basin with 25 Tcf of recoverable gas, is a very low-quality reservoir which has experienced rapid development in response to improvements in gas price and technology.

Of course, nothing in nature is as regular as the triangle indicates. The concept is valid, although if all gas resources were known their distribution according to quality of reservoir would only vaguely resemble a symmetrical triangle. In general, there is no doubt that much larger amounts of gas are present in low-quality reservoirs than in high-quality reservoirs.

GAS FIELDS—NORTH AMERICA

Figure 2 is a map of the western part of North America (excluding Alaska) showing sedimentary basins, gas fields, and tar-sand deposits. Several important relations are noted.

1. Western Canada and the United States are geologically continuous and have essentially similar rock systems.

2. The incidence of gas fields in western Canada is impressive. It is obviously a more favorable area for gas than western United States.

3. The amount of drilling in western Canada is less than half that of western United States. It is far less than in the highly developed United States Mid-Continent area. Canada has been only lightly explored.

4. A significant geologic difference is readily apparent. In Canada the original sedimentary basin is relatively undisturbed. The rocks of the Alberta syncline dip gently and thicken gradually westward. The original stratigraphic oil and gas traps are essentially in place. In the United States, the shelf in front of the overthrust belt from Montana to New Mexico has been broken into numerous separate basins and mountain uplifts. The reservoir rocks have been tilted in various directions, exposed to erosion, and subjected to completely altered hydrodynamic gradients in many places. Many traps were destroyed and much oil and gas was lost in the redistribution of fluids.

5. It is not generally appreciated that western Canada is one of the great hydrocarbon areas of the world. The Athabasca and adjoining areas of tar sands contain about 1,000 billion bbl of oil in place in Cretaceous sandstones. This substantially exceeds the proved recoverable oil reserves of the entire Persian Gulf region. The theme of this paper is that these same Canadian Cretaceous rocks, where greatly thickened and deeply buried, generated immense quantities of gas. Recognizing their apparent prolific capacity to generate oil, it is perhaps not difficult to imagine them as a source also for one of the world's great gas accumulations.

Figure 2 identifies three of the largest sandstone gas fields in western North America. The San Juan basin contains 25 Tcf of recoverable gas in Cretaceous sandstones. It is the largest gas field in the U.S. Rocky Mountains and the second largest field in North America. The geologic characteristics of this giant accumulation should obviously be of prime importance in guiding us to other large fields. Another large field in the U.S. Rocky Mountains is Wattenberg (1.3 Tcf). The largest field in Canada is Milk River with 9 Tcf of reserves. (The three largest carbonate-reservoir fields contain only about 2.3 Tcf each).

Table 1 lists the salient reservoir parameters of the three giant sandstone fields. Note that porosity and permeability are abnormally low. Reservoir pressure is also low.

Several other geologic characteristics are similar in each of the giant fields and distinguish them from ordinary gas traps. The structural position is similar. The San Juan field lies across the synclinal axis of the San Juan basin. Wattenberg is on the synclinal axis of the Denver basin. Milk River is far down the northeast plunge of the Sweet-

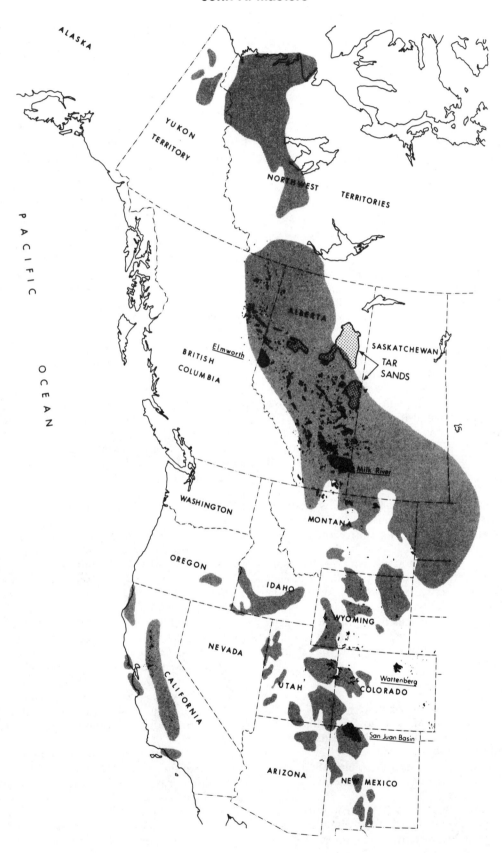

FIG. 2—Western North America sedimentary basins, gas fields, and tar sand deposits.

Deep Basin Gas Trap, Western Canada

Table 1. Reservoir Parameters of Largest Fields in Cretaceous Sandstone

	San Juan Basin (1)*	San Juan Basin (2)*	Wattenberg (3)*	Milk River (4)*
Reservoir	Mesaverde	Dakota	Dakota	Milk River
Average porosity (%)	10	7	9.5	15
Average permeability (md)	1.5	0.15	0.3	1
Water saturation (%)	34	35	44	45
Average pay (ft)	80	60	25	60
Original reservoir pressure (psi)	1,360	3,000	2,750	445
Average depth (ft)	5,400	7,000	8,000	1,100
Field size (sections)	3,140	1,100	970	7,000
Recoverable reserves (Tcf)	11**	7	1.3	9
Average well prod. (MCFD)	300	250	225	100
Well spacing (acres)	320	320	320	320

*References: (1) Allen (1955) and Pritchard (1973); (2) Deischl (1973); (3) Matuszczak (1973); (4) Suffield Block Study Committee (1972).

**Infill drilling to change spacing from 320 acres to 160 acres may increase reserves 70% to 19 Tcf (Long, 1977).

grass arch. Each of these fields shows a curious upside-down relation of water to gas. In each the reservoir rock is fairly porous and water saturated updip, and grades transitionally downdip into less porous, gas-saturated rock. This relation in the San Juan basin has been interpreted to be a hydrodynamic trap (Berry, 1959). It must be significant that three of the largest sandstone gas fields in western North America are in low-porosity Cretaceous sandstones, in downdip structural locations, with porous water-filled reservoir rock updip. These three fields contain more than half the total sandstone gas reserves of the region. The rest of the gas is distributed in over 1,000 separate fields. It is apparent that very large reserve additions might be made by exploring specifically for low-porosity fields.

POROSITY

Recognizing that a number of large fields are present in the United States in low-porosity sands, we may examine the distribution of proved gas reserves in Canada according to porosity (Energy Resources Conservation Board, 1974). Figure 3A is a graph of sandstone gas reserves in 1973 by intervals of porosity. The gas reserves in sandstones in Canada were shown to have a near normal distribution with the bulk of the reserves in porosity of 20 to 25%. There were practically no gas reserves measured in Canada in sandstone porosity less than 13%.

Figure 3B shows the addition of Milk River reserves in 1973-74. Note that the histogram is now skewed sharply to the left. The normal distribution shape is gone. The Milk River added very large reserves of gas in 13 to 18% porosity sandstone. This graph suggests the likelihood of large quantities of gas being present in sandstones with less than 13% porosity. Whether or not they are "reserves" depends on economics.

A study of United States gas reserves demonstrates the presence of very large amounts of gas in sandstones ranging in porosity from 5 to 13% in addition to, of course, large amounts of gas in sandstones of higher porosity. Figure 3C shows approximately 50 Tcf of gas in known fields in the United States, all in sandstones of less porosity than the lowest grade in which any reserves are measured in Canada. The implication is clear that Canada probably has large gas supplies in these lower porosity rocks which have not yet been identified. Their development is a question of economics.

SAN JUAN BASIN

Before considering Canadian basins, it is instructive to examine the trapping mechanism in the San Juan basin. This field contains most of the gas reserves of the entire U.S. Rocky Mountains. It is a huge stratigraphic trap covering 3,600 sq mi (9,325 sq km) across the axis of the San Juan basin, a Laramide structural basin in northwestern New Mexico. The producing zone includes almost the entire Cretaceous section with four main pay sands and numerous lesser sands over an interval of 5,000 ft (1,525 m).

SANDSTONE GAS RESERVES - ALBERTA

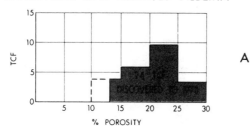

A

SANDSTONE GAS RESERVES - ALBERTA
(NO RESERVES RECOGNIZED LESS THAN 13%)

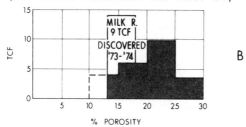

B

SANDSTONE GAS RESERVES - U.S.
(IN SANDS LESS THAN 13% POROSITY)

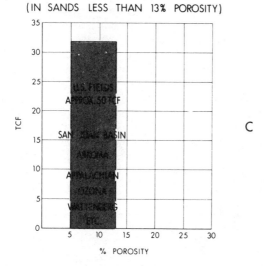

C

FIG. 3—**A.** Distribution of sandstone gas reserves by porosity. Approximate normal distribution curve. No reserves less than 13% porosity. From Energy Resources Conservation Board (1974). **B.** Curve is distorted by huge Milk River reserves. Still, essentially no reserves booked in Alberta less than 13% porosity. **C.** United States gas reserves in sandstones with less than 13% porosity. This illustration suggests that economics have permitted development in United States of large amount of lower quality reservoir rock. This grade of reservoir has not yet been tapped in Canada.

Figure 4 is an electric log cross section from southwest to northeast across the basin (Riggs, 1976). Structural dip is gently to the northeast. The following points are made with reference to the section.

1. The updip, water-wet sands with large SP deflections are thick, porous, and permeable.

2. The gas sands are thinner, shalier, and less porous and permeable.

3. The water-saturated section grades imperceptibly through a transition zone 5 to 10 mi (8 to 16 km) wide into a gas saturated zone. Wells in these areas will produce water, then water and gas, finally gas only.

4. The water in the transition zone is more saline than in the updip water zone. Silver (1968) said that the updip water is 3,000 to 5,000 ppm chlorides, whereas the "rind of saline waters around the Mesaverde gas reservoirs [is] . . . up to 33,000 ppm chlorides."

5. There is no evidence for a stratigraphic or structural barrier between the water and gas zones. There is only a small decrease in porosity and permeability through the transition zone, but enough perhaps to cause a significant increase in water saturation at the interface which would strongly influence permeability of the rock to gas. Unfortunately, the changes in porosity and permeability of the reservoir rock, although they have been observed by many geologists working in the basin, have never been measured and documented in the literature. Therefore, I cannot be quantitative about this observation although I will hazard an estimate from 20 log analyses that the change goes from about 20% porosity and 15 md permeability in the water zone to 10% porosity and 1 to 2 md permeability in the gas zone. The change through the transition zone appears to be quite gradual and of small magnitude.

6. Riggs (1976) said, "It appears that along the producing edges of any of the major pools, especially as you come updip toward water, that the selective good porosity and permeability intervals have taken on water and will not produce commercial gas. When exploring along the edges of these producing fields, one sometimes has to reverse his attitudes and look for tighter sand intervals to complete wells in rather than good porous intervals."

7. IMPORTANT: In the gas zone, all the rock exceeds 20 ohms of resistivity. The entire section is gas saturated; not only the main pay sands, but every silt zone and streak of sand in the entire section is gas charged and not just the Mesaverde rocks. The entire section from the Pictured Cliffs to the Dakota, 5,000 ft (1,525 m) of rock, is gas saturated wherever there is the thinnest stringer

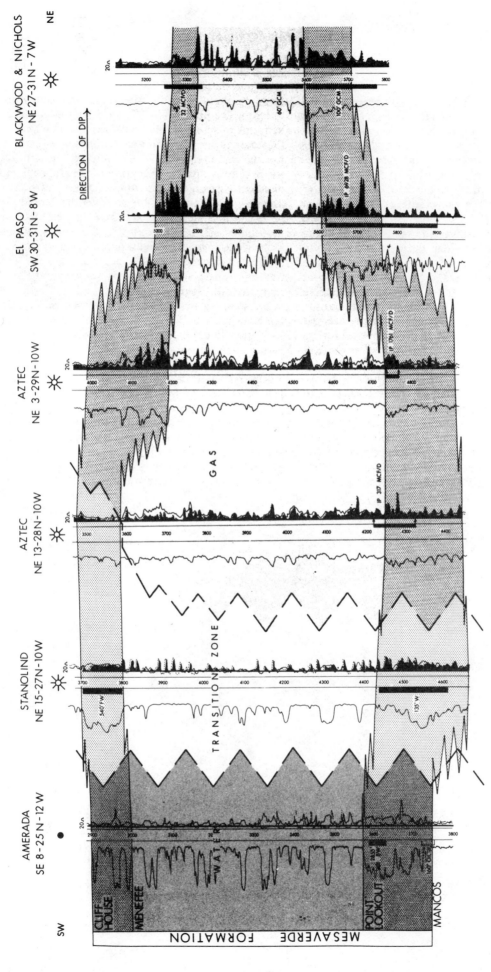

FIG. 4—Southwest-northeast electric log cross section of Mesaverde sands, San Juan basin, showing water-saturated area, transition zone, and gas-saturated area. Line of section shown on Figure 5. From Riggs (1976). Sands in gas area are shalier and less porous than those in water area.

503

John A. Masters

of porosity. Riggs (1976) observed, "Samples show that the entire Cretaceous section is gas saturated. Even the tightest shales when examined right off the shaker are bleeding gas under the microscope." Because of the pervasive gas saturation there is an unusual stacking of gas pay zones in this giant stratigraphic trap.

A map of the Mesaverde gas field in the San Juan basin (Fig. 5) shows structure, outcrop of the producing Mesaverde sandstones, the area of water saturation along the west and south sides, the area of tight, nonproductive sandstone along the northeast side, and the trace of the Figure 4 cross section. The gas-water transition zone along the southwest edge of the field, which was shown in the cross section, is indicated by the scattering of small pools south of the main accumulation. The water-saturated area, colored gray, extends at least 40 mi (64 km) up the gently dipping southwest flank of the basin. Note that there is about 500 ft (150 m) of gas column measured along the trace of the cross section. This is opposed to a water column of about 2,000 ft (610 m).

There is, at first glance, a significant change in these relationships at the northwest edge of the field. Here, one of the largest single pools of gas in North America extends to within 2 mi (3 km) of the outcrop. There is nearly 2,000 ft (610 m) of

gas column with its attendant buoyancy force trying to move updip. There is not a northeast-trending shale barrier between the gas and the outcrop. The producing sandstones extend unbroken and unchanged to the outcrop. Structurally, they are uplifted very sharply in the Hogback monocline, enough to put about 4,000 ft (1,220 m) of water column against the gas over the 2 mi (3 km) distance. The northwest end of the field on more detailed maps is "sawed off" in a nearly straight northeast line, essentially parallel with the +500-ft (+150 m) structure contour. Stratigraphic strike is perpendicular to this cutoff, parallel with the northwest-trending tight-sand line. The northwest edge of the field appears to be a spectacular example of the sealing capacity of either downdip water flow, water block in low permeability sands, or both.

Figure 6 is a potentiometric surface map of the Mesaverde sandstones in the San Juan basin taken from Hill et al (1961). It shows the gas accumulation entirely ringed by downdip water flow. The Petroleum Research Corporation geologists explained that "stratigraphic studies . . . indicate that the lenticular Mesaverde sands, which form the reservoirs for the Blanco (Mesaverde) gas field, increase both in number and in permeability updip (to the southwest) through the gas field. The updip edge of the field is not a consistently

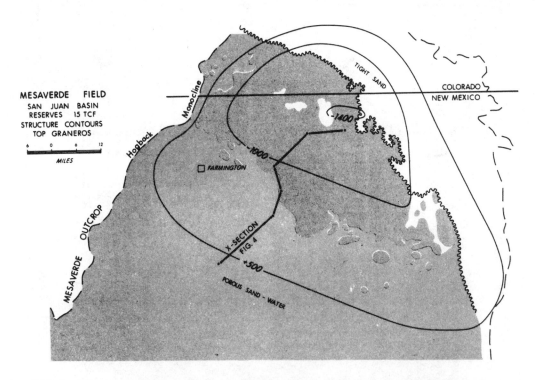

FIG. 5—Mesaverde gas trap sealed by water-saturated sands. Gas extends to within 2 mi of outcrop.

tight permeability barrier but a water-production zone. It has been proved that an exceedingly strong hydrodynamic gradient of such magnitude that it provides the trapping mechanism for this exceedingly large gas column exists along the south flank of the Blanco gas field" (Hill et al, 1961).

ALBERTA

Having identified certain geologic characteristics of a giant sandstone gas field, the next step is to consider the geology and distribution of gas in the Alberta basin. Space limitations make it difficult to present electric log cross sections because the size of the basin, both in width and depth, is so great. In order to maintain a log scale which is readable, I have prepared only a segment of a complete log section (Fig. 8) along with a diagrammatic section (Fig. 7) for clarification. They show the following.

1. The entire Mesozoic rock section, above the Paleozoic unconformity, is only 1,000 ft (300 m) thick in eastern Alberta. The section thickens gradually westward for 300 mi (480 km) and then abruptly into the Deep Basin to as much as 15,000 ft (4,570 m) where it is overridden by the Foothills overthrusts.

2. Thick sandstones in the west thin gradually eastward and many pinch out.

3. The sands in the eastern and central parts of the area, on the Alberta shelf, are moderately porous (20 to 25%) and permeable (over 50 md).

4. The sandstones on the shelf are characteristically water saturated.

5. Most of the gas and oil traps on the shelf are in updip lobes of porosity. They are conventional stratigraphic traps, gas on top of water, and their size is measured in billions of cubic feet of recoverable gas.

6. Resistivity of the entire rock section on the shelf, both shales and water-bearing sandstones, is about 5 ohms. A hydrocarbon pay sandstone is usually easily recognizable at 20 ohms or greater.

7. As the sandstones dip westward into the Deep Basin, they become less porous and permeable. This is caused by added clay content in these more rapidly buried sediments, greater compaction, and increased cementation and diagenesis.

8. As the porosity and permeability of the sandstones decrease downdip, the electrical resistivity shows that the section becomes gas saturated (the technical basis for this statement is described in a later section). West of a specific transition zone for each sand layer, the resistivity exceeds 20 ohms and in places reaches 200 ohms. It never falls back below 20. There is no longer any movable water in the rocks. They are entirely gas satu-

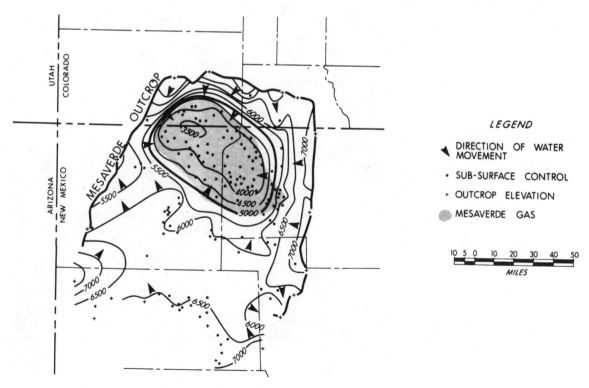

FIG. 6—Potentiometric surface map of Mesaverde sandstone. From Hill et al (1961).

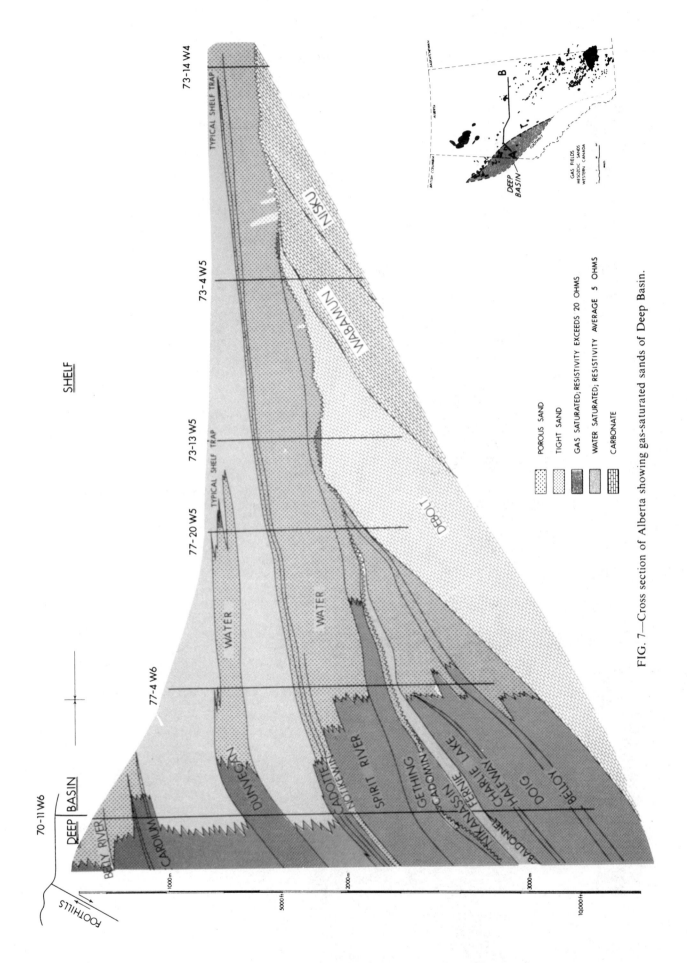

FIG. 7—Cross section of Alberta showing gas-saturated sands of Deep Basin.

POROUS SAND

TIGHT SAND

GAS SATURATED; RESISTIVITY EXCEEDS 20 OHMS

WATER SATURATED; RESISTIVITY AVERAGE 5 OHMS

CARBONATE

SHELF

DEEP | BASIN

FOOTHILLS

70-11 W6

77-4 W6

77-20 W5

73-13 W5

73-4 W5

73-14 W4

TYPICAL SHELF TRAP

TYPICAL SHELF TRAP

NISKU

WABAMUN

DEBOLT

WATER

WATER

BELLY RIVER

CARDIUM

DUNVEGAN

CADOTTE

NOTIKEWIN

SPIRIT RIVER

GETHING

CADOMIN

NIKANASSIN

FERNIE LAKE

CHARLIE LAKE

HALFWAY

BALDONNEL

DOIG

BELLOY

1000m

2000m

3000m

5000ft

10000ft

GAS FIELDS
MESOZOIC SANDS
WESTERN CANADA

DEEP
BASIN

BRITISH COLUMBIA

ALBERTA

SASKATCHEWAN

A

B

rated in the same way as the Cretaceous section of the San Juan basin. However, the Deep Basin section exceeds 15,000 ft (4,570 m) in thickness, covers a much larger area, and therefore contains an enormously larger trap volume.

9. From the Gething formation upward, the increase in resistivity related to gas saturation proceeds in steps. First, the lower Spirit River exceeds 20 ohms, then the upper Spirit River, then the Paddy-Cadotte, the Dunvegan, etc. On the highly exaggerated vertical scale of this cross section it appears that the change is occurring at progressively shallower depths. Actually, over a horizontal distance of 100 mi (160 km), the change generally occurs at a well depth of 2,500 to 4,500 ft (760 to 1,370 m). This amounts to virtually a level surface. It is certainly not an inverted gas below water contact. Instead, the Cretaceous sandstones, all rather similar rocks, probably reached a critical reduction in porosity and permeability at about the same depth of burial. The Jurassic and Triassic rocks, being of more variable composition, did not behave so uniformly in response to compaction.

10. Certain exceptions to the above generalizations seem to prove the rule. The Cadomin conglomerate maintains its good porosity and permeability deep into the basin. Consequently, it carries water far west of the gas-saturation boundary in the overlying beds. Only in parts of northeast British Columbia, close to the foothills, does it appear to be fully gas saturated. The beach facies of the Notikewin sand is porous and permeable as far west as we have seen it, and carries water. The Baldonnel, a porous Triassic dolomite present in northeast British Columbia, everywhere carries water and produces gas only on closed structures or in updip truncation traps.

11. With very limited exceptions the entire Mesozoic rock section in the Deep Basin is saturated with gas below a depth of about 3,500 ft (1,065 m). Within this area it is not possible to drill a dry hole; noncommercial wells, yes, but no completely dry holes. Every stringer of porosity holds gas.

12. Once west of the transition zone and into the "gas package" it is possible, indeed surprisingly common, to find "streaks" of good porosity and permeability. Rarely, we have found marine conglomerate layers, gas saturated, with permeabilities up to several darcys. Flow rates are very high. Commonly, we have found perfectly acceptable reservoir rock of 10 to 15% porosity and 1 to 10 md which will produce 1 to 3 MMcf of gas per day. These areas of acceptable porosity and permeability may extend over several townships and carry producible reserves of 1 Tcf or more. (Similar streaks of good-quality reservoir within the

gas package are well known in the San Juan basin.)

13. The cross-section well in 70-11 W6 represents the great Elmworth field. Tests have flowed gas at variable rates from different wells in the field from zones in the Cardium, Dunvegan, Paddy, Cadotte, Falher, Nikanassin, Halfway, and Belloy. The presence of 12 separate gas zones in this field (nearly every porous zone is gas saturated) demonstrates an unusual trap condition and requires a regional interpretation. It is certainly not a big anticline out on the Alberta shelf. It is a stratigraphic trap, albeit a most unusual one because it has multiple pays and the water is updip.

Because of the page size of the *Bulletin* it is not possible to reduce an electric log of a section several hundred miles long and 12,000 ft (3,600 m) deep and maintain readable data. For that reason, I presented in Figure 7 a diagrammatic section taken from an actual electric log section. The preceding comments and conclusions were derived from electric logs, not the diagrammatic section. Figure 8 shows a portion of the original electric log section. All resistivity less than 20 ohms (ohm meters) is gray; over 20 is red. Log analysis of several thousand wells tells us that, on average, 20 ohms is critical. Sand-shale sections exceeding 20 ohms of resistivity are generally gas saturated. Much resistivity on the logs considerably exceeds this number. The westernmost two logs have resistivities of 50 to 200 ohms. We are satisfied from regional data that the resistivity increases cannot be explained by decreasing water salinity, increasing rock density, cement or mineral content, or any other rock characteristic. The excess resistivity is caused by gas.

The numerous pay zones at Elmworth have been discussed. Log 11-15-70-11 W6 at the west end of the section is the Canadian Hunter discovery well. The portion of the log displayed on the section shows eight pay zones totaling 189 ft (57.6 m). In the entire log there is a total of 302 ft (92 m) of gas saturation in 12 pay zones. Most of the zones are of relatively poor reservoir quality but they are by no means beyond reach technologically or economically. Porosities average 10% and permeabilities are estimated from logs to average 0.5 md. These parameters are representative of a large part of the gas reservoirs in the Deep Basin.

The three Falher (Spirit River) pay zones represent some of the good porosity-permeability "streaks" which may be found inside the "gas package." Conglomerate is present in each zone with permeabilities which range from 50 md to several darcys. Drill-stem tests of these zones flowed 2 to 10 MMcf of gas per day.

Figure 9 is a map of the Cadotte sandstone showing the regional change in resistivity of a specific sand layer. In this sandstone the critical resistivity happens to be about 40 ohms, depending on porosity. All drill-stem test recoveries are shown. East of the 40-ohm line there are 66 water tests; west there are none. West of the line the recoveries are either mud or gas. The gas flows are as much as 2.4 MMcf per day. The gas fields east of the line are all conventional traps with gas on water. The trapping mechanisms in the large fields in the Pouce Coupe area are anticlines.

West of the 40-ohm line is a belt of gas-saturated Cadotte sandstone 150 mi (240 km) long by 40 mi (64 km) wide. Within this belt, wherever the sandstone is sufficiently porous and permeable, it will produce gas. Two low-resistivity wells appear as anomalies in the high-resistivity area: the well in 69-9 with resistivity of 40 and the well in 61-4 with resistivity of 25. However, because of the porosities and water resistivities both wells are calculated to be gas saturated.

Assuming only 10 ft (3 m) of effective porosity in the blanket layer of 50 to 100 ft (15 to 30 m) of sandstone, per section potential recoverable reserves would be about 3 Bcf and potential recoverable reserves of the entire belt would be 18 Tcf. It is not the purpose here to make a detailed analysis of the gas reserves of the Cadotte. It is enough to point out that the map shows a generally consistent increase of resistivity downdip to the southwest, and it indicates a large trap area with large potential reserves.

There are 10 to 20 porous zones in addition to

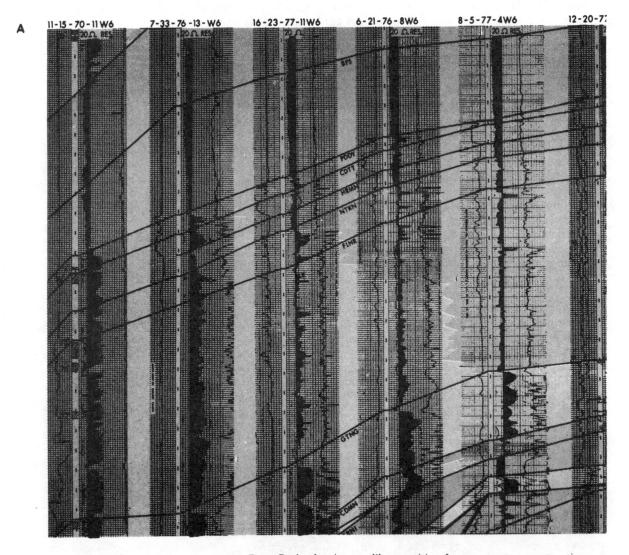

FIG. 8—Electric log section across Deep Basin showing steplike transition from water to gas saturation.

the Cadotte in the Deep Basin section. The general gas area of each may be defined by a resistivity map combined with gas and water shows. Further detailing of porosity may be obtained by logs, samples, and cores.

Figure 10 is an isopach map of the total gas-saturated section of the Deep Basin. It was constructed from 175 electric logs which were measured to determine the total net thickness of post-Mississippian section over 20 ohms resistivity. Part of the marine Cretaceous shale section never exceeds 20 ohms although every interbedded sandstone or siltstone layer does. The net gas section reaches a maximum thickness probably in excess of 10,000 ft (3,000 m). The reliability of this map was checked by a more detailed map which utilized 350 control wells that penetrated as

far as the Jurassic Fernie shale. The wedge shape and consistent westward increase in gas saturation was entirely confirmed. The data show a long, narrow wedge of gas-saturated rock over 400 mi (640 km) long and an average 60 mi (96 km) wide. At the northwest end the wedge of gas disappears as the bulk of the sandstone section changes to shale. To the southeast the wedge narrows to a thin strip in front of the foothills and probably extends well south of the map's boundary. I suspect that it could be traced, discontinuously perhaps, in a narrow belt in front of the foothills across southern Alberta, Montana, and south at least as far as the Cretaceous gas fields on the west side of the Green River basin in southwestern Wyoming. Conceptually, the Deep Basin trap may be 1,300 mi (2,100 km) or more in

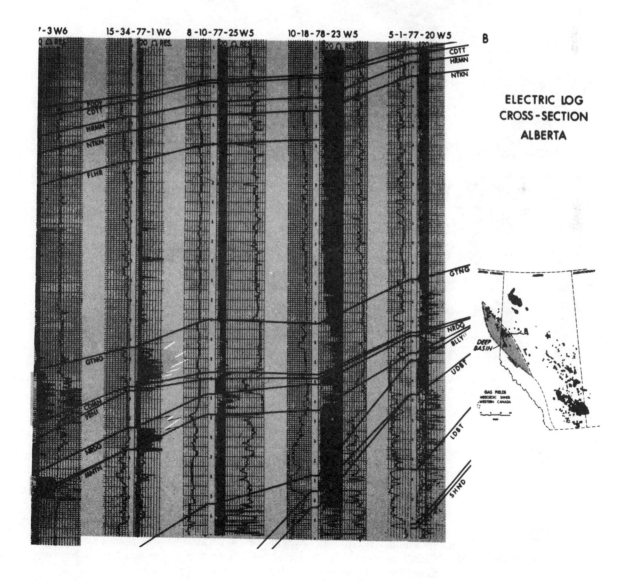

ELECTRIC LOG
CROSS-SECTION
ALBERTA

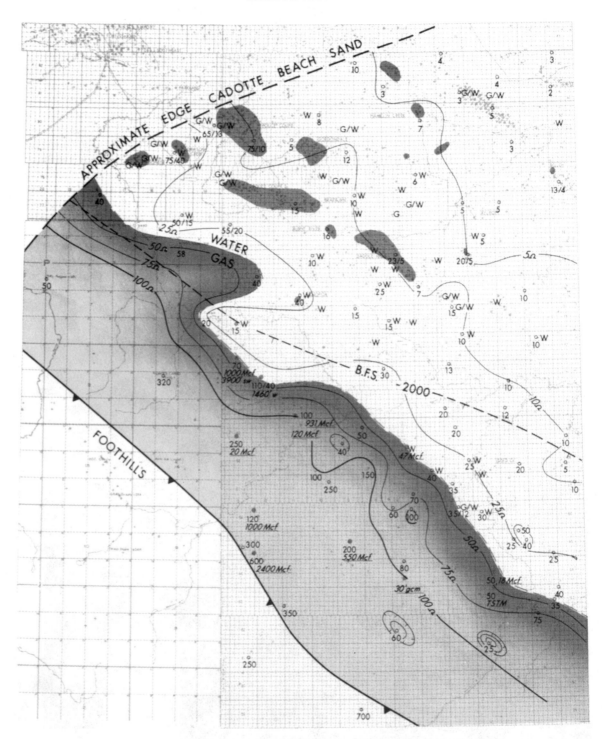

FIG. 9—Resistivity of Cadotte sand with formation tests and productive fields.

length.

The bulk volume of the entire Mesozoic rock section of the Deep Basin, in which virtually all porosity is gas saturated, is 30,000 cu mi (125,040 cu km). It compares to approximately 2,000 cu mi (8,336 cu km) for the San Juan basin. This tells us something about the incredible amount of gas which was generated in the Mesozoic rocks of western Canada and which has been effectively trapped beneath updip water.

WATER

This study did not include a thorough hydrodynamic evaluation. However, some significant information has been compiled. Figure 6, a potentiometric surface map of the San Juan basin Mesaverde sandstones, shows downdip water flow against the gas accumulation (Hill et al, 1961). The other producing zones in the San Juan basin have similar hydrodynamic conditions. These maps are of more than ordinary importance because they are part of the geologic pattern of the largest field in western North America.

Figure 11 is a potentiometric surface map of the Milk River sandstones in southeastern Alber-

ta, which comprise the largest gas field in Canada. The sandstones crop out at the south edge of the map. Water flow is downdip against the gas accumulation. Along the southwest edge of the field the gas saturated sandstones pass updip through a gas to water transition zone 6 to 8 mi (9.6 to 15 km) wide and then into free water. In the transition zone, the water is more saline than in the free water area (Suffield Block Study Committee, 1972; Map G-3). These same relations were observed in the San Juan basin Mesaverde field.

The eastern, updip edge of the Wattenberg field in the Denver basin is partly a permeability seal, but partly also a gas to water transition zone

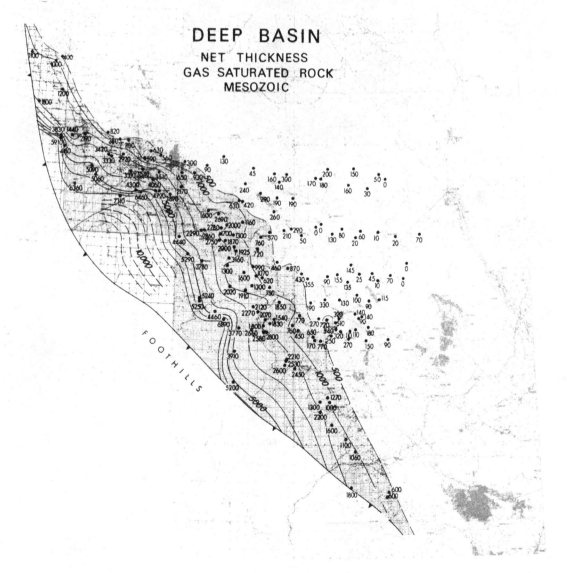

FIG. 10—Isopach of Mesozoic section which exceeds 20 ohms resistivity.

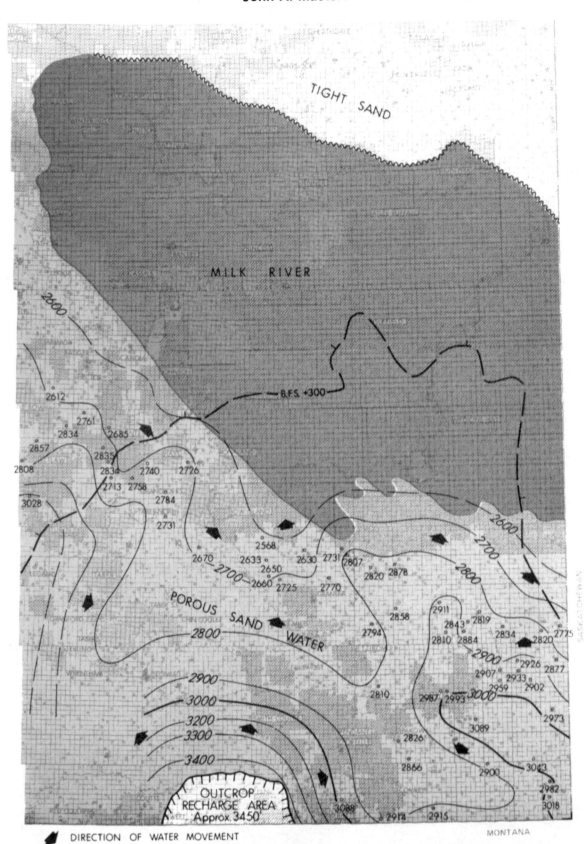

FIG. 11—Potentiometric surface map of Milk River sand and base of Fish Scales structure.

(R. A. Matuszczak, personal commun.). I do not know whether potentiometric mapping shows downdip water flow.

In the Deep Basin, hydrodynamic studies have seldom been used in exploration thus hydrodynamic information is lacking. However, a potentiometric surface map of the Cadotte sandstone (Fig. 12) was constructed as part of the example data on one of the porous zones in the Deep Basin. It shows clearly a westward, downdip water flow against the gas trap.

Hitchon (1964) published maps showing fluid flow in various formations in western Canada. While he demonstrated downdip flow in the Up-

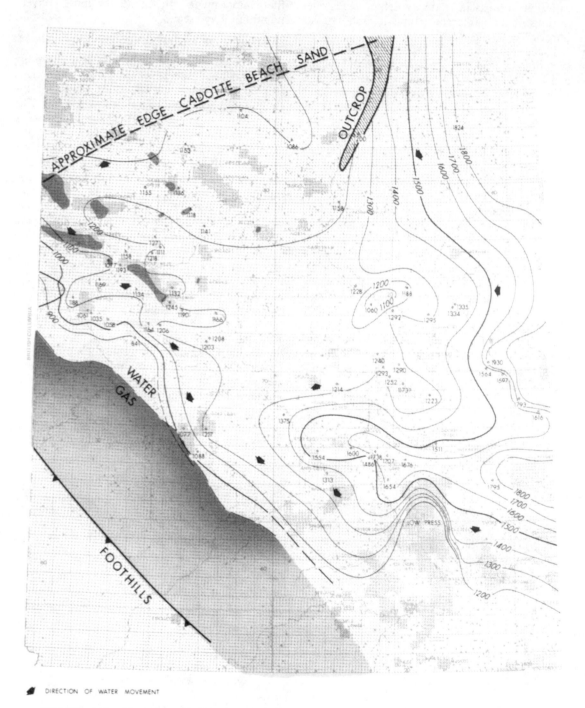

DIRECTION OF WATER MOVEMENT

FIG. 12—Potentiometric surface map of Cadotte sand and gas-water water line from resistivity map (Fig. 9).

per Cretaceous Belly River, Cardium and Viking, he showed updip flow in all beds below. However, the entire Lower Cretaceous section, 2,000 ft (610 m) thick, was lumped together so the map was highly generalized. At present, it is certain only that there is updip water in fairly porous and permeable sandstones which grades transitionally into downdip gas in relatively tighter sandstones in most of the Cretaceous, Jurassic, and Triassic porous zones (Fig. 13).

Exactly how the gas is trapped is an intriguing scientific question which will probably require considerable research to answer. Hydrodynamics seems to offer a solution by creating a back pressure against the buoyant force of the gas. However, there is some doubt that the Lower Cretaceous to Triassic beds show downdip water flow. Another possible solution is the effect that high water saturation has on permeability to gas in low-permeability rocks. High water saturation could cause a "water block" similar to the common formation damage around well bores in low-permeability sandstone. Figure 14 is a representative curve of relative permeability to gas for a low-permeability rock (1 md). At 10% water saturation, permeability to gas is the same as permeability to water with 100% water saturation. However at 40% water saturation (normal for Deep Basin gas reservoirs), permeability to gas is only three-tenths as much as water. At 65% water saturation, the rock is almost completely impervious to gas flow. This general relation is well known and has been described in many engineering papers (Keelan, 1972).

In Figure 13, we might imagine that at the gas-water interface the water saturation in the gas sandstones reaches 65%, at which point a rock with 1 md permeability would have no permeability to gas. A water block would be all the more effective with a large hydrostatic head behind it, and still more effective with a hydrodynamic head. The critical trapping mechanism would be fluid impairment rather than the more conventional lithologic seal.

SUMMARY OF TIGHT SANDSTONE GAS TRAPS

To summarize the regional aspects of downdip, tight sandstone gas traps, Figure 15 presents cross sections and maps, on the same scales, of the Deep Basin, San Juan basin, and Milk River. The Deep Basin has been described already in some detail. The San Juan basin cross section shows not only the Cliff House and Point Lookout gas sandstones of the Mesaverde (see Fig. 4), but also the other sands of the basin which make up a multipay "gas package" 5,000 ft (1,525 m) thick in which all porosity is gas saturated. Note that each gas sandstone is water saturated updip and the direction of water flow is downdip. Total reserves of the basin are 25 Tcf, although infill drilling to improve drainage in the Mesaverde may increase reserves another 8 Tcf (Long, 1977). The San Juan "gas package" contains about 2,000 cu mi (8,336 cu km) of rock in which all porosity is gas saturated. This compares with approximately 30,000 cu mi (125,000 cu km) in the mapped portion of the Deep Basin of western Canada.

The Milk River cross section shows the porous, beach sands of the Milk River at outcrop in southern Alberta. Slightly downdip to the north-

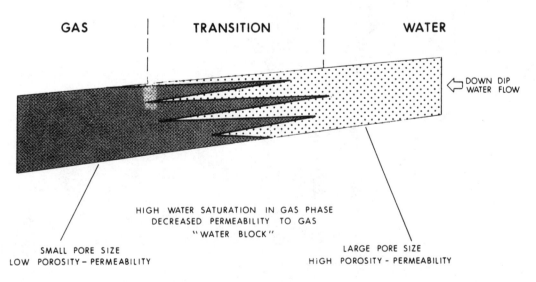

FIG. 13—Physical forces holding gas in place are not fully understood. Downdip water flow, or "water block," or both appear to be involved.

east they are the freshwater aquifer for hundreds of farm wells. Farther northeast they begin to change facies to offshore, marine, laminated thin sands and shales with an accompanying loss of porosity and permeability. It is in this facies that the great Milk River gas field is trapped. Like the San Juan basin field, water flow is downdip and there is no stratigraphic or structural barrier between the gas and water. There is only a gas to water transition zone about 6 to 8 mi (9.6 to 15 km) wide. Porosity and permeability in the water-saturated beach sands are about 25% and 100 md. In the gas phase these are reduced to about 14% and 1 md. Total Milk River reserves are in the range of 9 Tcf recoverable.

Milk River and San Juan basin are the two largest gas fields in western North America (ex-

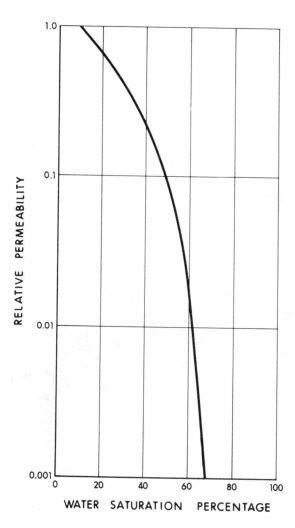

FIG. 14—Representative curve of relative permeability to gas of a low permeability (1 md) rock, taken from specific examples.

cluding Prudhoe Bay). The Deep Basin may be much larger than all three.

LOG ANALYSIS

Between 1973 and 1977 the price of gas in Canada increased nine-fold. There were, in that time period, about 25,000 dry holes which had already been drilled in western Canada, and logged and abandoned in view of the existing economics. Suddenly, for the first time in the history of the oil business, there was the opportunity to review thousands of well logs in the light of a nine-fold improvement in price, to say nothing of very substantial improvements in well completion technology. There was available perhaps $5 billion worth (at today's costs) of hard subsurface data, far superior to geophysics, to tell whether gas or oil was, or was not, present. The log analyses, combined with drill-stem tests, samples cuttings, mud logs, etc, would be reasonably definitive. It was critical to develop the highest level of electric log analysis technology.

The log analysis principles we use are similar to those used by the rest of the industry. Everyone is basically dependent on variations of the Archie saturation equation. We differ perhaps in the extent to which results are cumulated by formation, cross-checked against test and production results, and recalculated in light of regional relationships. We at Canadian Hunter may differ markedly in our attitude to log analysis in that we consider it our preeminent geophysical tool. Management and geologists both are acutely aware of its strengths, as well as its weaknesses, and the information is applied vigorously to exploration. The unusually productive results we have had are related in part to a thorough expertise in the subject, but in larger measure to the fact that apparently no one else in Canada had used the tool on a large exploration scale. We experienced the same burst of exploratory success which accompanied the introduction of other new exploration tools (e.g., surface mapping, core drilling, gravity, seismic, etc).

While it is true that in the last few years most companies have become reasonably sophisticated at analyzing the modern log suites, few have studied on a large scale the mass of old data accumulated from 1950 to 1970. During this time period most companies were directing their search toward the deep carbonate section. Little effort was made to test or analyze the overlying Mesozoic clastic section. In the light of increased prices for gas, Canadian Hunter log analysts concentrated on the clastic section.

The first stage of our exploratory analysis consisted of an experienced log analyst "scanning"

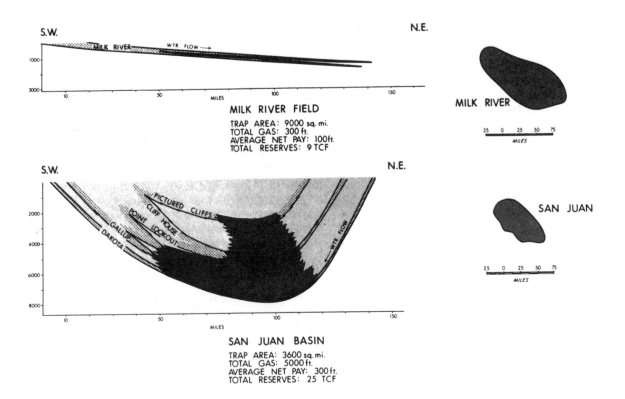

MILK RIVER FIELD

TRAP AREA: 9000 sq. mi.
TOTAL GAS: 300 ft.
AVERAGE NET PAY: 100 ft.
TOTAL RESERVES: 9 TCF

SAN JUAN BASIN

TRAP AREA: 3600 sq. mi.
TOTAL GAS: 5000 ft.
AVERAGE NET PAY: 300 ft.
TOTAL RESERVES: 25 TCF

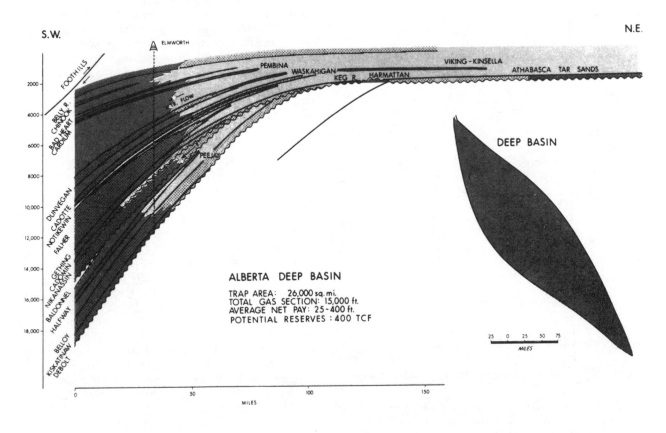

ALBERTA DEEP BASIN

TRAP AREA: 26,000 sq. mi.
TOTAL GAS SECTION: 15,000 ft.
AVERAGE NET PAY: 25-400 ft.
POTENTIAL RESERVES: 400 TCF

FIG. 15—Diagrammatic cross sections, on equal scales, of largest tight-sand gas traps.

well logs to look for zones of potential interest. This method enabled us to recognize almost instantly potential oil and gas zones, to process rapidly thousands of logs, and to select favorable wells for more detailed and thorough examination. Preparation of digitized logs and development of programs that would incorporate the sophisticated pattern recognition of an experienced analyst would have been too slow for scanning purposes.

The second stage of analysis consisted of a detailed examination of all logs on selected wells and calculation by intervals of porosity, water saturation, and recoverable hydrocarbon volume (Bcf per section). Permeability (K) was also calculated according to equations which combine porosity, water saturation, and a constant for the formation and area (Schultz, 1976). The results are useful in giving order of magnitude estimates. Permeability thickness (Kh) is found by multiplying permeability times thickness. The final step is to plot Kh versus test data and thereby develop an equation to estimate open-flow potential, or XOF.

The third stage of formation evaluation became the regional determination of petrophysical parameters for input into well-log calculations. Information from one well was no longer enough. However, the volume of data demanded computer recording and analysis.

Information on a particular zone for all wells in the surrounding region is collected. This includes porosities, water salinities, calculated saturations, and actual test data. The data for that zone are then assembled on a graph of porosity versus water saturation as shown in Figure 16. Water tests, hydrocarbon tests, and tight tests are indicated. From this data of log calculations, combined with test results, appropriate cutoff values are determined. These include a porosity cutoff, a water saturation cutoff, and a porosity times water saturation (ϕSw) cutoff. As shown in Figure 16 these cutoffs described the boundaries of values considered to be net pay. Presently, we have porosity-saturation charts on all significant pay zones in various geographic areas. Many of these charts have more than 100 data points.

We now use a detailed interval analysis as

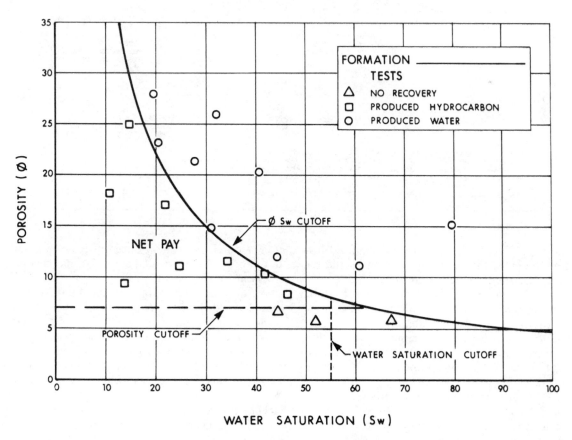

FIG. 16—Chart of porosity versus water saturation with test data to select approximate cutoffs for determination of net pay.

Table 2. Computer Print Out of Log Analysis from Digitized Logs

```
20/07/78              LOG ANALYSIS SUMMARY
                      WELL NAME: ELMWORTH
                      LOCATION: 11-15-70-11-W6
```

FORM	H	PORO WT	SW WT	PHIH SUM	KH SUM	BCF 0-25%	BCF 3-25%	BCF 7-25%	BCF 11-25%	BCF 15-25%	XOF mcf	
BLYR	19	21.	46.	4.1	83.7	1.5	1.5	1.5	1.5	1.5	324.	GROSS
	15	21.	45.	3.2	66.7		1.2	1.2	1.2	1.2	258.	NET ON 3% POROSITY
	15	21.	45.	3.2	66.7			1.2	1.2	1.2	258.	NET ON 7% POROSITY
BDHT	7	11.	77.	0.8	0.2	0.0	0.0	0.0	0.0	0.0	0.	GROSS
	0	0.	0.	0.0	0.0		0.0	0.0	0.0	0.0	0.	NET ON 3% POROSITY
	0	0.	0.	0.0	0.0			0.0	0.0	0.0	0.	NET ON 7% POROSITY
CRDS	18	20.	49.	3.6	50.0	2.3	2.3	2.3	2.3	2.3	586.	GROSS
	0	0.	0.	0.0	0.0		0.0	0.0	0.0	0.0	0.	NET ON 3% POROSITY
	0	0.	0.	0.0	0.0			0.0	0.0	0.0	0.	NET ON 7% POROSITY
DUNV	22	12.	59.	2.7	4.3	1.6	1.6	1.6	1.0	0.5	149.	GROSS
	11	12.	44.	1.3	3.3		1.6	1.6	1.0	0.5	115.	NET ON 3% POROSITY
	11	12.	44.	1.3	3.3			1.6	1.0	0.5	115.	NET ON 7% POROSITY
PDDY	11	11.	33.	1.2	4.9	2.1	2.1	2.1	1.0	1.0	260.	GROSS
	11	11.	33.	1.2	4.9		2.1	2.1	1.0	1.0	260.	NET ON 3% POROSITY
	11	11.	33.	1.2	4.9			2.1	1.0	1.0	260.	NET ON 7% POROSITY
CADT	23	10.	60.	2.2	0.4	1.6	1.6	1.6	0.0	0.0	24.	GROSS
	4	11.	39.	0.4	0.2		0.7	0.7	0.0	0.0	12.	NET ON 3% POROSITY
	4	11.	39.	0.4	0.2			0.7	0.0	0.0	12.	NET ON 7% POROSITY
FLAC	17	6.	32.	1.0	0.0	1.9	1.9	0.8	0.0	0.0	3.	GROSS
	17	6.	32.	1.0	0.0		1.9	0.8	0.0	0.0	3.	NET ON 3% POROSITY
	7	7.	43.	0.5	0.0			0.8	0.0	0.0	2.	NET ON 7% POROSITY
FLAS	35	7.	44.	2.3	0.2	3.8	3.8	2.0	0.0	0.0	14.	GROSS
	35	7.	44.	2.3	0.2		3.8	2.0	0.0	0.0	14.	NET ON 3% POROSITY
	16	8.	46.	1.3	0.2			2.0	0.0	0.0	10.	NET ON 7% POROSITY
FLAC	17	11.	36.	1.8	1.3	3.3	3.3	3.3	0.0	0.0	78.	GROSS
	17	11.	36.	1.8	1.3		3.3	3.3	0.0	0.0	78.	NET ON 3% POROSITY
	17	11.	36.	1.8	1.3			3.3	0.0	0.0	78.	NET ON 7% POROSITY
FLBS	49	6.	49.	3.0	0.2	4.5	4.5	2.3	0.0	0.0	10.	GROSS
	49	6.	49.	3.0	0.2		4.5	2.3	0.0	0.0	10.	NET ON 3% POROSITY
	19	7.	44.	1.4	0.1			2.3	0.0	0.0	9.	NET ON 7% POROSITY

(CONTINUED.................)

```
-WELL SUMMARY-

       570         45.7 154.8  66.2  66.2  42.2   7.0   5.2  2729.  GROSS
       376         29.2  86.2        36.5   4.4   2.6   0.0  2000.  NET ON 3% POROSITY
       189         18.8  85.3               4.4   2.6   0.0  1898.  NET ON 7% POROSITY
```

*** GROSS VALUES INCLUDE ALL INTERVALS WITH LESS THAN 65% WATER SATURATION
*** NET VALUES INCLUDE ALL INTERVALS WITH LESS THAN 65% WATER SATURATION
    THAT ARE UNDER THE POROSITY-WATER SATURATION CUTOFF FOR THAT ZONE
*** NET VALUES ARE ALSO DETERMINED BY POROSITY CUTOFFS I.E. GREATER THAN 3% POROSITY AND GREATER
    THAN 7% POROSITY
*** IN SANDS USE THE NET VALUES ON 7% POROSITY AND IN CARBONATES USE THE NET VALUES ON 3% POROSITY

shown in Table 2 generated from digitized logs. This provides the same information as the second-stage analysis described above, except it is done by computer so that information is stored for eventual recalculation with different parameters, retrieval of selected data, etc.

Table 3 is a log analysis summary which sums or averages the various measurements, applies porosity-water saturation cutoffs, and records recoverable Bcf per section in various categories of porosity. It is then possible to retrieve from the computer information such as "all wells in a selected area in Dunvegan with gas reserves in rock over 11% porosity, more than 5 ft thick, and with Kh greater that 2.5." The information will be vital to a resource assessment once adequate frac and test information are available to cross plot against reservoir measurements.

We are now developing the capability to provide detailed formation evaluation from well logs using advanced minicomputers. The new procedure involves digitizing the logs foot by foot and applying various corrections such as shale, hydrocarbon, borehole size, etc. Well logs recorded on magnetic tapes in the field will be processed in this system.

The final result will be a well calibrated analysis of available log data. Part or all of the procedure has been applied to more than 6,000 wells carefully selected to provide evaluation of critical

Deep Basin Gas Trap, Western Canada

Table 3. Summary of Log Analysis by Formation*

```
20/07/78
            WELL OPERATOR: CANADIAN HUNTER EXPLORATION
            WELL NAME: ELMWORTH

            LOCATION: 11-15-70-11-W6

            PROVINCE: ALTA      FIELD: ELMWORTH
            KB: 2350            FTD: 10202   IN TYLF
            DATE LOGGED: MAR 2/76      DATE ANALYZED: MAY/76/MAY 23/78
            LOGS RUN: DIL/BHCSG-C/ENL-FDC
            ANALYST: ETC             WELL STATUS: GAS WELL
```

INTERVAL	H	ACOU	NEUT	DENS	RT	PORO	SW	PHISW	PHIH	BCF	KH	RWA
BLYR												
1388- 1392	4	87.0	0.0	0.00	55.	22.1	43.	953.	0.89	0.3	15.8	2.70
1414- 1418	4	88.0	0.0	0.00	45.	22.9	46.	1054.	0.92	0.3	17.0	2.36
1421- 1426	5	90.0	0.0	0.00	60.	24.4	37.	913.	1.22	0.5	47.5	3.58
1590- 1596	6	81.0	0.0	0.00	50.	17.6	57.	1000.	1.05	0.3	3.4	1.54
BDHT												
2405- 2412	7	73.0	0.0	0.00	45.	11.5	77.	882.	0.80	0.0	0.2	0.59
CRDS												
2626- 2632	6	90.0	0.0	0.00	23.	24.4	47.	1142.	1.47	1.0	36.4	1.37
2632- 2636	4	86.0	0.0	0.00	30.	21.4	47.	1000.	0.85	0.6	10.9	1.37
2636- 2642	6	79.0	0.0	0.00	40.	16.0	54.	866.	0.96	0.6	2.2	1.03
2645- 2647	2	78.0	0.0	0.00	45.	15.3	53.	816.	0.31	0.2	0.6	1.05
DUNV												
4452- 4457	5	77.0	0.0	0.00	20.	15.8	74.	1162.	0.79	0.0	0.9	0.50
4670- 4676	6	70.0	0.0	0.00	80.	10.5	55.	581.	0.63	0.6	0.2	0.89
4695- 4698	3	72.0	0.0	0.00	180.	12.0	32.	387.	0.36	0.5	0.5	2.61
4713- 4715	2	79.0	0.0	0.00	70.	17.3	36.	621.	0.35	0.5	2.6	2.09
4883- 4889	6	69.0	0.0	0.00	55.	9.8	72.	701.	0.59	0.0	0.1	0.53
PDDY												
5834- 5838	4	68.0	0.0	0.00	200.	9.0	32.	292.	0.36	0.7	0.1	1.63
5866- 5870	4	68.0	0.0	0.00	100.	9.0	46.	412.	0.36	0.5	0.1	0.81
5870- 5873	3	77.0	0.0	0.00	110.	15.8	25.	393.	0.47	1.0	4.7	2.74
CADT												
5907- 5911	4	70.0	0.0	0.00	100.	10.5	39.	412.	0.42	0.7	0.2	1.11
5918- 5930	12	68.0	0.0	0.00	40.	9.0	72.	652.	1.08	0.0	0.1	0.33
5940- 5947	7	69.0	0.0	0.00	65.	9.8	52.	511.	0.68	0.9	0.1	0.62
FLAC												
6250- 6257	7	65.0	0.0	0.00	180.	7.1	43.	307.	0.50	0.8	0.0	0.91
6257- 6267	10	62.0	0.0	0.00	1600.	4.9	21.	103.	0.49	1.1	0.0	3.79
FLAS												
6267- 6271	4	67.0	0.0	0.00	1600.	6.8	19.	129.	0.27	0.6	0.1	7.37
6271- 6280	9	63.0	0.0	0.00	600.	4.4	39.	175.	0.40	0.7	0.0	1.17
6280- 6286	6	69.0	0.0	0.00	110.	8.0	50.	399.	0.48	0.7	0.0	0.70
6286- 6292	6	65.0	0.0	0.00	160.	5.6	56.	316.	0.34	0.4	0.0	0.50
6292- 6302	10	70.0	0.0	0.00	130.	8.6	44.	379.	0.86	1.4	0.1	0.95
FLAC												
6308- 6317	9	70.0	0.0	0.00	120.	10.9	35.	376.	0.98	1.8	0.8	1.42

*Net figures are after application of porosity—water saturation cut offs.

areas in western Canada. Although the results have been impressive, no one is more aware than we are of the limitations of the data, in particular lack of reliable water-resistivity information. The effect has been to create various degrees of reliability by formation and by geographic area. We are in a continuous program of checking and reanalyzing specific problem areas, but the overall picture has not changed appreciably. Even now, with all the mistakes and limitations, our geologists probably have more reliable log-derived data on porosity, pay thickness, water saturation, and reserves per section than has ever previously been assembled in Canada. In light of this, it is perhaps understandable that our geologic interpretations and potential gas reserve estimates differ appreciably from conventional

BCF MAPS

One of the map suites most useful for exploration has been Bcf per section maps of individual formations. Figure 17 is an example map of the Cadotte sandstone. It was derived from our log

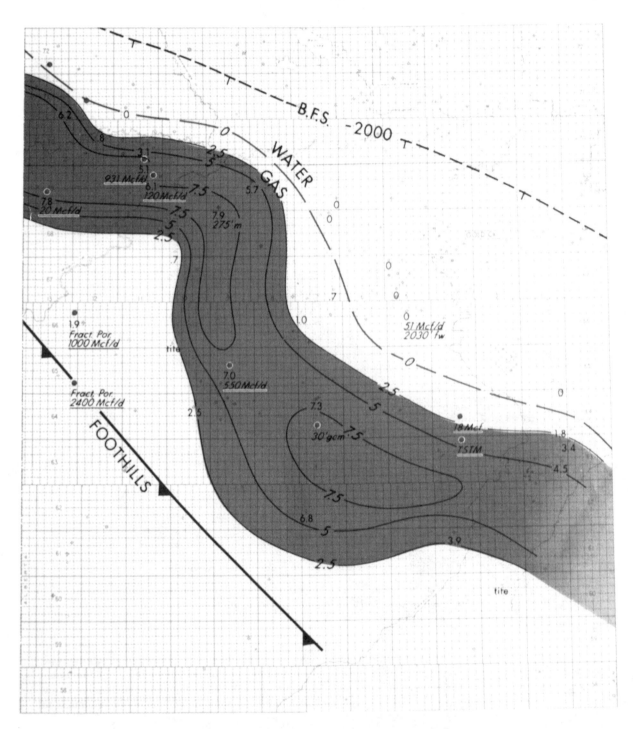

FIG. 17—Bcf per section map of Cadotte sand with formation tests.

analysis program which calculates porosity intervals, water saturation, pressure and temperature estimated from depth, and resultant Bcf per section after applying a recovery factor of 75%. By applying appropriate porosity and log-derived permeability cutoffs, we eliminate certain portions of the estimates and obtain a "recoverable reserve" number. Some geologists and engineers still may not agree with these numbers and, admittedly, they can be refined by more accurate data. Also there is room to differ on the porosity-permeability cutoffs. Therefore, our Bcf calculations may not be accurate by engineering standards in a local area. However, these data are internally consistent and are therefore an extremely valuable exploration tool for recognizing favorable areas, setting priorities, relative bid prices, etc. It is also superior data for assessing potential reserves. I would remind the reader that future reserve estimates are customarily done by assigning average recoveries per cubic mile of sediment or by projection of historic finding rates—both systems being subject to gross error.

The Cadotte sandstone map in Figure 17 shows the updip gas-water contact and a gas-saturated area covering 3,600 sq mi (9,325 sq km). Bcf per section are contoured to give a pattern to the data. As usual, care must be exercised in interpreting a contour map in areas of sparse well control. Nevertheless, the area of maximum gas is clearly indicated. The experienced explorationist will quickly see, however, that this map must be combined with information on porosity-permeability, drilling depths, land availability, etc, before the areas of maximum economic favorability can be determined. Nevertheless, this map alone is probably a more meaningful assessment of the gas potential of the Cadotte sandstone than is normally available by any other geologic or geophysical method. For the purposes of this analysis, we will consider only the area of the map within the 2.5 Bcf per section contours. The outlying portion on the northeast side is probably involved in the gas to water transition zone. The outlying portion on the southwest side, deep in the basin, is probably too tight for commercial production. Note that the wells along the foothills do not have significant matrix porosity although the logs do indicate fractures and there were significant gas flows.

As noted previously with regard to the Cadotte resistivity map (Fig. 9), west of the gas to water transition line there have been no water tests from the Cadotte. Drill-stem tests recover either gas or mud. The gas recoveries are substantial, giving good promise of commercial production following proper well stimulation. (Figure 18 will

show test recoveries from the early wells at Wattenberg as a comparison of reservoir quality and as an illustration of the kind of subtle information which leads to the discovery of giant tight sand gas fields.) The trend of Cadotte gas greater than 2.5 Bcf per section covers 1,600 sq mi (4,140 sq km). Note that this map has not been extended into British Columbia. At an average 6 Bcf per section there is indicated about 10 Tcf of recoverable gas. It should be noted that nothing has been said of the price necessary to recover these supplies. (I have also avoided using the term "reserves.") More detailed studies are necessary to arrive at estimates of the quantity of gas that can be recovered at $1.00/Mcf, $2.00/Mcf, etc. However, meaningful answers can be obtained. Suffice it to say that large quantities of gas are available at today's prices.

Before moving on to a discussion of total Bcf per section maps, we should consider Matuszczak's (1973) map showing the prediscovery wells at Wattenberg (Fig. 18). On the same scale as the Cadotte map of Figure 17, Matuszczak shows 22 wells in the Dakota J sandstone. A

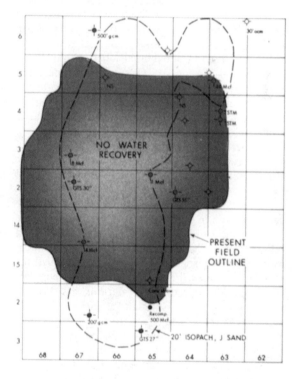

FIG. 18—Initial geologic outline of Wattenberg prospect, Colorado, showing drill-stem test recoveries (largest test was only 44 MCFD) from 21 wells. Present field outline superimposed. After Matuszczak (1973).

large area is indicated as water free with the test recoveries being either gas or mud. Many explorationists will be startled to see that the largest drill-stem test was only 44 Mcf per day. The outline of the present field, covering 970 sq mi (2,512 sq km) is superimposed. Recoverable reserves are 1.3 Tcf. Reservoir parameters are given in Table 1. Matuszczak (1973) said, "Hydraulic fracturing of the pay zone with sand, glass beads and water is a standard procedure. The J sandstone in this field has such low permeability that evaluation of a well is conclusive only after fracturing and testing through casing."

A similar map to Wattenberg could be constructed showing prediscovery wells in the San Juan basin Dakota field. There was a large area of only gas or mud recoveries, and gas flows were very small. All wells are hydraulically fractured to bring them onto production.

It is interesting, although not critically important, that the Basin Dakota, Wattenberg Dakota J, and Deep Basin Cadotte sandstones are all basal Late Cretaceous in age. These named fields are the larger of hundreds of oil and gas fields in the Dakota-Muddy-Viking-Cadotte continental-shoreline–shallow-water complex of sandstones which extends for 1,500 mi (2,415 km) from the San Juan basin in New Mexico to northwestern Alberta. Reservoir parameters of the Deep Basin Cadotte sandstones, from log analysis and test results, appear to be substantially better than Basin Dakota and Wattenberg Dakota. Only one small frac job has been done in the entire Cadotte gas trap area and the gas field awaits development. It is noteworthy that the Deep Basin Cadotte sandstone field alone could be a giant gas field.

Bcf per section maps by formation are useful for studying individual zones, but to see the whole picture, it is helpful to construct a total Bcf per section map (Fig. 19). This gross map is simply a cumulation of all the potentially recoverable gas measured in each well log. This is perhaps the first example of electric log analysis used on a regional scale as a reconnaissance geophysical tool. In many respects accuracy was sacrificed for speed, so long as the results were consistent. We were looking for economic relationships. We assumed constant pressure and temperature gradients over the whole region to simplify the gas-in-place calculations. Our water-salinity information was minimal. Some error was introduced by different depths of penetration, but it is significant only in the deeper parts of the basin where the entire Mesozoic section was not always drilled. The Bcf per section map was used primarily to define the area of maximum gas potential. Detailed exploration work within that area called

for much more accurate log analysis and gas-in-place calculations on each separate formation.

Again, space limitations prohibit printing a map covering all of Alberta and northeastern British Columbia. The well data would not be readable. However, Figure 19 shows the Bcf per section map covering part of the Deep Basin and adjoining shelf area. One may see the amount of well control, the Bcf per section values, the red circles for quantities of 2 to 7 Bcf, and red dots for greater than 7 Bcf. Note the much increased incidence of red dots in the shaded, Deep Basin area. Recognizing that the Deep Basin probably contains a continuous gas phase, it is sensible to attempt to contour Bcf per section although the resulting picture must be considered very cautiously if it is to be used for economic purposes.

Figure 20 is a simplified version of Canadian Hunter's total Bcf per section map which presents the basic statistical data derived from analysis of 4,731 logs. In the shelf area, we examined 4,187 logs and recognized gas (more than 2 Bcf) in 18% of them, averaging 6 Bcf per section. In the Deep Basin, we examined 544 logs and recognized gas in 84% of them, averaging 17 Bcf per section.

In large part, the log analysis was done quite independently of the geologic interpretation. A separate group of petrophysicists and computer experts worked on the logs and generally were relatively unaware of the location of the wells with respect to the geologists' concept of Deep Basin and shelf. Yet, the statistical results of 4,731 analyses outline with remarkable clarity the Deep Basin area and measure an abnormal incidence and quantity of gas in it. The geologic interpretation is essential to an adequate understanding of the potential of the area, but the log results alone are sufficient to find large quantities of gas.

The information on Figure 20 can be used to give a "broad brush" estimate of gas resources in the Deep Basin. Remember that our Bcf per section calculations are discounted to 75% "recoverable" and zones with estimated AOF's less than 250 Mcf per day are eliminated. Consider first, "proved" reserves. There are 472 wells in which we measure an average 17 Bcf per section. That amounts to 8 Tcf "proved" in the single sections around each well (although it does not yet meet the engineer's requirement of "tested"). "Probable" reserves would be the gas in a 1-mi (1.8 km) corridor around each well, or eight more sections, which amounts to an additional 64 Tcf. Total "proved" plus "probable" gas is 72 Tcf. (Total official proved gas reserves of Alberta and British Columbia are 59 Tcf; Canada Geol. Survey, 1977).

Recognizing that the separate wells are parts of

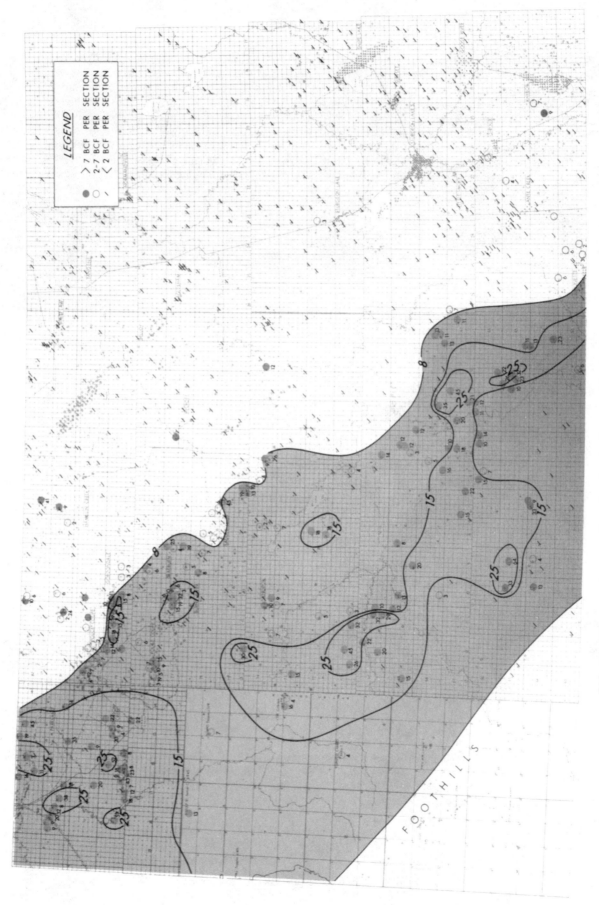

FIG. 19—Bcf per section map. Part of regional Alberta and British Columbia map.

a more or less continuous gas trap allows us to lump them together in an area 430 by 60 mi (690 by 96 km) totaling 26,000 sq mi (67,600 sq km). At an average 17 Bcf per section, the total gas resource calculates to be approximately 440 Tcf. This is still not a meaningful economic number, but it does present some measure of the resource base. It is now our task to determine what portions of the total can be economically recovered

at today's gas price of $1.00 after royalty, at $2.00 after royalty, at $3.00, etc. This will require considerable well fracturing and testing information.

In 1976 and 1977, Canadian Hunter drilled 36 wells in the Deep Basin, most of them classified as exploratory. Thirty wells were successfully completed. Flow rates in the completion zones ranged up to 15 MMcf per day and averaged 4 MMcf per day. Obviously, there is a significant

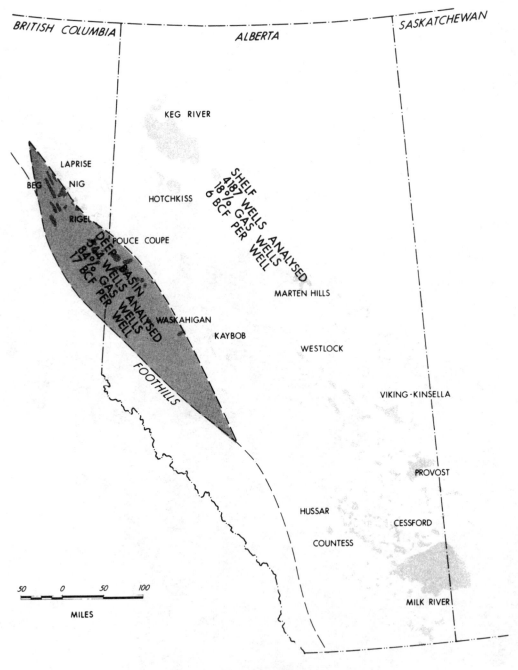

FIG. 20—Statistical data from regional Bcf per section map.

amount of good-quality reservoir rock within the "gas package." Based on our drilling experience, our opinion is that at today's net back price of $1.00/Mcf after royalty, the economically recoverable gas in the Deep Basin is in the range of 50 Tcf. At $2.00/Mcf the amount might increase to 150 Tcf. At present, however, we do not have sufficient frac and test information to calibrate against rock parameters in order to express anything more than "opinions" about the quantities of recoverable gas.

The Canada Geological Survey (1977) estimated a 50% probability that there is 38 Tcf remaining to be discovered in western Canada. That report stated, "It is likely that the larger pools, in most of the exploration plays, have already been found and exploration effort in the future will be devoted to searching for remaining smaller pools as well as improving the recovery from existing pools." Most major companies had apparently reached the same conclusion.

It is now apparent that western Canada has a supply of gas in the Deep Basin which has not been recognized in previous estimates of future reserves. It is large enough that it will probably significantly increase Canada's total resource base and have a significant impact on the economy of the country. These large supplies could displace part of Canada's foreign oil requirements and, exported to the United States, could resolve Canada's serious balance of payments deficit. In terms of total energy supply in North America, the Deep Basin gas will become a major factor.

FOOTHILLS

Of course, the great wedge of gas does not suddenly stop at the edge of the foothills. The entire gas-saturated Mesozoic section continues to thicken westward as it is caught up in the thrust slices and folding of the overthrust belt. In general, the porosity is much diminished in this area both by compaction from depth of burial and by silicification which seems to be a common process in overthrust provinces. However, there is also a significant amount of tectonic fracturing of the rocks which enhances permeability in a number of places.

Examination of all drill-stem tests in Mesozoic rocks of the foothills from 240 wells (through 1976) reveals the following: (1) only 49 drill-stem tests were taken in the Mesozoic section; (2) 25 tests recovered gas, 2 recovered water, the rest recovered mud; (3) 15 of the tests flowed greater than 1,000 Mcf per day.

Figure 21 shows the location of the gas tests along the 700 mi (1,100 km) foothills belt. All of these tests were drilled for Mississippian objectives. Consequently the structural location on the shallower Mesozoic beds is almost random— some high, some low, most on the flanks. Yet, practically no water was recovered; only gas or mud. Referring again to San Juan and Wattenberg, we know that this is the critical clue to an extensive tight-sand trap. Considering the foothills belt of Mesozoic sediment as simply the western extension of the Deep Basin it is, of course, logical to expect a fully gas-saturated section.

From the old cable tool wells, Turner Valley, the famous Mississippian field in the southern foothills, is recognizably a Cretaceous gas field. Logs of these wells recorded 357 separate gas shows (Conservation Board of Alberta, 1950). In 34 wells actual gas flows exceeded 500 Mcf per day. Seven of these wells flowed more than 4,000 Mcf per day. This gas does not come from the underlying Mississippian gas reservoir because it is sweet. Unfortunately, the Cretaceous sandstones are shallow and reservoir pressure is low so reserves at Turner Valley will not exceed a few hundred Bcf. But the field is a thoroughly documented example of an extensive Cretaceous sandstone gas accumulation in the foothills.

Six hundred miles (965 km) north is another, larger gas field at Grizzly on the Stoney Mountain anticline in British Columbia. Large reserves are present in Jurassic and Triassic sandstones with unevaluated shows in overlying Cretaceous sandstones. The paucity of water suggests a regional stratigraphic trap with the drilling merely concentrated on the structure by a geologic assumption that the field is structure-controlled. Fracturing related to the structure enhances permeability in the field area.

We believe that many more fields will be found in the foothills but at today's prices it is an economically marginal area in which to develop reserves. Well costs are very high and productivity is largely dependent on tectonic fracturing which, as a practical matter, is impossible to predict.

Considering the foothills as the western edge of the Deep Basin, it is logical that the Deep Basin area shown on Figure 21 probably extends south to the United States border and beyond.

CONCLUSIONS

1. Most natural resources are distributed in such a way that as the grade of the resource decreases the volume of it increases. This meaning is embodied in the expression "large, low-grade deposit."

2. Gas is distributed in the same way. As reser-

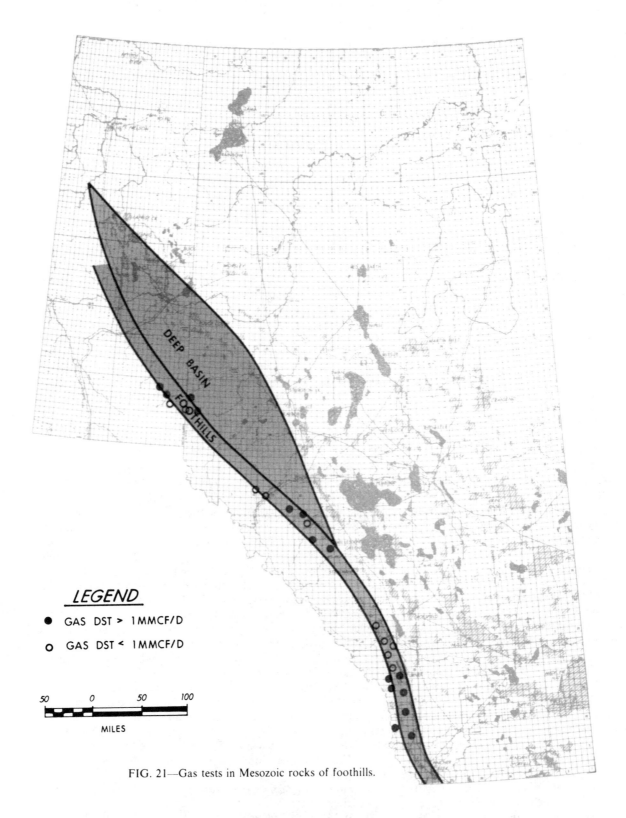

LEGEND

● GAS DST > 1MMCF/D

○ GAS DST < 1MMCF/D

50   0   50   100

MILES

FIG. 21—Gas tests in Mesozoic rocks of foothills.

voir quality deteriorates, larger and larger quantities of gas are contained in the rocks.

3. In western North America several giant sandstone gas fields occur in low porosity–low permeability Cretaceous sandstones, deep in the basins, and downdip from porous water-saturated reservoir rock.

4. In the San Juan basin the trapping mechanism is so pervasive that over a large area every porous zone in the entire Cretaceous section of 5,000 ft (1,525 m) is gas saturated.

5. The Mesozoic section of western Canada,

where it thickens rapidly into the Deep Basin in front of the Foothills overthrusts, increases significantly in electrical resistivity. Within this area there are numerous gas tests in many different sandstones with little or no recovery of formation water. The entire Mesozoic section appears to be gas saturated over an area 430 by 60 mi (690 by 96 km).

6. The gas area of the Deep Basin is separated from the water area of the shelf by a gas to water transition zone. As in the San Juan basin and Milk River field there is no shale barrier and no structural break.

7. The reservoir rocks of the Deep Basin, as in certain giant fields, become gradually less porous and permeable downdip from the water saturated rocks of the shelf. At a critical point the pore size apparently becomes so restricted that gas is trapped.

8. The trap occurs at the gas-water interface. It is proposed that high water saturation of the low-permeability rock reduces gas permeability to near zero, resulting in a "water block." This phenomenon is strengthened by either hydrostatic or hydrodynamic head of the updip water column.

9. Electric log analyses of most of the wells in the Deep Basin show 84% of them to carry recoverable gas, averaging 17 Bcf per section.

10. The entire Deep Basin area is indicated to contain potential recoverable gas resources of 440 Tcf.

11. Recoverable gas at $2.00/Mcf net after royalty may reach 150 Tcf.

12. Such large quantities of gas could have a powerful effect on Canada's national economy in addition to significantly increasing North America's total energy supply.

SELECTED REFERENCES

Allen, R. W., Jr., 1955, Stratigraphic gas development in the Blanco Mesaverde pool of the San Juan basin, *in* Geology of parts of Paradox, Black Mesa and San Juan basin: Four Corners Geol. Soc. Guidebook, p. 144-149.

Berry, F. A. F., 1959, Hydrodynamics and geochemistry of the Jurassic and Cretaceous Systems in the San Juan basin, northwestern New Mexico and southwestern Colorado: PhD thesis, Stanford Univ.

Burton, G. C., 1955, Sedimentation and stratigraphy of the Dakota Formation in the San Juan basin, *in* Geology of parts of Paradox, Black Mesa and San Juan basin: Four Corners Geol. Soc. Guidebook, p. 78-88.

Canada Geological Survey, 1977, Oil and natural gas resources of Canada, 1976: Energy, Mines and Resources Rept. EP77-1.

Century, J. R., 1977, Western Canadian oil and gas data base as working model for more effective North American resources management (abs.): AAPG Bull., v. 61, p. 775.

Conservation Board of Alberta, 1950, Schedule of wells drilled for oil and gas to 1949.

Deischl, D. G., 1973, The characteristics, history and development of the Basin Dakota gas field, San Juan basin, New Mexico, *in* Cretaceous and Tertiary rocks of the southern Colorado plateau: Four Corners Geol. Soc. Guidebook, p. 168-173.

Energy Resources Conservation Board, 1974, Reserves of crude oil, gas, natural gas liquids, and sulphur, Province of Alberta.

Fast, C. R., G. B. Helman, and R. J. Covlin, 1975, A study of the application of MHF to the tight Muddy "J" Formation Wattenberg field, Adams and Weld Counties, Colorado: Soc. Petroleum Engineers AIME Paper SPE 5624.

Gray, J. K., 1977, Future gas reserve potential Western Canadian sedimentary basin: 3d Natl. Tech. Conf. Canadian Gas Assoc.

Hill, G. A., W. A. Colburn, and J. W. Knight, 1961, Reducing oil finding costs by use of hydrodynamic evaluations, *in* Petroleum exploration, gambling game or business venture; Institute Economic Petroleum Exploration, Development, and Property Evaluation: Englewood, N.J., Prentice-Hall, p. 38-69.

Hitchon, B., 1964, Formation fluids, Chap. 15, *in* Geological history of western Canada: Alberta Soc. Petroleum Geologists, p. 201-217.

Independent Petroleum Association of America, 1976, The oil producing industry in your state.

Keelan, Dave K., 1972, Core analysis techniques and applications: Soc. Petroleum Engineers Paper 4160.

Last, Kloepfer Ltd., 1974, Suffield Evaluation Drilling Program: Rept. submitted to Suffield Evaluation Committee.

Long, M., 1977, Infill drilling expands in New Mexico: Oil and Gas Jour., March 21, p. 62-63.

Matuszczak, R. A., 1973, Wattenberg field, Denver basin, Colorado: Mtn. Geologist, v. 10, no. 3, p. 99-105; AAPG Mem. 24, p. 136-144.

Pritchard, R. L. A., 1973, History of Mesaverde development in the San Juan basin, *in* Cretaceous and Tertiary rocks of the southern Colorado plateau: Four Corners Geol. Soc. Guidebook, p. 174-177.

Reneau, W. E., Jr., and J. D. Harris, 1957, Reservoir characteristics of Cretaceous sands of the San Juan basin, *in* Geology of southwestern San Juan basin: Four Corners Geol. Soc. Guidebook, p. 40-43.

Riggs, E. A., 1976, Gas trapping mechanisms, Cretaceous section, San Juan basin, New Mexico: Unpub. rept. for Canadian Hunter.

Schultz, A. L., 1976, Estimate production from proposed wells with Kh maps: World Oil, April, p. 59-62.

Silver, Caswell, 1950, The occurrence of gas in the Cretaceous rocks of the San Juan basin: New Mexico Geol. Soc. Guidebook, 1st Field Conf., p. 109-123.

———— 1968, Principles of gas occurrence, San Juan basin, *in* Natural gases of North America: AAPG Mem. 9, p. 946-960.

Smith, M. B., et al, 1976, The azimuth of deep, penetrating fractures in the Wattenberg field: Soc. Petroleum Engineers AIME Paper SPE 6092, p. 24-35.

Suffield Block Study Committee, 1972, A resource evaluation, Suffield block: prepared for Province of Alberta.

American Association of Petroleum Geologists Memoir 38,
Elmworth—Case Study of a Deep Basin Gas Field, edited by
J. A. Masters, copyright © 1984 pp. 115-140.

Case History for a Major Alberta Deep Basin Gas Trap: The Cadomin Formation

Robert M. Gies
Canadian Hunter Exploration Ltd.
Calgary, Alberta

Abundant, high-quality geological data from the gas productive Elmworth region of the Alberta Deep basin reveal the existence of enormous gas accumulations found in an unconventional form of trap. This special form of "deep basin" gas trap defies conventional concepts of gas entrapment by turning them virtually upside-down. Gas is trapped in the deepest part of the basin, rather than on the flank of the basin. The gas/water contact occurs at the updip end of the accumulation, rather than the downdip end. Original gas accumulation pressures lie below the regional formation water pressure gradient, rather than above it as in conventional traps. Gas in at least one giant accumulation is in a dynamic state of updip migration, rather than in static equilibrium as found in conventional traps. The Lower Cretaceous Cadomin formation in the Alberta Deep basin provides some of the best available information on basic characteristics of deep basin gas traps.

Physical principles behind this form of gas entrapment were confirmed in fluid flow models designed to simulate the Cadomin gas-trapping conditions.

INTRODUCTION

In February 1976, a commercial gas discovery was made when a well was drilled near the small town of Elmworth, Alberta. As a consequence of that discovery, 700 follow-up wells were drilled throughout the surrounding region during the next 6½ years. The discovery well helped to confirm suspicions, based on log evaluations for a number of abandoned wells scattered throughout the area, that the Mesozoic section in the deepest portion of the Alberta sedimentary basin was virtually gas saturated, whereas the adjacent updip regions were mainly water bearing.

The follow-up drilling confirmed this unexpected geological phenomenon with the establishment of almost 6 tcf (170×10^9 cu m) of proven gas reserves. Ten gas plants have been built with a combined raw sweet gas processing capacity of 1.2 billion cu ft (34×10^6 cu m) per day, or the energy equivalent of 200,000 barrels of oil (31,800 cu m) per day.

Gas production is obtained from numerous formations extending throughout an elongate region covering several thousand square miles situated in the deepest part of the basin just in front of the northwest-trending disturbed belt (Fig. 1). This is not a unique gas entrapment situation. Most of the principal features are similar to the long established San Juan basin of New Mexico which contained an estimated 25 tcf (708×10^9 cu m) of gas in the deepest part of that basin.

In the beginning it was difficult to accept early indications of possible large gas accumulations trapped downdip from water, but as development progressed it became apparent from the results that, for reasons unknown at that time, our conventional concepts of gas entrapment, (that is, gas updip from water), were being turned virtually upside down. Gas/water contacts, for example, are developed at the updip termination of the deep basin gas accumulations whereas in conventional traps, the gas/water contacts are found at the downdip end. Most deep basin gas accumulations have original gas pressures that lie below the regional formation water pressure gradient, rather than above it as in conventional gas accumulations (Fig. 2). In at least one case, a deep basin accumulation appears to exist in a dynamic state of updip gas migration, while most conventional gas accumulations are found in a state of static equilibrium. Today we have developed an understanding of the basic physical principles behind this important form of hydrocarbon entrapment and as time goes on, we continue to learn new and important characteristics of deep basin traps.

Elmworth development has provided a vast quantity of high-quality, modern technical data of all kinds. Much of it is publicly available.

One of the best examples for illustrating the features of deep basin entrapment is found in the widespread Cadomin formation of Lower Cretaceous age (Fig. 3). An estimated 15 tcf (425×10^9 cu m) of gas in place are contained in the 3,800 sq mi (9,840 sq km) Cadomin gas accumulation. This paper presents a brief description of the geological setting for the Cadomin formation followed by a description and discussion of the gas and water it contains. The two-part approach emphasizes and demonstrates the physical compatibility that must exist between reservoir geology and reservoir fluid distribution (which includes fluid pressures). Experience has shown that failure to consider and integrate relevant data of all kinds can result in a seriously inaccurate concept of the trap. The geological setting of the reservoir involves not only structural considerations, but the equally important aspects of sediment types, sedimentary conditions, sediment distribution and diagenesis. The latter processes combine to establish present day reservoir "plumbing." The reservoir plumbing is a combination of variables, like porosity and permeability, together with reservoir continuity trends and fluid barriers (reservoir discontinuities) and seals. Such geological conditions in turn exert major controls over the distribution of gas and water throughout the reservoir formation. Mutual dependence between reservoir fluid distribution and reservoir rock geometry means that we can use fluid data to provide valuable additional information about the reservoir's plumbing; especially in areas somewhat remote from the well bore.

DEPOSITIONAL SETTING OF THE CADOMIN

By early Cretaceous time an extensive mountain system had formed in the region of the province of British Columbia (Fig. 4). The climate must have been humid with abundant rainfall. At numerous points along the east- and northeast-facing mountain front, rivers emerged out onto the eastern plains. The sand and pebble debris they carried was deposited abruptly as the rivers left the confines of their mountain canyon channels. Near the mountains, alluvial fans of great thickness developed. Water flow across the fans redistributed some of the sediments downslope toward the northeast to develop broad alluvial plains. Water runoff from the alluvial plains eventually joined a large northwestward-flowing braided river system which transported both water and sediments derived from mountain sources, and from eastern drainage areas as well, in a direction more or less parallel to the mountain front. This large braided trunk river is named the Spirit River (McLean, 1977).

The overall drainage system, therefore, consisted of two complementary but distinctively different fluvial types. In this case the braided trunk river system was necessary to carry away water derived from the mountain rivers. The river was, no doubt, controlled in its position and direction of flow by topography and structural activity of the basin at the time. Recognition of the two interacting fluvial systems is fundamental to developing a proper understanding of the resultant reservoir

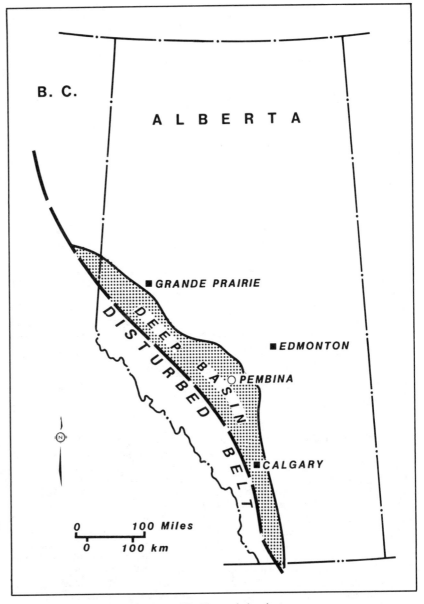

Figure 1. Index map showing location of the Elmworth deep basin.

plumbing patterns and consequent entrapment of the deep basin gas accumulation in the Cadomin.

As time passed, the two great fluvial systems competed for dominance of the region bordering the mountain front. Alluvial plains sediments, spread by stream flow downslope to the northeast, eventually were captured and transported northwestward by the Spirit River. The Spirit River channel migrated laterally

across the region beginning perhaps at its farthest eastward position, which is defined by a gently rising erosional feature called the Fox Creek Escarpment (Fig. 5). These and other geological interpretations of the Cadomin history throughout the region are founded largely on subsurface well control supplemented by outcrop investigations. Lateral channel migration may have been promoted by gradual basin subsidence in late Cadomin time which

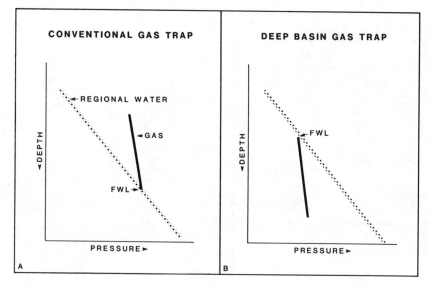

Figure 2. Pressure-depth relationships between conventional and deep basin gas traps.

saw the basin tilt to the southwest with the consequent lateral shifting of the Spirit River closer to the mountain front (Fig. 6). Remnant outliers of former alluvial plains sediments were cut off or became interbedded with Spirit River sediments throughout an extensive region. At the river's closest position to the mountain front, braided river sediments are overlain by a thin layer of alluvial plains sediments derived from the fans to the southwest.

In places, the Cadomin has been found from well control to be very thin or absent along northwest-trending belts parallel to the Spirit River channel courses. There is some evidence of clay channel fills and shale infilling, of possible late channel courses, by the basal Gething formation. This resulted in reservoir discontinuities or barriers along the Spirit River channel course. Such barriers act to influence present-day distribution of gas and water in the Cadomin reservoir.

Figure 7 is an isopach map of the total Cadomin interval. The map illustrates important Cadomin thickness trends throughout the region of interest. Thick isopachs in British Columbia reflect proximity to major alluvial fan sediment sources which spread north and eastward from the disturbed belt. These thick trends are dissected by a linear, isopach thin trend oriented northwestward which marks the position of the late Spirit River channel system.

Examination of Cadomin sediments in cores and drill cuttings permits the identification of the original type of fluvial conditions during deposition, that is, alluvial plain or braided river (Spirit River). The alluvial plains sediments consist of sandy conglomerates containing mainly chert and some quartzite pebbles that are characteristically poorly sorted with maximum pebble diameter of about 3 in (75 mm). Dark pebble colors predominate. The associated matrix sands are similarly dark in color and usually fine-grained chert sands with about 10% quartz grains (Fig. 8). Spirit River sediments also consist of sands and sandy conglomerates. However, the pebbles tend to be smaller, better sorted and light colors (white and pale apple green) predominate (Fig. 9). The associated matrix sands are also light colored, well sorted, coarse grained and mainly chert with up to 20 percent quartz. Quartzite pebbles are absent. Alluvial plains sediments that became reworked by the Spirit River are better sorted with coarser sand grains, although typical Spirit River light-colored clastics are the most abundant (Fig. 10).

Detrital clays have been identified in cores from the alluvial plains sediments. They line pore walls or partly infill some pores. These clays are thought to have been deposited in the pore system by muddy mountain river waters which infiltrated the alluvial plains sediments (sieve deposits). Clay rip up clasts, thin coaly

beds and shale beds tend to be characteristic of the Spirit River braided channel deposits.

Although the usual processes of burial diagenesis greatly reduced original porosity and permeability, distinct differences in the magnitudes of permeability are still governed by original sedimentary processes. As expected, the permeability of alluvial plains sandy conglomerate is very low (often less than a millidarcy at in situ conditions) due to small pore throat sizes of the fine-grained sand matrix. Even the occasional well-sorted, sand-free conglomerate beds may show reduced permeability due in part to clay sieve deposits. The well-sorted, coarse-grained braided river deposits, on the other hand, have permeabilities ranging up to several hundred millidarcys because of larger pore throats and less interstitial clays.

RESERVOIR FLUIDS

The second half of the Cadomin story concerns the reservoir fluids. Figure 11 is a map of the Elmworth region showing the gas and water distribution throughout the Cadomin. Geological control and reservoir fluid interpretations are based on examination of data from the 650 wells scattered throughout the area. Note the Cadomin structure contours which depict a very gently southwestward-dipping monocline. Cadomin sandy conglomerate was deposited throughout the entire map region. Updip areas (shown in white) are water bearing.

The Cadomin deep basin as accumulation (red) extends across 3,800 sq mi (9,840 sq km) in the downdip region of the map. There are also a few small, conventional type gas accumulations (orange), thought to be both structurally and stratigraphically trapped in the updip, predominantly water saturated region of the Cadomin. About 100 wells capable of initial gas flow rates in excess of 500,000 cu ft (14,160 cu m) per day have been completed to date. There are also a similar number of potential gas wells waiting on completion, or wells with lesser flow rates.

An important feature of the main gas accumulation is that it occupies a large proportion of the downdip, low-permeability alluvial plains sediments. The more permeable braided river deposits predominate throughout the updip region which is shown to be mainly water bear-

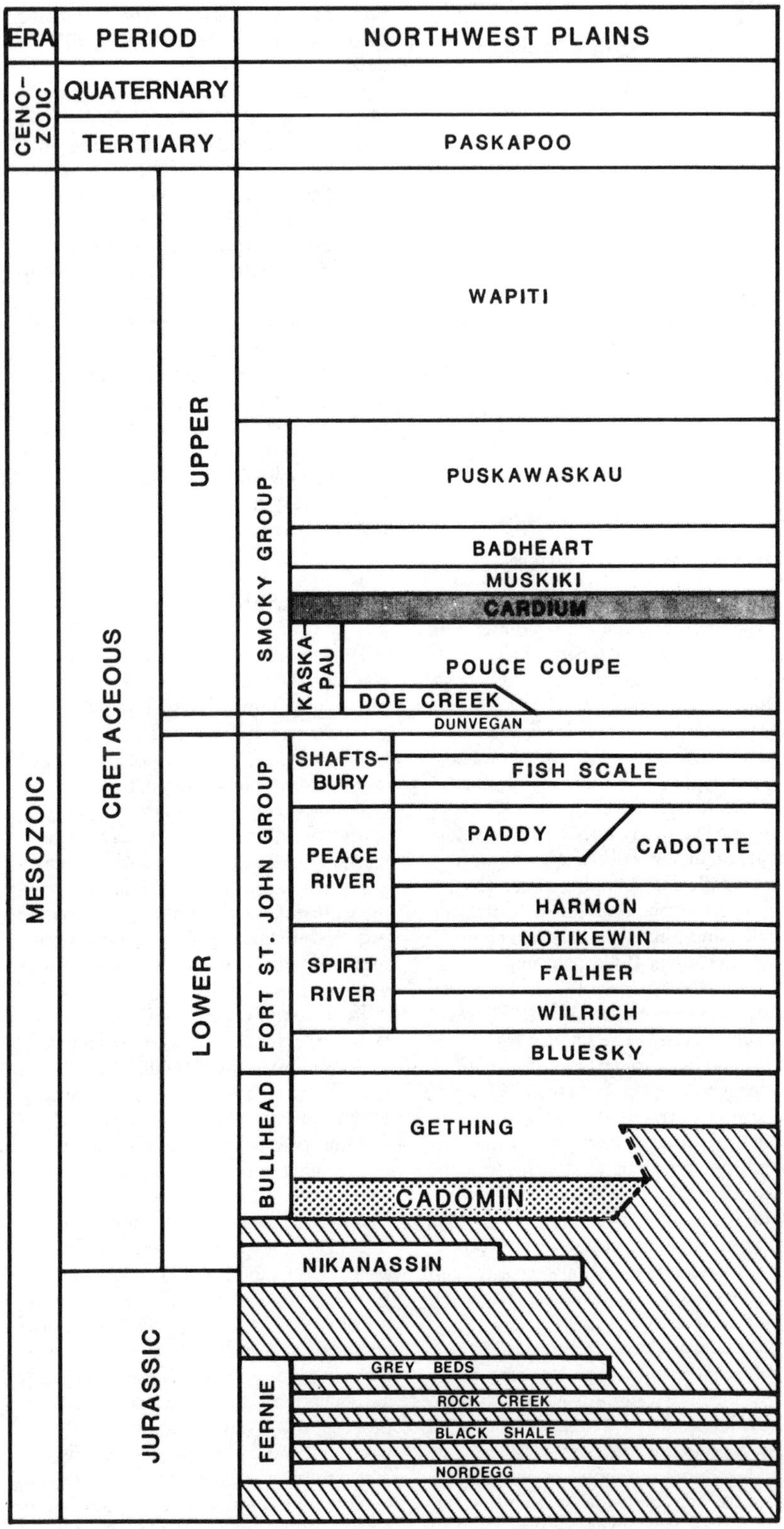

Figure 3. Table of stratigraphic names in the Elmworth deep basin-Mesozoic section.

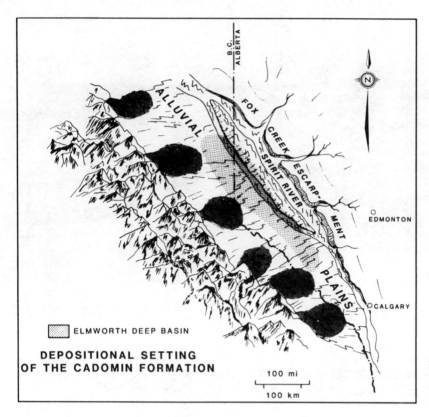

ELMWORTH DEEP BASIN

**DEPOSITIONAL SETTING
OF THE CADOMIN FORMATION**

100 mi

100 km

Figure 4. Sandy conglomerate derived from alluvial fans was deposited across broad alluvial plains to the east where the conglomerate was reworked by the northward flowing Spirit River system.

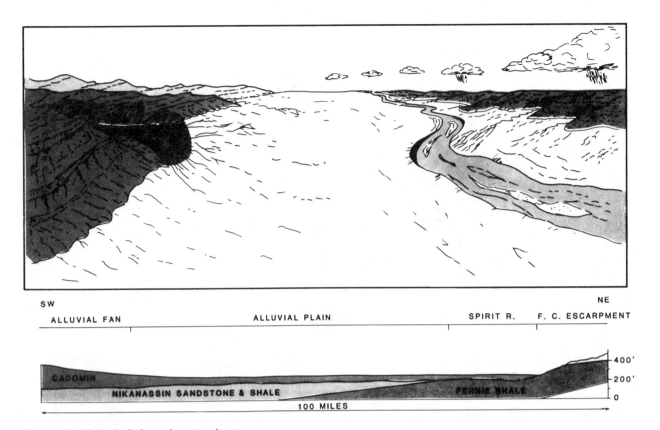

SW

ALLUVIAL FAN   ALLUVIAL PLAIN   SPIRIT R.   F. C. ESCARPMENT

NE

CADOMIN

NIKANASSIN SANDSTONE & SHALE   FERNIE SHALE

400'

200'

0

100 MILES

Figure 5. Model of early Cadomin depositional setting.

SW

NE

ALLUVIAL PLAIN

ALLUVIAL FAN ─────────────────────────────────── F.C. ESCARPMENT

SPIRIT R.

CADOMIN

NIKANASSIN

400'

200'

0

100 miles

Figure 6. Model of late Cadomin depositional setting.

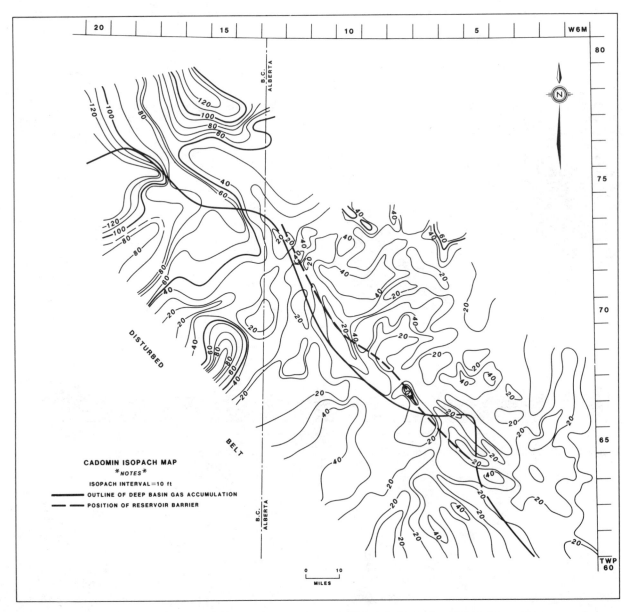

Figure 7. Isopach map of total Cadomin interval. Note northeast to east oriented "thicks" which reflect alluvial plains sedimentary trends versus linear northwestward "thin" along the course of a late Cadomin Spirit River channel system.

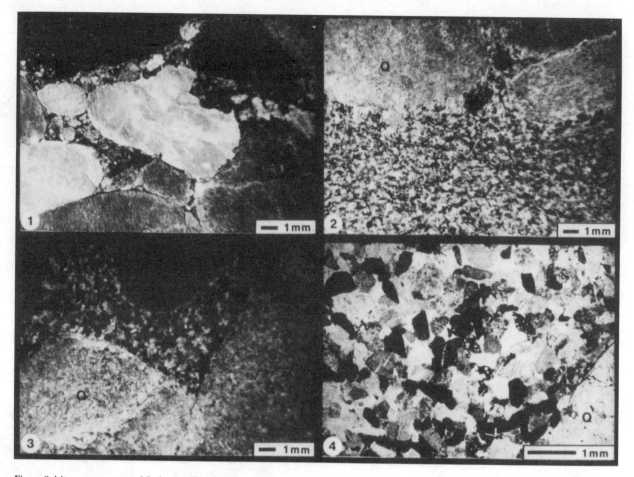

Figure 8. Microscopic views of Cadomin alluvial plain sandy conglomerate facies.

(1) 10-8-70-9W6, Cadomin Formation, 2,203m, 6.8×, ϕ = 1.2%, k = 1 md.
Typical dark-colored chert pebble conglomerate with medium grained, dark-colored chert and quartz sand matrix as seen through the binocular microscope with reflected light.

(2) 11-28-70-11W6, Cadomin Formation, 2,301m, 6.8×, ϕ = 2.2%, k = 0.065 md.
This binocular microscope view illustrates the fine-grained, tight-sand matrix frequently found in the alluvial plain conglomerate facies. Note the pink quartzite pebble (Q).

(3) 11-7-68-12W6, Cadomin Formation, 2,830.5m, 6.8×, ϕ = 1.9%, k = >0.01 md.
This binocular microscope view also illustrates the typical, dark colored, tight, fine sand matrix found in many alluvial plain conglomerates. Note pink quartzite pebbles (Q).

(4) Some visible porosity (blue color) can be seen in thin section through the polarizing microscope within the quartz-rich sand matrix of the same sample described in (3) above. Additional, unseen microporosity also is present among and within some of the sand grains. Note pervasive porosity infilling quartz overgrowth cement, and quartzite pebble (Q).

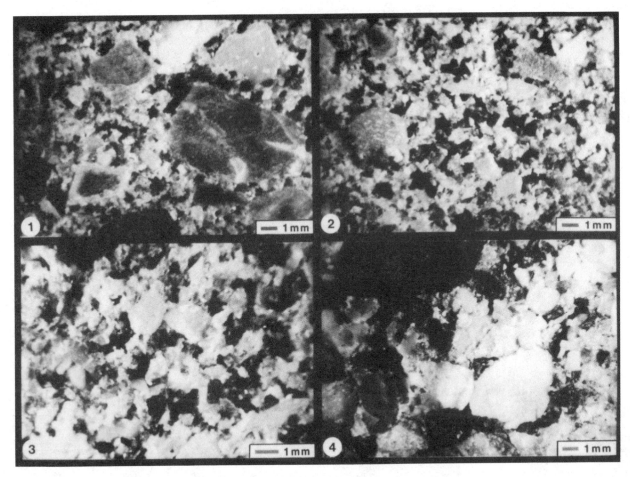

Figure 9. Binocular microscope views of Cadomin Spirit River sand conglomerate facies.

(1) 10-8-70-9W6, Cadomin Formation, 2,195.6m, 6.8×, ϕ = 4.5%, k = 20 md.
This view and photos 2 and 3 are from the same well which also contains alluvial plain facies as seen in Figure 7. Note that light chert colors predominate (white, pale green, and light brown), the sand matrix is coarse grained and matrix porosity is obvious. Quartzite pebbles are very rare or absent.

(2) 10-8-70-9W6, Cadomin Formation, 2,197m, 6.8×, ϕ = 3.6%, k = 18 md.
The porous, light-colored chert grain texture of this sandy conglomerate is very distinctive.

(3) This is an enlarged view (10.2×) of photo 2 illustrating the well sorted, essentially silt and clay free texture of the porous Spirit River facies.

(4) 7-11-65-2W6, Cadomin Formation, 2,342m, 6.8×, ϕ = 2.2%, k = 70 md.
Similar chert mineralogy and clean, porous texture are seen here in this example of a coarser-grained sandy chert some 56 mi (94 km) southeast of the 10-8 well (examples in photos 1-3).

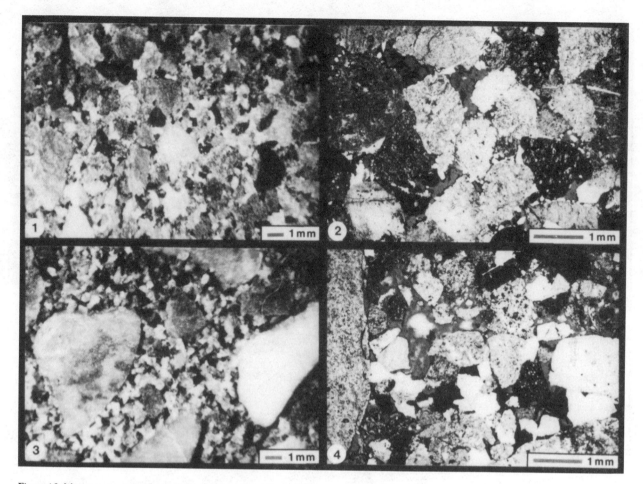

Figure 10. Microscopic views of mixed alluvial plain, Spirit River Cadomin conglomerate facies.

(1) 6-16-68-11W6, Cadomin Formation, 2,631.4m, 6.8×, ϕ = 10.6%, k = 1,367 md.
The binocular microscope view shown here reveals a well sorted, coarse-grained but dark-colored chert sand matrix.

(2) This thin section view (21×) from the same sample seen in photo 1 illustrates an unusual example of porosity and permeability enhancement due to chert dissolution and pore enlargement by circulating formation waters. At a later stage, druzy quartz developed on pore walls as silica saturation with respect to quartz increased.

(3) 10-21-68-4W6, Cadomin Formation, 2,015.7m, 6.8×, ϕ = 10.6%, k = 940 md.
Dark chert colors predominate (relative to typical Spirit River facies) but coarse, well sorted chert sands form the conglomerate matrix.

(4) The thin section view (21×) illustrates abundant porosity in the chert sand grain matrix. Quartz grains are not very abundant but where present show evidence of porosity destroying quartz cement overgrowths. Porosity creating leaching, on the other hand, is apparent among the chert sand grains.

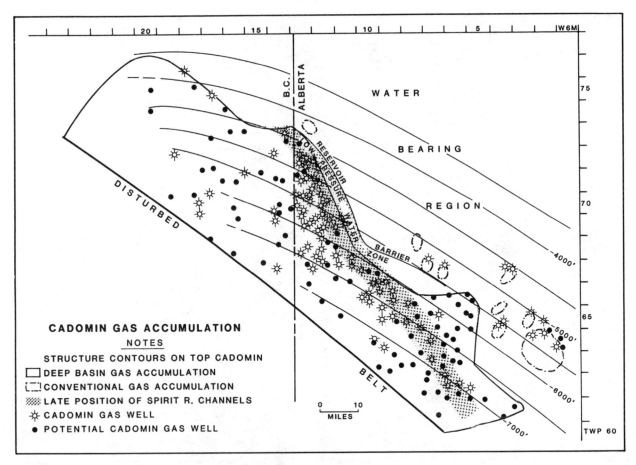

Figure 11. Map showing gas- and water-bearing regions in the Cadomin formation.

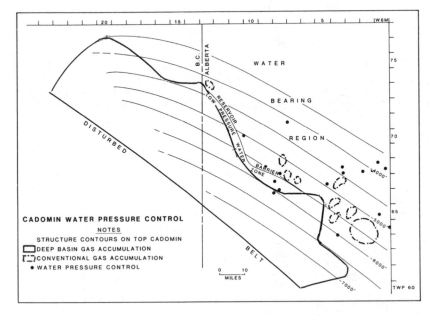

ing. But, there is also an important commingled and interbedded zone of both fluvial types that is 6 mi (9.6 km) or more in width running along the downdip, southwest side of the water/gas contact line. This is represented on Figure 11 by a shaded band which shows the position of the most southwestern extent of the Spirit River channel system.

Reservoir boundaries for this extensive gas field are important from the standpoint of understanding the gas-trapping mechanism. The downdip boundary occurs where the reservoir rock becomes either too tight to be productive, or it is cut off by thrust faulting along the edge of the disturbed belt. There is no evidence for a defined downdip gas/water contact.

Figure 12. Map showing distribution of Cadomin water pressure control points.

Table 1. Typical examples of Cadomin formation water properties.

No.	Well	DST No.	Interval	Recovery	pH	Ca	Mg	Na+k	HCO₃	SO₄	Cl	CO₃
						\multicolumn{7}{c}{Ion Concentration mg/ℓ}						
1	6-7-67-8W6	6	2,665-2,681.6 m	432 m of salt water - sample from downhole sampler	6.7	3,680	1,115	31,301	810	21	57,400	0
2	11-35-67-9W6	3	2,537-2,550 m	1,185 m of salt water - sample from downhole sampler	5.9	4,709	1,103	35,042	434	6	65,200	0
3	7-1-68-7W6	6	2,306-2,310 m	56 m mud, 224 m salt water - sample taken at top of tool	7.0	1,277	340	26,280	968	16	43,100	0
4	11-35-68-7W6	15	7,324-7,343 ft	1,560 ft salt water - sample taken at bottom of recovery	6.6	1,437	540	30,652	1,051	7	50,761	0
5	11-36-70-11W6	4	2,251-2,262 m	189 m salt water - sample taken at bottom of recovery	7.3	1,020	438	27,801	1,108	27	45,200	0

Notes: Analyses No.'s 1 and 2 are from wells located in the low-pressure water-bearing Cadomin south of the reservoir barrier shown in Figure 10 whereas analyses 3 and 4 are from nearby wells located north of the reservoir barrier and represent typical Cadomin water properties seen in many wells in the regional water-bearing portion of the Cadomin. Note that samples from the low-pressure region south of the barrier show distinctly higher concentrations of calcium, magnesium and chlorine ions. Analysis No. 5 represents Cadomin water properties north of the barrier but 25 mi (42 km) west of analyses 3 and 4.

In the updip region, a water/gas contact is defined from well control at about 4,265 ft (1,300 m) subsea in Township 73 Range 13. This is not believed to be a reservoir discontinuity or barrier, but simply a contact in low-permeability sediments where pressures in the water are in balance with gas phase pressures. It is like a conventional gas/water contact only turned upside-down. The vertical gas column height, defined by these updip and downdip limits, is about 2,600 ft (792 m).

An important reservoir barrier or discontinuity exists over a great distance in the Cadomin and it is closely linked to the gas accumulation because it acts as a lateral boundary. It is defined from well logs, fluid pressures, water analyses and fluid type control. Figure 11 shows the discontinuity oriented in a northwest-southeast direction running oblique to the structural contours. On the northeast side, the Cadomin is, for the most part, permeable but water bearing. But on the west and southwest side, conditions are more complex. We believe that the discontinuity is the result of the northwest Spirit River channel development during late Cadomin times (Fig. 4). The few wells that penetrated the barrier, or discontinuity trend, show that the Cadomin is unusually thin or is absent, with Lower Gething sediments resting directly on the underlying Nikanassin formation. Note that

Cadomin water occupies the pore system along both sides of the discontinuity trend, although the water-saturated zone is narrow along the southwest side. Water pressure measurements show that pressures on the northeast side are 160 psi (1,158 kPa) greater than water pressures on the opposite side at common subsea depths. Water analyses data also show a significant change in ion concentrations on opposite sides of the discontinuity trend, but show consistent properties for waters among a number of control points selected along either side (Table 1).

Original reservoir fluid pressures often provide considerable insight into the nature of hydrocarbon traps. The deep basin Cadomin gas accumulation is no exception. Figure 12 is a map of the basin showing well locations where 24 high-quality original formation water pressures were obtained (Table 2). In Figure 13 it is seen that 19 of these values fall along a straight line in a standard pressure-depth plot. The remaining five lower pressure points are from wells located in the water-bearing Cadomin located along the southwestern side of the discontinuity trend. Figure 14 is a map of the same area showing locations of 45 wells where good original Cadomin gas pressure values were obtained (Table 3). In all cases, pressures used were carefully qualified and calculated using standard Horner extrapolation

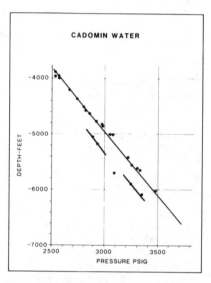

Figure 13. Pressure-depth profile for original Cadomin formation water pressures updip from the deep basin gas accumulation.

methods. These gas pressures were from drill-stem tests supplemented by post-completion pressure build-up tests. The resulting pressure-depth distribution is presented in Figure 15. Note that the gas pressure values appear to define a curved line, although the point spread increases with depth. This turned out to be a sur-

Table 2. Original formation water pressures, Cadomin formation, Elmworth area.

Point	Well	KB Elev. (ft)	DST Inter. Top Base (ft)		REC Depth (ft SS.)	Extr. ISIP (psig)	Extr. FSIP (psig)	Water Recovery (ft)	Remarks
A	7-20-64-1W6	2,845	7,889	7,905	−5,053	3,077	3,077	1,594	
B	11-1-64-5W6	2,936	8,982	9,012	−6,050	3,478		558	
C	7-11-65-2W6	2,843	7,675	7,713	−4,846	2,982	2,982	2,707	
D	10-23-65-5W6	2,586	8,234	8,300	−5,655	3,312		361	
E	10-20-67-5W6	2,342	7,254	7,355	−4,926	2,979	2,903	950	
F	6-7-67-8W6	2,851	8,743	8,798	−5,901	3,247	3,247	1,312	
G	11-35-67-9W6	2,623	8,321	8,364	−5,708	3,094	2,932	3,766	Limited reservoir
H	10-2-67-9W6	2,796	8,888	8,931	−6,128		3,347		Fm. wtr. on comp.
I	6-8-68-1W6	2,408	6,355	6,371	−3,950	2,665	2,552	2,650	
J	11-15-68-2W6	2,273	6,233	6,263	−3,942	2,532	2,539	4,711	
K	6-26-68-2W6	2,214	6,096	6,132	−3,897	2,539	2,533	4,500	
L	11-11-68-3W6	2,343	6,503	6,528	−4,225	2,671	2,612	1,400	
M	11-9-68-4W6	2,220	6,803	6,873	−4,571	2,836	2,830	4,790	
N	10-21-68-4W6	2,196	6,558	6,637	−4,373		2,752	1,000	Pkr. failed
O	6-8-68-8W6	2,750	8,205	8,268	−5,459		3,226	690	
P	6-20-68-9W6	2,552	8,123	8,222	−5,550		3,258	771	8 psi added for depth
Q	11-36-70-11W6	2,326	7,385	7,421	−5,062	3,048	3,048	620	
R	6-25-71-8W6	2,394	6,414	6,450	−4,006		2,574	560	
S	7-6-71-11W6	2,481	7,671	7,736	−5,193	2,987	2,932	184	Limited reservoir
T	10-28-71-11W6	2,647	7,388	7,451	−4,748		2,922	2,038	
U	7-13-71-12W6	2,700	7,779	7,812	−5,060	2,894	2,690	841	
V	7-11-72-12W6	2,679	7,380	7,407	−4,683		2,868	466	
W	10-24-72-12W6	2,588	7,086	7,112	−4,531	2,807	2,804	2,350	
X	11-18-79-16W6	2,683	4,957	5,016	−2,274	1,833	1,833	1,076	Falls on extrap. wtr. pressure gradient
Y	10-31-68-10W6	2,437	8,130	8,153	−5,696	3,335		2,300	

prise because a steeply inclined straight line is usually the case for gas accumulations. Careful re-examination of the pressure data showed nothing apparently wrong and, therefore, the curved gas pressure gradient line represents factual information that must be integrated with other geological data on a physically sound basis.

Figure 16 is a comparison of the pressure-depth relationship for both the Cadomin gas and water. Note that pressures in both fluids are about equal at a subsea depth of about 4,200 ft (1,280 m), which corresponds to the updip water/gas contact position previously described. Gas pressures at all other subsea depths are consistently less than water pressures at corresponding depths throughout the widespread water-bearing region of the Cadomin.

Figure 17 is a comparison between the curved line, representing actual Cadomin gas pressures, and a straight line for the gas-pressure gradient one would expect to see for an accumulation in static equilibrium. The wide deviation between the two is of considerable geological interest and importance. Gas-pressure gradients along the curved line range from about 0.06 psi/

ft (1.356 kPa/m) at shallow depths, to 0.65 + psi/ft (14.7 kPa/m) in the downdip area. This compares to the normal static pressure gradient of 0.055 psi/ft (1.243 kPa/m) for the kind of dry gas present in the Cadomin.

The comparison gives rise to the question: What is the explanation for the wide deviation found between the expected straight line, and the actual curved gas-pressure gradient line? There are two possibilities. One is that the pressure points forming the curve are from several pools that each have progressively higher pressures with increasing depth, and, therefore, they only appear to fall along a curve. This would require a straight line with a pressure gradient of about 0.055 psi/ft (1.243 kPa/m) to fit points in each of these individual pools. The second possibility is that there is only a single continuous Cadomin gas accumulation, but it is not in static equilibrium. Rather, it is in a dynamic state of updip gas migration. Although the concept of a dynamic gas accumulation is unusual, it is the preferable interpretation because it fits the geological data without any apparent technical weaknesses. A major difficulty

with the first possibility, that is, separate gas pools, is that there is no apparent mechanism for gas to have originally displaced the formation water from the water-wet Cadomin reservoir, and end up in an array of separate gas pools with gas pressures significantly below extrapolated water pressures for the same depths. If separate gas pools existed, then we should expect to find some evidence for lateral seals, or reservoir discontinuities, required to separate the pools. It is difficult to make such a geological case for these, based on the available well control.

The case for a single dynamic gas accumulation fits the geological and reservoir fluid data far better. In order to explain the curved gas gradient line, it is perhaps more easily understood by examining some of the fundamental principles of subsurface fluid hydrodynamics. Figure 18 illustrates the simplest example. A gently dipping porous reservoir is completely saturated with fresh water. Three wells have penetrated the aquifer at points A, B, and C. Water rises to the same position, relative to the horizontal datum, in each well. Note that the height of the water column in each well is a function of pressure in the

Table 3. Original formation gas pressure, Cadomin, Elmworth area.

Well	DST Interval Top	DST Interval Base	REC Depth (ft SS.)	Extr. ISIP (psig)	Extr. FSIP (psig)	Rate mmcf/d	Remarks
7-24-63-3W6	8,655	8,674	−5,652	3,112	3,097		Rec 820 ft oil & 1,800 ft XW
7-8-64-2W6	8,306	8,375	−5,427		3,160	1.74	
10-29-64-3W6			−5,407	3,091		0.74	P.B.U.*
6-2-65-3W6	8,110	8,131	−5,209	3,067	3,055	4.80	
7-32-65-3W6	7,930	7,972	−5,088	2,761	2,747	0.84	
7-7-65-4W6	8,566	8,694	−5,587	3,276	3,276	4.00	
5-12-65-5W6	8,484	8,543	−5,705	3,391		1.20	P.B.U.
6-32-65-7W6			−6,113	3,207		3.95	P.B.U.
5-32-65-9W6	10,446	10,463	−6,626	3,414	3,382	0.04	
10-27-65-10W6			−6,840	3,457		0.82	P.B.U.
7-13-66-9W6	9,144	9,183	−6,239	3,249	3,107	0.27	
10-19-66-9W6			−6,394	3,282		1.00	P.B.U.
12-31-66-9W6			−6,291	3,302		0.68	P.B.U.
9-10-66-10W6	9,527	9,580	−6,606		3,210	0.04	
10-26-66-11W6	9,390	9,432	−6,659	3,483	3,498	1.50	
10-33-67-7W6	7,792	7,884	−5,314	3,150	3,140	2.75	
10-4-67-10W6			−6,421	3,348		0.38	P.B.U.
10-12-67-10W6	9,228	9,255	−6,179	3,363	3,337	1.72	
7-15-67-10W6			−6,269	3,285		0.42	P.B.U.
6-21-67-10W6	9,160	9,199	−6,278	3,398	3,362	5.10	
10-25-67-11W6			−6,292	3,118		1.00	P.B.U.
10-3-68-4W6	6,880	6,915	−4,479		2,855	13.91	
6-11-68-8W6	8,093	8,148	−5,457	3,153	3,096	6.48	
7-4-68-11W6	9,160	9,200	−6,260	3,137		0.10	P.B.U.
6-16-68-11W6			−6,113	3,084		2.60	P.B.U.
7-1-68-12W6			−6,398	3,106		0.21	P.B.U.
11-7-68-12W6	9,261	9,291	−6,456		3,142	0.01	
11-14-68-12W6			−6,212	3,158		0.50	P.B.U.
7-14-68-13W6	9,324	9,363	−6,550		3,202	0.02	Poor Extrap.
6-28-68-13W6			−6,489	3,285		0.44	P.B.U.
10-7-69-12W6	8,707	8,743	−6,042	3,022	3,045	0.10	
6-26-69-12W6	8,284	8,338	−5,802	3,002		0.10	
6-28-69-12W6	8,336	8,372	−5,812		3,102	0.01	
10-34-69-13W6			−5,844	3,003		0.80	P.B.U.
11-1-70-12W6	7,989	8,038	−5,574		2,894	0.06	
11-7-70-12W6			−5,637	2,878		5.07	P.B.U.
11-33-70-12W6			−5,323	2,887		0.46	P.B.U.
11-1-70-13W6	8,392	8,438	−5,740	2,974	2,949	0.07	
10-34-70-13W6	8,202	8,287	−5,351		2,932	0.01	
10-25-70-14W6	8,441	8,497	−5,657		2,892	0.01	
7-5-71-12W6			−5,286	2,793		1.00	P.B.U.
6-18-71-12W6	7,949	7,982	−5,170	2,832	2,827	0.07	
11-4-71-13W6	8,150	8,195	−5,387	2,853	2,851	0.02	
11-11-71-13W6			−5,279	2,831		1.50	P.B.U.
11-19-71-13W6	8,104	8,159	−5,248		2,802	0.06	
10-22-71-13W6	7,841	7,874	−5,105	2,824	2,814	0.83	
13-33-71-13W6	7,815	7,871	−5,072		2,838	1.40	
10-35-71-13W6			−4,999	2,838		1.35	P.B.U.
6-6-72-12W6	7,647	7,664	−4,936	2,794	2,794	0.21	
7-8-72-13W6	7,759	7,803	−4,970	2,789	2,785	0.21	
10-12-72-13W6	7,568	7,615	−4,838	2,767	2,601	0.73	
7-27-72-13W6	7,540	7,558	−4,678		2,754	0.20	
6-5-74-13W6	6,934	7,000	−4,164	2,658	2,617	0.20	
b-82-B/93-P-1	9,268	9,393	−5,965	3,297	3,297	0.01	
a-43-D/93-P-1	9,892	9,934	−6,249		3,624	0.01	
d-68-K/93-P-1	8,543	8,587	−5,308	2,804	2,866	0.04	ISIP not extrap.
b-24-B/93-P-6	9,170	9,258	−5,523	3,362		1.04	
a-57-C/93-P-7			−5,633	3,686		1.60	P.B.U.
a-43-B/93-P-8	7,881	7,906	−4,865		2,705	0.04	
b-46-H/93-P-8			−4,436	2,687		2.50	P.B.U.

*P.B.U. signifies pressure measurement obtained from post-completion pressure build-up analysis.

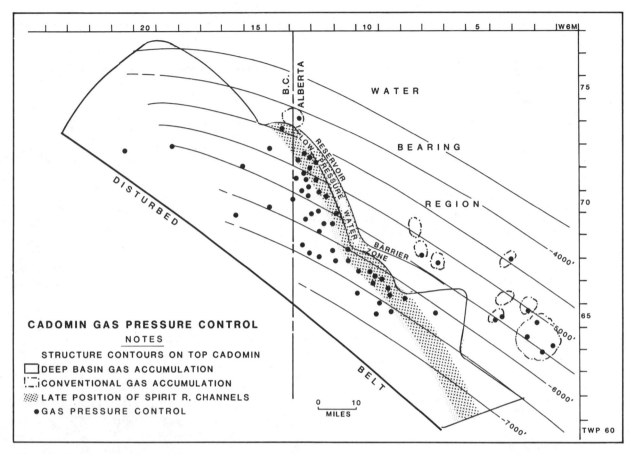

Figure 14. Map showing distribution of Cadomin gas pressure control points.

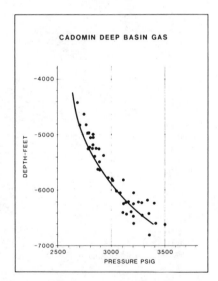

Figure 15. Pressure-depth plot for original Cadomin formation gas pressures throughout the deep basin gas accumulation.

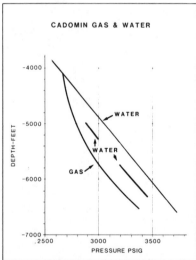

Figure 16. Pressure-depth comparison of Cadomin deep basin gas and regional Cadomin water systems.

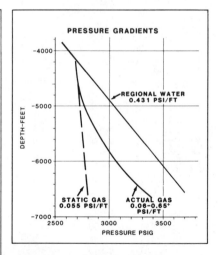

Figure 17. Pressure-depth comparison between actual curved gas pressure gradient line for Cadomin deep basin gas accumulation and the corresponding gradient line for a static gas accumulation.

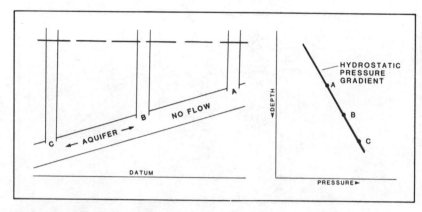

Figure 18. Model of a gently dipping aquifer with three wells located at positions A, B, and C. Hydrostatic conditions prevail in the water phase.

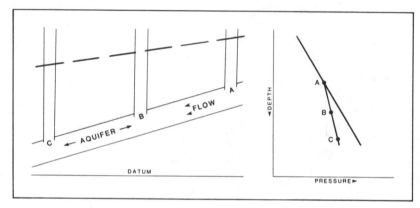

Figure 19. Downdip water flow in aquifer leads to energy loss in dynamic water phase and therefore a steepening of the apparent pressure gradient line.

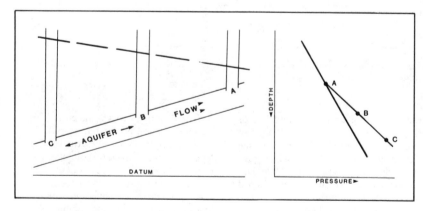

Figure 20. Updip water flow results in energy loss in updip direction and a flattening of the pressure gradient.

aquifer at that point. In this first case, hydrostatic conditions exist. As shown in the adjacent pressure-depth plot, the pressure increases with vertical depth in the water phase at a rate of 0.433 psi/ft (9.79 kPa/m). If, however, the water is in a dynamic state, that is, it is moving slowly through the reservoir, then, from the principles behind Darcy's law for fluid flow through porous media, there will be some energy loss in the water due to frictional resistance to flow through the pore system. The rate of energy loss along the flow path is largely a function of the permeability and rate of water flow. The higher the flow rate and the lower the permeability, the greater is the loss of energy per unit length along the flow path. In the case of downdip water flow, the energy loss, due to frictional resistance to flow, is reflected by a pressure loss in the water at downdip positions. The resultant apparent pressure gradient line, as would be established by making a series of measurements along the water flow path, is steeper than the gradient line for a static system (Fig. 19). Similarly, if water flow was in an updip direction, the energy loss in the moving water is reflected by an increase in the rate of pressure loss in the water phase at updip positions. The resultant pressure gradient line for updip water flow is, therefore, greater (flatter) than the static case (Fig. 20).

The foregoing simplified examples all assumed constant reservoir transmissibility throughout. Figure 21 shows what can happen if, for example, reservoir transmissibility progressively increases updip (perhaps due to increasing permeability, reservoir thickness, or both). Water movement is updip. Note that the water pressure gradient becomes a curve that steepens in the updip direction.

The same principle applies to updip migration of gas in the real world. Slow gas movement through the low-permeability Cadomin reservoir results in an energy loss to the gas phase and this·is reflected as a pressure loss updip along the gas migration path. Because the reservoir transmissibility to gas improves updip, the rate of energy loss along the gas migration path decreases, and therefore the apparent pressure gradient curve becomes steeper as it approaches the gradient for gas in static equilibrium (Fig. 17). Corresponding average in situ permeabilities, derived from pressure build-up analyses, show low values generally in the downdip regions in the order of 0.1 md, whereas in the updip areas it is 10 times greater at 1.0 md.

Pressure curves (Fig. 16) indicate that pressures are about equal at the updip position where gas is thought to be in contact with the regional water pressure system. The low gas pressures measured in downdip regions simply reflect the pressure in the gas phase along a gradient that is steeper than that for the regional formation water where both are in balance at some updip position. If the gas was in static equilibrium, then the pressure differences between formation waters northeast of the barrier and the gas southwest of it, would be much greater than they really are, as indicated by Figure 17. Such a high pressure difference would encourage formation water in the northeast to move across the barrier, if possible, and enter the much lower-pressured gas-saturated region.

Water/Gas Contact Phenomena

A question of considerable interest that must be answered is: How is it possible for nature to maintain a gas below water contact in the reservoir, as suggested from the Cadomin situation, without a seal to separate them? The explanation to this very puzzling phenomenon became apparent only after making an exhaustive series of model experiments designed to simulate the field observations.

Figure 22 is a diagram representing one of the models designed to duplicate fluid mechanisms involved. The model is constructed from transparent plexiglass, to permit direct observations of events taking place in the pore system. It consists of a vertical cylinder containing a sand pack having a total length of 28.3 inches (72 cm) and a diameter of 2.24 inches (5.7 cm). The sand pack rests on a porous screen support at its base. Sand permeabilities are high relative to the actual Cadomin reservoir. Permeability of the fine sand (small dot pattern) is 30 darcys while the coarse sand (open circles) is 1,100 darcys for an average sand column permeability of 40 darcys in this case. Sand porosity is 38%. A manometer, tied into the base of the sand column, will be used to measure water pressure. The bypass tube allows short-circuiting of the water from the top of the sand column, through a valve, to the base.

A typical experimental run begins with the total system water saturated to a little above the top of the sand column. The

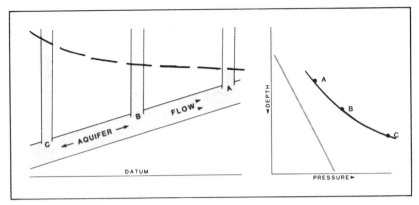

Figure 21. Updip flow in an aquifer with increasing transmissibility updip results in a pressure gradient curve that steepens updip.

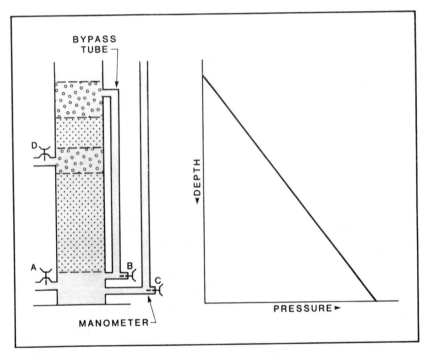

Figure 22. Water-saturated sand column in model designed to investigate gas entrapment principles.

water is colored for better observation of flow. At this point, the valves B and C are open and the water level in the manometer tube is level with the water level just above the sand column. Valves B and C are then closed. An air pump is attached to valve A, which is then opened and air is slowly injected at the base of the sand column. This action simulates generation of gas in deeply buried downdip areas from

organic rich source rocks, and entry of gas into adjacent low permeability sands.

As air injection continues, the water level at the top of the sand column is seen to gradually rise (Fig. 23). It does so because the injected air (unshaded) slowly displaces the mobile water phase upward and out of the sand column. Soon gas bubbles are seen escaping at the upper water surface. Also, the color of the sand

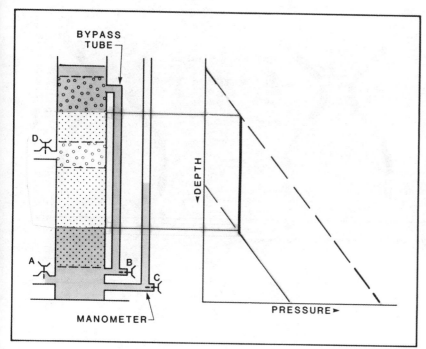

Figure 23. Following downdip gas injection gas accumulates at low pressure beneath water at coarse/fine sand contract.

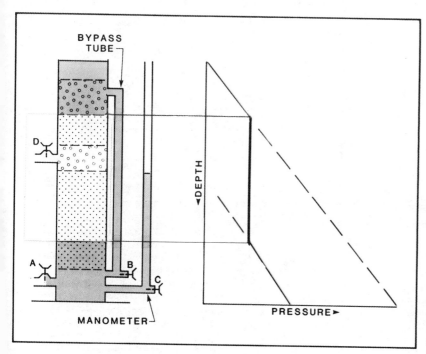

Figure 24. Model conditions following bleed-off of small quantity of gas from coarse sand lens through valve D.

column becomes lighter as the colored water is displaced out of the pore system. Air injection is continued until the water level at the top no longer continues to rise. Valve A is then closed and manometer valve C is then opened.

The water level in the manometer is observed to fall a considerable distance below its original position opposite the top of the sand column. This reaction obviously signifies that the water pressure at the base of the base of the sand column has dropped. (Note, the chamber below the sand column remains filled with water that is in contact with the base of the sand column.) Gas bubbles cease to rise to the water surface and a state of static equilibrium is soon reached in the system wherein the shortened water column height in the manometer remains fixed and stable as long as one wishes to observe it. Meanwhile the water wet sand, as seen through the transparent walls of the plexiglass column, remains coated with a water film and some isolated sand pores below the uppermost coarse/fine sand contact, remain completely water saturated. The top coarse sand segment is also totally water saturated. However, the middle coarse sand segment is water wetted but obviously gas saturated. This segment simulates locally developed permeable areas within the overall low-permeability Cadomin sand.

The model experiments suggest that water is unable to flow downward through the sand column and thereby displace gas out of the pores - even though there is a continuous water-wetting film throughout the gas-bearing sand. This water-wetting film and the isolated water-filled pores that occur throughout the generally gas-filled portion of the column are unable to transmit water flow, and do not transmit water-phase pressures in accordance with the normal static water pressure gradient of 0.433 psi/ft (9.79 kPa/m) from the top water surface to the water-filled chamber linked to the manometer at the base of the column.

If the valve connected to the short, coarse, gas-filled segment is opened momentarily to bleed off gas, and then closed again, the water level in the manometer drops abruptly and slowly rises again to about its previous position. At the same time, the water level above the top of the sand column is observed to fall slightly (Fig. 24).

This little experiment revealed that

with a reduction in gas pressure, some of the water was able to flow downward through the gas column in the fine sands, and accumulate at the base of the column. Initial reduction in gas pressure, due to momentary gas bleed off, reduced water pressure at the base of the column as evidenced by the fall in the manometer water column. Gradual rise again in the manometer indicated repressuring of the gas by water accumulating at the base of the column below the gas accumulation. As the gas pressure rose, however, downward water flow decreased until a state was reached where flow ceased and the manometer column stabilized roughly 12 inches (30.5 cm) below the top of the sand column.

Figure 25 is a simplified diagram that illustrates principles underlying the model reactions described above. This indicates what happens on a microscopic scale. When the non-wetting gas pressure is low, as in the left side of the diagram, a continuous, mobile water phase can exist separating the bound, immobile water film coating pore walls of the water-wet rock minerals, from gas that occupies the center of the pores. Water is able to flow around the gas in this situation. This is what happened when gas pressure in the model experiment was momentarily reduced by bleeding some of it off. As gas pressure increased, due to water accumulation at the base of the gas column, thereby compressing the gas above, the gas deformed into the pore throat displacing and disrupting continuity of the mobile water film. This is depicted in the right half of Figure 25.

The low gas pressure at equilibrium, as evidenced by the manometer water level, duplicates the pattern of field pressure observations. Gas pressures in the gas column below the water-saturated permeable sand, are less than water pressures if projected along a static water pressure gradient tied to the water surface above the column. Even though there is a bottom gas/water contact, there is still no continuously interconnected mobil water phase linking bottom water through the gas column with the upper water-saturated sands. Therefore, there are no buoyancy forces active to displace gas upward through the permeable, fine sand column.

It is important to recognize that when low pressure gas occupied the fine sand column, the relative permeability to water had become zero. Even though a water

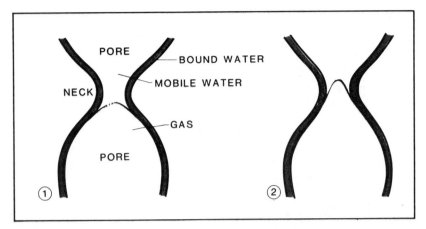

Figure 25. Simplified model of gas and water distribution on a microscopic scale across a pore throat.

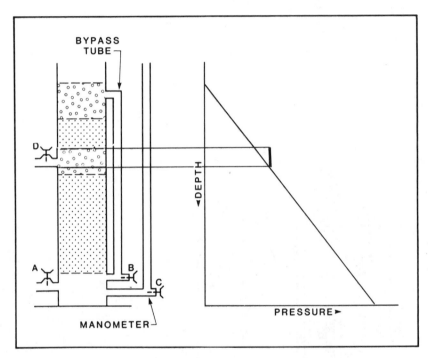

Figure 26. Model conditions after opening bypass control valve B.

film coated all pore walls and many isolated pores were still water filled, this water film was incapable of transmitting water flow or water pressures because water simply cannot flow along the water film unless its thickness reaches a certain minimum value.

A similar situation is believed to be the case of the updip water/gas contact in the Cadomin. At this position, the gas phase

pressure in the pores balances and disrupts continuity of the mobile water phase, thereby preventing water in updip sands to flow downdip and displace the gas. Two other conditions may also be developed: (1) the Cadomin may become more permeable immediately updip from the water/gas contact, and (2) the lateral barrier to the northeast may end at that position.

Returning to the model experiments

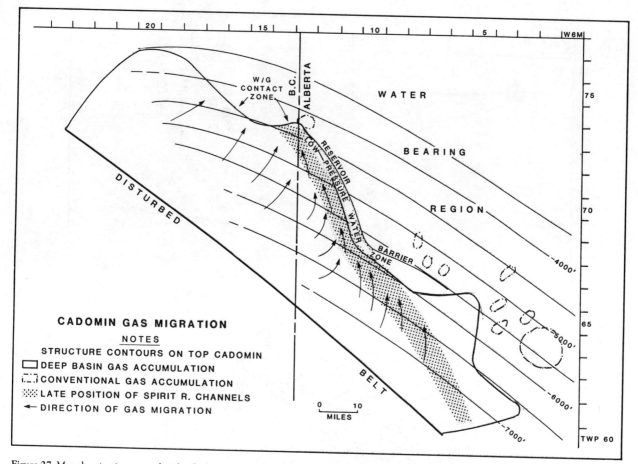

Figure 27. Map showing interpreted updip Cadomin gas migration paths as derived from detailed pressure analysis.

(Fig. 26), if the bypass tube valve B is opened, a continuous path of mobile water is now provided from the top to the base of the trapped gas column. Two things happen very rapidly: (1) the water level in the manometer rises immediately to a position opposite the top water surface in the column, and (2) gas bubbles begin escaping across the top water surface as the gas is gradually displaced by water flowing downward through the bypass tube into the base of the sand column. Eventually a new equilibrium position is reached, with a short gas column remaining trapped in the middle coarse sand interval. If pressures could be measured accurately, it would be observed that the middle gas column pressures are now slightly greater than the water pressures for corresponding depths.

Gas Migration

A question worth considering is: If, as it appears, gas is migrating updip in the Cadomin, why did it not all escape long ago? Based on average reservoir parameters and applying Darcy's Law, it is possible to calculate the rough average rate of gas escape across the updip water/gas contact. The calculated rate is in the order of 100,000 standard cu ft (2,800 cu m) of gas per day across a maximum contact length of 62 mi (100 km). Thus, the entire 15 tcf (425 × 10⁹ cu m) of estimated gas in place could escape in about 400,000 years. The most likely explanation to account for the fact that the accumulation exists despite updip gas leakage, is that gas is still being injected into the reservoir in downdip regions. The most likely source of gas is the organic-rich Cretaceous section in general, and the basal Gething section in particular. This section contains numerous coalbeds throughout the region as well as organic-rich shales and siltstones. The Cretaceous coals in the Elmworth deep

basin had enormous gas-generating capability in addition to other organic-rich shale source rocks.

The Cadomin deep basin gas accumulation appears to be a dynamic situation that has persisted for millions of years. Gas, generated in adjacent source rocks, moves vertically into the Cadomin layer and then slowly migrates updip, escaping eventually into the permeable, water-saturated region at the water/gas contact. Updip gas migration is a response to an unbalanced pressure gradient in the accumulation caused by comparatively low pressures at the updip end and higher gas pressures due to gas generation (maturation pressures) in the downdip region. Migration is not likely due to gas buoyancy in water. Low transmissibility throughout most of the gas reservoir tends to retard updip gas flow and thereby allows gas pressure to build up by the process of gas generation downdip, even though the rate of gas generation is

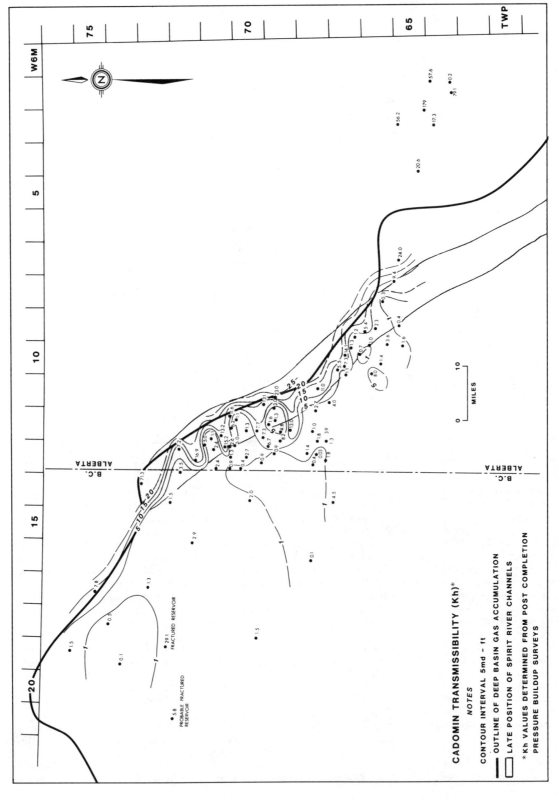

Figure 28. Cadomin reservoir transmissibility (Kh) map showing low Kh values downdip and higher values along the updip edge of the deep basin gas accumulation.

very low. This mechanism probably would not work effectively in relatively higher transmissibility reservoirs because the rate of gas generation could not keep pace with natural gas movement out of the downdip reservoir when water displacement originally took place and therefore buoyancy forces could provide rapid updip migration of generated gas. Because the extent of the low permeability Cadomin is so great in the downdip regions, gas can accumulate from a wide region. Based on the calculated average rate of gas migration across the water/gas contact stated earlier, the average rate of gas migration into the Cadomin from adjacent, active source rocks can be calculated. This rate works out to an average of 30 cu ft of gas (0.85 cu m) per acre (4,047 sq m) per year entering the Cadomin. Constant flushing of gas from downdip entry to an updip escape across the water/gas contact is believed to have resulted in exceptionally low residual water saturations in the Cadomin, as well as in other reservoirs. Under such high gas-saturated conditions, microporosity in the reservoir gains importance since the total gas volume and relative premeability to gas in such pores increase. Recent petrographic research suggests that microporosity is much more abundant than previously thought.

Much work remains to be done at Elmworth to assess indications of possible lower water saturations and higher porosities (due to the microporosity component) than the values in current use. For example, the assigned water saturations in gas-bearing sands usually fall in the 30 to 50+ percent range. Due to the enormous volumes of gas-bearing sandstones at Elmworth, a small reduction in average water saturation translates into a substantial increase in gas-in-place volumes and recoverable gas reserves. We speculate that additional future oil base core control and material balance estimates, based on several years field production history, could lead to reductions in current water saturation values by as much as 50%. Water saturations in the sands may actually lie in the 15 to 25% range.

The significance of microporosity as a contributor to long-term gas production has probably been highly underrated. Recent developments in petrographic techniques permit us, for the first time, to see and more fully appreciate previously unknown characteristics of micropore systems in rocks. This technology is under-

going further testing and improvement. The ability to see micropore networks and to evaluate their interconnection and association with macropore systems (which are necessary for economic recovery rates) should lead to more accurate petrophysically-based predictions of field production performance.

Detailed inspection of the scattered, downdip original gas pressures suggests gas migration paths through the Cadomin reservoir as depicted in Figure 27. In other words, gas would tend to move through the reservoir from locally high pressure areas to lower pressure areas (which, in turn, reflect lower to higher reservoir transmissibilities) assuming lateral reservoir continuity. Thus, gas apparently is migrating, in response to pressure gradients, out of lower permeability, relatively high-pressure alluvial plains sediments, into the comparatively higher permeability, northwestward-oriented Spirit River sediments, and thence updip to the water/gas contact. It is also along this latter trend that gas wells having the highest flow rates have been completed.

Figure 28 is a map showing the Cadomin deep basin gas reservoir transmissibility (Kh) trends. Transmissibility data are derived from analysis of post completion pressure build-up measurements which we consider to be the most reliable source of in situ information. The map clearly depicts improving reservoir transmissibility along the updip edge of the accumulation along the course of the late Spirit River channels.

The narrow, elongate, water-saturated, low pressure zone adjacent to the southwest side of the northwest-oriented barrier (Fig. 27), is indicated to be in pressure balance with the gas accumulation along its length. This is probably water that was bypassed when Cadomin gas originally moved updip into the region. The piston-like displacement of gas, as seen in the model experiments, could have bypassed the water, much in the same manner that injection water during an oil field waterflood will bypass some parts of the pool, due to reservoir continuity and transmissibility configurations that favor such movement.

Gas Accumulation Process

Figure 29 is a simplified diagram representing a gently dipping aquifer comprising low-permeability sands downdip,

giving way to higher-permeability sands updip. No permeability barrier exists at the contact to separate the two kinds of sands. In the beginning the sand is totally water saturated as in the previously described model experiment. Note the hydrostatic water-pressure gradient as depicted to the right of the section.

In Figure 30, gas begins to enter the downdip end as a result of large-volume gas generation from adjacent, downdip mature source rocks. Note that the gas pressure at the water/gas contact exceeds the water pressure in order to be able to move through the constricted pore system as a non-wetting fluid phase. Because gas pressures exceed water pressures, water is displaced updip out of the system by the gas.

In Figure 31, continuing gas influx has further displaced much of the original mobile water phase. Gas pressures always equal or exceed water pressures at the advancing water/gas contact front. Note, however, that downdip in the mobile gas saturated reservoir the pressure gradient for the gas now lies below the extrapolated water pressure gradient. As the water/gas contact continues to move updip, the water pressure at the advancing front decreases in accordance with the water-pressure gradient line shown on the graph.

In Figure 32, the moving gas front has now reached the more permeable sands and it cannot progress updip beyond that contact position. The change to a more permeable pore system updip sets the limit for the downdip gas accumulation as in the case for the Cadomin gas. As gas moves into the more permeable sand, the restriction to updip movement is almost eliminated because of much larger pore throats which allow gas to pass readily through them. At this point, buoyancy forces take over and in response, the gas is propelled rapidly updip such that the rate of gas influx across the contact cannot keep up with gas movement updip.

Figure 33 indicates what happens in this situation when the continuous gas supply being injected at the downdip end of the reservoir is shut off. Gas migration ceases and the gas pressure gradient steepens as the accumulation reaches a state of static equilibrium. At this point, the gas is trapped at low pressure relative to the regional water pressure gradient. There are no upward-acting buoyancy forces applied to the gas accumulation because hydraulic continuity of a mobile water phase extend-

ing through the reservoir from the updip
to the downdip end does not exist.

Exploration Considerations

From the practical point of view of
exploration for and exploitation of deep
basin type hydrocarbon accumulations,
one should bear in mind the following
considerations:

The intimate association of reservoir
rocks with thick, rich, mature source rocks
is a most important factor, and probably is
a prerequisite. Such reservoirs are well
positioned to capture hydrocarbons as they
are expelled from adjacent source rocks.
The reservoirs must be capable of retaining their hydrocarbon charge. Elmworth
experience suggests that if the reservoir is
very permeable, as in the case of a beach
conglomerate deposit, then it must be
bounded by a very effective shale seal or
low permeability sandstone. Proximity to
prolific gas source rocks and low reservoir
permeability is an important part of the
mechanism that allows regionally extensive sands to retain their gas charges
despite some continuous gas leakage from
the sands updip. Without locally developed reservoir "sweet spots" in these sands
however, the prospects for obtaining economic gas productivity are severely diminished. Geologists should consider the
various reservoir conditions that could
produce locally developed good reservoir
rock quality in a generally poor quality
reservoir. These would include: conglomerate or sand beaches, densely fractured
zones, areas enhanced by diagenetic mineral dissolution processes, local sand mineralogy which inhibits reservoir-destroying
diagenetic processes (that is, quartz cementation), late dissolution of calcite cements
deposited in sediments soon after burial,
marine transgressive (reworked) deposits,
etc.

If we recognize that enormous volumes
of gas have been generated in the past and
that most of it has probably migrated
updip out of the deep basin, then one
would tend to assume that the adjacent
updip region would be a good area to
explore for conventional structurally and
stratigraphically trapped gas accumulations. In practice, however, this may not
be the case in the Elmworth region. Adjacent updip regions of the Mesozoic contain abundant, generally more permeable
sands interbedded with potential trap seals
such as shales and coals. But the area has a

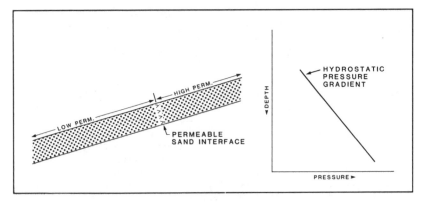

Figure 29. Gas migration model, stage 1, water-filled potential reservoir.

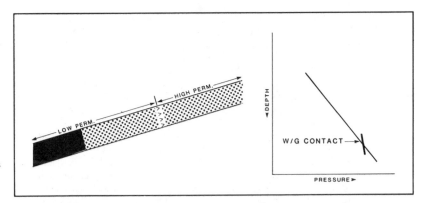

Figure 30. Gas entry begins in downdip region.

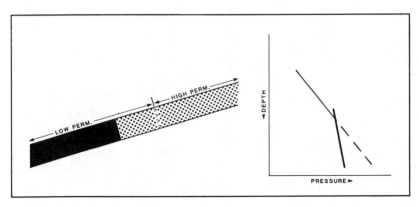

Figure 31. Gas accumulation expands updip as downdip gas influx continues. Note that gas
pressure gradient is greater than that for a static gas accumulation, due to updip energy loss in
the dynamic gas phase.

serious disadvantage in that it is more
remote from the gas-generating source
rocks and therefore there is a greater
chance for dispersal and loss of updip

migrating gas before it reaches some
potential traps. Much of the gas may in
fact seek an updip migration route that
offers permeable reservoir rock and conti-

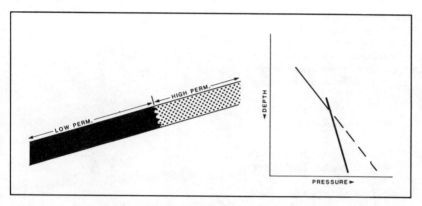

Figure 32. Updip development of the dynamic gas accumulation ends at contact with comparatively high-permeability reservoir.

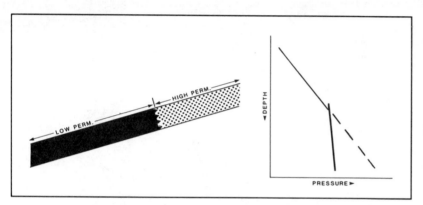

Figure 33. When downdip gas influx ceases gas pressure gradient steepens as accumulation reaches a state of static equilibrium.

nuity more or less directly to the updip outcrop. Also, the high mobility of gas through permeable water-bearing sands can lead to a rapid migration rate to the outcrop. Experience suggests that gas accumulations in adjacent updip sands tend to be small in areal extent and often have a gas/water contact in the interval. Thus, the advantage of high initial gas productivity, because of permeable sands, is offset by bottom water influx. Due to the comparatively limited rate of gas influx to updip traps, the confining seals, unlike in the deep basin, must be much more effective to allow development and retention of large gas accumulations. In this case, reservoir seals containing a continuous micropore system through the seals may not trap gas. Large conventional gas over water traps usually require extremely effective updip and lateral seals having very

low permeability. This is because the capillary displacement pressure (that is, gas pressure minus regional water pressure at any given subsea depth) which acts to displace gas into the water-wet pore system, increases very rapidly with increasing gas column height (ref. Fig. 2A). The resistance of fine capillary pore systems in trap seals to gas entry decreases with increasing capillary displacement pressures. Deep basin type gas traps, on the other hand, can exist with comparatively less effective (more permeable) updip seals because the capillary displacement pressure exerted by the accumulation at its updip edge is always low (Fig. 2B). These less rigid seal requirements favor development of very large deep basin type accumulations at Elmworth. The absence of large updip gas accumulations (in consideration of enormous volumes of gas that

evolved from deep basin source rocks), suggests there must be effective updip sand continuity which allowed most of the gas to escape through basin edge outcrops.

One last consideration is that the principles of deep basin gas entrapment should be equally valid for oil as well. An example of such a trap may exist at Elmworth in the Cardium formation of Upper Cretaceous age (Fig. 3). The widespread Cardium marine shoreline sandstone virtually blankets the region. It is known to contain high-quality, light-gravity oil throughout an extensive downdip region. The oil probably was derived from underlying organic rich, mature marine shale oil source rock within the Kaskapau. Updip the Cardium becomes water bearing similar to the Cadomin reservoir example. Exploitation of this vast oil resource in the Cardium has not been extensively pursued because the initial well drilling completion attempts experienced only limited economic success. Perhaps greater success would result through highly concentrated efforts to develop the kinds of geological, drilling/completion and supplemental recovery technology most appropriate to the specific reservoir conditions that exist in the Cardium. This would require a dedicated and resourceful management-technical team of experts who could pursue the costly drilling and complex experimental research programs required.

An important analog to the Cardium oil accumulation at Elmworth is the giant Pembina oil field located in the deep basin south of Elmworth and 70 mi (112 km) southwest of Edmonton (Fig. 1). The Pembina pool covers an area of 700,000 acres (283,000 ha). In parts of the field, oil productivity is enhanced by thin, permeable coarse sands and conglomerates at the top of the sandstone reservoir (Kerr, 1980). The oil is a high quality crude with API gravity ranging from 36 to 40°. Recoverable oil reserves, based on detailed study of 4,400 wells and a long production history, are estimated by the Alberta Energy Resources Conservation Board staff at 1.3 to 1.5 billion barrels (207 to 239 million cu m). Two oil reservoir features of particular geological interest and significance are: (1) the absence of a bottom water zone, and (2) unusually low connate water saturations of approximately 10% (based on 24 oil-base cored wells) as described by Purvis and Bober (1979). Both of these features suggest that deep basin trapping processes were active at Pembina.

CONCLUSIONS

Deep basin type gas traps represent a special class of hydrocarbon traps. This Cadomin example of the deep basin gas trapping mechanism points out and explains the unusual conditions that can occur. Other deep basin accumulations at Elmworth show similar relationships but with different features related to different reservoir plumbing developments.

The main points are as follows:

1. There must be a close match between a reservoir's plumbing and its fluid distribution (including fluid pressures);

2. The Cadomin deep basin gas accumulation is indicated to be dynamic where gas losses at the updip water/gas contact are in rough balance with downdip gas influx from adjacent, active source rocks; and

3. The gas accumulation is extensive. It occupies low-permeability rocks out of necessity (although locally high-permeability beds and areas are incorporated which thereby enhance gas well productivity), there is no downdip gas/water contact, and the accumulation is characterized by low pressures relative to projected regional formation water pressures.

REFERENCES

Kerr, W.C., 1980, A geological explanation for the variation in fluid properties across the Pembina Cardium field: Journal of Canadian Petroleum Technology, v. 19, n. 2, p. 76–84.

McLean, J. R. 1977, The Cadomin Formation; stratigraphy, sedimentology, and tectonic implications: Bulletin of Canadian Petroleum Geology, v. 25, n. 4, p. 792–827.

Purvis, R.A., and W.G. Bober, 1979, A reserves review of the Pembina Cardium oil pool: Journal of Canadian Petroleum Technology, v. 18, n. 3, p. 20–34.